Sustainable Environmental Waste Management Strategies

Series Editors

Alok Prasad Das, Department of Life Sciences, Rama Devi Women's University, Bhubaneswar, Odisha, India

Eric van Hullebusch, Équipe ACE, Institut de Physique du Globe de Paris, PARIS CEDEX 05, France

Sustainable waste management is an absolute imperative in a world that is already dealing with the effects of climate change. This book series seeks to publish an extensive collection of scientific books targeting researchers, scientists, professionals, academics, students, and anyone else involved in the wide, diverse field of sustainable waste management. The series accepts monographs, edited volumes, textbooks, and conference proceedings. Topics covered include but are not limited to environmental pollution and remediation, sources of environmental waste, environmental and human health, sustainable waste management, circular economy, nature-inspired environmental solutions, pollution prevention, and energy-efficient technologies for waste management. The series puts a particular emphasis on developing countries/ the majority world but accepts proposals globally as long as the topic is in scope.

Shreya Ghosh · Alok Prasad Das · Chandra Mohan ·
Eric van Hullebusch
Editors

Advances in Organic Waste Conversion Through Industrial Biotechnology

Springer

Editors
Shreya Ghosh
TCS iON
Kolkata, West Bengal, India

Chandra Mohan
School of Basic and Applied Sciences
K. R. Mangalam University
Gurugram, Haryana, India

Alok Prasad Das
Department of Life Sciences
Rama Devi Women's University
Bhubaneswar, Odisha, India

Eric van Hullebusch
Institut de Physique du Globe de Paris
Université Paris Cité
Paris, France

ISSN 3005-1622 ISSN 3005-1630 (electronic)
Sustainable Environmental Waste Management Strategies
ISBN 978-3-032-11376-4 ISBN 978-3-032-11377-1 (eBook)
https://doi.org/10.1007/978-3-032-11377-1

This Springer imprint is published by the registered company Springer Nature Switzerland AG
The registered company address is: Gewerbestrasse 11, 6330 Cham, Switzerland

Preface

The rapid expansion of industry and cities today has resulted in a significant buildup of organic waste, creating environmental and social challenges. Traditional disposal methods like landfilling, incineration, and open dumping are becoming unsustainable. These approaches lead to greenhouse gas emissions, resource waste, and ecosystem damage. In this context, industrial biotechnology has emerged as a powerful and eco-friendly tool for transforming organic waste into valuable products, aligning waste management practices with the principles of sustainable development and the circular economy.

This book, *Advances in Organic Waste Conversion Through Industrial Biotechnology*, provides an in-depth review of the most recent advancements in the field. It compiles leading research and highlights new technological innovations aimed at transforming organic waste into valuable products. The chapters span a multitude of methodologies—including microbial and enzymatic breakdowns, as well as the operation of biorefineries—which facilitate the conversion of waste into resources such as biofuels, biopolymers, biofertilizers, and other biobased chemicals. Notably, the book emphasizes the application of sophisticated approaches like metabolic engineering and synthetic biology, along with strategic process optimizations intended to enhance conversion efficiency while simultaneously minimizing environmental impacts. This book brings together insights from researchers, practitioners, and industry experts to present a balanced perspective on viewing organic waste as a renewable resource for industrial change instead of an environmental burden. It serves as a reference for students, scholars, policymakers, and professionals seeking sustainable waste management solutions. We hope this book will encourage creativity, foster cross-disciplinary cooperation, and facilitate the wider application of waste conversion technologies driven by biotechnology. In the end, the advancements discussed here highlight how important industrial biotechnology is to improving the environment, lowering reliance on fossil fuels, and advancing the development of a more robust and circular bioeconomy.

Kolkata, India — Shreya Ghosh
Bhubaneswar, India — Alok Prasad Das
Gurugram, India — Chandra Mohan
Paris, France — Eric van Hullebusch

Acknowledgements

The completion of this book, *Advances in Organic Waste Conversion Through Industrial Biotechnology*, truly owes much to the unwavering support, encouragement, and valuable contributions of numerous individuals and institutions. We are deeply appreciative of the contributing authors, whose perseverance, expertise, and thoughtful research have greatly enhanced this book with priceless insights and fresh perspectives. Their willingness to share information and work has made this collection comprehensive and impactful. Acknowledgments are extended to the organizations and institutions that facilitated the authors in their research endeavors. The authors express gratitude for the backing and resources provided by their overseeing entity, which proved instrumental in the successful completion of the project amid an intellectually enriching setting. Furthermore, the authors acknowledge the indispensable guidance and assistance rendered by the publishing house, particularly by its production and editorial teams, throughout the evolution of the book. The primary dedication of this book is directed toward aspiring environmental advocates and researchers committed to combating environmental waste. It is the aspiration of the authors that this publication will offer encouragement and inspiration to individuals engaged in similar pursuits.

Contents

Biorevolution: Turning Waste into Wealth with Cutting-Edge Biotechnologies for Sustainable Waste Management

Suprava Sahoo, Jyotirmayee Lenka, Swagat Kumar Rout, Sumitra Panda, Nikita Panda, and Basudeba Kar

1 Introduction

Human civilization has long been defined by its relationship with nature—both in the resources it extracts and the waste it generates. The rapid industrialization and population expansion of the modern era have intensified this dynamic, leading to an ever-growing burden of waste accumulation and an increasing dependence on finite fossil fuels. The pressing need for sustainable alternatives has thus ushered in a new paradigm, one that transcends mere waste disposal and embraces the principles of regeneration, circularity, and environmental stewardship. At the heart of this transformation lies biotechnology—a field that not only seeks to remediate environmental damage but also to unlock the latent potential of organic waste streams as a resource for a cleaner, more energy-efficient world.

Traditional waste management has long been centered on containment and elimination—landfills, incineration, and chemical treatments designed to remove waste from sight but often at the cost of long-term environmental degradation (Rao & Sharma, 2018). However, as scientific understanding deepens, so too does our appreciation for the intricate cycles of matter that sustain life on Earth. Biotechnology offers a profound shift in perspective: waste is no longer seen as an endpoint but as a beginning, a valuable input in a larger biochemical economy. This shift aligns with nature's own principles, where decomposition and renewal are seamlessly woven into the fabric of existence, ensuring balance and sustainability (Rao & Sharma, 2018).

S. Sahoo (✉) · J. Lenka · S. K. Rout · S. Panda · N. Panda · B. Kar
Centre for Biotechnology, Siksha O Anusandhan (Deemed to Be University), Ghatikia, Bhubaneswar, Odisha, India
e-mail: supi.sos2000@gmail.com

S. Ghosh et al. (eds.), *Advances in Organic Waste Conversion Through Industrial Biotechnology*, Sustainable Environmental Waste Management Strategies,
https://doi.org/10.1007/978-3-032-11377-1_1

Among the most revolutionary advancements in environmental biotechnology are mixed culture biotechnology (MCB) processes, which integrate microbial ecosystems to break down and transform organic matter into valuable byproducts (Kleerebezem & van Loosdrecht, 2007). Unlike conventional industrial processes that rely on pure microbial cultures with stringent conditions, MCB embraces diversity—an ecological approach that fosters resilience, adaptability, and efficiency. These processes not only mitigate pollution but also generate high-value products such as biofuels, organic acids, bioplastics, and essential nutrients, closing the loop in material and energy cycles.

A striking example of this transformation is anaerobic digestion, a process that mimics natural microbial ecosystems to convert organic waste into biogas—a sustainable energy source that reduces dependency on fossil fuels (Kleerebezem & Macarie, 2003). Similarly, solvent-producing fermentation and bioplastic synthesis harness microbial metabolism to generate environmentally friendly alternatives to petrochemical-derived products. By leveraging the principles of ecological selection, these biotechnologies offer not only waste remediation but also resource recovery, embodying the philosophy of waste-to-wealth (Seghezzo et al., 2006).

Presently, more than 50 countries have devised strategies for the bioeconomy (Svazas et al., 2019), recognizing its potential to revolutionize energy production, industrial processes, and environmental sustainability. Currently, the global production of energy, electricity, heat, chemicals, and fuels is predominantly reliant on fossil fuels, which contribute to 81% of the total primary energy supply, while nuclear energy accounts for 5% and sustainable energy sources provide 14%, with biomass contributing approximately 70% of this share (Dwyer & Teske, 2018). However, the industrial sector—particularly the chemical industry—consumes nearly 50% of its energy resources as raw materials rather than for direct energy generation (Dwyer & Teske, 2018). Given these challenges, the utilization of biomass feedstock and its transformation into food, feed, and biobased products—such as bioplastics, biofuels, and bioenergy—offer a viable alternative to fossil energy (Popp et al., 2021). Nevertheless, the increasing demand for biomass as an industrial feedstock may impact global food security by diverting agricultural production toward biobased materials, thereby altering the economic contribution of agriculture to gross domestic product (GDP) (Popp et al., 2021). This issue is further exacerbated by the finite availability of arable land, which constitutes only 18% of the Earth's surface area (Anekwe et al., 2022). Currently, biomass is primarily used for animal feed, followed by food production, energy generation, and chemical synthesis—collectively supplying approximately 13% of the global energy demand, with other renewable alternatives contributing an additional 5% (Popp et al., 2021). A critical aspect of transitioning toward a sustainable energy future is the "decarbonization" of the economy, which aims to reduce greenhouse gas (GHG) emissions by minimizing reliance on fossil fuel combustion. In this context, biotechnology plays a pivotal role in supporting the shift toward renewable energy sources, as it enables the development of biobased alternatives driven by consumer demand and policy initiatives (Kircher, 2019). The market viability of biochemical feedstocks—including lignocellulose, sugars, starch, and oils—depends on multiple factors, such as their cost, availability, and processing complexity (Kircher, 2019). However, biomass-based energy solutions often require additional processing steps, which increase both production costs and labor requirements. The advancement of biotechnological innovations in the energy and chemical industries has significantly reduced dependency on petroleum and other fossil fuels, yielding substantial environmental benefits. This shift has contributed to the convergence of the bioeconomy and circular economy, as economic incentives align with sustainability-driven practices (Aguilar et al., 2018). The development of a circular bioeconomy necessitates the internationalization of biotechnology-driven processes, particularly in bioenergy synthesis (Aguilar et al., 2018). Biotechnology, in essence, involves harnessing biological processes, microbial species, or engineered systems to create products that advance human welfare. In its broadest application, it includes genetic engineering and microbial manipulation for industrial and environmental applications. A key priority in modern biotechnology is the design and transfer of hybrid genes into microbial hosts, allowing for the expression of genes that are either partially or entirely absent in the original organism. These advancements have enabled breakthroughs in biofuel production, bioenergy generation, and sustainable industrial processes (Kleerebezem & van Loosdrecht, 2007).

Building upon the growing discourse on sustainable waste management, it is evident that unprecedented technological and economic advancements have significantly influenced the production, consumption, and disposal of various materials. With rapid population growth and urbanization, municipal solid waste (MSW) management faces increasing waste volumes, greater complexity, and heterogeneity. Addressing these challenges necessitates a continuous evaluation of emerging technologies to enhance waste processing and resource recovery. Waste-to-energy (WtE) has long been recognized as a macro-level strategy for sustainable MSW management (Moya et al., 2017), positioning MSW as a substantial renewable resource in the twenty-first century (Tsui & Wong, 2019). WtE incineration, widely adopted across developed and developing nations (Di Matteo et al., 2017), has undergone significant technological advancements, making it a mature waste treatment option. However, criticisms persist regarding its potential to encourage wastefulness, its lower priority in the waste management hierarchy compared to recycling, and its environmental risks, including the release of toxic substances and greenhouse gases (Xin-Gang et al., 2016). Consequently, public skepticism and social opposition have often challenged its political viability. In response, the concept of waste biorefinery has gained traction for its potential to generate biofuels and platform chemicals from the organic fraction of MSW, particularly food waste (Krook & Baas, 2013). The large-scale recovery of energy, reinforcement of material cycling, and economic implications of advancing a bioeconomy offer promising alternatives. While WtE incineration provides an immediate solution, the long-term success of MSW management within a circular economy depends on reducing waste disposal and maximizing resource recovery (Silva et al., 2017). Today, sustainable MSW management extends beyond environmental protection to achieving broader sustainability goals through energy recovery and material valorization. The transition toward a circular bioeconomy necessitates a broader perspective on WtE technologies, emphasizing not only electricity and heat generation but also the production of waste-derived biofuels and bioproducts (Malinauskaite & Jouhara, 2019). This shift urges government authorities and industries to explore socially acceptable and environmentally friendly WtE solutions, reinforcing MSW management with the highest sustainability potential. A well-implemented sustainable MSW framework will facilitate effective material recycling, expand renewable energy supplies, and contribute to biodiversity preservation (Malinauskaite et al., 2017). As global sustainability efforts intensify, the intersection of waste biotechnology, energy recovery, and circular economy principles will continue to redefine waste management strategies, making the optimization of biobased processes a critical priority for achieving long-term environmental and economic resilience.

Yet, despite the promise of these innovations, the challenges remain vast. Municipal solid waste, food waste, and agro-industrial residues continue to accumulate at an unprecedented scale, straining landfills and contaminating ecosystems with hazardous byproducts. The indiscriminate disposal of organic waste in open dumps and water bodies exacerbates climate change through methane emissions, contributes to groundwater contamination through leachate percolation, and poses public health risks. Traditional composting methods, while beneficial in recycling nutrients, cannot single-handedly resolve the complexities of modern waste streams. A multi-pronged biotechnological approach is thus imperative—one that synergistically integrates microbial bioprocesses, enzymatic degradation, and genetic engineering to enhance efficiency and sustainability.

2 Exploring the Many Faces of Waste

Waste generation is an inevitable byproduct of human activity, impacting ecosystems, public health, and global sustainability efforts. Various types of waste—such as industrial, domestic, hazardous, and organic—require distinct management strategies to mitigate their environmental consequences. Modern biotechnological advancements offer promising solutions for sustainable waste management.

1. **Municipal Solid Waste (MSW)**

Municipal solid waste (MSW) primarily consists of household, commercial, and institutional waste. It includes plastics, metals, paper, organic matter, and other general refuse. Improper disposal of MSW, particularly in landfills, leads to environmental hazards such as methane emissions, groundwater contamination, and soil degradation. Composting has emerged as an effective strategy for managing biodegradable fractions of MSW, reducing landfill waste while generating nutrient-rich soil amendments (Rao & Sharma, 2018). **Figure 1 Waste-to-Energy Pathways.**

2. **Industrial Waste**

Industrial activities generate vast amounts of waste, including chemical byproducts, heavy metals, and synthetic materials. Industries such as chemical manufacturing, food production, and metal processing produce hazardous residues that, if not

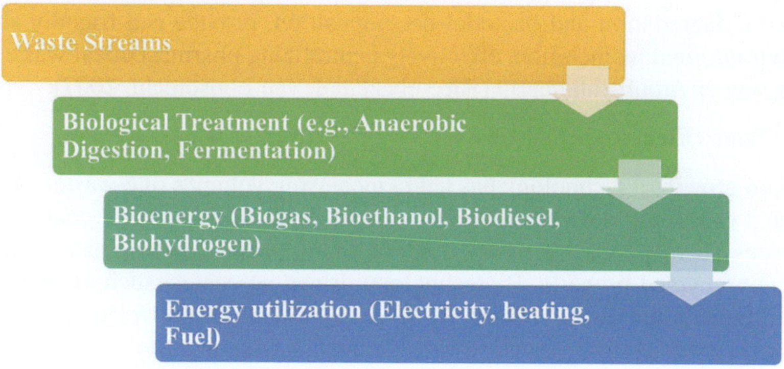

Fig. 1 Waste-to-energy pathways

managed properly, contribute to air and water pollution. Advances in bioremediation leverage microorganisms like *Deinococcus radiodurans*, which detoxify heavy metals and radioactive waste, offering a sustainable alternative to conventional waste treatments (Rao & Sharma, 2018).

3. **Agricultural, Agro-Processing, and Food Waste**

Agricultural and food processing industries generate significant amounts of organic waste, including crop residues, husks, manure, and food scraps. If not managed properly, these wastes contribute to soil and water pollution, methane emissions, and nutrient imbalances. However, biotechnology-based solutions such as biofertilizers, composting, and bioenergy generation can transform agricultural waste into valuable resources. Techniques like vermicomposting, using earthworms such as *Eisenia foetida*, and anaerobic digestion help convert organic waste into nutrient-rich biofertilizers and biogas, promoting sustainability and supporting circular economy models (Ezejiofor et al., 2014).

4. **Fruit and Vegetable Waste**

Fruit and vegetable waste constitutes a major portion of global food loss, with over 492 million tons wasted annually. This waste contributes to greenhouse gas emissions, nutrient depletion, and land pollution. Innovative approaches such as microalgae-based biorefineries can help upcycle fruit waste into biofuels, bioplastics, and nutraceuticals (Kamm et al., 2006).

5. **Pharmaceutical and Biomedical Waste**

Hospitals, laboratories, and pharmaceutical industries generate hazardous biomedical waste, including infected materials, chemicals, radioactive substances, expired drugs, syringes, and other contaminated materials. If not properly managed, this waste can lead to antibiotic resistance, toxic contamination of water bodies, and severe health and ecological risks. Advanced biodegradation techniques, such as enzymatic degradation and bacterial decomposition, provide eco-friendly alternatives to traditional incineration, effectively neutralizing pharmaceutical waste while minimizing environmental harm (Kleerebezem & van Loosdrecht, 2007).

6. **E-Waste (Electronic Waste)**

The rapid growth of technology has led to increasing volumes of e-waste, including discarded electronic devices, batteries, and circuit boards. E-waste contains toxic substances such as lead, mercury, and cadmium, which contaminate water and soil if not recycled properly. Biotechnology-driven processes, such as bioleaching and bioremediation using bacteria like *Acidithiobacillus ferrooxidans*, provide sustainable methods for recovering valuable metals from e-waste while minimizing environmental damage (Lederer, 2024).

7. **Plastic Waste and Marine Pollution**

Plastic waste is a growing global concern due to its long-lasting presence in the environment and its detrimental impact on marine ecosystems. Taking hundreds of years to degrade, plastics contribute to severe pollution and the bioaccumulation of microplastics in food chains, posing risks to both wildlife and human health. Innovative solutions such as biodegradable plastic alternatives and microbial degradation, using bacteria capable of breaking down synthetic polymers, offer promising strategies for mitigating plastic pollution (Tsui & Wong, 2019).

8. **Slaughterhouse and Meat Industry Waste**

The meat processing industry produces large quantities of animal byproducts, wastewater, and organic residues, which contribute to environmental pollution when improperly managed. Microbial treatment of slaughterhouse waste using specialized bacteria can convert organic waste into biofuels, fertilizers, and animal feed, supporting sustainable resource recovery (Rao & Sharma., 2018).

9. **Wastewater and Sewage Waste**

Urban wastewater and sewage contain high concentrations of organic matter, nitrogen, and pathogens, leading to waterborne diseases and eutrophication of aquatic ecosystems. Advanced biotechnology-based wastewater treatments, such as activated sludge systems, trickling filters, and membrane bioreactors (MBRs), enhance water purification while promoting water recycling (Ezejiofor et al., 2014; Rao & Sharma, 2018).

10. **Waste from Biofuel Production**

With the rising demand for bioenergy, waste generated from biomass processing, bioethanol production, and biodiesel conversion is increasing. These residues can be repurposed through anaerobic digestion, gasification, and fermentation technologies to produce additional energy sources such as biogas and biohydrogen (Lee et al., 2019).

11. **Waste from Meat and Dairy Industries**

Slaughterhouses and dairy industries generate organic waste, wastewater, and animal byproducts. If not properly treated, this waste leads to high biochemical oxygen demand (BOD) in water bodies, soil contamination, and greenhouse gas emissions. Biotechnological solutions such as enzymatic treatment and microbial digestion can aid in managing these waste streams efficiently (Ezejiofor et al., 2014).

12. **Construction and Demolition Waste**

Rapid urbanization has led to increased concrete, brick, wood, and metal waste from construction activities. Unmanaged disposal contributes to landfill overflow and resource depletion. Recycling and repurposing materials through green construction techniques can mitigate environmental impact (Liu et al., 2024).

13. **Textile and Dyeing Industry Waste**

The fashion industry generates chemical-laden wastewater, synthetic fibers, and fabric scraps that contribute to water pollution. Advances in enzyme-based textile treatment and circular fashion models can help reduce this waste burden (Lederer, 2024).

The diversity of waste types necessitates tailored management strategies that incorporate circular economy principles and biotechnological innovations. Sustainable waste management not only mitigates environmental harm but also unlocks opportunities for resource recovery, energy generation, and climate resilience. With continued advancements in biotechnology, waste can be transformed from an ecological burden into a valuable asset for future sustainability.

3 Ecological and Climate Impact of Waste Accumulation

3.1 Global Trends in MSW Generation

Currently, over one billion metric tons of municipal solid waste (MSW) are discarded globally, with estimates suggesting this number will rise to 2.2 billion metric tons by 2025 due to rapid urbanization, economic development, and changing consumption patterns, especially in developing countries (Chen et al., 2020). Waste generation is an inevitable consequence of human activity, contributing to an estimated 7–9 billion tons of global waste each year, with MSW accounting for roughly 2 billion tons in 2016 (Chen et al., 2020). MSW largely consists of household waste, though it can also include commercial and industrial waste, depending on the reporting criteria. Despite common predictions, uncertainties remain in waste generation figures due to regional differences in waste composition, collection methods, and disposal practices (Seberini, 2020). MSW poses considerable environmental, health, and economic challenges at local, regional, and global levels, making effective waste management strategies crucial. **Figure 2: Anaerobic Digestion Process for Biogas Production.** The environmental and health impacts of MSW vary based on disposal methods,

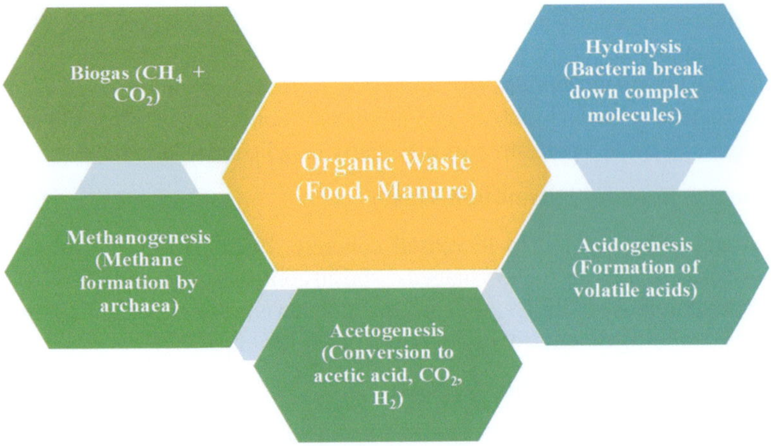

Fig. 2 Anaerobic digestion process for biogas production

with improper waste management leading to pollution, greenhouse gas emissions, and public health risks, while effective integration into circular economy models can unlock economic value (Vergara & Tchobanoglous, 2012).

3.2 Projected Waste Treatment in a Circular Framework

Although efforts to integrate waste into a circular economy—through recycling, composting, and waste-to-energy (WtE) initiatives—are increasing, progress remains slow (European Commission (EC), 2019). The total waste treated under circular methods is projected to double from 180 Mt (Million Metric Ton) in 2015 to 356 Mt by 2050 (Kaza et al., 2018). However, the recycling and composting rates show only marginal improvements:

- Recycling rates for metals and glass remain nearly stagnant at 34% in 2015 and 35% in 2050, highlighting challenges in collection systems and recycling infrastructure.
- Organic waste composting sees slight growth, increasing from 13% in 2015 to 17% in 2050, suggesting a need for better waste diversion strategies.
- Plastic waste leakage into marine environments continues to be a pressing issue, with an annual discharge rate of 8.76 million kg per year by 2050, worsening ocean pollution and threatening marine biodiversity.

4 Environmental Impact of Waste

4.1 Plastic Waste

Plastic waste is an increasing global issue because of its long-lasting presence in the environment and its negative impacts on wildlife and human health. It does not break down easily and often enters the food chain when organisms ingest it (Vergara & Tchobanoglous, 2012). This results in major ecological disruptions, bioaccumulation, and possible toxic effects on human populations. Additionally, plastics contribute to marine pollution, with an estimated 8 million tons entering the ocean annually (Wagner & Lambert, 2018).

4.2 Greenhouse Gas (GHG) Emissions from Waste

Fugitive emissions from waste treatments account for approximately 3–4% of global GHG emissions (Monni et al., 2006). Landfills and open dumps release methane, a greenhouse gas 21 times more potent than carbon dioxide, further exacerbating climate change. The open burning of waste, a common practice in developing nations, contributes significantly to air pollution and respiratory diseases due to the emission of hazardous pollutants (Monni et al., 2006).

The waste sector is a significant contributor to climate change due to methane (CH_4) and carbon dioxide (CO_2) emissions from landfills, dumps, and incineration (Chen et al., 2020). Key estimates include:

- GHG emissions from MSW are expected to almost double, from 1323 Mt CO_2e (Million Metric Tons of Carbon Dioxide Equivalent) in 2015 to 2383 Mt CO_2e by 2050.
- Landfills and dumps remain the dominant sources of emissions, as organic waste decomposition releases methane—a greenhouse gas 25 times more potent than CO_2 over a century.
- Waste-to-energy and composting contribute minimally to emissions, but their expansion remains insufficient to offset landfill methane outputs.

4.3 Leachate and Nutrient Accumulation and Pollution:

Waste leachate contains harmful substances, including nitrogen compounds, which contaminate soil and water sources. This contamination leads to nutrient imbalances in aquatic ecosystems, causing eutrophication, algal blooms, and subsequent depletion of oxygen levels, which negatively affects marine and freshwater biodiversity

(Guo et al., 2010). Nitrogen pollution from waste leachates has been linked to long-term health risks in humans, including waterborne diseases and metabolic disorders (Kjeldsen et al., 2002).

The accumulation of nitrogen (N) from MSW in Solid Waste Disposal Systems (SWDS) presents a growing environmental concern:

- Annual nitrogen deposition in landfills is projected to increase from 4.4 Mt N in 2015 to 5.8 Mt N by 2050.
- By 2050, an estimated 302 Mt N will have accumulated in landfills and dumps globally, contributing to long-term soil and groundwater pollution.
- Leachate contamination from unmanaged waste sites further exacerbates water quality degradation, increasing risks to both human and ecological health.

4.4 Social and Economic Impact of Waste:

Food waste has profound environmental consequences, contributing significantly to resource depletion, greenhouse gas emissions, and pollution. Despite sufficient global food production, nearly a billion people suffer from hunger and malnutrition, while vast amounts of edible food are discarded, particularly in developed nations (Hoornweg et al., 2013). Food waste not only deepens social inequalities but also puts significant pressure on natural resources, as food production relies heavily on water, land, and energy (Seberini, 2020). Disposing of food waste in landfills generates methane, a powerful greenhouse gas, which accelerates climate change. Furthermore, the overconsumption and waste of food contribute to harmful agricultural practices, such as deforestation, soil degradation, and biodiversity loss (Seberini, 2020). The economic burden of food waste extends beyond the immediate cost of discarded food, driving up demand and causing food prices to rise, which impacts lower-income groups the most. Additionally, the hidden environmental costs—such as pollution from transporting food and the depletion of natural resources—bring the total economic loss of global food waste to an estimated 2,600 billion dollars annually (Food and Agriculture Organization of the United Nations, 2019). Recognizing food waste as a significant public policy issue, the United Nations has included it in the Sustainable Development Goal (SDG) "Responsible Consumption and Production," which aims to reduce global per capita food waste at retail and consumer levels and minimize food loss along production and supply chains by 2030 (Principato et al., 2018). Addressing food waste is crucial for reducing environmental degradation, conserving essential resources, and mitigating climate change, aligning with global sustainability goals.

4.5 Environmental Consequences of Poor MSW Management:

Without improved waste management strategies, the global rise in MSW will lead to:

- Increased air and water pollution, as unregulated dumps and poorly managed landfills release toxic leachate and airborne pollutants.
- Threats to biodiversity, particularly from microplastic contamination in aquatic ecosystems.
- Rising public health risks, with unmanaged waste contributing to vector-borne diseases and hazardous air quality in urban areas.
- Depletion of natural resources, as inefficient recycling and waste recovery lead to continued reliance on virgin materials (Rao & Sharma, 2018).

If current trends persist, the environmental impact of MSW is expected to increase considerably by 2050, highlighting the urgent need for improvements in waste management systems globally.

5 Closing the Loop: Sustainable Waste Management Strategies for a Circular Economy and Bioenergy Generation

Waste management is essential for environmental sustainability, focusing on reducing, reusing, and recycling waste materials. Various strategies have been introduced to reduce waste buildup and its environmental effects, such as biotechnological solutions, zero-waste efforts, improvements in municipal solid waste management, waste-to-energy processes, and circular economy concepts. Below is a detailed look at these approaches.

1. **Biotechnological Approaches to Waste Management**

Biotechnology is crucial in contemporary waste management, using microbial processes to decompose waste materials and turn them into valuable products. This method minimizes the reliance on chemical treatments and supports environmentally friendly solutions. Figure 3: Biohydrogen Production via Dark Fermentation.

Microbial Degradation of Waste

- Some bacteria, like *Thauera aromatica*, are known for their ability to break down harmful substances like toluene, which is often present in industrial waste (Rao & Sharma, 2018).
- Other bacteria are capable of breaking down petroleum-based pollutants, helping to reduce pollution from oil spills and industrial discharges.

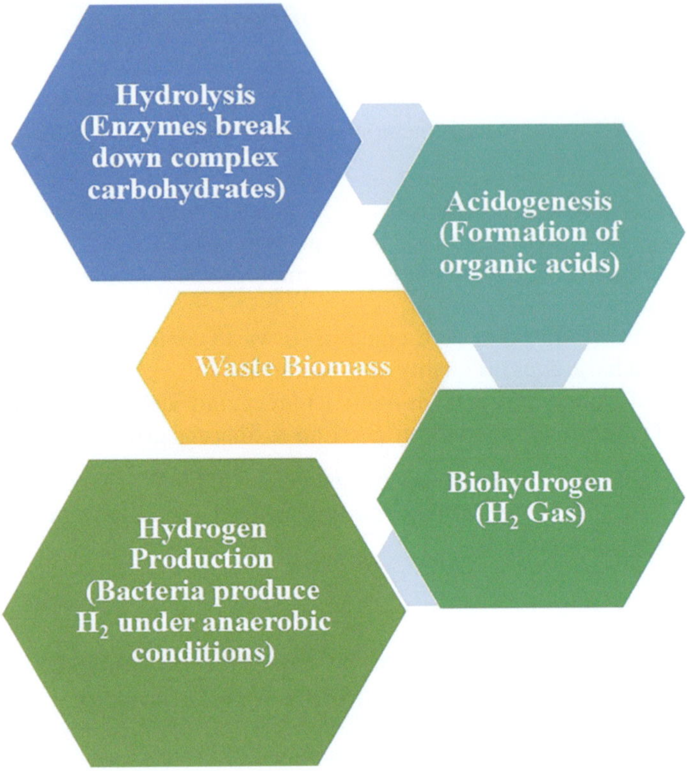

Fig. 3 Biohydrogen production via dark fermentation

Vermiculture and Organic Waste Processing

- Vermiculture uses earthworms, such as *Eisenia foetida*, to decompose organic waste materials. This process transforms food scraps, agricultural byproducts, and other biodegradable waste into nutrient-rich compost, enhancing soil fertility (Lim et al., 2016; Rao & Sharma, 2018).

Bioremediation and Biofiltration Techniques

- Bioremediation uses natural microorganisms to clean up contaminated environments, effectively removing heavy metals and toxic chemicals from soil and water (Barbosa et al., 2007).
- Biofiltration relies on microbial biofilms to filter pollutants from air and water, helping reduce emissions from industrial activities (Barbosa et al., 2007).

Fermentation and Bioprocessing

- Bioreactors help ferment organic waste, turning it into biofuels, bioplastics, and other valuable products.

- This method aligns with the circular economy by repurposing waste as a resource instead of discarding it (Safarudin et al., 2018).

2. **Zero-Waste Management Strategies**

The zero-waste approach aims to reduce waste creation by using efficient production techniques, sustainable packaging, and thorough recycling programs Seberini (2020).

Green Supply Chain Management (GSCM)

- GSCM focuses on using eco-friendly raw materials, energy-efficient production methods, and minimizing waste throughout the supply chain.
- Companies that implement GSCM lower their environmental impact while staying profitable (Seberini, 2020).

Innovative Recycling and Upcycling

- Advanced sorting technologies help separate recyclable materials from mixed waste.
- Upcycling turns waste into higher-value items, like converting discarded plastics into building materials or repurposing textile waste into new fabrics (Seberini, 2020).

Waste Audits and Policy Integration

- Regular waste audits help industries and municipalities assess waste generation patterns and implement targeted reduction strategies.
- Government policies promoting Extended Producer Responsibility (EPR) ensure that manufacturers take responsibility for the disposal and recycling of their products.

3. **MSW Management**

MSW includes household, commercial, and institutional waste. Managing this waste efficiently is crucial for reducing landfill dependency and mitigating environmental pollution (Seberini, 2020).

Integrated Waste Collection and Sorting

- Waste collection systems should incorporate source separation, allowing organic, recyclable, and hazardous waste to be processed separately (Vergara & Tchobanoglous, 2012).
- Smart waste bins with sensors can enhance collection efficiency by alerting authorities when they are full.

Landfill Diversion and Alternative Treatments

- Traditional landfills contribute to greenhouse gas emissions and groundwater contamination.

- Alternative treatments such as anaerobic digestion, composting, and incineration with energy recovery provide more sustainable disposal options (Wilson et al., 2015).

4. **Waste-To-Energy Technologies**

WtE processes convert waste materials into usable energy, reducing landfill accumulation while generating electricity or fuel.

Thermochemical Processes

- Incineration: Controlled combustion of waste materials to generate heat and electricity. Modern incineration facilities use advanced filtration systems to reduce air pollution (Tsui & Wong, 2019).
- Gasification: Converts solid waste into syngas (a mixture of hydrogen and carbon monoxide), which can be further refined into biofuels.
- Pyrolysis: It uses high temperatures in an oxygen-free environment to create bio-oil and biochar, which can be used as fuel and soil additives, respectively (Tsui & Wong, 2019).

Biochemical Processes

- Anaerobic Digestion: Bacteria break down organic waste without oxygen, producing biogas that can be used to generate electricity (Lee et al., 2024).
- Microbial Fuel Cells: Microorganisms produce electricity as they break down organic waste, offering an innovative approach to sustainable energy generation (Lee et al., 2024).

5. **Circular Economy Approaches**

The circular economy model focuses on reducing waste by creating products that are built to last, can be reused, and are easy to recycle (Vergara & Tchobanoglous, 2012).

Eco-Design and Sustainable Manufacturing

- Products should be made from materials that are simple to recycle and take apart when they reach the end of their life cycle.
- Sustainable manufacturing methods, like 3D printing with biodegradable materials, contribute to minimizing waste production (Vergara & Tchobanoglous, 2012).

Food Waste Upcycling

- Waste from agriculture and food processing can be repurposed into biofuels, animal feed, and biodegradable packaging.
- Using fruit waste as a nutrient source for microalgae cultivation provides an eco-friendly way to produce biofuels and bioplastics. (Vergara & Tchobanoglous, 2012).

Food Waste Reduction Programs

- Food waste is a major contributor to environmental harm worldwide. Solutions to reduce it include improving food supply chain efficiency, raising consumer awareness, and redistributing excess food. Innovations like smart packaging and AI-based inventory management systems also play a role in cutting food waste in the retail and hospitality industries (Lee et al., 2024).

Enhanced Waste Governance and Policy Frameworks

- Effective regulations and government policies are key to achieving sustainable waste management. Measures like landfill taxes, extended producer responsibility (EPR), and stricter environmental standards encourage businesses to adopt eco-friendly waste practices. Incorporating informal waste pickers into formal systems boosts efficiency and fosters social inclusion (Wilson et al., 2015).

Industrial Symbiosis and Resource Recovery

Industrial symbiosis promotes the idea of businesses exchanging waste materials as resources for each other. This practice improves resource efficiency and supports a more sustainable industrial ecosystem. For example, turning waste from food processing industries into biofuels or bioplastics helps create a low-carbon economy. By adopting these strategies, waste management systems can shift toward a more sustainable and efficient model, lowering environmental impacts while offering economic and social advantages (Bogner et al., 2008).

- Businesses in an industrial area can work together by repurposing each other's waste as raw materials. For instance, waste heat from factories can be transferred to nearby facilities that need thermal energy (Bogner et al., 2008).

Smart Waste Management Systems

The integration of artificial intelligence (AI), the Internet of Things (IoT), and Big Data Analytics into waste management has revolutionized waste collection and processing.

- Automated Waste Sorting: AI-based sorting systems enhance the separation of recyclable materials from mixed waste, improving recycling rates (Lee et al., 2024).
- Smart Waste Bins: These bins, equipped with sensors, alert authorities when full, minimizing unnecessary pickups and streamlining collection routes (Lee et al., 2024).
- Predictive Analytics: Data-driven methods help predict waste generation patterns, enabling cities to allocate resources more efficiently (Wilson et al., 2015).

Plasma Gasification for Hazardous and Non-Recyclable Waste

Plasma gasification is a WtE technology that operates at extreme temperatures (over 5,000 °C) in an oxygen-free setting, converting waste into syngas and slag (Tsui & Wong, 2019).

- Processing Non-Recyclable Plastics and E-Waste: This approach efficiently handles complex waste like electronic waste and non-recyclable plastics while minimizing harmful emissions (Tsui & Wong, 2019).
- Syngas Applications: The generated syngas can be used for electricity production, hydrogen fuel generation, or synthetic fuel creation, contributing to cleaner energy solutions (Tsui & Wong, 2019).

11. **Upcycling of Waste into High-Value Products**

Upcycling transforms waste into valuable materials and products without breaking them down as in traditional recycling.

- Food and Agricultural Waste: Byproducts from fruit and vegetable processing can support microalgae growth, leading to the production of biofuels, bioplastics, and nutritional supplements (Lee et al., 2024).
- Construction and Demolition Waste: Reclaimed concrete, wood, and metals can be reused in sustainable building materials, reducing reliance on new resources (Lee et al., 2024).

12. **Extended Producer Responsibility (EPR) and Sustainable Product Design**
    - EPR policies: These policies require manufacturers to manage the disposal of their products, encouraging sustainable packaging and better product design (Tsui & Wong, 2019).
    - Biodegradable Packaging: The development of compostable bioplastics and edible packaging helps reduce plastic waste, particularly in the food and beverage industries.
    - Modular Electronics: Designing electronics with easily replaceable and recyclable components extends product lifespan and minimizes e-waste (Tsui & Wong, 2019).

Decentralized Waste Treatment

Small-scale composting and anaerobic digestion plants at community levels reduce transportation costs and landfill dependency (Vigil et al., 2015).

- Public Awareness Campaigns: Educational initiatives encourage responsible waste disposal, recycling participation, and sustainable consumption behaviorsive-Based Recycling Programs: Pay-as-you-throw (PAYT) schemes charge residents based on the amount of waste they generate, encouraging waste reduction and recycling. Bioremediation for Landfill Waste and Industrial Pollution
- Bioremediation uses microorganisms and plants to degrade pollutants in waste-contaminated sites (Barbosa et al., 2007; Vigil et al., 2015).
- Phytoremediation: Plants like sunflowers and poplars absorb heavy metals and contaminants from landfill leachates and industrial waste sites. Engineered microbial strains enhance the breakdown of persistent organic pollutants (POPs), hydrocarbons, and plastic waste (Barbosa et al., 2007). Certain bacteria and fungi can

break down PET plastics and other synthetic polymers, offering an eco-friendly alternative to traditional recycling. Optimization of Landfill Management and Methane Recovery

- Although landfills are the least preferred waste disposal method, optimized landfill management reduces environmental impacts (Vigil et al., 2015).

- **Sanitary Landfills with Gas Recovery:**

Modern landfills capture methane emissions for energy production, reducing greenhouse gas output. Covering landfills with compost and biochar reduces methane leaks by enhancing microbial methane oxidation. Instead of opening new landfill sites, existing ones can be expanded vertically using engineered landfill caps to extend their lifespan. Blockchain for Waste Traceability and Circular Economy

Blockchain technology ensures transparent tracking of waste materials across supply chains (Vigil et al., 2015).

- **Verifiable Recycling Credits:**

Blockchain-based systems can authenticate recycling data, reducing fraud in extended producer responsibility (EPR) programs. Industries can trade recyclable materials on blockchain-secured platforms, promoting resource efficiency and industrial symbiosis. Waste Valorization through Hydrothermal Processing. Hydrothermal carbonization and liquefaction convert wet organic waste into biochar, bio-oil, and syngas (Guerrero et al., 2013).

- **Sludge and Agricultural Residues:**

These technologies effectively convert wastewater sludge, manure, and crop residues into renewable biofuels. Supercritical water gasification generates hydrogen-rich syngas, providing a cleaner alternative to fossil fuels (Guerrero et al., 2013).

- **Anaerobic Digestion and Gas Reforming:**

Biogas from food and agricultural waste can be used to produce green hydrogen, a sustainable fuel option. Plasma reactors improve hydrogen production from waste-based syngas, contributing to the growth of hydrogen energy systems (Hasan & Ammenberg, 2019).

13. **Blockchain for Waste Traceability and Circular Economy**

Blockchain technology facilitates transparent and trackable monitoring of waste materials across the entire supply chain (Tsui & Wong, 2019). Figure 3: Biohydrogen Production via Dark Fermentation and Fig. 4.

- **Verifiable Recycling Credits**: Blockchain technology can verify recycling data, helping to prevent fraud in extended producer responsibility (EPR) programs (Tsui & Wong, 2019).
- **Material Trading Platforms**: Industries can exchange recyclable materials through secure blockchain platforms, enhancing resource efficiency and collaboration (Tsui & Wong, 2019).

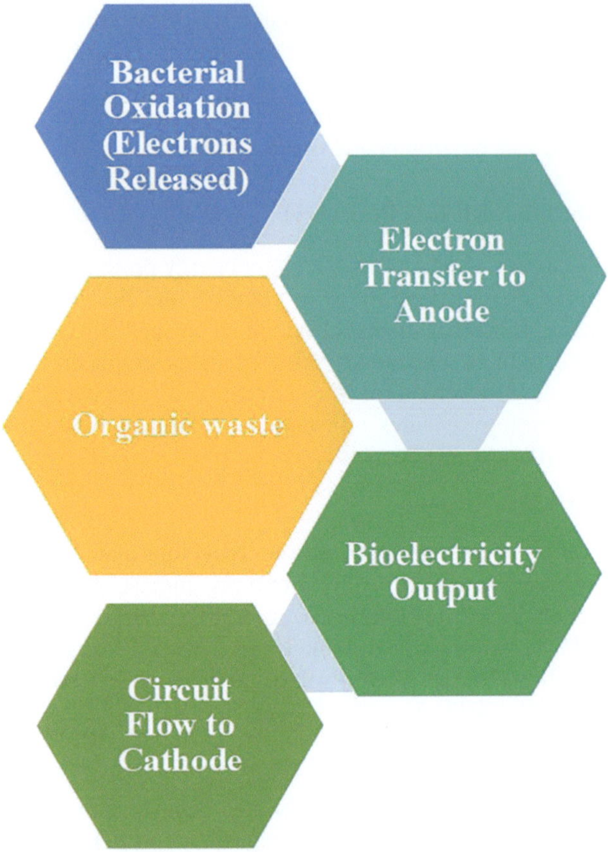

Fig. 4 Microbial fuel cell (MFC) for bioelectricity generation

The future of waste management will depend on the use of advanced biotechnologies, smart digital systems, and strong policy measures. **Table 1: Waste disposal technologies and their biotechnological approaches.** These strategies will help maximize waste utilization, support circular economy efforts, and minimize environmental harm. By combining innovation, regulatory improvements, and community involvement, waste can be repurposed into a valuable resource, paving the way for a low-carbon, zero-waste future. A sustainable approach to waste management requires multiple strategies, including biotechnological advancements, zero-waste practices, better MSW handling, waste-to-energy solutions, and circular economy models. By adopting these methods, industries and local governments can cut down on environmental damage while generating economic benefits. Prioritizing resource recovery and waste reduction will create a more sustainable future for both cities and rural areas. **Bioenergy Generation Methods Using Waste represented in Table** 2.

Table 1 Waste disposal technologies and their biotechnological approaches

Waste type	Biotechnology used	Process description	End product
Organic waste	Anaerobic digestion	Microorganisms break down organic material in oxygen-free conditions	Biogas (CH_4, CO_2)
Solid waste	Composting	Aerobic microbial decomposition of organic matter	Organic compost
Plastic waste	Biodegradation using engineered microbes	Enzymes break down plastics into biodegradable components	Biodegradable compounds
E-waste	Bioleaching	Bacteria and fungi extract metals from electronic waste	Recovered metals
Hazardous waste	Bioremediation	Use of bacteria/fungi to detoxify pollutants	Detoxified soil/ water

Table 2 Bioenergy generation methods using waste

Energy type	Biotechnology used	Process description	Feedstock
Bioethanol	Microbial fermentation	Yeast/bacteria convert sugars into ethanol	Lignocellulosic biomass, food waste
Biodiesel	Transesterification using microbes	Microbial enzymes catalyze conversion of oils to biodiesel	Waste cooking oil, algae
Biogas	Anaerobic digestion	Decomposition of organic matter to methane-rich gas	Organic waste, sewage sludge
Biohydrogen	Dark fermentation	Anaerobic bacteria produce hydrogen from organic substrates	Agricultural waste, industrial effluents
Bioelectricity	Microbial fuel cells (MFCs)	Electrochemical reactions via bacteria generate electricity	Organic wastewater

6 Future Perspectives

The future of waste management will be driven by ongoing advancements in biotechnologies and bioenergy solutions. In the coming years, a number of promising trends and innovations are set to transform the industry.

1. **Advancements in Microbial Biotechnology**

- Advances in genetic engineering and synthetic biology will lead to the development of specialized microbes that can break down plastics, pharmaceuticals, and hazardous waste more efficiently.
- Improved bioremediation methods, including the use of engineered bacteria and fungi, will make cleaning up contaminated sites more effective (Barbosa et al., 2007).

2. **Next-Generation Waste-to-Energy Technologies**

- Enhancements in gasification and pyrolysis will boost energy conversion efficiency while lowering greenhouse gas emissions.
- Hybrid waste-to-energy systems that merge thermochemical and biochemical methods will improve bioenergy production.
- More efficient microbial fuel cells and bioelectrochemical systems will become practical for decentralized energy generation.

3. **Smart Waste Management Systems**

- The integration of AI, machine learning, and the Internet of Things (IoT) will play a bigger role in waste management, making sorting, tracking, and processing more efficient.
- Sensor-equipped smart bins will optimize waste collection and help cut down operational costs.

4. **Expansion of Circular Economy Practices**

- Product design will prioritize durability, easy recyclability, and natural decomposition.
- Manufacturers will shift to zero-waste processes, reusing by-products and minimizing waste.
- Agricultural and food processing waste will increasingly be converted into biofuels, bioplastics, and high-value products.

5. **Policy and Regulatory Enhancements**

- Stronger regulations on waste disposal will be enforced by governments, along with incentives to encourage sustainable waste management practices.
- Global partnerships and financial support will help developing regions adopt advanced waste management technologies.

6. **Public Awareness and Behavioral Change**

- Public education programs and community-driven initiatives can foster responsible waste disposal and encourage recycling habits.
- Locally organized efforts, such as zero-waste movements and small-scale composting projects, will work alongside government strategies to improve waste management.

7 Conclusion

Waste management has experienced a substantial transformation in recent years, driven by a growing global emphasis on sustainability, resource recovery, and environmental conservation. Conventional waste disposal techniques such as landfilling and open incineration are gradually being phased out in favor of advanced biotechnological and bioenergy-based methods. These modern approaches not only mitigate environmental pollution but also yield valuable by-products, including biofuels, biofertilizers, and industrial raw materials.

Biotechnological interventions have demonstrated effectiveness in breaking down organic and hazardous waste through processes such as microbial degradation, anaerobic digestion, biofiltration, and enzymatic treatment. Specific microbial strains, such as *Thauera aromatica*, play a significant role in decomposing complex hydrocarbons, while bioreactors are instrumental in converting organic waste into biofuels and bioplastics. Additionally, vermiculture has gained prominence as an efficient technique for transforming organic waste into nutrient-rich compost.

Another major advancement in waste management is the integration of bioenergy production. Technologies like gasification, pyrolysis, and anaerobic digestion contribute to a sustainable energy economy by converting waste into biofuels, syngas, and biogas. MFCs are also garnering attention for their capacity to generate electricity through the decomposition of organic waste.

The transition to a circular economy - centered on resource efficiency, waste reduction, and product life cycle extension—has become a global priority. Sustainable practices such as GSCM, upcycling, and industrial symbiosis enable industries and municipalities to significantly lessen their environmental impact. Additionally, WtE facilities play a crucial role in addressing landfill overflow while contributing to renewable energy production.

Despite these advancements, certain challenges persist, including the high costs of implementing advanced waste management technologies, regulatory complexities, and the need for more efficient waste sorting and collection systems. Developing nations, in particular, encounter difficulties in adopting cutting-edge waste management techniques due to financial and technological constraints. Future efforts should focus on making sustainable waste management solutions more accessible and affordable.

The future of waste management relies on successfully integrating biotechnological innovations and bioenergy production within a circular economy framework. Utilizing microbial biotechnology, advanced WtE technologies, AI-driven waste tracking, and supportive regulations can lead to a more sustainable and efficient waste management system. Addressing financial and infrastructural challenges in emerging economies will be crucial for achieving global sustainability goals. Ultimately, transitioning toward a waste-free society requires collaboration among governments, industries, research institutions, and individuals. With continuous advancements in biotechnology and bioenergy, waste can be transformed from an environmental challenge into a valuable resource, fostering a cleaner and more sustainable future.

References

Aguilar, A., Wohlgemuth, R., & Twardowski, T. (2018). Perspectives on bioeconomy. *New Biotechnology, 40*, 181–184.

Anekwe, I. M. S., Armah, E. K., & Tetteh, E. K. (2022). Bioenergy production: emerging technologies. *Biomass, biorefineries and bioeconomy, 225*.

Barbosa, V. L., Tandlich, R., & Burgess, J. E. (2007). Bioremediation of trace organic compounds found in precious metals refineries' wastewaters: A review of potential options. *Chemosphere, 68*(7), 1195–1203.

Bogner, J., Pipatti, R., Hashimoto, S., Diaz, C., Mareckova, K., Diaz, L.,... & Gregory, R. (2008). Mitigation of global greenhouse gas emissions from waste: conclusions and strategies from the Intergovernmental Panel on Climate Change (IPCC) Fourth Assessment Report. Working Group III (Mitigation). *Waste Management & Research, 26*(1), 11–32.

Chen, D. M. C., Bodirsky, B. L., Krueger, T., Mishra, A., & Popp, A. (2020). The world's growing municipal solid waste: Trends and impacts. *Environmental Research Letters, 15*(7), Article 074021.

Di Matteo, U., Nastasi, B., Albo, A., & Astiaso Garcia, D. (2017). Energy contribution of OFMSW (Organic Fraction of Municipal Solid Waste) to energy-environmental sustainability in urban areas at small scale. *Energies, 10*(2), 229.

Dwyer, S., & Teske, S. (2018). *Renewables 2018 global status report.*

European Commission (EC). (2019). *On the implementation of the circular economy action plan communication from the commission to the European Parliament, the Council, the European Economic and Social Committee and the Committee of the Regions* (p. 5). European Commission

Ezejiofor, T. I. N., Enebaku, U. E., & Ogueke, C. (2014). Waste to wealth-value recovery from agro-food processing wastes using biotechnology: A review. *British Biotechnology Journal, 4*(4), 418.

Food Loss and Food Waste. (2019). *Food and Agriculture Organization of the United Nations* [online]. Retrieved September 13, 2019 http://www.fao.org/food-loss-and-foodwaste/en/

Guerrero, L. A., Maas, G., & Hogland, W. (2013). Solid waste management challenges for cities in developing countries. *Waste Management, 33*(1), 220–232.

Guo, J. S., Abbas, A. A., Chen, Y. P., Liu, Z. P., Fang, F., & Chen, P. (2010). Treatment of landfill leachate using a combined stripping, Fenton, SBR, and coagulation process. *Journal of Hazardous Materials, 178*(1–3), 699–705.

Hasan, A. M., & Ammenberg, J. (2019). Biogas potential from municipal and agricultural residual biomass for power generation in Hazaribagh, Bangladesh–A strategy to improve the energy system. *Renewable Energy Focus, 29*, 14–23.

Hoornweg, D., Bhada-Tata, P., & Kennedy, C. (2013). Environment: Waste production must peak this century. *Nature, 502*(7473), 615–617.

Kamm, B., Gruber, P. R., & Kamm, M. (Eds.). (2006). *Biorefineries-industrial processes and products*. Wiley-VCH.

Kaza, S., Yao, L., Bhada-Tata, P., & Van Woerden, F. (2018). *What a waste 2.0: A global snapshot of solid waste management to 2050*. World Bank Publications.

Kircher, M. (2019). Bioeconomy: Markets, implications, and investment opportunities. *Economies, 7*(3), 73.

Kjeldsen, P., Barlaz, M. A., Rooker, A. P., Baun, A., Ledin, A., & Christensen, T. H. (2002). Present and long-term composition of MSW landfill leachate: A review. *Critical Reviews in Environmental Science and Technology, 32*(4), 297–336.

Kleerebezem, R., & Macarie, H. (2003). Treating industrial wastewater: anaerobic digestion comes of age: anaerobic treatment systems offer important advantages over conventionally applied aerobic processes for removing organic pollutants from water-based streams. (Cover Story). *Chemical Engineering, 110*(4), 56–65.

Kleerebezem, R., & van Loosdrecht, M. C. (2007). Mixed culture biotechnology for bioenergy production. *Current Opinion in Biotechnology., 18*(3), 207–212.

Krook, J., & Baas, L. (2013). Getting serious about mining the technosphere: A review of recent landfill mining and urban mining research. *Journal of Cleaner Production, 55*, 1–9.

Lederer, S. (2024). *Waste management-municipal solid waste management in emerging economies with the waste-management planning software WaPla* [Doctoral dissertation]. Technische Universität Wien.

Lee, S. Y., Sankaran, R., Chew, K. W., Tan, C. H., Krishnamoorthy, R., Chu, D. T., & Show, P. L. (2019). Waste to bioenergy: A review on the recent conversion technologies. *BMC Energy, 1*(1), 1–22.

Lee, A., Lan, J. C. W., Jambrak, A. R., Chang, J. S., Lim, J. W., & Khoo, K. S. (2024). Upcycling fruit waste into microalgae biotechnology: Perspective views and way forward. *Food Chemistry: Molecular Sciences, 8*, Article 100203.

Lim, S. L., Lee, L. H., & Wu, T. Y. (2016). Sustainability of using composting and vermicomposting technologies for organic solid waste biotransformation: Recent overview, greenhouse gases emissions and economic analysis. *Journal of Cleaner Production, 111*, 262–278.

Liu, D., Yousaf, Z., & Rosak-Szyrocka, J. (2024). Environmental performance through green supply chain management practices, green innovation, and zero waste management. *Sustainability, 16*(24), 11173.

Malinauskaite, J., & Jouhara, H. (2019). The trilemma of waste-to-energy: A multi-purpose solution. *Energy Policy, 129*, 636–645.

Malinauskaite, J., Jouhara, H., Czajczyńska, D., Stanchev, P., Katsou, E., Rostkowski, P., & Spencer, N. (2017). Municipal solid waste management and waste-to-energy in the context of a circular economy and energy recycling in Europe. *Energy, 141*, 2013–2044.

Monni, S., Pipatti, R., Lehtilä, A., Savolainen, I., & Syri, S. (2006). Global climate change mitigation scenarios for solid waste management. *Espoo, Technical Research Centre of Finland. VTT Publications, 603*, 51–55.

Moya, D., Aldás, C., López, G., & Kaparaju, P. (2017). Municipal solid waste as a valuable renewable energy resource: A worldwide opportunity of energy recovery by using waste-to-energy technologies. *Energy Procedia, 134*, 286–295.

Popp, J., Kovács, S., Oláh, J., Divéki, Z., & Balázs, E. (2021). Bioeconomy: Biomass and biomass-based energy supply and demand. *New Biotechnology, 60*, 76–84.

Principato, L., Pratesi, C. A., & Secondi, L. (2018). Towards zero waste: An exploratory study on restaurant managers. *International Journal of Hospitality Management, 74*, 130–137.

Rao, R. K., & Sharma, P. K. (2018). *Indian Journal of Agriculture and Allied Sciences.*

Safarudin, A., Millati, R., Taherzadeh, M. J., & Niklasson, C. (2018). Inhibition of patchouli oil for anaerobic digestion and enhancement in methane production using reverse membrane bioreactors. *Renewable Energy, 129*, 748–753.

Seberini, A. (2020). Economic, social and environmental world impacts of food waste on society and Zero waste as a global approach to their elimination. In *SHS web of conferences* (vol. 74, p. 03010). EDP Sciences.

Seghezzo, L., Cuevas, C. M., Trupiano, A. P., Guerra, R. G., Gonzalez, S. M., Zeeman, G., & Lettinga, G. (2006). Stability and activity of anaerobic sludge from UASB reactors treating sewage in subtropical regions. *Water Science and Technology, 54*(2), 223–229.

Silva, T. F., Soares, P. A., Manenti, D. R., Fonseca, A., Saraiva, I., Boaventura, R. A., & Vilar, V. J. (2017). An innovative multistage treatment system for sanitary landfill leachate depuration: Studies at pilot-scale. *Science of the Total Environment, 576*, 99–117.

Svazas, M., Navickas, V., Krajnakova, E., & Nakonieczny, J. (2019). Sustainable supply chain of the biomass cluster as a factor for preservation and enhancement of forests. *Journal of International Studies, 12*(2), 309–321.

Tsui, T. H., & Wong, J. W. (2019). A critical review: Emerging bioeconomy and waste-to-energy technologies for sustainable municipal solid waste management. *Waste Disposal & Sustainable Energy, 1*(3), 151–167.

Vergara, S. E., & Tchobanoglous, G. (2012). Municipal solid waste and the environment: A global perspective. *Annual Review of Environment and Resources, 37*(1), 277–309.

Vigil, M., Marey-Pérez, M. F., Huerta, G. M., & Cabal, V. Á. (2015). Is phytoremediation without biomass valorization sustainable?—Comparative LCA of landfilling vs. anaerobic co-digestion. *Science of the Total Environment, 505*, 844–850.
Wagner, M., & Lambert, S. (2018). *Freshwater microplastics: emerging environmental contaminants?* (p. 303). Springer Nature.
Wilson, D. C., Rodic, L., Modak, P., Soos, R., Carpintero, A., Velis, K.,... & Simonett, O. (2015). *Global waste management outlook*. UNEP.
Xin-Gang, Z., Gui-Wu, J., Ang, L., & Yun, L. (2016). Technology, cost, a performance of waste-to-energy incineration industry in China. *Renewable and Sustainable Energy Reviews, 55*, 115–130.

Biotechnology for Waste Disposal and Alternative Bioenergy Generation

Sonit Bhusan Patanaik, Rabindra Nath Padhy, and Kanishka Uthansingh

1 Introduction

Biotechnology is proving to be very important in tackling the leading global issues of waste and energy. The combination of biological processes and new technological methods in biotechnology helps change waste into useful products, in line with what the circular bioeconomy believes (Ashokkumar et al., 2022). The approach combines different methods to help decrease pollution and make new alternative bioenergy sources such as biogas, bioethanol, biodiesel, and biohydrogen (Rajmohan et al., 2021). The main approach in biotech waste management is to treat agricultural and food residues along with municipal solid waste in the same way as microbes and enzymes to produce useful energy. They save waste for our landfills and help reduce greenhouse gas emissions by replacing fossil fuels with clean bioenergy (Dhanya et al., 2020; Reddy Koteswara et al., 2025; Swain et al., 2025). Most bioenergy projects use anaerobic digestion, composting, or fermentation to bring about this change (Awasthi et al., 2020).

Sustainable bioenergy development is often driven by microalgae and other photosynthetic organisms. By cultivating these microbes on wastewater from industry, we can clean the water and produce biomass that can be used later for energy (Ahmad et al., 2022). Thanks to algal biotechnology, nutrient-charged waste can be efficiently cleaned up, and its mass can be utilized to create biodiesel and biogas (Arora et al., 2021; Bala et al., 2024; Das et al., 2023a; John & Das, 2024). Taking advantage of

S. B. Patanaik · K. Uthansingh (✉)
Multi-Speciality Research Unit, KIMS Cancer Center, KIMS Hospital, KIIT University, Bhubaneswar, India
e-mail: uthansinghtintu@gmail.com; kanishka.uthansingh@kims.ac.in

R. N. Padhy
Central Research Laboratory, Institute of Medical Sciences and SUM Hospital, Siksha O Anusandhan Deemed to be University, Bhubaneswar, India

S. Ghosh et al. (eds.), *Advances in Organic Waste Conversion Through Industrial Biotechnology*, Sustainable Environmental Waste Management Strategies,
https://doi.org/10.1007/978-3-032-11377-1_2

algae in the waste sector boosts waste handling, strengthens the energy infrastructure, and supports environmental protection (Moreira et al., 2023).

Microbial fuel cells (MFCs) are an important advance in biotechnology that lets waste be treated while generating electricity using microorganisms. Since this technology supports sustainability and clean energy, it is an important tool in the field of the circular bioeconomy (Naha et al., 2023). Electroactive microorganisms in MFCs change organic substances in waste or sludge, sending electrons to an external power source, which creates electricity. MFCs stand out because they are used to treat strong organic materials like residues from food, industry, and human excreta. Experiments have established that MFCs are capable of degrading difficult organic materials in municipal wastes while generating electricity from them (Kumar et al., 2022a; Verma et al., 2022). Therefore, MFCs are preferred by those setting up wastewater treatment in small communities and energy recovery systems that are not connected to the power grid.

These materials also support flexible technological uses in addition to their green features. Adding AnTPs and MBRs to the system can enhance how much is treated, boost energy gain, and reduce how much is spent and discharged (Ardakani & Gholikandi, 2020). Lately, discovering new proton exchange membranes in materials science has brought significant improvements to MFCs in their efficiency and their ability to produce energy (Shabani et al., 2020). Researchers are also studying MFCs in places other than water, such as soils and sediments, where SMFCs are becoming popular for low-maintenance systems powering sensors placed far from human access (Bose et al., 2021). In industrial uses, MFCs treat high-organic-load wastewater while producing energy, which results in both lower costs and fewer carbon emissions (Brunschweiger et al., 2020).

2 Biotechnological Approaches to Waste Disposal

Waste management in the modern era calls for ways that result in less pollution, recycling waste, and developing useful products using that waste. Biotechnology has many tools and microbes, enzymes, and teams that all support each other to create bioenergy and useful products from waste. We look at biotechnologies that are used for waste disposal and potential new directions for these methods.

(i) Microbial Decomposition and Bioremediation

Microbial decomposition supports nearly all biotechnology processes for waste disposal. Microorganisms such as bacteria, fungi, and archaea have the natural ability to transform various organic wastes into simpler, non-poisonous molecules or substances that store energy (Das et al., 2025). This technique depends on these bacteria to purify contaminated regions, including soil, sediment, and waste from industry (Shah, 2020).

Using enzymes, microbes split big polymers and biochemicals present in complex waste into smaller pieces, known as monomers, which are finally broken down into minerals. To control municipal waste, agricultural leftovers, and industrial effluents, this process is required. Through this process, unique microbes help produce biogas, methane, and stabilized digestate from organic waste that is totally deprived of oxygen (Rao et al., 2010). Substituting conventional fossil fuels with biogas can help the planet become more energy sustainable.

In the past few years, genetic engineering in microbial biotechnology has helped increase how fast degradation happens, the resilience of microbes to tough environments, and the types of chemicals they work with (Das et al., 2024a; Mishra et al., 2024). Different microbial species are being combined so that pollutant-breaking tasks are split among them (Shah, 2020).

Going forward, AI and cameras might be added to control microbial removal, letting bioremediation adapt more quickly to new levels of pollution and environmental factors, helping to make waste processing more environmentally friendly.

(ii) **Enzymatic Waste Degradation**

Microbes are responsible for everything, with enzymes spearheading the efforts to help make waste breakdown faster and easier. Enzymatic degradation is favored for its highly precise effect, easy handling under non-harsh conditions, and minimal formation of other pollutants (Shah, 2020). Cellulases, proteases, lipases, and ligninases in the enzyme group help break apart cellulose, proteins, fats, and lignin found in organic waste. Treating recalcitrant materials such as lignocellulosic agricultural residues with enzymes is necessary so that the materials can be converted into sugar (Phwan et al., 2018). They can be used to make bioethanol, which is an important renewable fuel (Ibeto et al., 2011; Tse et al., 2021). Biotechnology has made it possible to discover and build enzymes that stay stable even when conditions in a factory become very hot, the pH changes, and there are inhibitors. By using immobilization techniques, you can take advantage of an enzyme two or three more times, lowering costs and raising the efficiency of enzyme-based reactions (Al Makishah, 2017).

Industry trends suggest that the production of biofuels, bioplastics, and chemical products from waste involves multiple-enzyme systems that act in series (Ashokkumar et al., 2022). Using enzymes in these processes helps cut energy use, which benefits the environment.

(iii) **Wastewater treatment and biotechnological innovations**

Often considered a problem for the environment, wastewater can now be used to produce both bioenergy and bioproducts with the help of new biotech methods.

Wastewater is mainly treated with microbes and algae.

a. Anaerobic digestion is often used to deal with tough industrial and municipal wastewater at a high concentration. When microbial communities break down organic matter, the produced biogas mainly consists of methane and CO_2, which can be regarded as renewable energy (Dhanya et al., 2020; Rao et al., 2010). This solution lowers organic pollution and produces energy afterward, which finishes the waste management cycle.

b. Microalgae are being studied as a fast-improving approach for treating wastewater while producing additional biomass, which can transform into biodiesel, bioethanol, biohydrogen, or several other bioproducts (Ahmad et al., 2022; Moreira et al., 2023). Algae take the nutrients from wastewater, safeguard against eutrophication, and reduce spending on chemical fertilizers (Jayaseelan et al., 2021).

Microalgal biorefineries use wastewater and make biofuels simultaneously, applying principles from circular bioeconomy. It is still hard to grow these facilities and improve how biomass is farmed and worked on (Bhatia et al., 2021). Because of advances in reactor design, better microalgae genetics, and automation, the earlier challenges are being solved rapidly (Huang et al., 2022a; Li & Zhu, 2021).

Other biotechnological innovations include:

- Microbial fuel cells convert wastewater organics directly into electricity while treating water.
- Enzymatic treatments complement microbial degradation bybreaking down complex compounds that microbes struggle with alone (Shah, 2020).
- Integration with nanotechnology for enhanced pollutant removal and energy recovery (Ashokkumar et al., 2022).

With increasing regulatory pressures and environmental concerns, the drive for sustainable wastewater management is pushing the adoption of these integrated biotechnologies globally. The future will likely see smart, modular biorefineries that treat wastewater on-site, recover multiple bioproducts, and operate with minimal environmental footprints, contributing significantly to renewable energy targets and circular economy goals.

3 Biotechnology in Alternative Bioenergy Generation

Biotechnology plays a key role in transition because it empowers the sustainable production of biofuels from a diverse array of biological resources such as plants, algae, and microbial systems. Plant biotechnology has provided new possibilities for bioenergy production due to biomass increase, stress resistance, and conversion improvement. Manipulation of energy crops like sugarcane, maize, and switchgrass through genetic engineering provides the option of specific expression of traits that enhance the biofuel yield while minimizing environmental impact (Joshi & Nookaraju, 2012). The importance of engineering transgenic feedstocks with higher cellulose content, but lower levels of lignin, has been highlighted to make enzymatic hydrolysis more efficient during the production of bioethanol (Moon et al., 2010).

Biotechnological interventions in agriculture can also provide for energy needs in developing countries. Biotechnology integration in traditional farming systems, in developing nations, can raise bioenergy returns without threatening food safety (Ruane et al., 2010). Among these is the application of molecular markers, tissue culture, and genetic modification to select breeds, high-yield and energy-rich food grains. Another promising way of bioenergy generation is microbial biotechnology, including, for example, the use of mixed microbial cultures. Mixed cultures provide flexibility and resilience in processing complex organic feeds, such as agricultural residues and wastewater, to produce biofuels such as biogas and hydrogen (Kleerebezem & van Loosdrecht, 2007). These systems usually do not need as strict sterility and nutrient conditions as monocultures, and therefore, they are more economically viable for large-scale application.

Algal biotechnology is an energy frontier because it has high lipid contents and the fastest growth rate, ideal for biodiesel production. Strategies that have been studied to improve the productivity of algal biomass and lipid accumulation are strain selection, metabolic engineering, and photobioreactor design (Wilhelm & Jakob, 2011). However, even because of technical breakthroughs, issues of scalability and energy input–output balance continue to be faced. Advances in fermentation technology have only expanded the horizon of the generation of such second-generation biofuels. By employing engineered microbes, agricultural residues and non-food lignocellulosic biomass can be converted to ethanol and other fuels, thus avoiding the food vs fuel debate (Sivakumar et al., 2010; Tabatabaei et al., 2015). Such approaches reflect the global pressure on carbon-free energy systems. A detailed review indicates that despite the numerous sources of bioenergy, with those sources ranging from crops and algae to waste, biotechnology is the common link that ensures efficient and sustainable conversion of the biomass into usable fuels (Rasool & Hemalatha, 2016).

(i) **Anaerobic digestion for biogas production**

Anaerobic digestion (AD) is a building block of the biotechnology used to produce alternative bioenergy, which provides a sustainable solution to converting organic waste to biogas. This process is biological and takes place without oxygen, in which microorganisms cause the breakdown of biodegradable substances, such as agricultural residues, food wastes, and algae, to biogas, mainly consisting of carbon dioxide and methane. Other than the production of renewable energy, AD also takes care of waste management and environmental conservation. Recent reviews emphasize the importance of AD technology to the global shift to renewable energy. Obileke et al. (2021) focus on its flexibility in terms of substrate utilization as well as its scalability, meaning it is practical for urban and rural applications. Design of AD systems is being optimized toward improving efficiency through reactor design innovations, microbial consortia selection, and pretreatment technologies. One of the greatest benefits of anaerobic digestion is its ability to adapt to any given feedstock. Another study highlighted the high biogas yields that could be obtained from a broad variety of organic wastes in the Indian scenario, to buttress the potential of AD to play a major role in the rural energy demands (Rao et al., 2010). Similarly, the biogas generation obtained from AD is especially appropriate for decentralized energy production in off-grid/ agricultural societies (Martins das Neves et al., 2009).

The combination of algal biomass with anaerobic digestion systems is a new and promising trend. Another experiment confirmed that the microalgae, namely Chlorella *sp.*, can be a good feedstock owing to high biomass productivity and high lipid content (Prajapati et al., 2014). Not only does this improve biogas output level, but it also sustains a closed-loop system when integrated with wastewater treatment and nutrient recycling. AD is now being envisaged in the context of biorefineries, going beyond energy production. As reported by other authors, modern AD-based biorefineries can coproduce biogas and value-added products, such as fertilizers, biohydrogen, and biochemicals, thus increasing economic viability and sustainability (Surendra et al., 2015). These integrated systems imply the transition from the single-product energy systems to multipurpose bio-industrial platforms. Improvement of the anaerobic digestion process efficiency can also be achieved through microbial engineering and synthetic biology. It additionally touches on the latest growth in manipulation of the microbial community and confinement of genetically tailored strains for boosting methane production (Balagurusamy et al., 2021). At the same time, even more extensive applications of the anaerobic process that go beyond the traditional methane-based outputs, toward controlled fermentation systems that provide other biobased chemicals, without necessarily generating biogas (Kleerebezem et al., 2015). Although beneficial, anaerobic digestion is hampered by several technical and economic barriers: slow microbial kinetics, system sensitivity to changes in the surrounding environment, and the cost of infrastructure. However, consistent research and policy support are tackling such limitations. Other studies emphasized the critical role of government incentives, rural development policies, and public–private partnership in scaling up deployment of AD technologies (Korres et al., 2013).

(ii) **Fermentation and bioethanol production**

Fermentation is an essential biotechnological process for the generation of bioethanol in which sugars extracted from biomass are converted to ethanol by microorganisms. First-generation bioethanol production is mainly based on food crops, such as corn and sugarcane, but second-generation bioethanol has received much attention because it utilizes non-food biomass, including agricultural residues, forest residues, and even microalgae. Second-generation bioethanol has several environmental and economic benefits since it escapes the "food vs. fuel" conundrum, while lignocellulosic biomass gets used, which must undergo pretreatment and a saccharification process to convert complex carbohydrates to fermentable sugars (Rastogi & Shrivastava, 2017; Robak & Balcerek, 2018). However, the following work has been done in understanding how disruption can convert cyclodextrin. Challenges such as high pretreatment cost and enzyme requirements have remained hurdles in scaling up this technology (Muktham et al., 2016).

The rapid growth and large quantities produced by microalgae have inspired more recent efforts to consider them as bioethanol production feedstock (Harun et al., 2010). Improving the microbial fermentation process for turning sugars into ethanol includes choosing better strains, optimizing process conditions, and carrying out fermentation on solid substrates (Karimi et al., 2021). In addition, an alternative way to produce bioethanol is known as syngas fermentation, where microbes convert gases like carbon monoxide, hydrogen, and carbon dioxide into alcohol (Kennes et al., 2016). Even so, there are problems in bioethanol production, such as how quickly fermentation occurs, the costs involved, and duplicating the process for commercial use (Sivakumar et al., 2010; Vohra et al., 2014).

(iii) **Biodiesel production from waste oils and microbial sources**

Recently, turning waste oil and microscopic organisms into biodiesel has become a main way to produce bioenergy sustainably. Waste oils, which include vegetable and animal fat, are currently being processed by transesterification into biodiesel. The process turns the triglycerides in these oils into biodiesel by using methanol or ethanol and with a catalyst, according to Azócar et al. (2010). Not only are waste oils recycled to benefit the environment, but they also do away with environmental contamination (Das et al., 2024b; Mohanty et al., 2024). Also, Yarrowia lipolytica and Rhodococcus are microbes that can now be used in biodiesel processing because they accumulate large amounts of lipids from lignocellulosic biomass and residues (Chintagunta et al., 2021). Such microbes ensure biofuel can be made without disturbing food crops, making them a good alternative for use in biofuel production (Christophe et al., 2012).

Biodiesel production from oleaginous microorganisms is becoming more popular, possessing both eco-friendliness and scalability compared to other biodiesel sources. According to a new study, agricultural residues and industrial by-products are now being used as feedstocks in the production of microbial biodiesel (Muniraj et al., 2015). Besides, changes in seasonal production for plant-based feedstocks do not strongly impact microbial lipid production systems (Rittmann, 2008). Increased yield of biodiesel is possible because metabolic engineering and genetic engineering have made microorganisms produce more (Srivastava, 2019). Moreover, linking microbial biodiesel production with the production of ethanol and biogas opens beneficial ways to create integrated biofuel plants (Bardhan et al., 2019). Turning waste oil and microorganisms into biodiesel could improve the production process's sustainability, reduce costs, and make it more environmentally safe.

(iv) **Microbial fuel cells for energy recovery**

These special microorganism technologies, called MFCs, handle waste by using the chemical activity of microorganisms to generate electricity. The main goal of circular bioeconomy fits perfectly well with such technology, which can help manage waste sustainably and produce renewable energy at the same time (Naha et al., 2023). What makes MFC useful is that electroactive bacteria oxidize common organic materials from wastewater, transfer electrons outside, and produce an electric current. In MFCs, converting all types of toxic organic materials, including food residues, industrial wastewater, and waste itself, is a definite selling point. According to the studies, MFCs can oxidize the complex waste compounds and produce enough electricity to be used from part of the chemical energy released (Kumar et al., 2022b; Verma et al., 2022). That is why MFCs are a top choice for wastewater treatment in smaller, remote areas and for energy recycling in places not linked to the grid.

Besides being environmentally and energy efficient, MFCs are also flexible in their technological use. When MFs are linked with other processes, the results include higher efficiency, more energy, less money spent on operations, and less waste being

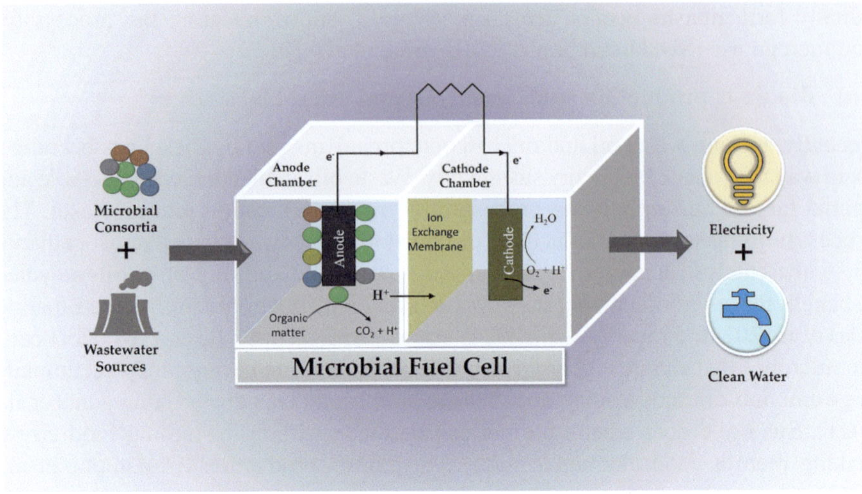

Fig. 1 Microbial fuel cells in treatment and product recovery from wastewater (Malik et al., 2023)

created (Ardakani & Gholikandi, 2020). Moreover, the use of PEMs in MFCs has improved their performance and raised the energy output they produce (Shabani et al., 2020). There is ongoing research into using MFCs for more unusual purposes in sediments and soils, and in this context, sediment microbial fuel cells (SMFCs) offer a low-maintenance option for powering sensors far from or under much water. In industries, MFCs are also used to process wastewater with a high organic load, and while doing so, they generate energy that reduces both cost and carbon emissions (Brunschweiger et al., 2020).

Just as MFCs, microbial consortia made of different and interacting types of microbes help to break down the complex wastes present in wastewater. As a result, these microbes use the organic pollutants as food, releasing electrons into the anode, which pass to an external circuit where they produce electricity. At the same moment, this reaction helps remove contaminants from the wastewater. By integrating electricity and water treatment processes, there is a chance to recycle clean water and harvest power from the wastewater sources (Fig. 1) (Malik et al., 2023).

4 Advancements in Microbial and Enzyme Engineering for Bioenergy

The rising worldwide need for clean and sustainable energy has accelerated the need for improved biotechnological techniques to turn waste products into useful energy resources. One of the main thrusts in this area is microbial and enzymatic systems engineered for higher efficiency and robustness in bioenergy production (Joshi &

Mishra, 2022; Keasling et al., 2021). Drawing from synthetic biology, metabolic engineering, and advances from bioprocesses, researchers have transformed the process of transforming organic and lignocellulosic waste into the production of biofuels and high-value bioproducts, significantly (Akdemir et al., 2022; Saini et al., 2022a). Three major advances, genetic modification of microorganisms, enzymatic hydrolysis of lignocellulosic biomass, and mixed microbial cultures, have emerged as prime strategies within this biotechnology revolution (Areniello et al., 2023; Singh et al., 2023; Xu et al., 2023).

(i) **Genetically modified microorganisms for enhanced biofuel production**

Genetic engineering has allowed for the development of very effective microbial platforms that are designed to maximize biofuel yields, substrate consumption, and resistance to process inhibitors. Microalgae, bacteria, and yeasts have all been greatly tweaked to steer metabolic flux toward the precursors for biofuels such as lipids for biodiesel or ethanol from sugars (Das et al., 2023a; Dey et al., 2024). Microalgae, on the other hand, have had their lipid accumulation drastically increased through strategies such as CRISPR-Cas mediated gene editing, overexpression, and knockout of competitive metabolic pathways (Khoo et al., 2023; Xue et al., 2021).

Likewise, genetically produced Saccharomyces cerevisiae strains had been generated to enhance ethanol production from industrial products, including but not limited to sugarcane molasses, with deregulated genes such as the PHO4 gene that regulates phosphorus metabolism and energy conversion efficiency (Wu et al., 2020). Genetically altered marine microbes have also been used to optimize saline waste streams, thereby propelling new career paths in marine biotechnology to produce biofuels (Singh et al., 2023).

Such engineered strains often incorporate metabolic pathways enabling the spatial conservation of substrate integration alongside the product formation, thus shortening processing time and costs massively (Keasling et al., 2021). The exploitation of synthetic biology tools, making it possible to assemble modular, inducible, and high-throughput pathways, makes the scalability of biofuel production even more achievable (Mirsalami & Mirsalami, 2025). Meanwhile, simultaneous use of bioprospecting and metabolic engineering continues to reveal new microbial organisms that could be used to produce biofuels from unconventional feedstocks, Adegboye et al. (2021), Joshi and Mishra (2022).

(ii) **Enzymatic hydrolysis of lignocellulosic biomass**

Lignocellulosic biomass sources, including crop residues and woody materials, constitute a plentiful but inadequately exploited resource to produce bioenergy. However, its construction complexity, particularly the presence of recalcitrant lignin, presents limitations in the bioconversion. Hydrolysis has been predominantly achieved by enzymatic means, especially through cellulase and hemicellulase, which have become identified as a viable, eco-friendly option to the more dangerous chemically based treatments. Latest strategies for improving enzymatic hydrolysis emphasize pretreatment process, enzyme formulation, and processes (Li et al., 2022; Saini et al., 2022b).

Pretreatment techniques, including those based on sulfite delignification and glycerol organosolv processes, have been demonstrated to enhance enzyme accessibility through alteration of the lignin structure, overriding the biomass crystallinity (Huang et al., 2022a; Song et al., 2024). Also, the addition of surfactants and bioadditives has enhanced the stability of the enzyme and the substrate-enzyme interaction, leading to faster saccharification of complex polymers (Xu et al., 2021).

Enzyme recyclability and process economics have both been enhanced because of cellulase immobilization, which hence renders hydrolysis more commercially viable (Xu et al., 2023). Modern bioreactor designs and modeling methods have allowed for fine control of reaction parameters so that the rates and hydrolysis yield of the process could be made more precise (Olivieri et al., 2021; Rohrbach & Luterbacher, 2021). In addition, enzymatic hydrolysis is now being used more often as part of the downstream fermentation process to produce ethanol and prebiotic oligomers (Saini et al., 2022a; Ávila et al., 2023).

(iii) **Mixed microbial cultures for efficient bioconversion**

Ecologically and functionally, mixed microbial cultures (MMCs) provide a flexibility to convert different types of waste materials into biofuels and biochemicals. Unlike monocultures, MMCs have a synergistic metabolic network that is capable of copolymerization of many substrates and detoxification of inhibitory compounds, making them quite adaptable to real-world heterogeneous waste streams (Akdemir et al., 2022).

The applications of MMCs cover a spectrum, ranging from biogas and bioethanol to bioplastics and microbial proteins. Cocultures of yeast and fungi have been successfully applied in the conversion of residues of food processing, such as yellow wine waste, to high-value microbial proteins (Zhu et al., 2020). Similarly, customized consortia of Bacillus with enzyme-producing fungi have made possible the effective hydrolytic/fermentative conversion of kitchen waste, such as kitchen bagasse, to biosurfactant, like surfactin (Pan et al., 2021). Engineering of cocultures of Escherichia coli and Bacillus subtilis has also shown successful bioconversion of plant hydrolysates into biofuels, which proves the utility of synthetic consortia (Kim et al., 2020).

In new reactor systems, MMCs were utilized to convert biogas into PHAs, biodegradable plastics with extensive applications even under harsh conditions such as high salinity (Carvalho et al., 2022; Cattaneo et al., 2022). The incorporation of bottom-up synthetic ecology strategies has also afforded isometric design of microbial consortia (Lin, 2022) optimized for lignocellulose degradation and energy recovery.

5 Digital Approaches to Treating Waste and Producing Biofuels from Microalgae

Microalgae are an extremely versatile type of photosynthetic microorganism that are being developed at the crossroads of environmental biotechnology and renewable energy systems. These organisms function on two levels: processing different types of wastewaters while at the same time producing biomass that can be used to produce biofuels and other value-added bioproducts. This portion details modern capital investments in the application of microalgae for purposes of carbon capture, treatment of wastewater, and sustainable bioresource exploitation.

(i) **Algal bioreactors for the mitigation of CO_2 and biofuel synthesis**

Algal bioreactors have received high attention as biotechnological solutions to reduce anthropogenic CO_2 emissions while producing renewable biofuels. The ability of microalgae to oxidize CO_2 during photosynthesis and turn it into lipids, carbohydrates, and proteins that can be precursors to produce biodiesel, bioethanol, and biohydrogen (Hoang et al., 2023a; Marangon et al., 2023). Scalable closed photobioreactors and open raceway ponds are the most employed systems, and their design vitally determines biomass productivity and CO_2 sequestration capabilities.

Novel cultivation techniques, such as nutrient modulation, light modification, and carbon addition, report increased lipid accumulation and a better yield of biofuels (Vijayalakshmi et al., 2020; Wang et al., 2022). These systems are becoming increasingly interfaced with industrial exhaust streams for in situ CO_2 capture-providing carbon–neutral or even carbon-negative biofuel production platforms (Ali et al., 2022a).

Furthermore, studies have proved that the integrated use of algal bioreactors, along with nutrient sources in the form of wastewater, increases sustainability as well as economic viability (Ali et al., 2022b; Srimongkol et al., 2022). This method synergistically positions waste CO_2 and nutrients toward biomass production, hence lowering the cost of operations and leaving a minimal environmental footprint.

(ii) **Algae-based wastewater treatment systems**

MbWT systems based on microalgae represent an eco-friendly way of eliminating contaminants, including nitrogen, phosphorus, heavy metals, and organic carbon, from wastewater of municipal, agricultural, and industrial backgrounds (Abdelfattah et al., 2023; Kumar et al., 2022b). These systems are based on the high nutrient uptake potential of microalgae in synergy with oxygen-evolving effects of photosynthesis to promote nitrification, denitrification, and biological phosphorus removal.

Organisms such as Chlorella vulgaris, Scenedesmus obliquus, and Spirulina platensis have shown satisfactory pollutant removal efficiencies, and they are manageable in terms of incorporation in tertiary wastewater treatment programs (Ali et al.,). Hybrid systems containing either Microalgae plus bacteria or constructed wetlands have displayed greater resilience and adaptability in managing fluctuating wastewater compositions (Liu & Hong, 2021; Mehariya et al., 2021).

Algal-based systems also allow biomass recovery afterward, and this biomass can be refined into biofuels or other bioproducts, thus supporting zero-waste and circular bio-economy assumptions (López-Sánchez et al., 2022; Moreira et al., 2023). Notably, biofilm reactor and immobilized microalgae technologies have been developed for valid configurations that are favorable for controlling biomass yield and sustainability in different operating conditions (Han et al., 2022).

(iii) **High-value bioproducts from algal biomass**

The bioproducts gained from biomass production from microalgae in wastewater treatment systems, apart from biofuels, are rich sources of high-value products. Such are carotenoids (astaxanthin, β-carotene), polyunsaturated fatty acids (PUFAs), phycobiliproteins, polysaccharides, and antioxidants that may be applied in food, pharmaceutical, cosmetic, and nutraceutical industries (Catone et al., 2021; Peter et al., 2021).

Microalgal biorefineries currently utilize cascade extraction and fractionation approaches to obtain several bioproducts from a stream of biomass, thus increasing economic viability and reducing waste (Gondi et al., 2022; Pessôa et al., 2021). Furthermore, new advances in algomics-omics-guided exploration of algal metabolism are making it possible to target biosynthesis of compounds of interest and broaden valorization strategies (Kadri et al., 2023).

Studies indicate that the combination of wastewater treatment with valuable compound extraction, such as isoprene or biofuels, can offer return on investment far greater than biofuels exclusively, assuming genetic improvement of strain and metabolite stimulation of cultivation conditions (Liu & Hong, 2021; Srimongkol et al., 2022). This makes micro-algal biotechnology an environmental, as well as an industrial, platform for sustainable biomanufacturing.

6 Integration of Waste-To-Energy (WTE) Technologies in the Bio-Economy

The incorporation of waste-to-energy (WTE) technologies into the bio-economy is becoming a revolutionary approach to realizing sustainability intentions by optimizing the valorization of waste, providing renewable energy sources, and implementing a circular resource approach (Ray et al., 2024). Not only do these technologies decrease the volume of waste that is landfilled, but they are also turning organic and inorganic waste streams into valuable products (biogas, bio-oil, syngas, and biochar). This transformation from linear waste management to a circular resource use basis is central to the transition toward a low-carbon and resource-efficient economy.

(i) **Circular economy and resource recovery strategies**

Viewing waste as useful, not harmful, is an important part of a circular economy and waste-to-energy technology transformation. Using this WTE model, energy, nutrients, and key compounds are recovered from waste streams, demonstrating the ideas of resource efficiency and zero waste (Bhuyan, 2024; Yaashikaa et al., 2020).

Biogas and compost are produced from organic waste using AD and composting; such products are then used for renewable energy and soil improvement. As a result, nutrients are recycled and farmers need less chemical fertilizer (Kapoor et al., 2020). The pyrolysis of household waste creates biochar that stores carbon and helps the soil, and the syngas can be used to create electricity and heat (Sajid et al., 2022; Sarker et al., 2025).

WTE systems are integrated into urban systems, contributing to clean waste disposal and generating chances for employment as well as new forms of energy in less developed countries (Hoang et al., 2023a; Vasileiadou, 2024). Besides promoting stable material circulation and environmental protection in the long run, public–private collaborations allow the policy-driven WTE supply chains based on circular economy legislation to succeed (Atstaja et al., 2024; Fernando et al., 2022).

The diagram shows how various organic wastes are recycled using biotechnology to produce useful bioenergy products. Materials including agricultural leftovers, municipal garbage, factory wastewater, used oils and sewage water are treated via biological processes such as microbial decomposition, enzymatic decomposition, anaerobic digestion, fermentation, power from microbial cells and treatment using microalgae. They allow waste to be converted into various things, including biogas, bioethanol, biodiesel, electricity, treated water and bioproducts which is good for the environment and provides alternative energies (Fig. 2).

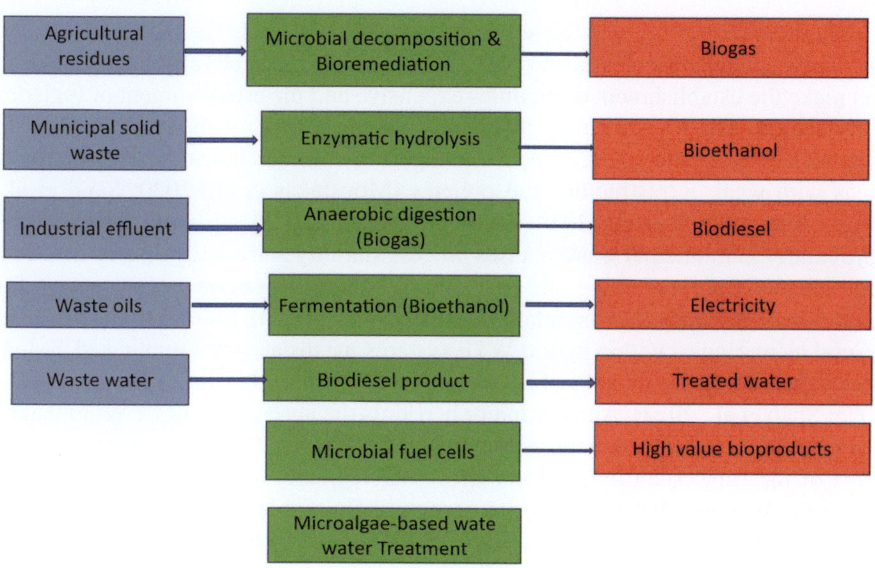

Fig. 2 Waste Recycling Pathways in Biotechnology for Bioenergy

(ii) **Hybrid approaches: combining biological, chemical, and thermochemical processes**

Hybrid WTE systems are being increasingly deployed to improve process efficiency and to increase product diversification. These systems combine various biological (anaerobic digestion, fermentation), thermochemical (gasification, pyrolysis, hydrothermal liquefaction), and chemical processes to optimize the valorization of waste from complex, heterogeneous feedstocks (Mishra et al., 2023; Shah et al., 2022). Food waste may undergo enzymatic hydrolysis for fermentation to bioethanol, while residual solids may be subjected to anaerobic digestion to produce biogas, creating a multi-product biorefinery model (Duan et al., 2021). Pyrolysis, subsequent catalytic upgrading is utilized to transform plastic-rich wastes into synthetic fuels, but hydrothermal liquefaction (HTL) is particularly useful for high moisture biomass and produces biocrude and recovers nutrients in one process (Mahadevan et al., 2025; Taghipour et al., 2023).

These integrated systems can be operated either cascaded or in parallel to provide flexibility in the processing of diverse waste streams, which improves the energy yield (yield from the waste is deposited or withdrawn) and eliminates residue. They also facilitate the recovery of coproducts such as, but not limited to, biochar, hydrogen, acetic acid, and platform chemicals that can be used as starting materials for other industrial processes, thereby closing material loops and enriching the circular bioeconomy (Ali et al., 2020; Shah et al., 2021).

(iii) **Scaling up and commercialization challenges**

Although WTE technologies have shown great promise from the lab to pilot scale, they have several critical issues to overcome before they can be fully scaled up. This can make the establishment of modules expensive and the establishment of logistics complicated. The logistics may therefore be inconsistent, or the quality of waste feedstock may be inconsistent, hence hampering commercialization, especially in areas with weak infrastructure and policies (Moustakas et al., 2020; Sajid et al., 2022). Metchnikoff et al. (2020) notes that the variability in waste composition, particularly in municipal waste, causes process stability and energy output, meaning that advanced preprocessing and real-time monitoring systems are necessary for operational efficiency to be maintained (Hoang et al., 2023a). Moreover, the fusion of several technologies within hybrid WTE systems means a more complicated system design that requires knowledgeable technical staff to run and maintain (Rezania et al., 2023; Shah et al., 2022). Another major barrier is the absence of the development of markets for secondary products, including digestate, biochar, and recovered nutrients. Most of the time, such by-products encounter regulatory provisions or consumer rejection, meaning that they are not economically viable (Fernando et al., 2022; Taghipour et al., 2023). To work out these problems, governments and industries need to work collectively toward deeper policies, incentives, and awareness campaigns to facilitate WTE commercialization and attract private sector investment.

Nevertheless, increased use of digital technologies such as IoT, AI, and blockchain provides new opportunities in the optimization of WTE operation and supply chains.

Smart WTE systems can adapt based on feedstock, calibrate the emissions, and realize the best energy production, which makes it easier to scale up and more economical (Atstaja et al., 2024; Singh et al., 2024).

7 Conclusion

At the meeting point of caring for nature and advancing energy, biotechnology supports us in turning global waste into something useful. This chapter has demonstrated how new approaches in biotechnology are turning waste handling and alternative energy into modern solutions. Thanks to these advances, less pollution occurs, and we can move more toward a sustainable circular bioeconomy, ensuring waste serves to produce clean energy and valuable bioproducts. Connecting different microbes and new biotechnology is increasing both the efficiency and the feasibility of this process, allowing waste to be transformed into useful products. While microalgae are often overlooked, such systems are showing great promise by cleaning wastewater, trapping carbon dioxide, and generating biofuels, as well as important chemicals, making them valuable allies against climate change. However, the process is not done yet, as the flexibility of raw materials, economic problems, and improving the production methods must be addressed by the evolving next-generation biotechnology.

8 Future Perspectives

Targeting better efficiencies of bioconversions requires progress in enzyme design and using digital and data-driven methods for optimization in bioreactors. Soon, combining bio, chem, and thermo reactors function together on a single site could be what really changes the game for energy and the environment. Imagine that our waste-to-energy technology is so advanced that landfills are abandoned and fossil fuels are no longer important; however, they just become part of history books. Essentially, biotechnology goes beyond managing waste and bioenergy; it changes the path we take with our environment. By introducing and supporting interdisciplinary innovation and circular economy, we move toward building a cleaner, smarter, and more sustainable world. What really matters with waste and energy is that what we throw away is moved forward into a cleaner, biopowered future.

References

Abdelfattah, A., Ali, S. S., Ramadan, H., El-Aswar, E. I., Eltawab, R., Ho, S. H., Elsamahy, T., Li, S., El-Sheekh, M. M., Schagerl, M., & Kornaros, M. (2023). Microalgae-based wastewater treatment: Mechanisms, challenges, recent advances, and future prospects. *Environmental Science and Ecotechnology, 13*, Article 100205.

Adegboye, M. F., Ojuederie, O. B., Talia, P. M., & Babalola, O. O. (2021). Bioprospecting of microbial strains for biofuel production: Metabolic engineering, applications, and challenges. *Biotechnology for Biofuels, 14*(1), 5.

Ahmad, A., Banat, F., Alsafar, H., & Hasan, S. W. (2022). Algae biotechnology for industrial wastewater treatment, bioenergy production, and high-value bioproducts. *Science of the Total Environment, 806*, Article 150585.

Akdemir, H., Liu, Y., Zhuang, L., Zhang, H., & Koffas, M. A. (2022). Utilization of microbial cocultures for converting mixed substrates to valuable bioproducts. *Current Opinion in Microbiology, 68*, Article 102157.

Al Makishah, N. H. (2017). Bioenergy: Microbial biofuel production advancement. *International Journal of Pharmaceutical Research and Allied Sciences, 6*(3).

Ali, J., Rasheed, T., Afreen, M., Anwar, M. T., Nawaz, Z., Anwar, H., & Rizwan, K. (2020). Modalities for conversion of waste to energy—Challenges and perspectives. *Science of the Total Environment, 727*, Article 138610.

Ali, S. S., El-Sheekh, M., Manni, A., Ruiz, H. A., Elsamahy, T., Sun, J., & Schagerl, M. (2022a). Microalgae-mediated wastewater treatment for biofuels production: A comprehensive review. *Microbiological Research, 265*, Article 127187.

Ali, S. S., Mastropetros, S. G., Schagerl, M., Sakarika, M., Elsamahy, T., El-Sheekh, M., Sun, J., & Kornaros, M. (2022b). Recent advances in wastewater microalgae-based biofuels production: A state-of-the-art review. *Energy Reports, 8*, 13253–13280.

Ardakani, M. N., & Gholikandi, G. B. (2020). Microbial fuel cells (MFCs) in integration with anaerobic treatment processes (AnTPs) and membrane bioreactors (MBRs) for simultaneous efficient wastewater/sludge treatment and energy recovery-A state-of-the-art review. *Biomass and Bioenergy, 141*, Article 105726.

Areniello, M., Matassa, S., Esposito, G., & Lens, P. N. (2023). Biowaste upcycling into second-generation microbial protein through mixed-culture fermentation. *Trends in Biotechnology, 41*(2), 197–213.

Arora, K., Kumar, P., Bose, D., Li, X., & Kulshrestha, S. (2021). Potential applications of algae in biochemical and bioenergy sector. *3 Biotech, 11*(6), 296.

Ashokkumar, V., Flora, G., Venkatkarthick, R., SenthilKannan, K., Kuppam, C., Stephy, G. M., Kamyab, H., Chen, W. H., Thomas, J., & Ngamcharussrivichai, C. (2022). Advanced technologies on the sustainable approaches for conversion of organic waste to valuable bioproducts: Emerging circular bioeconomy perspective. *Fuel, 324*, Article 124313.

Atstaja, D., Cudecka-Purina, N., Koval, V., Kuzmina, J., Butkevics, J., & Hrinchenko, H. (2024). Waste-to-energy in the circular economy transition and development of resource-efficient business models. *Energies, 17*(16), 4188.

Ávila, P. F., de Mélo, A. H., & Goldbeck, R. (2023). Cello-oligosaccharides production from multi-stage enzymatic hydrolysis by lignocellulosic biomass and evaluation of prebiotic potential. *Innovative Food Science and Emerging Technologies, 85*, Article 103335.

Awasthi, M. K., Sarsaiya, S., Patel, A., Juneja, A., Singh, R. P., Yan, B., Awasthi, S. K., Jain, A., Liu, T., Duan, Y., & Pandey, A. (2020). Refining biomass residues for sustainable energy and bio-products: An assessment of technology, its importance, and strategic applications in circular bio-economy. *Renewable and Sustainable Energy Reviews, 127*, Article 109876.

Azócar, L., Ciudad, G., Heipieper, H. J., & Navia, R. (2010). Biotechnological processes for biodiesel production using alternative oils. *Applied Microbiology and Biotechnology, 88*, 621–636.

Bala, K., Verma, S., Dogra, A., Das, A. P., & Kumar, R. (2024). Sustainable management for urban plastic waste generation. *Sustainable Management of Urban Plastic Waste Through Circular Economic Approaches*, 211–236. https://doi.org/10.1201/9781003465447-13

Balagurusamy, N., & Chandel, A. K., (Eds.). (2021). *Biogas production: From anaerobic digestion to a sustainable bioenergy industry*. Springer Nature.

Bardhan, P., Gupta, K., & Mandal, M. (2019). Microbes as bio-resource for sustainable production of biofuels and other bioenergy products. In *New and future developments in microbial biotechnology and bioengineering* (pp. 205–222). Elsevier.

Bhatia, S. K., Mehariya, S., Bhatia, R. K., Kumar, M., Pugazhendhi, A., Awasthi, M. K., Atabani, A. E., Kumar, G., Kim, W., Seo, S. O., & Yang, Y. H. (2021). Wastewater based microalgal biorefinery for bioenergy production: Progress and challenges. *Science of the Total Environment, 751*, Article 141599.

Bhuyan, P. (2024). Waste-to-energy transition: importance in circular bioeconomy and a tool to reduce greenhouse gas emission. In *Advances in energy from waste 2024* (pp. 503–524). Woodhead Publishing.

Bose, D., Santra, M., Sanka, R. V., & Krishnakumar, B. (2021). Bioremediation analysis of sediment-microbial fuel cells for energy recovery from microbial activity in soil. *International Journal of Energy Research, 45*(4), 6436–6445.

Brunschweiger, S., Hofmann, T., & Glas, K. (2020). Industrial wastewater treatment with simultaneous energy recovery using microbial fuel cells–A review. *Brewing Science,* 73.

Carvalho, J. M., Marreiros, B. C., & Reis, M. A. (2022). Polyhydroxyalkanoates production by mixed microbial culture under high salinity. *Sustainability, 14*(3), 1346.

Catone, C. M., Ripa, M., Geremia, E., & Ulgiati, S. (2021). Bio-products from algae-based biorefinery on wastewater: A review. *Journal of Environmental Management, 293*, Article 112792.

Cattaneo, C. R., Rodríguez, Y., Rene, E. R., García-Depraect, O., & Muñoz, R. (2022). Biogas bioconversion into poly (3-hydroxybutyrate) by a mixed microbial culture in a novel Taylor flow bioreactor. *Waste Management, 150*, 364–372.

Chintagunta, A. D., Zuccaro, G., Kumar, M., Kumar, S. J., Garlapati, V. K., Postemsky, P. D., Kumar, N. S., Chandel, A. K., & Simal-Gandara, J. (2021). Biodiesel production from lignocellulosic biomass using oleaginous microbes: Prospects for integrated biofuel production. *Frontiers in Microbiology, 12*, Article 658284.

Christophe, G., Kumar, V., Nouaille, R., Gaudet, G., Fontanille, P., Pandey, A., Soccol, C. R., & Larroche, C. (2012). Recent developments in microbial oils production: A possible alternative to vegetable oils for biodiesel without competition with human food? *Brazilian Archives of Biology and Technology, 55*, 29–46.

Das, A., Dey, S., Bala, K., Gnanasekaran, R., & Das, A. P. (2024). Microbial bioremediation technology for sustainable treatment and management of synthetic microfiber waste. *Sustainable Microbial Technology for Synthetic and Cellulosic Microfiber Bioremediation*, 221–238. https://doi.org/10.1007/978-3-031-62600-5_12

Das, A., Dey, S., Gacem, A., Yadav, K. K., Nath, H., Ram, D. K., & Das, A. P. (2025). Revolutionizing solid hydrocarbon sourcing: Plastic's journey from waste to upcycled treasure. *Environmental Science and Engineering, Part, F218*, 331–351. https://doi.org/10.1007/978-3-031-81615-4_14

Das, A. P., Mallavarapu, M., & Ghosh, S. (2023a). Ecotoxicity of mining pollutants on the environment and their remediation. *Environmental Chemistry and Ecotoxicology, 5*, 165–167. https://doi.org/10.1016/J.ENCECO.2023.08.002

Das, A. P., Mohanty, A. K., van Hullebusch, E. D., Figueiredo, G., Das, A. P., Mohanty, A. K., van Hullebusch, E. D., & Figueiredo, G. (2023b). Bioremediation as an emerging technology for the removal of synthetic microplastic pollutants from marine ecosystem. *MarPB, 194*, Article 115297. https://doi.org/10.1016/J.MARPOLBUL.2023.115297

Das, A. P., Nayak, B., & Behera, I. D. (2024b). Applications of microorganisms in advanced geomicrobiological and biotechnological exploration. *Geomicrobiology Journal, 41*(4), 309–311. https://doi.org/10.1080/01490451.2024.2327059/ASSET//CMS/ASSET/DE2A1F67-08C1-475A-ABC7-2B3F4C9E8BCF/01490451.2024.2327059.FP.PNG

Dey, S., Dash, A., & Das, A. P. (2024). Bioremediation as a sustainable and circular solution for mining waste management and bioenergy generation. *Sustainable Management of Mining Waste and Tailings: A Circular Economy Approach*, 107–119. https://doi.org/10.1201/9781003442455-6/BIOREMEDIATION-SUSTAINABLE-CIRCULAR-SOLUTION-MINING-WASTE-MANAGEMENT-BIOENERGY-GENERATION-SUDESHNA-DEY-AKANSHYA-DASH-ALOK-PRASAD-DAS

Dhanya, B. S., Mishra, A., Chandel, A. K., & Verma, M. L. (2020). Development of sustainable approaches for converting the organic waste to bioenergy. *Science of the Total Environment, 723*, Article 138109.

Duan, Y., Chen, H., Liu, T., Awasthi, S. K., Pandey, A., Zhang, Z., & Awasthi, M. K.: Food waste biorefinery: Case study in China for enhancing the emerging bioeconomy. In *Waste biorefinery 2021* (pp. 421–438). Elsevier.

Fernando, Y., Tseng, M. L., Aziz, N., Ikhsan, R. B., & Wahyuni-TD, I. S. (2022). Waste-to-energy supply chain management on circular economy capability: An empirical study. *Sustainable Production and Consumption, 31*, 26–38.

Gondi, R., Kavitha, S., Kannah, R. Y., Kumar, G., & Banu, J. R. (2022). Wastewater based microalgae valorization for biofuel and value-added products recovery. *Sustainable Energy Technologies and Assessments, 53*, Article 102443.

Han, M., Zhang, C., Li, F., & Ho, S. H. (2022). Data-driven analysis on immobilized microalgae system: New upgrading trends for microalgal wastewater treatment. *Science of the Total Environment, 852*, Article 158514.

Harun, R., Danquah, M. K., & Forde, G. M. (2010). Microalgal biomass as a fermentation feedstock for bioethanol production. *Journal of Chemical Technology and Biotechnology, 85*(2), 199–203.

Hoang, A.T., Nguyen, H.P., Pandey, A., Lichtfouse, E., Veza, I., Ahmed, S.F., Ağbulut, Ü., Duong, X. Q., & Nguyen, X. P. (2023). Waste to energy in circular economy. In *Renewable energy in circular economy 2023* (pp. 113–133). Springer International Publishing.

Hoang, A. T., Sirohi, R., Pandey, A., Nižetić, S., Lam, S. S., Chen, W. H., Luque, R., Thomas, S., Arıcı, M., & Pham, V. V. (2023a). Biofuel production from microalgae: Challenges and chances. *Phytochemistry Reviews, 22*(4), 1089–1126.

Huang, C., Li, R., Tang, W., Zheng, Y., & Meng, X. (2022b). Improve enzymatic hydrolysis of lignocellulosic biomass by modifying lignin structure via sulfite pretreatment and using lignin blockers. *Fermentation, 8*(10), 558.

Huang, H., Zhong, S., Wen, S., Luo, C., & Long, T. (2022a). Improving the efficiency of wastewater treatment and microalgae production for biofuels. *Resources, Conservation and Recycling, 178*, Article 106094.

Ibeto, C. N., Ofoefule, A. U., & Agbo, K. E. (2011). A global overview of biomass potentials for bioethanol production: A renewable alternative fuel. *Trends in Applied Sciences Research, 6*(5), 410.

Jayaseelan, M., Usman, M., Somanathan, A., Palani, S., Muniappan, G., & Jeyakumar, R. B. (2021). Microalgal production of biofuels integrated with wastewater treatment. *Sustainability, 13*(16), 8797.

John, A., & Das, A. P. (2024). Government strategies for achieving circular economy for plastic waste management in Southeast Asia. *Sustainable Management of Urban Plastic Waste Through Circular Economic Approaches*, 72–82. https://doi.org/10.1201/9781003465447-6

Joshi, C. P., & Nookaraju, A. (2012). New avenues of bioenergy production from plants: Green alternatives to petroleum. *Journal of Petroleum and Environmental Biotechnology, 3*(07).

Joshi, S., & Mishra, S. (2022). Recent advances in biofuel production through metabolic engineering. *Bioresource Technology, 352*, Article 127037.

Kadri, M. S., Singhania, R. R., Haldar, D., Patel, A. K., Bhatia, S. K., Saratale, G., Parameswaran, B., & Chang, J. S. (2023). Advances in algomics technology: Application in wastewater treatment and biofuel production. *Bioresource Technology, 387*, Article 129636.

Kapoor, R., Ghosh, P., Kumar, M., Sengupta, S., Gupta, A., Kumar, S. S., Vijay, V., Kumar, V., Vijay, V. K., & Pant, D. (2020). Valorization of agricultural waste for biogas based circular economy in India: A research outlook. *Bioresource Technology, 304*, Article 123036.

Karimi, F., Mazaheri, D., Saei Moghaddam, M., Mataei Moghaddam, A., Sanati, A. L., & Orooji, Y. (2021). Solid-state fermentation as an alternative technology for cost-effective production of bioethanol as useful renewable energy: A review. *Biomass Conversion and Biorefinery,* 1–7.

Keasling, J., Garcia Martin, H., Lee, T. S., Mukhopadhyay, A., Singer, S. W., & Sundstrom, E. (2021). Microbial production of advanced biofuels. *Nature Reviews Microbiology, 19*(11), 701–715.

Kennes, D., Abubackar, H. N., Diaz, M., Veiga, M. C., & Kennes, C. (2016). Bioethanol production from biomass: Carbohydrate vs syngas fermentation. *Journal of Chemical Technology and Biotechnology, 91*(2), 304–317.

Khoo, K. S., Ahmad, I., Chew, K. W., Iwamoto, K., Bhatnagar, A., & Show, P. L. (2023). Enhanced microalgal lipid production for biofuel using different strategies including genetic modification of microalgae: A review. *Progress in Energy and Combustion Science, 96*, Article 101071.

Kim, S. Y., Yang, Y. H., & Choi, K. Y. (2020). Bioconversion of plant hydrolysate biomass into biofuels using an engineered Bacillus subtilis and Escherichia coli mixed-whole cell biotransformation. *Biotechnology and Bioprocess Engineering, 25*, 477–484.

Kleerebezem, R., Joosse, B., Rozendal, R., & Van Loosdrecht, M. C. (2015). Anaerobic digestion without biogas? *Reviews in Environmental Science and Bio/technology, 14*, 787–801.

Kleerebezem, R., & van Loosdrecht, M. C. (2007). Mixed culture biotechnology for bioenergy production. *Current Opinion in Biotechnology, 18*(3), 207–212.

Korres, N. E., O'Kiely, P., Benzie, J. A., & West, J. S. (2013). *Bioenergy production by anearobic digestion.* Routledge.

Kumar, N., Sharma, G., Chandel, H., Shyam, K., Thakur, S., Vaswani, P., & Saxena, G. (2022b). Microalgae in wastewater treatment and biofuel production: Recent advances, challenges, and future prospects. *Omics Insights in Environmental Bioremediation, 27*, 237–271.

Kumar, S. D., Yasasve, M., Karthigadevi, G., Aashabharathi, M., Subbaiya, R., Karmegam, N., & Govarthanan, M. (2022a). Efficiency of microbial fuel cells in the treatment and energy recovery from food wastes: Trends and applications-a review. *Chemosphere, 287*, Article 132439.

Li, X., Shi, Y., Kong, W., Wei, J., Song, W., & Wang, S. (2022). Improving enzymatic hydrolysis of lignocellulosic biomass by bio-coordinated physicochemical pretreatment—A review. *Energy Reports, 8*, 696–709.

Li, Z., & Zhu, L. (2021). The scientometric analysis of the research on microalgae-based wastewater treatment. *Environmental Science and Pollution Research, 28*, 25339–25348.

Lin, L. (2022). Bottom-up synthetic ecology study of microbial consortia to enhance lignocellulose bioconversion. *Biotechnology for Biofuels and Bioproducts, 15*(1), 14.

Liu, X. Y., & Hong, Y. (2021). Microalgae-based wastewater treatment and recovery with biomass and value-added products: A brief review. *Current Pollution Reports, 7*, 227–245.

López-Sánchez, A., Silva-Gálvez, A. L., Aguilar-Juárez, Ó., Senés-Guerrero, C., Orozco-Nunnelly, D. A., Carrillo-Nieves, D., & Gradilla-Hernández, M. S. (2022). Microalgae-based livestock wastewater treatment (MbWT) as a circular bioeconomy approach: Enhancement of biomass productivity, pollutant removal and high-value compound production. *Journal of Environmental Management, 308*, Article 114612.

Mahadevan, V., Subbaiyan, N., Panchamoorthy, G. K., Jayaseelan, A., Palaniappan, S. K., & Siengchin, S. (2025). Hydrothermal liquefaction of confused waste to bio-oil: A study on elemental and energy recovery. *Energy Conversion and Management, 325*, Article 119353.

Malik, S., Kishore, S., Dhasmana, A., Kumari, P., Mitra, T., Chaudhary, V., Kumari, R., Bora, J., Ranjan, A., Minkina, T., & Rajput, V. D. (2023). A perspective review on microbial fuel cells in treatment and product recovery from wastewater. *Water, 15*(2), 316.

Marangon, B. B., Magalhães, I. B., Pereira, A. S., Silva, T. A., Gama, R. C., Ferreira, J., Castro, J. S., Assis, L. R., Lorentz, J. F., & Calijuri, M. L. (2023). Emerging microalgae-based biofuels: Technology, life-cycle and scale-up. *Chemosphere, 326*, Article 138447.
Martins das Neves, L. C., Converti, A., & Vessoni Penna, T. C. (2009) Biogas production: New trends for alternative energy sources in rural and urban zones. *Chemical Engineering and Technology: Industrial Chemistry-Plant Equipment-Process Engineering-Biotechnology, 32*(8), 1147–1153.
Mehariya, S., Goswami, R. K., Verma, P., Lavecchia, R., & Zuorro, A. (2021). Integrated approach for wastewater treatment and biofuel production in microalgae biorefineries. *Energies, 14*(8), 2282.
Mirsalami, S. M., & Mirsalami, M. (2025). Advances in genetically engineered microorganisms: Transforming food production through precision fermentation and synthetic biology. *Future Foods, 11*, Article 100601.
Mishra, K., Siwal, S. S., Nayaka, S. C., Guan, Z., & Thakur, V. K. (2023). Waste-to-chemicals: Green solutions for bioeconomy markets. *Science of the Total Environment, 887*, Article 164006.
Mishra, S., Dash, D., & Das, A. P. (2024). Aquatic microbial diversity on plastisphere: Colonization and potential role in microplastic biodegradation. *Geomicrobiology Journal, 41*(4), 312–323. https://doi.org/10.1080/01490451.2023.2209750
Mohanty, M., Tripathy, B., Koteswara Reddy, G., & Das, A. P. (2024). Greener technologies for establishing restraint over microplastic pollution. *Sustainable Microbial Technology for Synthetic and Cellulosic Microfiber Bioremediation*, 239–258. https://doi.org/10.1007/978-3-031-62600-5_13
Moon, H. S., Abercrombie, J. M., Kausch, A. P., & Stewart, C. N. (2010). Sustainable use of biotechnology for bioenergy feedstocks. *Environmental Management, 46*, 531–538.
Moondra, N., Jariwala, N. D., & Christian, R. A. (2021). Microalgae based wastewater treatment: A shifting paradigm for the developing nations. *International Journal of Phytoremediation, 23*(7), 765–771.
Moreira, J. B., Santos, T. D., Duarte, J. H., Bezerra, P. Q., de Morais, M. G., & Costa, J. A. (2023). Role of microalgae in circular bioeconomy: From waste treatment to biofuel production. *Clean Technologies and Environmental Policy, 25*(2), 427–437.
Moustakas, K., Loizidou, M., Rehan, M., & Nizami, A. S. (2020). A review of recent developments in renewable and sustainable energy systems: Key challenges and future perspective. *Renewable and Sustainable Energy Reviews, 119*, Article 109418.
Muktham, R., Bhargava, S. K., Bankupalli, S., & Ball, A. S. (2016). A review on 1st and 2nd generation bioethanol production-recent progress. *Journal of Sustainable Bioenergy Systems, 6*(3), 72–92.
Muniraj, I. K., Uthandi, S. K., Hu, Z., Xiao, L., & Zhan, X. (2015). Microbial lipid production from renewable and waste materials for second-generation biodiesel feedstock. *Environmental Technology Reviews, 4*(1), 1–6.
Naha, A., Debroy, R., Sharma, D., Shah, M. P., & Nath, S. (2023). Microbial fuel cell: A state-of-the-art and revolutionizing technology for efficient energy recovery. *Cleaner and Circular Bioeconomy, 5*, Article 100050.
Obileke, K., Nwokolo, N., Makaka, G., Mukumba, P., & Onyeaka, H. (2021). Anaerobic digestion: Technology for biogas production as a source of renewable energy—A review. *Energy and Environment, 32*(2), 191–225.
Olivieri, G., Wijffels, R. H., Marzocchella, A., & Russo, M. E. (2021). Bioreactor and bioprocess design issues in enzymatic hydrolysis of lignocellulosic biomass. *Catalysts, 11*(6), 680.
Pan, F. D., Liu, S., Xu, Q. M., Chen, X. Y., & Cheng, J. S. (2021). Bioconversion of kitchen waste to surfactin via simultaneous enzymolysis and fermentation using mixed-culture of enzyme-producing fungi and Bacillus amyloliquefaciens HM618. *Biochemical Engineering Journal, 172*, Article 108036.
Pessôa, L. C., Deamici, K. M., Pontes, L. A., Druzian, J. I., & de Jesus, A. D. (2021). Technological prospection of microalgae-based biorefinery approach for effluent treatment. *Algal Research., 60*, Article 102504.

Peter, A. P., Khoo, K. S., Chew, K. W., Ling, T. C., Ho, S. H., Chang, J. S., & Show, P. L. (2021). Microalgae for biofuels, wastewater treatment and environmental monitoring. *Environmental Chemistry Letters, 19*, 2891–2904.

Phwan, C. K., Ong, H. C., Chen, W. H., Ling, T. C., Ng, E. P., & Show, P. L. (2018). Overview: Comparison of pretreatment technologies and fermentation processes of bioethanol from microalgae. *Energy Conversion and Management, 173*, 81–94.

Prajapati, S. K., Malik, A., & Vijay, V. K. (2014). Comparative evaluation of biomass production and bioenergy generation potential of Chlorella spp. through anaerobic digestion. *Applied Energy, 114*, 790–797.

Rajmohan, K. S., Ramya, C., & Varjani, S. (2021). Trends and advances in bioenergy production and sustainable solid waste management. *Energy and Environment, 32*(6), 1059–1085.

Rao, P. V., Baral, S. S., Dey, R., & Mutnuri, S. (2010). Biogas generation potential by anaerobic digestion for sustainable energy development in India. *Renewable and Sustainable Energy Reviews, 14*(7), 2086–2094.

Rasool, U., & Hemalatha, S. (2016). A review on bioenergy and biofuels: Sources and their production. *Brazilian Journal of Biological Sciences, 3*(5), 3–22.

Rastogi, M., & Shrivastava, S. (2017). Recent advances in second generation bioethanol production: An insight to pretreatment, saccharification and fermentation processes. *Renewable and Sustainable Energy Reviews, 80*, 330–340.

Ray, S. K., Dey, S., Das, A. P., & John, A. (2024). Microplastic waste management through greener technologies : A review. *Sustainable Management of Urban Plastic Waste Through Circular Economic Approaches*, 237–261. https://doi.org/10.1201/9781003465447-14

Reddy Koteswara, G., Bhargavi, S., Yamuna, D., Anjani Devi, M., Jeevana Lakshmi Kumar, G., Manas, T., Adarsh, S., Bhavya Sai, V., Nath, H., Alreshidi, M. A., Togun, H., Ram, D. K., & Das, A. P. (2025). Recent advancements in the remediation of marine hydrocarbon and plastic pollutants. *Environmental Science and Engineering, Part F218*, 127–142. https://doi.org/10.1007/978-3-031-81615-4_7

Rezania, S., Oryani, B., Nasrollahi, V. R., Darajeh, N., Lotfi Ghahroud, M., & Mehranzamir, K. (2023). Review on waste-to-energy approaches toward a circular economy in developed and developing countries. *Processes, 11*(9), 2566.

Rittmann, B. E. (2008). Opportunities for renewable bioenergy using microorganisms. *Biotechnology and Bioengineering, 100*(2), 203–212.

Robak, K., & Balcerek, M. (2018). Review of second generation bioethanol production from residual biomass. *Food Technology and Biotechnology, 56*(2), 174.

Rohrbach, J. C., & Luterbacher, J. S. (2021). Investigating the effects of substrate morphology and experimental conditions on the enzymatic hydrolysis of lignocellulosic biomass through modeling. *Biotechnology for Biofuels, 14*(1), 103.

Ruane, J., Sonnino, A., & Agostini, A. (2010). Bioenergy and the potential contribution of agricultural biotechnologies in developing countries. *Biomass and Bioenergy, 34*(10), 1427–1439.

Saini, J. K., Kaur, A., & Mathur, A. (2022a). Strategies to enhance enzymatic hydrolysis of lignocellulosic biomass for biorefinery applications: A review. *Bioresource Technology, 360*, Article 127517.

Saini, R., Patel, A. K., Saini, J. K., Chen, C. W., Varjani, S., Singhania, R. R., & Di Dong, C. (2022b). Recent advancements in prebiotic oligomers synthesis via enzymatic hydrolysis of lignocellulosic biomass. *Bioengineered, 13*(2), 2139–2172.

Sajid, M., Raheem, A., Ullah, N., Asim, M., Rehman, M. S., & Ali, N. (2022). Gasification of municipal solid waste: Progress, challenges, and prospects. *Renewable and Sustainable Energy Reviews, 168*, Article 112815.

Sarker, T. R., Ethen, D. Z., Asha, H. H., Islam, S., & Ali, M. R. (2025). Transformation of municipal solid waste to biofuel and bio-chemicals–a review. *International Journal of Environmental Science and Technology, 22*(5), 3811–3832.

Shabani, M., Younesi, H., Pontié, M., Rahimpour, A., Rahimnejad, M., & Zinatizadeh, A. A. (2020). A critical review on recent proton exchange membranes applied in microbial fuel cells for renewable energy recovery. *Journal of Cleaner Production, 264*, Article 121446.

Shah, A. V., Singh, A., Mohanty, S. S., Srivastava, V. K., & Varjani, S. (2022). Organic solid waste: Biorefinery approach as a sustainable strategy in circular bioeconomy. *Bioresource Technology, 349*, Article 126835.

Shah, A. V., Srivastava, V. K., Mohanty, S. S., & Varjani, S. (2021). Municipal solid waste as a sustainable resource for energy production: State-of-the-art review. *Journal of Environmental Chemical Engineering., 9*(4), Article 105717.

Shah, M. P. (Ed.). (2020). *Microbial bioremediation and biodegradation.* Springer.

Singh, P., Fnu, K., & Encarnação, T. (2023). Genetic modification: a gateway to stimulate the industrial production of biofuels. In *Marine organisms: a solution to environmental pollution? Uses in bioremediation and in biorefinery* (pp. 237–260). Springer International Publishing.

Singh, K., Kumar, N., Bharti, A., Thakur, P., & Kumar, V. (2024). Waste to energy conversion: Key elements for sustainable waste management. In *Integrated waste management: A sustainable approach from waste to wealth 2024* (pp. 91–117). Springer Nature Singapore.

Sivakumar, G., Vail, D. R., Xu, J., Burner, D. M., Lay, J. O., Ge, X., & Weathers, P. J. (2010). Bioethanol and biodiesel: Alternative liquid fuels for future generations. *Engineering in Life Sciences, 10*(1), 8–18.

Song, G., Sun, C., Madadi, M., Dou, S., Yan, J., Huan, H., Aghbashlo, M., Tabatabaei, M., Sun, F., & Ashori, A. (2024). Dual assistance of surfactants in glycerol organosolv pretreatment and enzymatic hydrolysis of lignocellulosic biomass for bioethanol production. *Bioresource Technology, 395*, Article 130358.

Srimongkol, P., Sangtanoo, P., Songserm, P., Watsuntorn, W., & Karnchanatat, A. (2022). Microalgae-based wastewater treatment for developing economic and environmental sustainability: Current status and future prospects. *Frontiers in Bioengineering and Biotechnology, 10*, Article 904046.

Srivastava, R. K. (2019). Bio-energy production by contribution of effective and suitable microbial system. *Materials Science for Energy Technologies, 2*(2), 308–318.

Surendra, K. C., Sawatdeenarunat, C., Shrestha, S., Sung, S., & Khanal, S. K. (2015). Anaerobic digestion-based biorefinery for bioenergy and biobased products. *Industrial Biotechnology, 11*(2), 103–112.

Swain, S., Purnima Sudha, M., Dey, S., Badu, Y., Bhoi, S. R., Alreshidi, M. A., Qasim, M. T., Nath, H., Ram, D. K., & Das, A. P. (2025). Sources of micro and nano plastic hydrocarbon pollutants in marine bodies and their treatment. *Environmental Science and Engineering, Part, F218*, 45–67. https://doi.org/10.1007/978-3-031-81615-4_3

Tabatabaei, M., Karimi, K., Kumar, R., & Horváth, I. S. (2015). Renewable energy and alternative fuel technologies. *BioMed Research International, 2015*, Article 245935.

Taghipour, A., Foukolaei, P. Z., Ghaedi, M., & Khazaei, M. (2023). Sustainable multi-objective models for waste-to-energy and waste separation site selection. *Sustainability, 15*(22), 15764.

Tse, T. J., Wiens, D. J., & Reaney, M. J. (2021). Production of bioethanol—A review of factors affecting ethanol yield. *Fermentation, 7*(4), 268.

Vasileiadou, A. (2024). From organic wastes to bioenergy, biofuels, and value-added products for urban sustainability and circular economy: A review. *Urban Science, 8*(3), 121.

Verma, M., Verma, M. K., Singh, V., Singh, J., Singh, V., & Mishra, V. (2022). Advancements in applicability of microbial fuel cell for energy recovery from human waste. *Bioresource Technology Reports, 17*, Article 100978.

Vijayalakshmi, S., Anand, M., & Ranjitha, J. (2020). Microalgae-based biofuel production using low-cost nanobiocatalysts. In *Microalgae cultivation for biofuels production 2020* (pp. 251–263). Academic.

Vohra, M., Manwar, J., Manmode, R., Padgilwar, S., & Patil, S. (2014). Bioethanol production: Feedstock and current technologies. *Journal of Environmental Chemical Engineering, 2*(1), 573–584.

Wang, X., Qin, Z. H., Hao, T. B., Ye, G. B., Mou, J. H., Balamurugan, S., Bin, X. Y., Buhagiar, J., Wang, H. M., Lin, C. S., & Yang, W. D. (2022). A combined light regime and carbon supply regulation strategy for microalgae-based sugar industry wastewater treatment and low-carbon biofuel production to realise a circular economy. *Chemical Engineering Journal, 446*, Article 137422.

Wilhelm, C., & Jakob, T. (2011). From photons to biomass and biofuels: Evaluation of different strategies for the improvement of algal biotechnology based on comparative energy balances. *Applied Microbiology and Biotechnology, 92*, 909–919.

Wu, R., Chen, D., Cao, S., Lu, Z., Huang, J., Lu, Q., Chen, Y., Chen, X., Guan, N., Wei, Y., & Huang, R. (2020). Enhanced ethanol production from sugarcane molasses by industrially engineered Saccharomyces cerevisiae via replacement of the PHO_4 gene. *RSC Advances, 10*(4), 2267–2276.

Xu, C., Alam, M. A., Wang, Z., Chen, H., Zhang, J., Huang, S., Zhuang, W., & Xu, J. (2021). Mechanisms of bio-additives on boosting enzymatic hydrolysis of lignocellulosic biomass. *Bioresource Technology, 337*, Article 125341.

Xu, C., Tong, S., Sun, L., & Gu, X. (2023). Cellulase immobilization to enhance enzymatic hydrolysis of lignocellulosic biomass: An all-inclusive review. *Carbohydrate Polymers, 321*, Article 121319.

Xue, J., Balamurugan, S., Li, T., Cai, J. X., Chen, T. T., Wang, X., Yang, W. D., & Li, H. Y. (2021). Biotechnological approaches to enhance biofuel producing potential of microalgae. *Fuel, 302*, Article 121169.

Yaashikaa, P. R., Kumar, P. S., Saravanan, A., Varjani, S., & Ramamurthy, R. (2020). Bioconversion of municipal solid waste into bio-based products: A review on valorisation and sustainable approach for circular bioeconomy. *Science of the Total Environment, 748*, Article 141312.

Zhu, W., He, Q., Gao, H., Nitayavardhana, S., Khanal, S. K., & Xie, L. (2020). Bioconversion of yellow wine wastes into microbial protein via mixed yeast-fungus cultures. *Bioresource Technology, 299*, Article 122565.

Biodegradable Polymers: A Green Solution to Environmental Plastic Pollution

Raj Kumar, Bhagat Singh, Arushi Dogra, Abhiranjan Kumar, Sarfaraz Ali, Alok Prasad Das, and Kiran Bala

1 Introduction

Rising as a major environmental concern, plastic pollution is typified by non-degradable plastic garbage accumulating in ecosystems, oceans, and landfills. Over 8.3 billion tons of plastics have been generated since the 1950s; just around 9% have been recycled, while the rest have been burned or thrown away. Conventional plastics, generated from petroleum, are polymers with carbon–carbon backbones that withstand natural breakdown and cause ongoing waste in the environment spanning millennia. This tenacity fuels ecological degradation, harm to animals, and microplastic pollution of food systems. Often called "bioplastics" when generated from biological sources, biodegradable polymers are meant to break down more

R. Kumar
Department of Zoology, S.N. Sinha College Tekari, Gaya M.U., Bihar, India

B. Singh
Department of Zoology, Niilm University, Kaithal Haryana, India

A. Dogra
Faculty of Life Sciences and Biotechnology, South Asian University, Delhi, India

A. Kumar
B.M.D. College, Dayalpur, Vaishali (Hajipur), India

B.R.A.B.U. Bihar, Muzaffarpur, Bihar, India

S. Ali
Department of Biotechnology, Magadh University, Bodh-Gaya, India

K. Bala (✉)
Department of Zoology, Deshbandhu College, University of Delhi, New Delhi, India
e-mail: kbala@db.du.ac.in

A. P. Das
Department of Life Sciences, Rama Devi Women's University, Bhubaneswar, Odisha, India

S. Ghosh et al. (eds.), *Advances in Organic Waste Conversion Through Industrial Biotechnology*, Sustainable Environmental Waste Management Strategies,
https://doi.org/10.1007/978-3-032-11377-1_3

easily via natural processes. These materials either have chemical structures fit for microbial breakdown or come from renewable biomass (bio-based). The idea behind biodegradable plastics is appealing: materials that perform the same purposes as conventional plastics but can break down into benign end products following use. As part of a larger approach to solve plastic pollution, the appeal of a "green" plastic has attracted major interest in science, industry, and policy. This chapter investigates biodegradable polymers as a green fix for plastic pollution of the environment.

1.1 Definition of Biodegradable Polymers

Biodegradable polymers, quite unlike the plastics we're used to, can be broken down naturally. We're talking about bacteria and fungi doing the work, turning these polymers into water, carbon dioxide, and biomass. It's a sharp contrast to those conventional plastics hanging around for ages, seriously messing with the environment (Reddy Koteswara et al., 2025; Swain et al., 2025). Biodegradable polymers come from renewable sources, think starch or polylactic acid (PLA), but also synthetic ones, like polyhydroxyalkanoates (PHA) (Kothawade & Padwal, 2025). Because they can break down in certain conditions, they're great for things like packaging and even biomedical devices, you know, where cutting down on plastic waste is a big deal (Kumar et al., 2024). With everyone looking for sustainable solutions, getting what biodegradable polymers are all about, and where they fit in, is key to tackling the plastic pollution problem (Stoica et al., 2024). An informative breakdown of the biodegradation process really shows how beneficial these materials are over their life (Das et al., 2024c; Reddy Koteswara et al., 2025). Unlike conventional plastics (e.g., polyethylene, polystyrene) which survive due to stable carbon–carbon bonds, biodegradable plastics are designed to have connections (such as ester, amide, or glycosidic bonds) that bacteria can split by enzymatic action and hydrolyzed. Eventually, this mechanism breaks down the polymer into simpler molecules that fit within natural biogeochemical cycles, so producing ideally no hazardous residue. It is crucial to separate related phrases sometimes used in talks of sustainable plastics: Unlike fossil fuels, bio-based plastics are plastics derived from renewable biological sources such as plant biomass. A bio-based plastic is not always biodegradable; for instance, bio-based polyethylene stays non-biodegradable yet chemically equivalent to conventional polyethylene. On the other hand, some biodegradable plastics such as polycaprolactone, PCL have fossil-fuel derived origin.

1.2 Importance of Addressing Environmental Issues

The exploration of biodegradable polymers is driven by the global issue of plastic pollution. The versatility, durability, and affordability of plastics have enabled their infiltration into every aspect of contemporary life since the onset of mass production.

Regrettably, these identical attributes render plastics persistent pollutants. (Geyer et al., 2017) estimated that 8.3 billion metric tons of virgin plastics were produced, resulting in 6.3 billion tons of plastic waste, of which 79% was deposited in landfills or the natural environment, 12% was incinerated, and just 9% was recycled. Without intervention, it is projected that 12 billion tons of plastic waste will accumulate in landfills or the environment by 2025, according to current trends. The durability of conventional plastics enables items such as packaging, bottles, and bags, utilized for few minutes, to persist for millennia. Plastic waste is now ubiquitous, found in surface waterways, deep-sea layers, remote mountain peaks, and Arctic ice. Air, water, and human organs have disclosed the presence of microplastics (particles smaller than 5 mm) resulting from direct emissions (e.g., synthetic fibers) or the degradation of larger items (Bala et al., 2024; Das et al., 2025; Dey et al., 2024). The environmental degradation is significant: plastics can facilitate the transmission of invasive species or toxins; soil quality deteriorates; marine organisms experience entanglement or poisoning.

It's hard to overstate the urgency of environmental issues, especially when you consider plastic pollution and how far its effects reach into ecosystems and our own health. The world is on track to produce 34 billion metric tons of primary plastic by 2050, so how we handle waste and the accumulation of plastics is clearly a big challenge. Studies show that regular plastics contribute to greenhouse gas emissions and cause long-term damage to the environment. However, alternatives like biodegradable polymers offer a potential way to lessen these effects (Abdullayeva and Yamil, 2022). To really know how well bioplastics work, we have to understand their lifecycle and how they break down (Parameswaranpillai et al., 2021). Using biodegradable materials could substantially decrease the amount of plastic waste in water environments and encourage a more sustainable future that benefits both biodiversity and public health (Das et al., 2024a; Ghosh & Das, 2020; John & Das, 2024; Mishra & Das, 2022). Therefore, tackling environmental issues with new materials and responsible habits is vitally important for protecting ecological integrity.

2 Understanding Biodegradable Polymers

Grasping the nature of biodegradable polymers sheds light on how they might help with today's plastic pollution issue, notably when we think about how they're made and what happens to them. These polymers break down naturally, mostly thanks to tiny organisms, which really cuts down on the bad effects that regular plastics have on the environment. Studies show that how well these materials break down depends on things like the environment and the tiny organisms that live in different places, making them pretty adaptable (Agarwal et al., 2024). Also, coming up with ways to test how well things break down has been super important in figuring out how these materials affect the environment, which makes them more believable as a sustainable option (Kothawade & Padwal, 2025).

2.1 Types of Biodegradable Polymers

Depending on their composition and origin, biodegradable polymers can be categorized. This impacts their breakdown and utilization. PLA and PHA are popular study topics for aliphatic polyesters. Due to their renewable nature, they can be used for packaging and medical instrument manufacture. Polysaccharides, like starch and cellulose, are important natural products. Overall, it's biodegradable and environmentally friendly. Not only are these polymers environmentally friendly, but they also present difficult recycling and recovery issues, making it crucial to understand how they work and how they affect the environment.

2.1.1 Polylactic Acid

Polylactic acid is a popular biodegradable material. Fermented maize starch or sugarcane to lactic acid produces this thermoplastic aliphatic polyester. Following polymerization PLA is biodegradable and biobased in industrial composting. PLA is suitable for packaging (bottles, food containers), disposable cutlery, textile fibers, and 3D printing filament since its mechanical properties mirror those of PET. It has a low melting point (~150–170 °C), is translucent, semi-crystalline, or amorphous. PLA's bio-based nature and smaller carbon footprint make it more attractive than fossil polymers, especially if generated using renewable energy. Industrial PLA composts at temperatures above 55 °C with significant humidity and microbial activity. Hydrolysis of PLA's ester bonds leads to microbial digestion of lactic acid monomers, resulting in months of CO_2, water, and biomass. PLA does not biodegrade in soil or sea since it requires prolonged high temperatures. PLA can behave like plastic in soil or water for years unless it is finely fragmented and certain bacteria are present.

2.1.2 Polyhydroxyalkanoates

Polyhydroxyalkanoates (PHAs) are biodegradable polyesters produced spontaneously by bacteria as energy storage molecules. Polyhydroxybutyrate (PHB) and copolymers like PHB-Valate (PHBV) are common monomers in PHA. Since they biodegrade in marine and soil environments, these stand out. PHA polymers are hydrophobic semi-crystalline thermoplastics. PHB is fragile but similar to polypropylene. Adding hydroxyvalate units to monomers increases flexibility and toughness. Use cases include packaging films, mulch films, throwaway products, and medical devices. Biocompatible PHA. Microorganisms digest waste oils or carbohydrates into PHA granules, which are removed. General biodegradability distinguishes PHAs. They biodegrade well in compost, soil, and marine environments with the right microbes. Then what? PHA is a natural polymer that bacteria break down with enzymes. Complete biodegradation yields water, biomass, and methane or carbon dioxide. In warm compost, PHAs break down in weeks; in cooler marine water, it may take months or longer, but is faster than regular plastics and leaves no microplastics.

2.1.3 Other Aliphatic Polyesters

From fossil fuels—1,4-butanediol, adipic acid, terephthalic acid synthetic biodegradable polyester PBAT. Although not bio-based, it is totally biodegradable and mixed for flexibility and speed. Characteristics and Use: PBAT is a flexible, tear-resistant polymer that biodegrades like LDPE. It's used in mulch films, compostable shopping and garbage bags, and PLA mixing to make it less brittle. PBAT is soft and melts at ~ 110 °C, making it suitable for ordinary plastic equipment. Because bacteria attack its aliphatic segments (adipate), PBAT biodegrades efficiently in compost. Terephthalate units also dissolve. The terephthalate content may limit marine breakdown, yet it is biodegradable (EN 13,432, ASTM D6400) in soil and marine environments. Haider et al. (2018 in Angewandte Chemie) found that soil can mineralize PBAT entirely in a year. PBAT may biodegrade in seawater, but not as quickly as PHAs, according to marine research.

2.1.4 Starch-Based Plastics

Biodegradable plastics can be manufactured from corn, potato, cassava, etc. starch. TPS is plasticized starch made process able with glycerol and water. Starch is also important in composite biodegradable polymers. Uses and traits: Hydrophilic pure TPS has low mechanical strength. Starch can be combined with PLA, PCL, or PBAT to make faster-biodegrading, cheaper polymers. Agricultural films, foam, biodegradable bags, "peanuts," and loose-fill packing are used. Blends save money and hasten biodegradation since bacteria eat starch. Naturally biodegradable starch is easily digested by bacteria (Mishra et al., 2024; Mohanty et al., 2024b). Blended starch biodegrades quickly to provide a porous structure that lets bacteria target leftover polymers like PBAT. Compostable starch-based plastics. However, high-starch materials can break apart if wet, which aids biodegradation. Depending on formulation, industrial composting can degrade starch plastics into CO_2 and water within weeks to months.

2.1.5 Cellulose and Natural Polymer Derives

Cellulose, a common plant cell wall polymer, is biodegradable. Low-degree substitution cellulose acetate is used in several biodegradable products. Films of fiber: Older biodegradable packaging films include cellophane and regenerated cellulose. Modern forms may use cellulose coatings or casings. Highly chemically modified cellulose (such cellulose acetate with considerable substitution) breaks down slowly. Chitosan, a biodegradable polysaccharide made from crab shells, is used in films and coatings for therapeutic or specialty uses. Proteins: Biodegradable protein-based plastics including soy protein, gelatin, and casein are being studied, however they often have niche uses or require considerable modification.

2.2 *Mechanisms of Biodegradation*

Understanding biodegradation's mechanisms helps explain how biodegradable polymers reduce plastic pollution. Microorganisms break down polymers into simpler pieces during biodegradation, releasing nutrients into the environment. The biodegradable substance, its chemical makeup, and its surroundings affect how this happens. Polyhydroxyalkanoates (PHAs) breakdown swiftly in the water because microbial groups produce enzymes to break them down (Huang et al., 2020). Microbial activity is highly affected by temperature, moisture, and oxygen, which might affect biodegradation. Biodegradable solutions like these can help combat traditional plastics' rising concerns (Bartolo et al., 2021). The lifespan study of biodegradable polymers shows why these mechanisms are important for sustainability. If they are to reduce plastic pollution, biodegradable polymers must biodegrade efficiently in real life. This section covers biodegradation processes, ecosystems (industrial compost, soil, marine), end products, and timescales. It also examines circumstances when biodegradation fails or produces microplastics (Das et al., 2024b; Mohanty et al., 2024c; Tripathy et al., 2024).

2.2.1 Biodegradation Mechanism

Most polymers breakdown physiologically in two phases: Break lengthy polymer chains into water-soluble oligomers and monomers. Biologically, bacteria release lipases and esterases that break down polymer chains, or abiotically, UV-induced oxidation breaks down ester bonds. Small molecules are metabolized by microorganisms, changing polymer carbon into cellular biomass and metabolic waste. Methane (CH_4) is produced in anaerobic conditions, while CO_2 and water are the major outputs under aerobic conditions. This process has been represented in Fig. 1.

Polymers react differently to enzymes. PLA, PHAs, starch, and cellulose are depolymerized by enzymes such as proteinase K, amylases, and cellulases. Temperature, humidity, oxygen, pH, and microorganisms affect biodegradation. Composting facilities provide 50–70 °C, humidity, oxygenation, and a strong microbial community for quick biodegradation. Soil with temperate temperatures, microbiological composition, and moisture content degrades more slowly. Cold marine regions, especially the open ocean, have lower microbial numbers and delayed disintegration. Sunlight helps plastic surface layers photodegrade, making them more accessible to bacteria, but biofouling reduces its efficiency. Complete vs. partial biodegradability: Complete biodegradation produces no polymer waste by fully mineralizing CO_2 (and CH_4 under anaerobic conditions) and biomass. A partially biodegradable

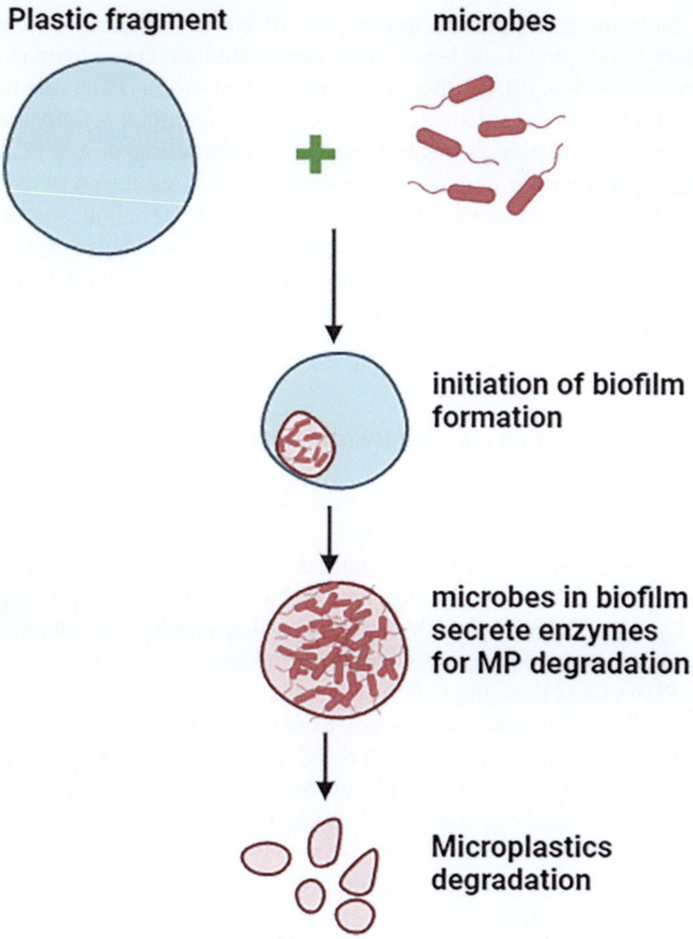

Fig. 1 Mechanism of plastic degradation

plastic object may break down into micro-fragments or change its chemical makeup without mineralization, releasing microplastics. Certification testing may require a 90% conversion to CO_2 within a specified timeframe to demonstrate complete biodegradation under ideal conditions.

2.2.2 Industrial Compost Industrial Composting

Municipal composting facilities break down organic waste aerobically. The ASTM D6400 and EN 13,432 standards define biodegradable plastic as achieving at least ~ 90% conversion to CO_2 within 180 days in industrial composting conditions (58 °C, regular aeration, rotation, presence of other organic materials) compared to cellulose.

Unrecognizable material particles in compost after passing through a 2 mm filter. Tests on plant development and heavy metal limits demonstrate no harm to compost. The right polymers: Pla, PBAT, PBS, starch blends, and various PHAs are biodegradable plastics that meet these characteristics. Increased microbial activity and warmth accelerate their degradation. A heated compost windrow transforms PLA cups or PBAT bags to CO_2 within months. Compost (humus) is produced without visible plastic particles, with the carbon liberated as CO_2. CO_2 release indicates that carbon is not stored, but rather part of the biogenic carbon cycle, obtained from living plants and not contributing fossil carbon to the atmosphere. Compost improves soil fertility. Composting helps biodegradable plastics avoid landfills and incineration by using carbon naturally.

2.2.3 Home Composting and Soil Degradation

Overall, home composting is less controlled than industrial composting and runs at lower temperatures (maybe 30–40 °C at the pile's heart). Some biodegradable polymers may not break down in household compost after months or years. Home compostable polymers: Many products are certified to a home composting standard (e.g., OK Compost Home by TÜV Austria), which requires 12 months at lower temperatures. PLA seldom survives such circumstances, even in very small, thin structures. PHAs and other starch derivatives do.

In uncontrolled environments like gardens or agricultural fields, soil biodegradation depends on biota and climate. Starch blends break down quickly in a few months because soil microbes consume it. PHAs biodegrade in soil within months to a year in soils rich in plants and organic waste because microorganisms break them down. In gentler soils, PBAT and PBS decompose slowly, but in active soil situations like cultivated fields, decomposition is commonly observed annually (Kanungo et al., 2023; Satapathy et al., 2024).

PLA breaks down slowly in soil, but shards may last for years without warm, moist soil and bacteria. Studies reveal that PLA pieces stay in soil for a year in temperate climates. PCL degrades easily in soil due to its softer, low melting polymer. Cellulose-based cellophane and paper degrade in months.

2.2.4 Determinants Affecting Degradation Speed

Polymer Chemistry: Essential is the molecular architecture that is, the backbone connections, existence of easily hydrolysable bonds. Usually, amorphous areas degrade faster than crystalline areas. Copolymers can improve amorphous content, best shown by PBAT's random structure in contrast to the extremely crystalline character of high-PHB content PHA, which shows considerable crystallinity and may undergo delayed degradation until certain crystallites are disturbed.

Additives: Plasticizers, including glycerol or citrate in starch/PCL combinations, could improve microbial access and flexibility. On the other hand, if antioxidants or some stabilizers remain unconsumed, they could prevent degradation.

Physical Form: Because their surface area encourages microbial invasion, thin films degrade faster than thick sections. Whereas a thick injection-molded PLA fork needs a longer period for degradation, a 50-micron film could break down quickly.

Environmental Microorganisms: Appropriate microorganisms are absolutely vital. Compost has a varied community ready to tackle many problems. Particular environments devoid of particular microbes, including open ocean gyres devoid of PHA degraders in the absence of PHA, can hinder activities.

Temperature: The Q10 effect suggests that higher temperatures speed microbial metabolism and responses. Whereas it is somewhat slow in soil at 20 °C, PLA hydrolysis quickly in compost at 58°Cp.

Moisture/Oxygen: Especially aerobic activities depend on this for microbial activity. Minimal deterioration will occur in a dried habitat—desert soil, arid temperature, or a dry portion of compost. Anaerobic conditions give way to slower anaerobes such methanogens.

pH: Hydrolysis can be sparked by an acidic or basic surroundings. For example, base-catalyzed hydrolysis causes some polymers to break down more quickly in somewhat alkaline compost.

Time: Many biodegradable plastics may eventually break down in different environments given enough length (years or decades); nevertheless, the aim is to achieve breakdown within human-relevant timescales (months, or at most, years) to so reduce pollution.

2.3 Applications in Various Industries

Biodegradable polymers are becoming increasingly valued for their adaptability in different sectors, providing new ways to tackle environmental plastic issues. For instance, in dentistry, switching to biodegradable options instead of poly (methyl methacrylate) (PMMA) is very important because regular acrylic polymers greatly add to plastic waste after they're used (Naveed & Ms, 2024). Agriculture is also seeing more biodegradable films and coatings made from polymers such as polylactic acid (PLA) being used; these improve crop protection while lessening environmental harm. Moreover, the packaging sector is adopting these materials to cut down on using traditional plastics, which encourages sustainable methods and a smaller carbon footprint (Kumar et al., 2024). Also, progress in microbial bioremediation indicates that specific microorganisms might effectively break down plastic waste, which enhances using biodegradable materials to handle pollution completely

(Agarwal et al., 2024). The significant ability of biodegradable polymers in these areas highlights their importance in creating a more sustainable future.

2.4 Current Research and Developments

The growing plastic pollution crisis may find some relief in recent progress made with biodegradable polymers, signaling a move toward more sustainable materials. Research is currently focused on biopolymers from various origins-like carbohydrates, proteins, and lipids-all geared toward creating environmentally friendly replacements for traditional plastics made from petroleum. Also, existing recycling methods are being examined, and new systems, such as block chain, are being developed to make waste management more accountable. This all syncs up with global efforts addressing the problem of single-use plastics, something public demand for "greener" solutions is really pushing. What's more, detailed studies have started looking at how plastics break down and affect different ecosystems, offering crucial understanding for mitigation strategies (Napper & Thompson, 2023). When these findings are integrated into policy and industry, they could lead to more sustainable waste management and ways to reduce pollution. Images, like, are very important for visually capturing these advances and helping people better understand the potential of biodegradable polymers.

3 Environmental Impact of Plastic Pollution

Plastic pollution's environmental impact remains a critical issue. It poses significant challenges to ecosystems and, indeed, human health on a global scale. Plastics, coming as they do mainly from non-renewable petroleum sources, contribute to broad environmental degradation. This includes greenhouse gas emissions and a long-term ecological damage resulting from their persistence in nature. Now, biodegradable polymers represent a promising alternative, generally speaking. However, they also face their own serious challenges linked to their lifecycle and waste management (Priyadarshini et al., 2024; Sahu et al., 2024). Current literature, in most cases, indicates that many biodegradable plastics don't achieve effective disposal within existing waste management frameworks, often causing confusion among consumers and inefficient processing systems, which, in turn, exacerbate pollution levels.Furthermore, even though bioplastics are often presented as a cleaner option, they might still generate considerable emissions, not to mention land-use changes, that threaten ecological sustainability (Abdullayeva & Yamil, 2022). It is important to understand that visual representations such as enhance our understanding by illustrating the processes involved in managing plastic waste and the necessary transitions to sustainable solutions. It has been summarized in Fig. 2.

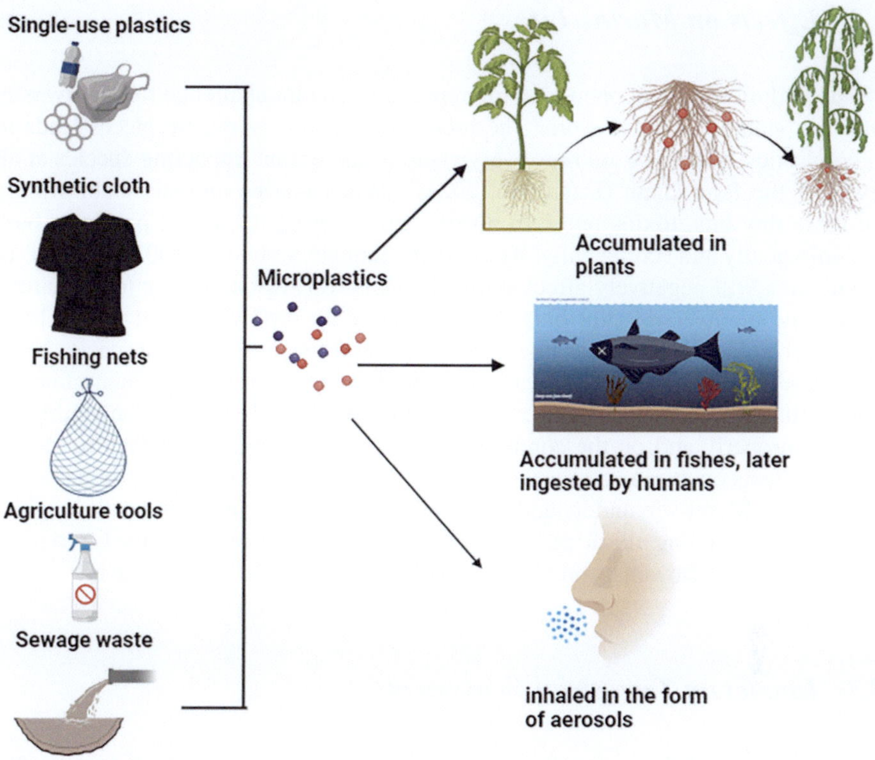

Fig. 2 Impact of plastics

3.1 Statistics on Plastic Waste

Plastic waste statistics are truly alarming; conventional plastics aren't biodegradable, which really makes the environmental crisis worse. Globally, we produce about 300 million tons of plastic each year. This contributes a lot to the estimated 8 million tons of plastic waste that ends up in our oceans every year. This pollution is a threat to marine ecosystems. It can pose serious risks to biodiversity and human health. Research indicates plastic's persistence in the environment is linked to ineffective recycling only about 9% of plastic waste actually gets recycled (Macheca et al., 2024). So, biodegradable alternatives, maybe from corn starch, could be a more sustainable solution. This could reduce our reliance on fossil fuels and help tackle plastic pollution (Samuel et al., 2024). If you look at visual information, like a full lifecycle analysis of plastic waste, you can see how much our current waste management hurts the environment.

3.2 Effects on Marine Life

Plastic pollution's effect on marine life represents a major challenge for biodiversity and how stable ecosystems are. The breakdown of bigger plastic pieces leads to microplastics, and these are now everywhere in the ocean, impacting species at all levels of the food chain (Das et al., 2024a). When marine animals unknowingly eat these tiny bits, toxins build up inside them, causing damaging problems both physiologically and ecologically. By 2050, it's thought around 12,000 metric tons of plastic junk will negatively affect marine habitats worldwide, adding to the danger for marine biodiversity. Not only does this mess up food webs, but it also hurts important biological processes like reproduction and how well different species grow. Biodegradable polymers could be a fix here, as they seem to cause less environmental trouble than regular plastics. Still, how well they work really depends on getting rid of them properly and on the conditions in the environment. If we want to reduce the bad effects on marine ecosystems, facing these problems headon is vital (Ncube et al., 2020). Moreover, understanding how these materials live and break down gives us a start for lessening plastic pollution's widespread influence on marine life, a point made clear in the flowchart on plastic waste's environmental consequences.

3.3 Impact on Terrestrial Ecosystems

While plastic waste's detrimental effects often overshadow terrestrial ecosystems, it's crucial to acknowledge its pervasive impact on these environments. The gradual degradation of plastics leads to the release of harmful substances into soil and water, thereby affecting both plant and animal life, even if such effects aren't as obvious as oceanic plastic pollution. Furthermore, microplastics make their way into terrestrial food chains, impacting everything from tiny insects to larger mammals. Now, consider biodegradable polymers, like those derived from sodium alginate: these offer a potentially significant shift. In most cases, they break down much faster and can even boost soil health by adding organic matter and promoting beneficial microbial activity. With the ability to help foster more resilient ecosystems that can cope with climate change and various pollutants, biodegradable polymers may well represent a practical path towards reducing the environmental burden associated with conventional plastics (Islam, 2025; Agarwal et al., 2024; Balushi et al., 2024). The interplay of biodegradable materials and natural ecosystems could indicate important advancements in pursuing sustainable waste management approaches. This interaction highlights the capacity of biodegradable alternatives to decrease soil pollution and support environmental recovery, which could ultimately improve both the environment and human wellbeing.

3.4 Human Health Concerns

Conventional plastics, used so widely, introduce notable concerns related to human health, mostly tied to chemical additives. These additives have a tendency to leach into our food and water. While intended to improve how plastics function, these substances present real risks; they can move from the plastic products into the environment and, as a result, potentially expose us through food contact (Hahladakis et al., 2017). The health impacts of pollutants like microplastics, along with the chemicals they carry, are especially alarming; research indicates that they may disrupt endocrine functions and possibly lead to ongoing health problems. As we become more conscious of these dangers, biodegradable polymers emerge as a possible solution (Irshad et al., 2024). They reduce harmful leachates and help the environment. Because biodegradable polymers assist in breaking down materials safely, without leaving harmful stuff behind, they can lower the risks linked to plastic pollution, which helps keep the public healthier (Patra JK et al., 2018). The move toward these polymers really highlights how important the link between material science and human health truly is.

3.5 Economic Costs of Plastic Pollution

Plastic pollution's economic toll goes far beyond just environmental harm; it places financial strains on public health, waste disposal, and the services our ecosystems provide. Because plastic waste lingers for so long, municipalities face significant cleanup costs. Cities, for instance, spend millions each year on waste management programs, which, generally speaking, see limited success thanks to plastic's resilient nature (Kumar et al., 2024).

Furthermore, plastic compromises marine ecosystems, thereby threatening both fisheries and tourism both of which depend on healthy, clean environments. Now, biodegradable polymers offer a potentially compelling fix. Such materials could cut down on the economic consequences tied to plastic pollution by lowering landfill expenses and boosting sustainability in how we manage waste (Kameri-Mbote & Kaguru, 2025). Investing significantly in green technology and in creating biodegradable alternatives is vital. Not only will this curb the harmful effects of plastics, but it will also foster economic resilience in industries that are affected (Agarwal et al., 2024). A comprehensive strategy to integrate these kinds of solutions could really tackle the urgent economic implications of plastic pollution.

3.6 Long-Term Environmental Consequences

Plastic pollution's long-term effects on the environment are seriously concerning, especially in the oceans, where plastic trash is everywhere. Some figures show that globally, about 6300 Mt of plastic waste has been produced, and a big chunk of it ends up in our oceans, impacting the health of both ecosystems and biodiversity Thompson and De Falco (2020). This enduring pollution doesn't just endanger sea creatures through ingestion and entanglement but also throws food chains off balance, causing broader ecological consequences. Plus, microplastics created when bigger plastic pieces break down have shown up in different water environments, highlighting just how far plastic contamination spreads and how it might get into the food we eat (Browne, 2007). Given this reality, biodegradable polymers are emerging as a potential sustainable alternative, and insights into their environmental effects, as explored in, could really help us move toward lessening plastic pollution and nurturing healthier ecosystems.

3.7 Role of Public Awareness and Education

Generally speaking, public awareness and education are crucial when we talk about the challenges of plastic pollution and getting people to use biodegradable polymers more often. Good environmental education can really help people understand how bad plastic waste is and why biodegradable options are better, which can lead to more responsible choices. You see, research indicates that when communities are more aware, they tend to be more willing to recycle and dispose of waste properly, suggesting that focused educational campaigns can actually change how society views plastic usage (Ibrahim et al., 2025). Also, integrating these education initiatives into school programs can help create a future generation that understands environmental stewardship better, specifically about the life cycle of materials and the importance of bioplastics in cutting down on pollution (Rahman et al., 2025; Mohanty et al., 2024a, 2024b, 2024c; Ray et al., 2024). As the lifecycle flowchart of plastic waste depicted in illustrates, understanding these concepts really drives home the need for change. Ultimately, public engagement is essential for successful environmental interventions.

4 Advantages of Biodegradable Polymers

Biodegradable polymers provide a potentially effective answer to the widespread problem of plastic pollution; compared to conventional petroleum-based plastics, they offer several environmental advantages. A crucial benefit is that they tend to break down faster in different environments. This, in turn, reduces the build-up of synthetic plastic waste. This waste is significantly harmful to both terrestrial and aquatic ecosystems (N/ A, 2022). For example, bioplastics like poly (lactic acid) (PLA) can be a viable alternative in applications, spanning from packaging to various consumer goods. Consequently, this supports a move toward a circular economy. In addition, numerous biodegradable polymers come from renewable resources, like corn or sugarcane. In most cases, this helps to lower our reliance on fossil fuels while also lowering greenhouse gas emissions linked to producing plastics (Bartolo et al., 2021). These materials break down into harmless by-products, which can enrich the soil rather than contributing to toxic microplastics. This showcases an important step toward sustainable materials management, especially when considering current severe environmental challenges. Thus, the use of biodegradable polymers has significant potential in tackling the pressing issue of plastic pollution.

4.1 Reduction of Plastic Waste

Plastic waste is a major ecological problem, pushing us to find new ways to lessen its effects. Biodegradable polymers are a key answer, giving us a more sustainable option than regular plastics. These polymers, made from materials like starch and cellulose, can help lower plastic pollution by breaking down more effectively (Agarwal et al., 2024).

Also, recent research shows that microbes are very important in breaking down plastics, proving that some microorganisms can handle plastic waste well (Sharma et al., 2025). Circular economy strategies, such as those used by new African businesses that turn waste into valuable products, highlight how biodegradable materials can change plastic waste management. As these environmentally friendly polymers keep getting better, they not only offer an environmental fix but also a way to create sustainable industrial practices that value ecological health over pollution. Generally speaking, these methods show real promise in tackling the growing problem of plastic waste, in most cases.

4.2 Lower Carbon Footprint

Pursuing a lower carbon footprint when it comes to plastic production and its uses, the incorporation of biodegradable polymers is a promising tactic. These biodegradable alternatives, unlike your typical petroleum-based plastics, tend to come from renewable resources—think corn starch or even algae. This, in turn, really cuts down on greenhouse gas emissions during the whole product lifecycle. Also, studies have shown that materials which are biodegradable, like polylactic acid or PLA for short, have the potential to reduce both energy consumption and carbon emissions, specifically throughout the phases of production and disposal (Anggraini, 2025). Integrating these kinds of polymers into manufacturing is not only a response to the crucial issue of plastic pollution, but it also falls in line with global sustainability efforts, offering a feasible route to cutting down on carbon footprints across various industries. These materials make use of natural degradation processes, which lessens our dependence on landfills and brings down the environmental impacts we see tied to plastic waste, making them an important piece of the puzzle in paving the way for a more environmentally responsible future.

4.3 Compatibility with Natural Ecosystems

Biodegradable polymers, when introduced into natural ecosystems, represent more than just a sustainable choice compared to standard plastics; they signal a significant move toward environmental balance. These polymers are formulated to break down via natural biological actions, which reduces their longevity in ecosystems, unlike synthetic plastics. As an example, studies show that substances such as polylactic acid (PLA) and polyhydroxyalkanoates (PHAs) generally show good interaction with both soil and water environments. Upon breakdown, these materials can return helpful nutrients back into the ecosystem, supporting soil health and decreasing pollution. These interactions are exceptionally important because they lessen the negative environmental effects typically linked to persistent plastic waste, which may release harmful toxins and microplastics into our food supply (Naser et al., 2021). In most cases, successful decomposition, which various studies have affirmed, adds to the viability of using these materials across varied environments. This reinforces the idea that biodegradable polymers offer a pathway to a more sustainable and ecologically thoughtful approach to materials science (Mrozik et al., 2021; Yaradoddi et al., 2020). Image6 clearly shows this transition, highlighting the potential for biodegradable materials to solve plastic pollution problems, while it also displays their compatibility with natural ecosystems.

4.4 Potential for Composting

Composting biodegradable polymers shows real promise for better waste management. Because regular plastics stick around for ages, biodegradable ones break down naturally, helping reduce waste. When these polymers compost, they enrich the soil, boosting its quality and helping plants grow, which is super important for farming. Making stuff like poly (methyl methacrylate) really shows we need compostable options to cut down on pollution (Ahmad et al., 2024). Studies suggest that biodegradable films from renewable sources are similar to regular plastics but break down easier (Naveed & Ms, 2024). Still, we need good plans for collecting and treating these materials to make composting actually work (Kumar et al., 2024). So, viewing composting as a way to handle these polymers at the end of their life could really boost their sustainability.

4.5 Versatility in Applications

Biodegradable polymers are quite versatile, useful across a spectrum of applications and notably important for tackling plastic pollution. Materials like polylactic acid (PLA) have shown their value in packaging, agriculture, and even medicine, due to their renewable sources and natural decomposition abilities in several settings. Consider packaging: biodegradable polymer solutions can really cut down on waste and improve sustainability. Compostable bags and films are great examples, breaking down well in composting setups (Kumar et al., 2024). Moreover, in medical applications, these polymers can be used in sutures or drug delivery. This minimizes long-term waste, all while meeting healthcare needs (Naveed & Ms, 2024). This adaptability highlights the potential for reducing plastic pollution alongside diverse functions (Ebelechukwu & Asekun, 2024). Ultimately, more and more biodegradable polymers are being adopted; this signals an industry-wide move to more sustainable practices, paving the way for a greener future.

4.6 Contribution to Circular Economy

Generally speaking, the move toward a circular economy plays a crucial role when we consider the ongoing problem of plastic pollution, and this is especially true thanks to how we can innovatively use biodegradable polymers. These polymers, designed to be kinder to the environment, can cut our dependency on standard plastics, materials that traditionally use a production method ending in major waste issues. In most cases, the circular economy pushes for products that are designed sustainably—think reuse, recycle, and biodegradation—helping us hold onto resources and cut environmental damage. The paradigm shift helps both ecosystems and economic stability, particularly in the food and beverage sectors. Research suggests that using biodegradable materials improves waste management, especially when combined with recycling and composting (Arijeniwa et al., 2024; Terzioğlu et al., 2024). The incorporation of biodegradable polymers in our economy is a transformation, exemplifying how we can reduce waste and foster more sustainable approaches when it comes to the environment.

5 Constraints and Difficulties

Although biodegradable polymers offer great potential to address plastic waste, wide acceptance is hampered. A major issue is the unequal rate of breakdown of these polymers; environmental elements, such as temperature and the surrounding microorganisms, significantly influence this and thereby reduce their practical efficacy (Kothawade & Padwal, 2025). Many biodegradable plastics also fail criteria for compostability. Lots of them so still wind up in landfills, breaking down gradually and releasing methane gas (Kothawade & Padwal, 2025). As you can see in countries like India where the markets are still young, high production costs and restricted raw material sources hold down the spread of bioplastics even further. Moreover, improper disposal of biodegradable plastics could lead to contamination similar to that of conventional plastics, so stressing the need of effective waste management. To fully grasp the possibilities of biodegradable polymers, their lifetime must be closely controlled.

5.1 Biodegradable Polymer's Production Cost

The production costs of biodegradable polymers define their economic practicality mostly, which influences the larger shift from conventional plastics to greener options. The cost loads can seem heavy since these manufacturing techniques sometimes rely on complex technology and biomass-derived ingredients. For instance, although biopolymers such as PLA and PHA have many environmental benefits, their manufacturing and processing expenses usually remain above those of petroleum-based plastics. As economies of scale haven't yet occurred yet, thereby reducing market competitiveness (Gonz Eález-López et al., 2023). Scaling up manufacturing further complicates this financial dilemma. Furthermore influencing investment and running expenses is regulatory issues, which calls for a careful balance between sustainability goals and economic viability (Biodegradable Materials and Their Applications 2022). Developing plans to increase the acceptance of biodegradable polymers in different industrial environments depends on an awareness of these economic issues.

5.2 Performance Relative to Conventional Polymers

Although they are seen as a hopeful replacement for conventional plastics in the attempt to reduce environmental damage, biodegradable polymers have some performance-related issues that might hinder their general acceptance. Mechanical characteristics are a big problem; often, biodegradable versions lack the strength or durability of regular plastics, which would restrict their employment in industries including packaging and construction (Osman et al., 2023). Furthermore, depending on the surroundings, their rate of breakdown may vary really greatly. This may lead to less dependability of the substance over time and during consumption. Furthermore complicated by the complex nature of biodegradation is recycling, especially in light of the sectors still mostly dependent on traditional plastics (Chen et al., 2022). Thus, it's crucial to address these performance-related issues even if biodegradable polymers help reduce the quantity of plastic garbage we produce. By doing this, we can contribute to ensure that, as we strive to reduce plastic pollution, they can be really a sustainable choice. Emphasizing the several facets of biodegradability in practical situations, the image which describes the lifetime effects of materials—do highlights these elements.

5.3 Misconceptions Regarding Biodegradability

It's common to misinterpret precisely how "biodegradable" plastics actually are, which can lead to some regrettable presumptions about their favorable environmental effects. Many people, for instance, tend to think that anything labeled biodegradable will break down anywhere. Research indicates, however, that the degree of degradation of these materials relies considerably on variables such temperature, moisture level, and species of microorganisms present. When these conditions are not optimal, the breakdown process may not function as predicted and generate non-biodegradable by-products (Kumar et al., 2024). Furthermore, the word "biodegradable" could let buyers believe it's acceptable to discard these products into landfills. To really break down, many biodegradable polymers actually require specific industrial composting facilities. These misinterpretations help us to explain why we still fight with plastic pollution instead of finding solutions. This scenario emphasizes the need of increased education on suitable disposal techniques and clearer labels. Fostering really sustainable practices and ensuring that biodegradable polymers perform as they are meant as ecologically responsible substitutes for conventional plastics dependent on addressing these misconceptions (Samuel et al., 2024).

5.4 Discharge and Recycling Infrastructure

Dealing with plastic pollution truly depends on effective methods of disposal and recycling biodegradable polymers. Getting these materials to fit in with our present waste systems is absolutely crucial since they can truly help to reduce the environmental concerns generated by ordinary plastics. Still, there are some really large obstacles to overcome. For example, we must establish some basic guidelines for waste collecting and handling so that we may distinguish between biodegradable and non-biodegradable plastics. Given that single-use plastics are increasingly found in food packaging and improper disposal of them could compromise the whole purpose of employing bioplastics, this is especially true (Ncube et al., 2020). Examining the lifespan of plastic garbage and stressing the need of improved recycling sites helps one to see some really large gaps in our waste management, thereby making recycling difficult. Furthermore, new technologies in composting are constantly emerging, thus spreading the knowledge will help us to change to a more sustainable way of living including biodegradable elements into our daily life.

5.5 Regulatory and Policy Obstacles

Generally speaking, the sustainable change to biodegradable polymers, as opposed to conventional plastics, often runs across difficulties, largely from a tangle of policy and regulatory red tape that can truly impede advancement. Policy can be a bit behind the times, especially when it comes to knowing what makes these materials unique and why they're good for the environment; regulators sometimes don"t quite get it. This lack of clarity makes it more difficult to certify fresh biodegradable materials; obstacles new products confront trying to enter the market can compromise their chances versus regular plastics. Still major challenges include also consumer acceptance of biodegradable solutions and how businesses purchase materials. These problems highlight how urgently fully defined rules and coordinated policies supporting development on biodegradable technology are needed. As shown in, creative ideas like lifespan evaluations can help highlight the possibilities of biodegradable substitutes, therefore providing strong justification for their wider adoption and help to reduce plastic pollution.

5.6 Consumer Behavior and Market Acceptability

Consumer behavior is closely related to differing degrees of environmental awareness and determines how successfully biodegradable polymers find acceptance on the market. Studies recently have shown us that what customers believe about plastic substitutes—stuff like biodegradable possibilities—really rely on knowing the sustainability aspects, you know, like can it be recycled, can it be composted? Though not everyone understands why these things are good, or even relevant, the demand for greener materials has surely increased. From those who just lack interest to those who know all about the environment, we encounter all kinds of people. We have to get the information out in a way that everyone understands the environmental pluses if we truly want people to accept biodegradable materials, particularly with all the concerns about plastic pollution worsening Kautish et al. (2021). Though consumer decisions are somewhat complex, they emphasize the need of carefully adjusting marketing to several populations (Chalannavar et.al., 2025). For example, shows the challenges consumers have with biodegradable items and what we may do about them, thereby providing a good image of their interaction with them.

6 Conclusion

In the end, considering biodegradable polymers highlights their main role as a sustainable alternative to conventional plastics, so addressing the relevant problem of plastic pollution endangering the environment. These materials not only show great promise because they break down better by natural processes but also have less negative effects on the environment, so reducing the consequences connected with daily plastic waste. Research indicates that these polymers may be deftly applied into other sectors, like medical and packaging, such new concepts meant to boost their features and application (Kothawade & Padwal, 2025). Reevaluating lifetime strategies and waste management plans is still essential, though, as research on the problems of attaining correct biodegradability in various environments have highlighted (Naveed & Ms, 2024). Starting the use of biodegradable polymers will enable us to assist a circular system that prioritizes environmental preservation while reducing plastic pollution broad effects (Stoica et al., 2024). Working together in research, development, and practical use of these essential resources can help to shape the future and open the path toward a more permanent one. The lifetime effect of biodegradable polymers is rather well shown for visual clarity.

Generally speaking, the study of biodegradable polymers emphasizes their promise as a viable approach to cope with plastic pollution, therefore emphasizing its relevance in the broader perspective of sustainable materials management. Emphasizing the purposes of catalytic degradation and sophisticated recycling techniques (Chen et al., 2022), important factors include the need for working biodegradable processes that may transform synthetic plastics into environmentally beneficial alternatives. Furthermore, the rising demand for biopolymers mostly polylactic acid (PLA) showcases a shift toward renewable sources able to replace petroleum-based materials (Ilyas et al., 2021). Cutting waste and maximizing resource utilization depend on including a circular economy model into biocomposite development, therefore promoting a sustainable approach including waste management and recycling (Shanmugam et al., 2021). By means of biodegradable polymers, one can significantly reduce plastic pollution, thereby fostering ecological balance and enabling structural changes in the food packaging sector and more generally (Ncube et al., 2021).

References

Abdullayeva, M., & Yamil, N. G. A. (2022). Environmental impact assessment of petroleum-based bioplastics. *World Science, 5(77)*. https://doi.org/10.31435/rsglobal_ws/30092022/7867

Agarwal, T., Atray, N., & Sharma, J. G. (2024). A critical examination of advanced approaches in green chemistry: microbial bioremediation strategies for sustainable mitigation of plastic pollution. *Future Journal of Pharmaceutical Sciences, 10*(1). https://doi.org/10.1186/s43094-024-00645-x

Ahmad, N. A. M., Ali, N. A., Afzal, N. A., Tanveer, N. N., & Mustafa, N. A. (2024). Corn waste a sustainable solution for plastic pollution: A mini review. *NUST Journal of Engineering Sciences, 17*(1), 17–27. https://doi.org/10.24949/njes.v17i1.817

Anggraini, W. (2025). Development of biodegradable materials for consumer product packaging: environmental impact and consumer preferences. *Jurnal Pijar MIPA, 20*(2), 304–310. https://doi.org/10.29303/jpm.v20i2.8588

Arijeniwa, V. F., Akinsemolu, A. A., Chukwugozie, D. C., Onawo, U. G., Ochulor, C. E., Nwauzoma, U. M., Kawino, D. A., & Onyeaka, H. (2024). Closing the loop: A framework for tackling single-use plastic waste in the food and beverage industry through circular economy- a review. *Journal of Environmental Management, 359*, Article 120816. https://doi.org/10.1016/j.jenvman.2024.120816

Bala, K., Anshul, Verma, S., Dogra, A., Das, A. P., Kumar, R., & Nandini. (2024). Sustainable management for urban plastic waste generation. In *Sustainable management of urban plastic waste through circular economic approaches* (pp 211–236). https://doi.org/10.1201/9781003465447-13

Balushi, A. A., Amara, Y., Balushi, K. A., Mannai, S. E., Saber, R. B., Masmoudi, A. S., Cherif, A., & Souissi, Y. (2024). Harnessing the plastic-degrading potential of pseudomonas species for environmental sustainability. *IOP Conference Series Earth and Environmental Science, 1401*(1), Article 012009. https://doi.org/10.1088/1755-1315/1401/1/012009

Biodegradable materials and their applications. (2022). In *Wiley eBooks*. https://doi.org/10.1002/9781119905301

Browne, M. A. (2007). Environmental and biological consequences of microplastic within marine habitats. Pearl. https://pearl.plymouth.ac.uk/fose-theses-other/314/

Chen, J., Wu, J., Sherrell, P. C., Chen, J., Wang, H., Zhang, W., & Yang, J. (2022). How to build a microplastics-free environment: Strategies for microplastics degradation and plastics recycling. *Advanced Science, 9*(6). https://doi.org/10.1002/advs.202103764

Das, A., Dey, S., & Das, A. P. (2024b). Ecotoxicological impacts of synthetic microfiber pollutants and development of sustainable mitigation strategies. *Environmental Chemistry and Ecotoxicology*. https://doi.org/10.1016/J.ENCECO.2024.12.003

Das, A., Dey, S., Gacem, A., Yadav, K. K., Nath, H., Ram, D. K., & Das, A. P. (2025). Revolutionizing solid hydrocarbon sourcing: Plastic's journey from waste to upcycled treasure. *Environmental Science and Engineering, Part, F218*, 331–351. https://doi.org/10.1007/978-3-031-81615-4_14

Das, A., Dey, S., Bala, K., Gnanasekaran, R., & Das, A. P. (2024a). Microbial bioremediation technology for sustainable treatment and management of synthetic microfiber waste. In *Sustainable microbial technology for synthetic and cellulosic microfiber bioremediation* (pp. 221–238). https://doi.org/10.1007/978-3-031-62600-5_12

Das, A. Prasad., Behera, I. Dipamitra., & Das, N. Prasad. (2024c). *Renewable energy generation and value addition from environmental microfiber pollution through advanced greener solution*. p. 278.

Dey, S., Dash, N., Ray, S. K., Priyadarsini, S., Chhotaray, C., & Das, A. P. (2024). Occurrence and quantification of tire wear plastics pollutants from Highway City Dust of Bhubaneswar City of Odisha. In *Sustainable microbial technology for synthetic and cellulosic microfiber bioremediation* (pp. 77–89). https://doi.org/10.1007/978-3-031-62600-5_5

Di Bartolo, A., Infurna, G., & Dintcheva, N. T. (2021). A review of bioplastics and their adoption in the circular economy. *Polymers, 13*(8), 1229. https://doi.org/10.3390/polym13081229

Ebelechukwu, F. C., & Asekun, F. O. (2024). Assessment and characterization of plastic pollution in two coastal sites in Lagos Nigeria: Implications for environmental management and public awareness. *Open Journal of Environmental Research (ISSN 2734–2085), 5*(1), 43–59. https://doi.org/10.52417/ojer.v5i1.621

Geyer, R., Jambeck, J. R., & Law, K. L. (2017). Production, use, and fate of all plastics ever made. *Science Advances, 3*(7). https://doi.org/10.1126/sciadv.1700782

Ghosh, S., & Das, A. P. (2020). Microbial metagenomics: current advances in investigating microbial ecology and population dynamics. In S. K. Nayak & B. B. Mishra (Eds.), *Frontiers in soil and environmental microbiology* (1st ed., pp. 247–254). CRC Press.

González-López, M. E., De Jesús Calva-Estrada, S., Gradilla-Hernández, M. S., & Barajas-Álvarez, P. (2023). Current trends in biopolymers for food packaging: a review. *Frontiers in Sustainable Food Systems, 7*. https://doi.org/10.3389/fsufs.2023.1225371

Hahladakis, J. N., Velis, C. A., Weber, R., Iacovidou, E., & Purnell, P. (2017). An overview of chemical additives present in plastics: Migration, release, fate and environmental impact during their use, disposal and recycling. *Journal of Hazardous Materials, 344*, 179–199. https://doi.org/10.1016/j.jhazmat.2017.10.014

Haider, T. P., Völker, C., Kramm, J., Landfester, K., & Wurm, F. R. (2018). Plastics of the Future? The impact of biodegradable polymers on the environment and on society. *Angewandte Chemie International Edition, 58*(1), 50–62. https://doi.org/10.1002/anie.201805766

Ibrahim, M. A., Saba, A. O., Mohammed, M., & Ojewole, A. E. (2025). Public awareness and perception of the ecological impacts and responsibility for plastic pollution in Nigeria. *Environmental & Socio-economic Studies, 13*(1), 15–27. https://doi.org/10.2478/environ-2025-0002

Ilyas, R., Sapuan, S., Harussani, M., Hakimi, M., Haziq, M., Atikah, M., Asyraf, M., Ishak, Razman, Nurazzi, N., Norrahim, M., Abral, H., & Asrofi, M. (2021). Polylactic acid (PLA) biocomposite: processing, additive manufacturing and advanced applications. *Polymers, 13*(8), 1326. https://doi.org/10.3390/polym13081326

Irshad, M. A., Khan, A., Zafar, Z., Najib, M. S., & Ahmad, T. (2024). Biopolymers for a sustainable food system: sources, applications, and environmental impacts. *Journal of Food Process Engineering, 47*(11). https://doi.org/10.1111/jfpe.14770

Islam, N. F. A. S. (2025). The impact of plastic waste on ecosystems and human health and strategies for managing it for a sustainable environment. *International Journal of Latest Technology in Engineering Management & Applied Science, 14*(3), 706–723. https://doi.org/10.51583/ijltemas.2025.140300075

John, A., & Das, A. P. (2024). Government strategies for achieving circular economy for plastic waste management in Southeast Asia. In *Sustainable management of urban plastic waste through circular economic approaches* (pp. 72–82). https://doi.org/10.1201/9781003465447-6

Kameri-Mbote, P., & Kaguru, M. (2025). The role of indigenous and traditional knowledge in combatting plastic pollution. *Strathmore Law Journal*, 75–90. https://doi.org/10.52907/slj.v8i1.534

Kanungo, J., Sahoo, T., Swain, L. P., & Behera, I. D. (2023). *Toxicity of Persistent Hydrocarbon Pollutants, Sources and Sustainable Remediation Process*. 39–65. https://doi.org/10.1007/978-3-031-48220-5_2

Kautish, P., Sharma, R., Mangla, S. K., Jabeen, F., & Awan, U. (2021). Understanding choice behavior towards plastic consumption: An emerging market investigation. *Resources Conservation and Recycling, 174*, 105828. https://doi.org/10.1016/j.resconrec.2021.105828

Kothawade, S., & Padwal, V. (2025). Advancements in biodegradable Polymers: Sustainable solutions for environmental challenges. *Current Green Chemistry, 12*. https://doi.org/10.2174/0122133461347075241121063345

Kumar, S., Kumar, S., Kumar, S., Yadav, T., Dhapola, P. S., & Singh, P. K. (2024). Reducing environmental plastic pollution by designing polymer materials for managed End-of-Life. *Macromolecular Symposia, 413*(1). https://doi.org/10.1002/masy.202300146

Macheca, A. D., Mutuma, B., Adalima, J. L., Midheme, E., Lúcas, L. H. M., Ochanda, V. K., & Mhlanga, S. D. (2024). Perspectives on plastic waste management: Challenges and possible solutions to ensure its sustainable use. *Recycling, 9*(5), 77. https://doi.org/10.3390/recycling9050077

Mishra, S., Dash, D., & Das, A. P. (2024). Aquatic microbial diversity on plastisphere: Colonization and potential role in microplastic biodegradation. *Geomicrobiology Journal, 41*(4), 312–323. https://doi.org/10.1080/01490451.2023.2209750

Mishra, S., & Das, A. P. (2022). Treatment of the wastewater polluted with synthetic microfiber released from washing machine. In B. B. Das, H. Hettiarachchi, P. K. Sahu, & S. Nanda (Eds.), *Lecture notes in civil engineering* (vol. 207, pp. 109–117). Springer Science and Business Media Deutschland GmbH. https://doi.org/10.1007/978-981-16-7509-6_9/COVER

Mohanty, M., Mohanty, J., Dey, S., Dutta, K., Shah, M. P., & Das, A. P. (2024b). The face mask: A tale from protection to pollution and demanding sustainable solution. *Emerging Contaminants, 10*(2), Article 100298. https://doi.org/10.1016/J.EMCON.2023.100298

Mohanty, J., Dey, S., Bala, K., & Das, A. P. (2024). Recent advances in microplastic pollution for its sustainable management. In *Sustainable management of urban plastic waste through circular economic approaches* (pp. 27–48). https://doi.org/10.1201/9781003465447-3

Mohanty, M., Tripathy, B., Koteswara Reddy, G., & Das, A. P. (2024c). Greener technologies for establishing restraint over microplastic pollution. In *Sustainable microbial technology for synthetic and cellulosic microfiber bioremediation* (pp. 239–258). https://doi.org/10.1007/978-3-031-62600-5_13

Napper, I., & Thompson, R. (2023). Plastics and the environment. *Annual Review of Environment and Resources, 48*(1), 55–79. https://doi.org/10.1146/annurev-environ-112522-072642

Naser, A. Z., Deiab, I., & Darras, B. M. (2021). Poly(lactic acid) (PLA) and polyhydroxyalkanoates (PHAs), green alternatives to petroleum-based plastics: a review. *RSC Advances, 11*(28), 17151–17196. https://doi.org/10.1039/d1ra02390j

Naveed, N., S, B., & Ms, K. (2024). An outlook on the significance of biodegradable dental polymers to address plastic pollution and promote environmental sustainability. *Deleted Journal, 35*(3), 408–413. https://doi.org/10.21608/asdj.2024.292310.1295

Ncube, L. K., Ude, A. U., Ogunmuyiwa, E. N., Zulkifli, R., & Beas, I. N. (2020). Environmental impact of food packaging materials: A review of contemporary development from conventional plastics to polylactic acid based materials. *Materials, 13*(21), 4994. https://doi.org/10.3390/ma13214994

Ncube, L. K., Ude, A. U., Ogunmuyiwa, E. N., Zulkifli, R., & Beas, I. N. (2021). An overview of plastic waste generation and management in food packaging industries. *Recycling, 6*(1), 12. https://doi.org/10.3390/recycling6010012

Osman, A. I., Hosny, M., Eltaweil, A. S., Omar, S., Elgarahy, A. M., Farghali, M., Yap, P., Wu, Y., Nagandran, S., Batumalaie, K., Gopinath, S. C. B., John, O. D., Sekar, M., Saikia, T., Karunanithi, P., Hatta, M. H. M., & Akinyede, K. A. (2023). Microplastic sources, formation, toxicity and remediation: a review. *Environmental Chemistry Letters, 21*(4), 2129–2169. https://doi.org/10.1007/s10311-023-01593-3

Parameswaranpillai, J., Rangappa, S. M., Siengchin, S., & Jose, S. (2021). *Bio-based epoxy polymers, blends, and composites.* http://books.google.com/books?id=MecSEAAAQBAJ&dq=BioBased+Polymeric+Films&hl=&source=gbs_api

Priyadarshini, S., Jagatee, S., & Das, A. P. (2024). Synthetic fabrics and microfiber pollution–An assessment of their global impact. *Environmental Science and Engineering, Part, F2333*, 137–157. https://doi.org/10.1007/978-3-031-51792-1_8

Rahman, M. M., Jahan, I., U-Shah, F., Shahide, A. M. S., Shobuj, I. A., Hossain, M. T., Alam, E., & Islam, M. K. (2025). Public perception on plastic pollution: a web-based study in Dhaka City, Bangladesh. *BMC Public Health, 25*(1). https://doi.org/10.1186/s12889-025-21959-2

Ray, S. K., Dey, S., Das, A. P., & John, A. (2024). Microplastic waste management through greener technologies : A review. In *Sustainable management of urban plastic waste through circular economic approaches* (pp. 237–261). https://doi.org/10.1201/9781003465447-14

Reddy Koteswara, G., Bhargavi, S., Yamuna, D., Anjani Devi, M., Jeevana Lakshmi Kumar, G., Manas, T., Adarsh, S., Bhavya Sai, V., Nath, H., Alreshidi, M. A., Togun, H., Ram, D. K., & Das, A. P. (2025). Recent advancements in the remediation of marine hydrocarbon and plastic pollutants. *Environmental Science and Engineering, Part, F218*, 127–142. https://doi.org/10.1007/978-3-031-81615-4_7

Sahu, A., Dey, S., Mallick, K., Priyadarsini, S., Behera, I. D., & Das, A. P. (2024). A review on advanced technology for sustainable management of synthetic microplastic waste. In *Sustainable management of urban plastic waste through circular economic approaches* (pp. 262–279). https://doi.org/10.1201/9781003465447-15

Samuel, H. S., Ekpan, F. M., & Ori, M. O. (2024). Biodegradable, recyclable, and renewable polymers as alternatives to traditional petroleum-based plastics. *Asian Journal of Environmental Research., 1*(3), 152–165. https://doi.org/10.69930/ajer.v1i3.86

Satapathy, S., Kar, D., Priyadarsini, S., Das, A. P., & Kuanar, A. (2024). Application of microorganisms in emerging waste management strategies and remediation of synthetic and cellulosic microfibers. In *Sustainable microbial technology for synthetic and cellulosic microfiber bioremediation* (pp. 183–206). https://doi.org/10.1007/978-3-031-62600-5_10

Shanmugam, V., Mensah, R. A., Försth, M., Sas, G., Restás, Á., Addy, C., Xu, Q., Jiang, L., Neisiany, R. E., Singha, S., George, G., E, T. J., Berto, F., Hedenqvist, M. S., Das, O., & Ramakrishna, S. (2021). Circular economy in biocomposite development: State-of-the-art, challenges and emerging trends. *Composites Part C Open Access, 5*, 100138. https://doi.org/10.1016/j.jcomc.2021.100138

Sharma, B. K., Jha, A., Bhalani, D., Pillai, S., Pal, S., Patil, K., & Nagaraj, K. (2025). Advancements in bio-resource-based polymers and composites: sustainable alternatives to non-biodegradable plastics for a greener future: A review. *Current Green Chemistry, 12*. https://doi.org/10.2174/0122133461372269250503082219

Stoica, M., Bichescu, C. I., Crețu, C., Dragomir, M., Ivan, A. S., Podaru, G. M., Stoica, D., & Stuparu-Crețu, M. (2024). Review of bio-based biodegradable polymers: Smart solutions for sustainable food packaging. *Foods, 13*(19), 3027. https://doi.org/10.3390/foods13193027

Swain, S., Purnima Sudha, M., Dey, S., Badu, Y., Bhoi, S. R., Alreshidi, M. A., Qasim, M. T., Nath, H., Ram, D. K., & Das, A. P. (2025). Sources of micro and nano plastic hydrocarbon pollutants in marine bodies and their treatment. *Environmental Science and Engineering, Part, F218*, 45–67. https://doi.org/10.1007/978-3-031-81615-4_3

Terzioğlu, N., Ceschin, F., Jobling, S., & Tarverdi, K. (2024). Archetypes to categorise upstream packaging strategies for a circular economy. *Resources Conservation & Recycling Advances, 21*, Article 200211. https://doi.org/10.1016/j.rcradv.2024.200211

Thompson, R. C., & De Falco, F. (2020). Marine litter: Are there solutions to this environmental challenge? In *Springer water* (pp. 39–44). https://doi.org/10.1007/978-3-030-45909-3_8

Tripathy, B., Sahoo, P. P., Sundaray, H., & Das, A. P. (2024). In-vitro biodegradation of discarded marine microplastics across the eastern coast of the Bay of Bengal, India using Exiguobacterium sp. *Environmental Chemistry and Ecotoxicology, 6*, 236–247. https://doi.org/10.1016/J.ENCECO.2024.07.003

Yaradoddi, J. S., Banapurmath, N. R., Ganachari, S. V., Soudagar, M. E. M., Mubarak, N. M., Hallad, S., Hugar, S., & Fayaz, H. (2020). Biodegradable carboxymethyl cellulose based material for sustainable packaging application. *Scientific Reports, 10*(1). https://doi.org/10.1038/s41598-020-78912-z

Nanobiotechnology: An Approach to Climate Resilience and Sustainability

Basudeba Kar, Sumitra Panda, Jyotirmayee Lenka, Nikita Panda, and Suprava Sahoo

1 Introduction

The twin concerns of climate change and environmental sustainability are the twenty-first century's defining crisis. As global temperatures continue to climb, with the World Meteorological Organization announcing that 2023 was the warmest year on record, measuring 1.45 °C over pre-industrial levels (WMO, 2024), the need for creative technical solutions has never been greater. This global emergency has far-reaching consequences, including intensified weather events, biodiversity loss, agricultural disruption, water scarcity, and rising sea levels, all of which endanger human wellbeing and ecosystem stability (IPCC, 2022).

Nanobiotechnology: The combination of nanotechnology and biotechnology creates a new opportunity to solve these difficulties at the molecular and cellular levels. This interdisciplinary field integrates nanoscale material engineering (1–100 nm) with biological systems and processes to create technologies that replicate, augment, or interface with natural mechanisms (Fakruddin et al., 2012). Unlike traditional techniques, which frequently necessitate huge infrastructure or resource-intensive procedures, nanobiotechnological solutions can be beautiful, economical, and biomimetic, interacting with natural systems rather than fighting them.

Carbon capture and environmental cleanup are among the potential applications, which also include renewable energy generation and precision agriculture. For example, engineered nanoparticles can be created to selectively bind to and neutralize toxins in soil and water bodies, whereas nanobiocatalysts can convert industrial waste streams into useful products via energy-efficient pathways (Singh et al., 2020). Similarly, nanobiosensors embedded in agricultural systems offer real-time monitoring

B. Kar (✉) · S. Panda · J. Lenka · N. Panda · S. Sahoo
Centre for Biotechnology, Siksha O Anusandhan Deemed to be University, Bhubaneswar, India
e-mail: basudebakar@soa.ac.in

S. Ghosh et al. (eds.), *Advances in Organic Waste Conversion Through Industrial Biotechnology*, Sustainable Environmental Waste Management Strategies,
https://doi.org/10.1007/978-3-032-11377-1_4

of crop health and environmental conditions, allowing for more adaptable responses to shifting climatic patterns (Prasad et al., 2017).

This chapter investigates how nanobiotechnology can help with climate resilience and environmental sustainability across several dimensions. We begin by defining the fundamental principles of nanobiotechnology and the unique problems faced by climate change. We then investigate strategic frameworks for deploying nanobiotechnological solutions, outlining specific applications in energy, agriculture, water management, and environmental remediation. We present a detailed review of nanobiotechnology's potential to improve our climate adaptation and mitigation capabilities by examining case studies of successful implementations and critically analyzing technical and ethical issues.

2 Fundamentals of Nanobiotechnology

2.1 Defining the Field

Nanobiotechnology represents the intersection of nanotechnology and biotechnology, combining the principles of nanoscale engineering with biological systems and processes. While nanotechnology broadly encompasses the manipulation of matter at dimensions between 1 and 100 nm, nanobiotechnology specifically focuses on biological or biologically inspired systems at this scale (Roco, 2003). This hybrid discipline leverages both the unique properties that emerge at the nanoscale, such as enhanced surface area-to-volume ratios, quantum effects, and altered reactivity, and the exquisite specificity, self-assembly capabilities, and information-processing capacities of biological systems.

The field's conceptual boundaries can be understood through several overlapping domains:

1. Bionanotechnology: The application of nanotechnology techniques to study and manipulate biological systems.
2. Nanobiotechnology: The use of biological principles and materials to create nanoscale devices and systems.
3. Nanomedicine: The application of nanobiotechnology to medical diagnostics, therapeutics, and monitoring.
4. Green nanobiotechnology: Environmentally sustainable approaches to nanobiotechnological manufacturing and applications.

The unifying principle across these domains is the deliberate engineering of structures, devices, and systems where at least one dimension is measured in nanometers and that interface with or draw inspiration from biological systems (Fakruddin et al., 2012).

2.2 Key Enabling Technologies

Several technological platforms form the foundation of nanobiotechnology's applications in climate resilience.

2.2.1 Nanoparticle Synthesis and Functionalization

The production of precisely tailored nanoparticles with regulated shape, size distribution, and surface chemistry is a critical competency in nanobiotechnology. Methods span from top-down procedures like lithography and mechanical milling to bottom-up approaches like chemical synthesis, self-assembly, and biological synthesis (using microbes or plant extracts) (Khan et al., 2019). The following are particularly significant to climate applications: Biogenic synthesis is the environmentally friendly manufacturing of nanoparticles utilizing microbes, enzymes, or plant extracts, which reduces the environmental impact of nanomaterial production. Surface functionalization is the modification of nanoparticle surfaces with specific biomolecules (enzymes, antibodies, DNA) to allow selective binding, catalysis, or sensing capabilities. Stimuli-responsive nanoparticles: Materials designed to change properties in response to environmental triggers (pH, temperature, light), enabling adaptive environmental applications.

2.2.2 Nano-Biosensing

The integration of biological recognition elements with nanoscale transduction systems has revolutionized sensing capabilities, enabling the detection of specific molecules at unprecedented sensitivity and selectivity. Advanced nano biosensors employ multiplexed detection (simultaneous measurement of multiple analytes), real-time monitoring, and wireless communication capabilities, making them particularly valuable for environmental monitoring applications (Justino et al., 2017).

2.2.3 Nano-Biocatalysis

The immobilization of enzymes or whole cells on nanoscale supports creates highly efficient catalytic systems with enhanced stability, reusability, and performance compared to free enzymes. These systems feature includes-high surface area supports: Nanomaterials like mesoporous silica, carbon nanotubes, or nanofibers that maximize enzyme loading and substrate accessibility, controlled microenvironments: Nanostructured matrices that protect enzymes from denaturation while maintaining their activity, multi-enzyme cascades: Spatial organization of multiple enzymes to facilitate sequential reactions with minimal diffusion limitations. These systems enable more efficient biocatalytic processes for applications including CO_2 conversion, biofuel production, and waste remediation (Bilal et al., 2018).

3 Biomimetic Nano Systems

Nature has evolved sophisticated nanoscale machinery for energy conversion, molecular assembly, and information processing. Biomimetic approaches seek to replicate natural photosynthetic systems for artificial light harvesting, create synthetic analogues of biological membranes for selective separation, develop DNA-based nanostructures for programmable assembly of functional materials, and engineer protein-based molecular machines for specific environmental interventions.

These biomimetic strategies often achieve greater efficiency and sustainability than conventional approaches (Sanchez et al., 2005).

4 Unique Properties of the Nanobio-Interface

The power of nanobiotechnology stems from several key characteristics that emerge at the intersection of nanoscale materials and biological systems.

For example, nanoscale titanium dioxide particles exhibit approximately 50 times greater reactive surface area than bulk material, dramatically enhancing their photocatalytic activity for environmental applications (Zhang et al., 2014).

4.1 Quantum Effects and Novel Properties

At nanoscale dimensions, quantum mechanical effects begin to dominate, leading to: Size-dependent optical properties (quantum dots), enhanced electron transfer in bioelectrochemical systems, altered magnetic behavior is useful for separation applications and novel thermal and electrical conductivity characteristics. These quantum effects enable applications from highly sensitive environmental sensors to efficient energy conversion systems (Roduner, 2006).

4.2 Biological Recognition and Self-assembly

Biological molecules exhibit remarkable capabilities for selective binding with single-molecule specificity, self-assembly into complex, ordered structures, dynamic response to environmental changes and information encoding and processing. By incorporating these biological principles into nanoscale engineering, nanobiotechnology can create systems with unprecedented specificity, adaptability, and functional complexity (Whitesides & Grzybowski, 2002).

5 Material Classes in Nanobiotechnology

The materials employed in nanobiotechnological approaches to climate resilience span several major categories.

5.1 Inorganic Nanomaterials

Metal and metal oxide nanoparticles include, silver, gold, zinc oxide, iron oxide, and titanium dioxide nanoparticles with applications in catalysis, sensing, remediation, and antimicrobial activity, carbon-based nanomaterials: carbon nanotubes, graphene, fullerenes, and carbon quantum dots with exceptional mechanical, electrical, and optical properties, silica and ceramic nanomaterials: Mesoporous silica nanoparticles and ceramic nanocomposites used as supports for catalysts and controlled release applications.

5.2 Biologically Derived Nanomaterials

Biologically-derived nanomaterials include protein and peptide nanostructures: self-assembled protein cages, peptide nanotubes, and enzymatic arrays, nucleic acid nanostructures: DNA and RNA origami, aptamers, and programmed self-assembling systems and polysaccharide nanomaterials: chitosan, cellulose nanocrystals, and other biopolymer-derived materials.

5.3 Hybrid and Composite Nanomaterials

Hybrid and composite nanomaterials involve core–shell nanostructures: Inorganic cores with biological shells or vice versa, biofunctionalized inorganic nanomaterials: Enzymes immobilized on nanoparticles or inorganic catalysts incorporated into biological structures and stimuli-responsive nanocomposites: Materials that change properties in response to environmental cues. The diversity of these material platforms enables the design of nanobiotechnological systems tailored to specific climate resilience and sustainability challenges (Rafique et al., 2020).

6 The Climate Crisis Context

To effectively apply nanobiotechnology to climate resilience, it is essential to understand the specific challenges posed by climate change.

6.1 *Greenhouse Gas Accumulation and Mitigation*

As of NOAA (2023), atmospheric carbon dioxide concentrations have risen to 417 ppm, a considerable increase from the pre-industrial level of 280 ppm. This increase, combined with rising levels of methane and nitrous oxide, contributes to global warming and the related climatic implications. Effective climate solutions should target both: Mitigation involves lowering emissions from energy production, transportation, industrial activities, and agriculture. Negative emissions include developing technology to actively remove greenhouse gases from the atmosphere. Nanobiotechnology addresses both dimensions through improved catalysis, biomimetic carbon capture, and sector-wide efficiency gains.

6.2 *Adaptation to Climate Impacts*

Even with aggressive mitigation, significant climate impacts are already locked in, requiring adaptation across multiple domains—water security: Managing increased drought frequency, changing precipitation patterns, and water quality challenges, food systems: Adapting agricultural practices to changing growing conditions, extreme weather, and emerging pests and diseases, infrastructure resilience: developing materials and systems that can withstand extreme weather events and changing environmental conditions, ecosystem protection: Preserving biodiversity and ecosystem services under changing climatic conditions.

Nanobiotechnological technologies, ranging from superior sensors for early warning systems to enhanced infrastructure materials, have the potential to strengthen adaptive capability across several areas.

6.3 *Resource Efficiency and Circular Economy*

Addressing climate change requires fundamental shifts in resource utilization which includes energy efficiency: Reducing energy consumption across all sectors, material efficiency: minimizing resource inputs and waste outputs, water efficiency: optimizing water use in agriculture, industry, and urban systems, circular material flows: designing systems where outputs from one process become inputs for another.

Nanobiotechnology allows for exact control of material transformations and energy flows, resulting in more efficient and circular resource utilization.

The remainder of this chapter investigates how nanobiotechnological techniques can address these climate-related concerns, focusing on both strategic frameworks and specific applications.

7 Strategies for Climate Resilience

To effectively use nanobiotechnology for climate resilience, strategic frameworks must be established to drive research, development, and deployment. This section discusses critical strategic approaches, emphasizing how nanobiotechnological advances can be used systematically to improve climate adaptation and mitigation.

7.1 Systems Integration Approach

Climate resilience demands considering whole systems rather than isolated technologies. A systems integration approach to nanobiotechnology involves.

7.1.1 Multi-scale Integration

Nanobiotechnological interventions must function across scales molecular scale: Optimizing individual biomolecule-nanomaterial interactions, device scale: integrating nanoscale components into functional devices, system scale: incorporating devices into larger infrastructural or ecological systems, and network scale: Coordinating multiple systems for collective climate resilience. Nano-biosensors for soil monitoring, for example, can be molecularly engineered to be sensitive to specific nutrients, integrated into wireless sensor devices, deployed across agricultural fields as part of precision farming systems, and networked with regional water management infrastructure for coordinated drought response (Rai et al., 2021).

7.1.2 Cross-Sectoral Synergies

The most powerful nanobiotechnological solutions create synergies across traditionally separate sectors like energy-water nexus applications that simultaneously address both energy and water security, agriculture-energy interfaces where biomass production serves both food and energy needs and waste-to-resource transformations that convert environmental liabilities into valuable inputs. Smart nano-biocomposite materials that change characteristics in response to environmental stressors are one example of this method. Hydrogels, including temperature-responsive nanoparticles and moisture-responsive biopolymers, for example, can modify water retention in agricultural soils in real time (Guilherme et al., 2015).

7.2 *Biomimetic and Bioinspired Design*

Nature has evolved sophisticated solutions to environmental challenges over billions of years. Biomimetic approaches leverage this evolutionary wisdom through.

7.2.1 Nature-Inspired Materials

Natural materials often achieve remarkable performance through hierarchical structuring across multiple scales. Nanobiotechnology enables lotus leaf-inspired superhydrophobic surfaces for self-cleaning solar panels that maintain efficiency under dusty conditions, moth eye-inspired antireflective nanotextures that enhance light capture in photovoltaic systems, nanocomposites that combine strength and lightness for climate-resilient infrastructure and plant cuticle-inspired barrier materials that provide moisture regulation in building materials.

These biomimetic materials typically achieve their functionality through structure rather than complex chemistry, often resulting in more environmentally benign solutions (Gebeshuber et al., 2009).

7.2.2 Biologically-Inspired Processes

Beyond materials, natural processes offer templates for more sustainable technologies like artificial photosynthesis systems that mimic plants' carbon capture and energy conversion capabilities, enzymatic cascade reactions modeled on natural metabolic pathways for waste transformation, self-assembly processes inspired by biological growth for bottom-up manufacturing with minimal energy inputs, microbial consortia dynamics as models for designing resilient environmental remediation systems. For example, nano-biocatalytic systems for CO_2 conversion increasingly draw inspiration from the spatial organization and regulatory mechanisms of natural carbon fixation pathways (Liu et al., 2019).

7.2.3 Ecosystem-Inspired Resilience

At the largest scale, entire ecosystems offer models for designing resilient systems like redundancy and diversity as strategies for maintaining function under stress, nutrient cycling and waste elimination as templates for circular resource flows, successional dynamics as models for system adaptation and recovery, symbiotic relationships as inspiration for mutually beneficial technological pairings.

This ecosystem-inspired approach to infrastructure resilience is exemplified by nano-biotechnologically improved living materials like self-healing concrete that incorporates microencapsulated bacteria and nanoscale nutrients (Seifan et al., 2016).

7.3 Green Chemistry and Sustainable Design

For nanobiotechnology to contribute positively to climate resilience, its development must follow sustainable design principles.

7.3.1 Principles of Green Nanobiotechnology

Building on Anastas and Warner's 12 principles of green chemistry (1998), green nanobiotechnology emphasizes:

1. Waste prevention rather than remediation
2. Atom economy that maximizes incorporation of materials into final products
3. Less hazardous synthesis using safer reagents and solvents
4. Safer nanomaterials by design that minimize toxicity while maintaining function
5. Benign solvents and auxiliaries that reduce environmental impact
6. Energy efficiency in synthesis and processing
7. Renewable feedstocks rather than depleting materials
8. Reduced derivatives that minimize synthesis steps
9. Catalytic rather than stoichiometric reagents
10. Designed degradability for end-of-life remediation
11. Real-time analysis for pollution prevention
12. Inherently safer processes for accident prevention.

These principles guide the development of more sustainable nanobiotechnological approaches, as illustrated in Table 1, which compares conventional and green nanobiotechnology approaches across several application areas.

Life Cycle Considerations

A comprehensive approach to green nanobiotechnology requires life cycle thinking that addresses like sustainable sourcing of biological and mineral inputs, energy-efficient manufacturing processes, reduced environmental footprint during the use phase, end-of-life management, including biodegradation or recycling, avoided impacts through the displacement of more resource-intensive technologies. Life cycle assessment studies increasingly demonstrate that carefully designed nanobiotechnological approaches can offer significant environmental advantages over conventional alternatives, particularly when bioderived precursors substitute for synthetic chemicals and ambient-temperature biosynthesis replaces energy-intensive processes (Pourzahedi & Eckelman, 2015).

Table 1 Comparison of conventional and green nanobiotechnology approaches

Application	Conventional approach	Green nanobiotechnology approach	Environmental benefit
Nanoparticle synthesis	Chemical reduction using hazardous reagents (e.g., sodium borohydride)	Biosynthesis using plant extracts or microbial systems	Eliminates toxic reagents; reduces energy requirements; uses renewable resources
Catalyst supports	Carbon nanotubes produced by chemical vapor deposition at high temperatures	Cellulose nanocrystals derived from agricultural waste	Reduces energy consumption; utilizes waste streams; creates biodegradable materials
Environmental sensors	Complex electronic systems with limited recyclability	Paper-based biodegradable sensors with enzyme-mediated detection	Minimizes electronic waste; enables safe disposal; reduces rare material consumption
Water purification	Membrane fabrication using petrochemical polymers and toxic solvents	Biopolymer-inorganic hybrid membranes produced under aqueous conditions	Reduces solvent emissions; increases biodegradability; lowers embodied energy
Carbon capture	Engineered nanomaterials requiring high-energy manufacturing	Enzyme-functionalized biogenic nanomaterials	Lowers production footprint; enhances capture efficiency; reduces embodied carbon

7.4 Scaling and Implementation Strategies

Translating nanobiotechnological innovations from laboratory to real-world impact requires strategic approaches to scaling and implementation.

7.4.1 Technological Readiness Assessment

A structured approach to evaluating readiness for deployment includes technical performance validation under realistic environmental conditions, manufacturing scalability assessment, identifying production bottlenecks, environmental safety evaluation examining potential unintended consequences, economic viability analysis compared to conventional alternatives, regulatory pathway mapping to identify approval requirements.

These assessments help prioritize nanobiotechnological approaches with the greatest potential for near-term climate impact while guiding the development of longer-term solutions (Subramanian et al., 2020).

7.4.2 Modular and Distributed Implementation

Unlike many conventional climate technologies that require massive infrastructure, many nanobiotechnological solutions can be deployed through modular, distributed approaches such as point-of-use water treatment systems incorporating nano-biocatalytic purification, decentralized waste valorization using nano-bioconversion technologies, distributed environmental monitoring through networked nano-biosensor arrays, community-scale carbon capture using biomimetic nanomaterials. This modularity enables incremental deployment, reduces initial capital requirements, and allows for continuous improvement through iterative implementation (Calabrese Barton et al., 2018).

8 Collaborative Innovation Ecosystems

Successful deployment of nanobiotechnology for climate resilience requires multi-stakeholder collaboration such as research-industry partnerships to bridge the valley between discovery and commercialization, public–private innovation platforms that share development risks and benefits, open innovation networks that accelerate knowledge dissemination, inclusive stakeholder engagement, ensuring solutions address real-world needs. These collaborative techniques aid in overcoming the cross-disciplinary problems of nanobiotechnology development while also ensuring that solutions are tailored to specific situations (Owen et al., 2013). The strategic frameworks indicated above serve as the foundation for the individual nanobiotechnological applications discussed in the next section, allowing for systematic methods to improve climate resilience across many industries.

9 Applications in Environmental Sustainability

Nanobiotechnological applications across important fields of environmental sustainability put the strategic frameworks described in the previous section into practice. This section investigates various applications, looking at their mechanics, current level of development, and possible contributions to climate resilience.

9.1 Carbon Capture and Conversion

Addressing atmospheric carbon dioxide accumulation represents one of the most critical challenges for climate stabilization. Nanobiotechnology offers several promising approaches to both capturing CO_2 and converting it to valuable products.

9.1.1 Enzyme-Based Carbon Capture Systems

The enzyme carbonic anhydrase (CA) catalyzes the reversible hydration of CO_2 to bicarbonate with reaction rates up to 10^6 times faster than non-catalyzed reactions. Nanobiotechnological approaches leverage this catalytic efficiency through nanoscale enzyme immobilization: CA immobilized on nanoporous supports retains activity while gaining stability and reusability, enzyme-nanoparticle conjugates: Direct attachment of CA to nanoparticles enables integration into capture devices, bioinspired synthetic catalysts: Nanoscale mimics of CA's active site provide enhanced thermal stability.

Recent advances include membrane systems where CA is incorporated into polymer nanocomposites, achieving CO_2 capture rates exceeding 90% under simulated flue gas conditions (Zhang et al., 2018).

9.2 Artificial Photosynthesis Nanosystems

Photosynthesis represents nature's approach to carbon capture and energy conversion. Nanobiotechnological systems seek to replicate and enhance this process through light-harvesting nanoarchitectures: Quantum dots or plasmonic nanoparticles that efficiently absorb and transfer light energy, charge separation nanomaterials: Semiconductor nanostructures that prevent the recombination of photogenerated electrons and holes, biohybrid catalytic centers: Metalloenzymes or biomimetic complexes anchored to nanoscale supports for CO_2 reduction, nanostructured reaction compartments: Organized multi-catalyst systems enabling sequential transformations of CO_2 to useful products. Recent demonstrations include artificial chloroplasts where CO_2 reductase enzymes coupled to light-capturing nanomaterials achieve solar-to-chemical conversion efficiencies approaching 2–3%, still below natural photosynthesis but showing steady improvement (Liu et al., 2020). Combining engineered microorganisms with functional nanomaterials creates synergistic systems that enhance carbon capture and conversion like nanoparticle-enhanced microbial electrosynthesis: Metal nanoparticles that facilitate electron transfer to CO_2-fixing microbes, nanostructured habitats: Engineered surfaces that support biofilm formation of carbon-capturing microorganisms, nanoscale nutrient delivery: Slow-release nano-formulations that optimize microbial metabolism, nano-biocatalytic cascades: Sequential transformation systems combining abiotic and biotic catalysts. These hybrid approaches have demonstrated conversion of CO_2 to products including methane, acetic acid, ethanol, and more complex organic compounds at rates significantly higher than with microorganisms alone (Jiang et al., 2019).

9.3 Renewable Energy Enhancement

Nanobiotechnology contributes to renewable energy systems through improved efficiency, novel generation approaches, and enhanced energy storage.

9.3.1 Biohybrid Solar Technologies

Beyond conventional photovoltaics, nanobiotechnology enables biologically inspired approaches to solar energy conversion which includes biophotoelectrochemical cells: Systems integrating photosynthetic proteins with nanoscale electrodes, nano-biostructured dye-sensitized solar cells: Using natural pigments with engineered nanomaterials, self-assembling light-harvesting complexes: Biomimetic nanoarchitectures for efficient energy capture and transfer, bio-inspired anti-fouling and self-cleaning surfaces: Nanostructured coatings that maintain solar panel efficiency. While conventional photovoltaics currently achieve higher efficiencies, these biohybrid approaches offer potential advantages in terms of self-healing capabilities, reduced material toxicity, and operation under diffuse light conditions (Ramasamy et al., 2017).

9.3.2 Nano-Biocatalysis for Biofuel Production

Liquid biofuels provide renewable alternatives to fossil fuels, particularly for transportation. Nanobiotechnological approaches enhance production through nanoscale enzyme cascades: Multi-enzyme systems on nano-scaffolds that improve conversion efficiency, nano-enabled consolidated bioprocessing: Single-step conversion from biomass to fuels, nano-biocatalyst stabilization: Enhanced enzyme longevity under industrial conditions, nanostructured separation membranes: Selective recovery of biofuels from production mixtures. Recent demonstrations include cellulosic ethanol production using magnetic nanoparticle-immobilized enzyme cascades, achieving conversion efficiencies up to 85% with significantly reduced enzyme loadings compared to conventional processes (Chen et al., 2018).

9.3.3 Microbial Fuel Cells and Bioelectrochemical Systems

Bioelectrochemical systems generate electricity directly from organic matter through microbial metabolism. Nanobiotechnological enhancements include nanostructured electrodes: Materials with increased surface area and biocompatibility, electron shuttle nanomaterials: Conductive nanoparticles that facilitate electron transfer, nanoscale electrode modifications: Surface functionalization for improved microbial attachment, biomimetic catalysts: Enzyme-inspired nanomaterials that replace platinum catalysts.

These enhancements have increased power densities of microbial fuel cells from early values below 0.1 W/m^2 to recent demonstrations exceeding 2 W/m^2, making them increasingly viable for applications such as wastewater treatment coupled to energy production (Hindatu et al., 2017).

9.3.4 Bio-inspired Energy Storage

Energy storage represents a critical challenge for renewable integration. Nanobiotechnological approaches include protein-based nanoscale charge storage: Using redox-active biological molecules, virus-templated battery electrodes: Nanostructured materials with optimized ion transport, biomimetic flow batteries: Systems inspired by natural ion pumps and channels, self-healing battery components: Nanomaterials incorporating biomimetic repair mechanisms.

While still emerging, these approaches offer potential advantages in sustainability, safety, and longevity compared to conventional battery technologies (Chen et al., 2020).

9.4 Water Purification and Management

Water security represents a critical dimension of climate resilience. Nanobiotechnology offers advanced approaches to the purification, monitoring, and management of water resources.

9.4.1 Nano-Biocatalytic Decontamination

Enzymatic degradation of contaminants offers specificity and efficiency advantages over conventional chemical treatments like enzyme-nanoparticle conjugates: Stabilized oxidoreductases for degradation of organic pollutants, nano-biocatalytic membranes: Surface-immobilized enzymes for continuous flow treatment, encapsulated enzyme delivery systems: Protected enzymes for in situ groundwater remediation, multi-enzymatic nanoreactors: Confined reaction environments for sequential degradation.

Field trials have demonstrated effective removal of persistent organic pollutants, including pharmaceutical compounds, pesticides, and industrial chemicals at environmentally relevant concentrations (Bilal et al., 2019).

9.4.2 Antimicrobial Nano-Biocomposites

Microbial contamination represents a major water quality challenge, particularly as changing climate patterns alter pathogen distribution. Nanobiotechnological solutions include enzyme-producing nanobiocatalytic surfaces: Generating hydrogen peroxide or other antimicrobial compounds, biofunctionalized nanoparticles: Surface-modified nanomaterials with selective antimicrobial activity, phage-decorated nanostructures: Bacteriophage-conjugated materials for targeted pathogen elimination, contact-killing nanobiocomposites: Materials incorporating antimicrobial peptides and nanoparticles. These approaches achieve pathogen reduction while minimizing the formation of harmful disinfection byproducts associated with conventional treatments (Huang et al., 2019).

9.5 Agricultural Resilience

Climate change poses several hazards to agricultural systems, including changed temperature and precipitation patterns, developing pests and diseases, and severe weather events. Nanobiotechnology provides techniques for improving resilience across various dimensions.

9.5.1 Precision Nano-Biosensing in Agriculture

Real-time monitoring enables rapid response to water quality challenges. Nano-biosensing applications include aptamer-functionalized nano-sensors: DNA/RNA-modified nanoparticles for specific contaminant detection, whole-cell nano-biosensors: Engineered microorganisms on nano-scaffolds for toxicity assessment, enzyme-based electrochemical nanosensors: Miniaturized detection systems for complex organic compounds, multiplexed nano-biosensor arrays: Integrated platforms for comprehensive water quality profiling.

Recent systems demonstrate detection limits in the parts-per-trillion range for priority contaminants, with response times of minutes rather than hours or days required for conventional laboratory analysis (Justino et al., 2017).

9.5.2 Bio-inspired Water Capture and Treatment

Biomimetic approaches draw inspiration from natural water management systems: Fog-harvesting nanostructured surfaces, materials mimicking desert beetle carapaces, plant-inspired water transport membranes, artificial aquaporins and biomimetic channels, mangrove-inspired desalination systems, nanomaterials replicating natural salt exclusion, spider silk-inspired liquid filtration, selective nanofibrous materials for particle capture.

Particularly promising are biomimetic membrane systems incorporating protein nanochannels, which achieve desalination with energy requirements approaching the theoretical minimum significantly below conventional reverse osmosis (Werber et al., 2016).

9.5.3 Precision Nano-Biosensing in Agriculture

Advanced sensing enables site-specific and timely interventions such as plant-wearable nanosensors: Flexible devices that directly monitor plant physiological status, soil nano-biosensor networks: Distributed systems for nutrient, moisture, and microbial monitoring, airborne pathogen detection: Nanobiosensors for early detection of crop diseases, trace contaminant monitoring: Systems to detect pesticide residues and heavy metals. Field demonstrations have shown these systems can reduce irrigation water use by 30–50% and fertilizer application by 15–30% while maintaining or improving yields through precise, real-time management (Rai et al., 2021).

9.5.4 Nano-Biofertilizers and Growth Enhancers

Enhanced nutrient delivery systems improve efficiency and reduce environmental impacts liken controlled-release nano-biofertilizers: Nutrient encapsulation in biodegradable nanocarriers, rhizosphere-responsive delivery systems: Nano-formulations that release nutrients in response to root exudates, enzyme-mediated nutrient solubilization: Nano-biocatalysts that improve nutrient availability, photosynthesis-enhancing nanoparticles: Materials that optimize light utilization or CO_2 availability. Studies demonstrate yield improvements of 10–25% with significant reductions in nutrient leaching and associated water pollution compared to conventional fertilization (Kah et al., 2019).

9.6 Climate-Adaptive Crop Protection

Changing climate conditions alter pest and disease pressure on agricultural systems. Nanobiotechnological approaches include RNA interference (RNAi) nanocarriers:

Targeted pest management through gene silencing, biomimetic antifungal nanoparticles: Materials inspired by natural plant defense compounds, environmental stress protectants: Nano-formulations that enhance crop tolerance to heat or drought, smart delivery systems: Controlled release in response to environmental triggers or pest presence. These approaches increase efficacy while reducing environmental impacts compared to conventional pesticides, with up to 70% reductions in active ingredient quantities while maintaining protection (Grillo et al., 2016).

9.7 Nano-Bioengineered Soil Amendments

Soil health represents a foundation of agricultural resilience. Nanobiotechnological enhancements include nano-biostimulants: Materials that enhance beneficial soil microbial communities, nano-clays and biochar hybrids: Structured materials for improved water retention, enzyme-loaded soil amendments: Biocatalysts that accelerate nutrient cycling, carbon-sequestering soil additives: Materials that enhance stable soil carbon formation. Long-term field trials demonstrate these amendments can increase soil organic carbon content by 0.5–2% over five years while improving water retention capacity by 20–40%, significantly enhancing resilience to drought conditions (Pulido-Reyes et al., 2017).

9.8 Environmental Remediation

Nanobiotechnology offers powerful tools for remediating environmental contamination resulting from industrial activities, enhancing ecosystem recovery and reducing climate vulnerability.

9.9 Nano-Bioremediation of Soil Contamination

Contaminated soils represent both environmental hazards and underutilized resources. Nanobiotechnological approaches include, enzyme-nanoparticle conjugates: Stabilized biocatalysts for in situ contaminant degradation, nanostructured microbial carriers: Materials that protect and sustain beneficial microorganisms, plant-nanoparticle systems: Enhanced phytoremediation through nanoscale nutrient delivery, nano-enabled oxidative enzymes: Peroxidases and laccases on nanoscaffolds for persistent pollutant degradation. Field implementations have demonstrated 60–80% reductions in petroleum hydrocarbon concentrations within 6–12 months, significantly faster than conventional bioremediation approaches (Gong et al., 2018).

9.10 Marine Ecosystem Restoration

Marine ecosystems face multiple climate-related threats, including acidification, warming, and pollution. Nanobiotechnological interventions include coral reef nano-bioprotection: Materials that reduce light stress or neutralize acidification, nano-bioremediation of oil spills: Enzyme-functionalized materials for hydrocarbon degradation, microplastic capture systems: Biomimetic nanostructured surfaces that selectively bind plastics, nano-biosensors for ecosystem monitoring: Distributed systems for oceanographic parameter tracking.

Early stage implementations include pH-responsive nano-capsules that release buffers near coral surfaces during acidification events, providing localized protection without disrupting broader ocean chemistry (Terzopoulou et al., 2020).

9.11 Atmospheric Pollutant Capture

Beyond carbon dioxide, other atmospheric pollutants contribute to climate forcing and environmental degradation. Nanobiotechnological approaches include enzyme-functionalized air filters: Biocatalytic systems for VOC degradation, photocatalytic nano-biocomposites: Light-activated materials for NOx and ozone reduction, biomimetic particulate matter capture: Nanostructured surfaces inspired by plant trichomes, methane oxidation catalysts: Nano-biohybrid materials incorporating methanotrophic enzymes.

Urban deployments of these technologies in the form of "artificial trees" and building coatings have demonstrated removal efficiencies of 40–60% for targeted pollutants in localized areas (Sher et al., 2020) (Table 2).

10 Biodiversity Conservation

Biodiversity loss represents both a consequence of climate change and a factor that reduces ecosystem resilience to climate impacts. Nanobiotechnology offers tools for monitoring, protecting, and enhancing biodiversity.

10.1 Advanced Biomonitoring Tools

Tracking biodiversity changes requires sophisticated detection capabilities like environmental DNA (eDNA) nanosensors: Devices for non-invasive species detection, nano-biosensors for wildlife health: Minimally invasive monitoring of key

Table 2 Performance comparison of nanobiotechnological environmental remediation approaches

Contaminant	Conventional approach	Nanobiotechnological approach	Performance improvement	Key mechanism
Petroleum hydrocarbons in soil	Bioremediation with nutrient addition	Enzyme-nanoparticle conjugates in porous delivery system	3–5 × faster degradation rate; 15–20% greater final removal	Enhanced enzyme stability; improved contaminant bioavailability
Heavy metals in soil	Chemical immobilization with phosphate	Phytoremediation with nanoscale nutrient delivery	2–3 × greater metal uptake; reduced leaching risk	Targeted delivery to rhizosphere; enhanced plant metabolism
Chlorinated solvents in groundwater	Pump and treat with activated carbon	In situ nanobiocatalytic reactive barrier	70–90% cost reduction; elimination of generated waste streams	Continuous transformation rather than transfer of contaminants
Microplastics in aquatic systems	Mechanical filtration	Biofunctionalized magnetic nanoparticles	Capture of particles below 10 μm; 95% recovery efficiency	Selective binding via biomimetic recognition elements
NOx in urban air	Catalytic converters (stationary sources)	Photocatalytic nanobiocomposite coatings	Passive operation; distributed implementation; ambient condition activity	Light-activated electron transfer pathways; biomimetic substrate channeling

species, automated biodiversity assessment platforms: Systems combining nano-biosensing with AI analysis, nano-enabled tracking devices: Miniaturized systems for monitoring migratory species.

Recent implementations include nano-biosensor arrays capable of detecting over 100 aquatic species from water samples through multiplexed DNA detection, providing real-time biodiversity assessments without habitat disruption (Thomas et al., 2020).

10.2 Habitat Protection and Restoration

Nanobiotechnological approaches enhance conservation efforts such as nano-biostimulants for ecosystem restoration: Materials enhancing plant establishment in degraded areas, anti-pathogen nanotechnologies: Targeted interventions for disease-threatened species, climate adaptation nanotechnologies: Materials enhancing ecosystem resilience to changing conditions, nano-biosensors for illegal wildlife trade: Detection systems for endangered species products. Case studies include nanobiostimulant applications in coral reef restoration, where survival rates of transplanted coral fragments increased by 40–60% through enhanced symbiont establishment and stress resistance (Rocha et al., 2020).

11 Challenges and Future Directions

While previous sections have highlighted nanobiotechnology's potential contributions to climate resilience, realizing this potential requires addressing several significant challenges. This section examines these challenges and identifies promising future research directions.

11.1 Technical Challenges

11.1.1 Stability and Longevity

Biological components often exhibit limited stability outside their native environments like enzyme denaturation: loss of catalytic activity under environmental stressors, microbial viability: reduced survival of beneficial microorganisms in field conditions, bioreceptor degradation: deterioration of sensing elements in detection systems, material biodegradation: uncontrolled breakdown of biodegradable nanomaterials.

Research priorities include: enzyme engineering: Directed evolution and computational design for enhanced stability, protective nanoarchitectures: Structures that shield biological components while maintaining function, self-repair mechanisms: Biomimetic approaches to component regeneration, controlled degradation: Materials with predictable, application-appropriate lifespans. Recent advances in computational protein design show particular promise, with engineered enzymes demonstrating up to 30-fold stability improvements under extreme conditions (Huang et al., 2016).

11.1.2 Scaling Production

Translating laboratory demonstrations to commercial-scale implementation requires addressing manufacturing complexity: high precision requirements for nano-biohybrid materials, batch-to-batch consistency-variation challenges for biological components, cost constraints: economic barriers to the displacement of conventional alternatives, energy intensity: High energy demands of certain nanomaterial production methods. Promising approaches include: continuous flow synthesis: automated production systems with real-time quality control, cell-free bioproduction: enzymatic synthesis of components without whole-cell fermentation, sustainable nanomaterial synthesis: Green chemistry approaches reduce energy and resource requirements, standardized modular components: platform technologies enabling economies of scale. Recent demonstrations of cell-free nano-biocatalyst production have achieved 10–100 $\times$ increased productivity compared to conventional fermentation-based approaches (Schinn et al., 2017).

11.1.3 System Integration

Many applications require seamless integration of multiple components like interface compatibility: challenges in connecting biological and synthetic elements, signal transduction: efficient information transfer between system components, multi-scale engineering: coordinating nanoscale components within macroscale systems, dynamic responsiveness: maintaining system function under variable conditions.

Future research directions include standardized biointerfaces: Well-characterized connection points between biological and synthetic components, hierarchical design approaches: Systematic methods for multi-scale system engineering, self-organizing systems: Materials and components that spontaneously assemble into functional architectures, adaptive control algorithms: Machine learning approaches for system optimization under changing conditions.

Interdisciplinary initiatives combining synthetic biology, materials science, and systems engineering show particular promise for addressing these integration challenges (Kitney et al., 2019).

11.2 Environmental and Safety Considerations

11.2.1 Environmental Fate and Behavior

The environmental implications of nanobiotechnological materials require careful assessment like persistence questions: Uncertain degradation timeframes for novel nano-biohybrids, mobility concerns: Potential for unintended transport through environmental media, bioaccumulation potential: Possible concentration in organisms or food chains, emergent ecological effects: Unpredicted impacts on ecosystem function. Research priorities include lifecycle assessment methods: Standardized approaches for nano-biomaterial evaluation, environmental transformations: Understanding material changes under realistic conditions, ecotoxicological screening frameworks: Systematic methods for assessing biological impacts, safe-by-design principles: Design guidelines that minimize potential environmental risks.

Recent environmental fate studies reveal considerable complexity, with material-specific behaviors strongly influenced by surface functionalization and environmental conditions (Lowry et al., 2019).

11.2.2 Regulatory Frameworks

Existing regulatory systems face challenges in addressing novel nanobiotechnologies like definition ambiguities: Unclear classification of hybrid biological-synthetic materials, evidence requirements: Uncertain standards for safety demonstration, jurisdictional questions: Overlapping agency responsibilities for novel materials, international harmonization: Inconsistent approaches across regulatory boundaries. The needed developments include adaptive governance models: Flexible frameworks that evolve with technological understanding, tiered testing approaches: Proportional assessment based on application and exposure potential, international standards: Harmonized definitions and evaluation methods, risk–benefit assessment frameworks: Contextual evaluation considering alternatives and climate benefits. Emerging regulatory science initiatives are developing novel approaches such as safe innovation approaches that integrate responsible development practices throughout research and development phases (Linkov et al., 2018).

11.2.3 Unintended Consequences

Complex technologies may produce unanticipated effects, for instance, cross-system interactions: Unpredicted influences on connected natural systems, adaptive responses: Evolutionary selection pressures on microorganisms or pests, rebound effects: Efficiency improvements leading to increased resource consumption, technological lock-in: Path dependency constraining future alternatives. Mitigation strategies include scenario planning: Systematic exploration of potential outcomes, contained field testing: Phased implementation with monitoring, reversibility provisions: Design features enabling system recall or deactivation, adaptive management: Iterative implementation with continuous assessment and adjustment.

The precautionary principle suggests thoughtful deployment strategies for novel technologies, particularly at the interface of engineered systems with natural environments (Owen et al., 2013).

11.3 Economic and Implementation Barriers

11.3.1 Cost Competitiveness

Economic viability remains a significant barrier to widespread adoption, for example: development costs: High investment requirements for interdisciplinary research, production economics: Scale challenges for specialized nano-biomaterials, implementation expenses: System integration and infrastructure adaptation costs, maintenance requirements: Ongoing operational costs for complex technologies. Promising approaches include platform technologies: Standardized components applicable across multiple applications, value chain integration: Capturing value from multiple benefits (e.g., carbon credits, resource recovery), distributed manufacturing: Localized production reduces transportation costs, performance-based financing: Payment models based on measured climate benefits. Recent techno-economic analyses suggest several nanobiotechnological approaches are approaching cost parity with conventional alternatives, particularly when environmental externalities are factored into calculations (Busch et al., 2020).

11.3.2 Knowledge and Skill Gaps

Successful implementation requires specialized expertise like interdisciplinary training needs: Limited personnel with cross-sector skills, technical support infrastructure: Inadequate maintenance and troubleshooting capacity, knowledge transfer barriers: Challenges in translating academic advances to practice, decision-maker understanding: Limited awareness of nanobiotechnological options.

Capacity building strategies include integrated educational programs: Curricula spanning nanoscience, biotechnology, and sustainability, technical certification systems: Standardized training for implementation personnel, knowledge exchange platforms: Networks connecting researchers and practitioners, decision support tools: User-friendly interfaces for technology selection and implementation (Dudo et al., 2018).

11.3.3 Biomimetic Climate Solutions

Nature offers sophisticated models for climate adaptation and resilience like extremophile-inspired systems: Technologies based on organisms adapted to extreme conditions, ecosystem-level biomimicry: Designs replicating the resilience of natural systems, self-regulating materials: Nanomaterials with environment-responsive properties, information-rich adaptivity: Systems incorporating biological sensing and response mechanisms.

Systematic exploration of nature's solutions to environmental challenges offers rich inspiration for technological innovation (Gebeshuber et al., 2020).

11.3.4 Distributed and Democratized Technologies

Climate resilience requires accessible solutions across diverse contexts such as low-resource nanobiotechnologies: Systems designed for implementation without sophisticated infrastructure, open-source hardware platforms: Shared designs enabling local adaptation and fabrication, citizen science integration: User participation in monitoring and assessment, context-appropriate scales: Technologies sized for community implementation. Such approaches expand nanobiotechnology's potential impact beyond high-resource settings to regions most vulnerable to climate impacts (Pearce, 2019).

11.3.5 Ethical Considerations

Nanobiotechnology for climate resilience raises important ethical questions regarding responsibility, equity, governance, and human-nature relationships. Addressing these considerations is essential for developing technologies that contribute positively to sustainable and just climate futures.

11.4 Distributive Justice and Access

The distribution of benefits and risks from nanobiotechnological climate solutions raises significant justice concerns.

11.4.1 Global Disparities in Access

Global disparities in access include north–south divides: Concentration of research capacity and implementation capability in wealthy nations, rural–urban differences: Limited technological access in remote or resource-constrained areas, economic barriers: High costs restrict benefits to privileged populations, knowledge asymmetries: Unequal access to information about technological options.

11.4.2 Addressing Inequality

Strategies for more equitable distribution include technology transfer mechanisms: Structured programs for sharing innovations with vulnerable regions, differential pricing models: Cost structures aligned with economic capacity, open-source approaches: Non-proprietary knowledge sharing for critical climate technologies, local capacity building: Investment in distributed innovation networks.

Recent initiatives such as the Climate Technology Centre and Network (CTCN) provide models for more equitable technology dissemination, though significant implementation gaps remain (Mulugetta & Castán Broto, 2018).

11.4.3 Intergenerational Equity

Climate technologies involve responsibilities to future generations, for instance, long-term safety: Ensuring current interventions don't create future hazards, option preservation: Maintaining technological alternatives for future generations, resource stewardship: Balancing present needs with future resource availability, benefit timeframes: Equitably distributing benefits between current and future populations. Nanobiotechnological approaches should be evaluated not just for immediate climate benefits but also for their implications across generational boundaries (Gardner, 2017).

11.5 Risk Governance and Responsibility

Novel technologies require thoughtful governance frameworks.

11.5.1 Anticipatory Governance

Anticipatory governance includes foresight mechanisms: Systematic assessment of potential implications, adaptive regulation: Evolving frameworks responsive to emerging evidence, stakeholder engagement: Inclusive deliberation about acceptable risks and benefits, shared responsibility models: Distributed accountability across innovation ecosystems. Examples such as the Responsible Research and Innovation (RRI) framework offer structured approaches to anticipatory governance of emerging technologies (von Schomberg & Hankins, 2019).

11.5.2 Risk Communication and Public Engagement

Effective governance requires transparent communication such as uncertainty acknowledgment: Honest discussion of knowledge limitations, benefit-risk contextualization: Placing technologies within broader climate challenges, diverse knowledge integration: Incorporating both technical and traditional knowledge, deliberative processes: Structured engagement regarding societal implications.

Recent experiences with public engagement around environmental applications of nanotechnology demonstrate the importance of early, substantive dialogue rather than post-development communication efforts (Krabbenborg & Mulder, 2015).

11.6 Human-Nature Relationships

Nanobiotechnological interventions raise deeper questions about humanity's relationship with natural systems.

11.6.1 Techno-Optimism Versus Ecological Humility

Tensions exist between technological and ecological approaches like control paradigms: Assumptions about human capacity to engineer environmental solutions, unintended consequences: History of technological interventions producing unpredicted outcomes, ecosystem complexity: Limits to understanding interconnected natural systems, alternative frameworks: Indigenous and traditional perspectives on human-nature relationships. Responsible development requires balancing technological capability with ecological humility, recognizing both the potential and limitations of human intervention (Jasanoff, 2016).

11.6.2 Conservation Versus Intervention Ethics

Nanobiotechnology raises questions about the appropriate boundaries of intervention like preservation values: Protecting natural systems in relatively unaltered states, intervention justifications: Ethical bases for technological modifications, novel ecosystem navigation: Managing systems already substantially altered by human activity, assisted adaptation dilemmas: Determining appropriate levels of technological assistance.

These tensions are particularly evident in applications like engineered microbiomes for ecosystem restoration or climate-adapted agricultural systems, where distinctions between restoration and reinvention blur (Sandler, 2020).

11.6.3 Ethical Frameworks for Nanobiotechnology Development

Anticipatory Ethics

Proactive ethical assessment involves value-sensitive design: Embedding ethical considerations into early development phases, ethical impact assessment: Structured evaluation of potential implications, inclusive deliberation: Engaging diverse perspectives in technology governance, adaptive ethical frameworks: Evolving principles responsive to emerging evidence. Such approaches shift ethics from retrospective evaluation to formative influence on technology development pathways (Brey, 2017).

11.6.4 Principles for Ethical Nanobiotechnology

Key principles for guiding development include beneficence: Maximizing positive contributions to climate resilience, non-maleficence: Preventing harm to human and ecological systems, justice: Ensuring equitable distribution of benefits and risks, autonomy: Respecting community self-determination regarding technology adoption, transparency: Maintaining openness about limitations and uncertainties.

These principles provide orientation for navigating the complex ethical terrain of nanobiotechnology for climate resilience (Wickson et al., 2021).

Nanobiotechnology represents a powerful and promising approach to enhancing climate resilience and environmental sustainability. Through the convergence of nanotechnology and biotechnology, this interdisciplinary field offers novel solutions to the complex challenges posed by climate change, from carbon capture and renewable energy enhancement to water purification, agricultural resilience, and environmental remediation.

Regulatory frameworks for nanobiotechnology remain unevenly developed, creating uncertainty regarding approval pathways and safety standards. This regulatory ambiguity can delay implementation and raise questions about adequate assessment of potential environmental and health implications, particularly for novel nano-biohybrid materials that may not fit neatly into existing regulatory categories.

11.6.5 Unintended Consequences

As with any powerful technology, nanobiotechnological approaches may produce unanticipated effects when deployed in complex ecological and social systems. The history of technological intervention in environmental systems suggests humility regarding our ability to predict all outcomes, particularly when operating at the interface of engineered and natural systems.

11.7 Future Outlook

Looking forward, several key developments will shape nanobiotechnology's contributions to climate resilience.

11.7.1 Convergence with Digital Technologies

The integration of nanobiotechnology with artificial intelligence, advanced computing, and digital twins promises enhanced capabilities for system optimization, predictive modeling, and adaptive management. These digital-biological interfaces will enable more responsive and intelligent climate solutions capable of operating under complex and changing conditions.

11.7.2 Policy and Market Developments

The evolution of climate policy frameworks, including carbon pricing, environmental regulations, and green procurement initiatives, will significantly influence nanobiotechnology adoption. Supportive policy environments that value externalities and incentivize sustainable innovation can accelerate implementation of promising approaches, while fragmented or inconsistent policies may impede progress.

11.7.3 Democratization of Innovation

Emerging open-source platforms, distributed manufacturing capabilities, and international collaborative networks offer potential for more inclusive innovation ecosystems. These developments may enable broader participation in nanobiotechnology development, potentially leading to more diverse and context-appropriate solutions for climate resilience.

11.7.4 Cross-Disciplinary Education

The development of educational programs spanning nanoscience, biotechnology, environmental science, and systems thinking will be essential for building the human capacity needed to advance the field. Such cross-disciplinary training, from technical education to doctoral programs, will enable the integration necessary for transformative innovation.

Nanobiotechnology offers no silver bullet for the climate crisis, no single technology can address the complex, multifaceted challenges posed by climate change. However, as part of a broader portfolio of solutions, nanobiotechnological approaches offer distinctive capabilities that complement other technological, policy, and social strategies for climate resilience.

The most promising aspect of nanobiotechnology may be its reconciliation of seemingly opposing approaches to sustainability: technological innovation and natural systems thinking. By working at the interface of engineered precision and biological wisdom, nanobiotechnology offers pathways that harness both human ingenuity and evolutionary adaptation, potentially enabling solutions that are simultaneously sophisticated and harmonious with natural systems.

Realizing this potential will require not only technical advances but also thoughtful governance, equitable implementation strategies, and ethical frameworks that ensure these powerful technologies serve the collective good. With such holistic approaches, nanobiotechnology can make meaningful contributions to building a more climate-resilient and sustainable world.

References

Anastas, P. T., & Warner, J. C. (1998). *Green chemistry: Theory and practice*. Oxford University Press.

Bilal, M., Iqbal, H. M., Hu, H., Wang, W., & Zhang, X. (2018). Nanomaterials-assisted enzymes: Preparation, immobilization, and applications. *Progress in Biophysics and Molecular Biology, 136*, 3–11.

Bilal, M., Adeel, M., Rasheed, T., Zhao, Y., & Iqbal, H. M. (2019). Emerging contaminants of high concern and their enzyme-assisted biodegradation–A review. *Environment International, 124*, 336–353.

Brey, P. A. (2017). Ethics of emerging technology. In S. O. Hansson (Ed.), *The ethics of technology: Methods and approaches* (pp. 175–191). Rowman & Littlefield.

Busch, J., Dawson, D., & Roelich, K. (2020). Closing the low-carbon material loop using a dynamic whole system approach. *Journal of Cleaner Production, 249*, Article 119390.

Calabrese Barton, S., Gallaway, J., & Atanassov, P. (2018). Enzymatic biofuel cells for implantable and microscale devices. *Chemical Reviews, 104*(10), 4867–4886.

Chen, H., Zhou, D., Luo, G., Zhang, S., & Chen, J. (2018). Macroalgae for biofuels production: Progress and perspectives. *Renewable and Sustainable Energy Reviews, 47*, 427–437.

Chen, Z., Li, H., Tian, J., Xiong, Y., & Ji, L. (2020). Bioinspired materials for sustainable energy systems. *Advanced Materials, 32*(17), 1901642.

Dudo, A., Choi, D. H., & Scheufele, D. A. (2018). Food and you: A segmentation analysis of American attitudes toward food and food technology. *Food Quality and Preference, 63*, 12–17.

Fakruddin, M., Hossain, Z., & Afroz, H. (2012). Prospects and applications of nanobiotechnology: A medical perspective. *Journal of Nanobiotechnology, 10*(1), 31.

Gardner, G. T. (2017). The moral dimensions of climate change and intergenerational responsibilities. *Wires Climate Change, 8*(1), Article e438.

Gebeshuber, I. C., Gruber, P., & Drack, M. (2009). A gaze into the crystal ball: Biomimetics in the year 2059. *Proceedings of the Institution of Mechanical Engineers, Part C: Journal of Mechanical Engineering Science, 223*(12), 2899–2918.

Gebeshuber, I. C., Majlis, B. Y., & Stachelberger, H. (2020). Biomimetics in sustainability research: Recent advances in biomimetic membranes, sensors, tools and processes. *Sustainability, 12*(2), 4810.

Gong, X., Liu, Y., Li, S., & Yu, H. (2018). Application of iron-based nanomaterials in soil and groundwater remediation: A review. *Environmental Science: Processes & Impacts, 20*(9), 1203–1226.

Grillo, R., Abhilash, P. C., & Fraceto, L. F. (2016). Nanotechnology applied to bio-encapsulation of pesticides. *Journal of Nanoscience and Nanotechnology, 16*(1), 1231–1234.

Guilherme, M. R., Aouada, F. A., Fajardo, A. R., Martins, A. F., Paulino, A. T., Davi, M. F., et al. (2015). Superabsorbent hydrogels based on polysaccharides for application in agriculture as soil conditioner and nutrient carrier: A review. *European Polymer Journal, 72*, 365–385.

Hindatu, Y., Annuar, M. S. M., & Gumel, A. M. (2017). Mini-review: Anode modification for improved performance of microbial fuel cell. *Renewable and Sustainable Energy Reviews, 73*, 236–248.

Huang, P. S., Boyken, S. E., & Baker, D. (2016). The coming of age of de novo protein design. *Nature, 537*(7620), 320–327.

Huang, Z., Lin, Y., Fang, G., & Wang, G. (2019). Emerging antimicrobial peptides for water disinfection. *Journal of Water Process Engineering, 27*, 178–194.

IPCC. (2022). *Climate change 2022: Impacts, adaptation and vulnerability. Contribution of Working Group II to the sixth assessment report of the intergovernmental panel on climate change*. Cambridge University Press.

Jasanoff, S. (2016). The ethics of invention: Technology and the human future. W.W. Norton & Company.

Jiang, Z., Wang, B., Li, Y., Chan, J. K. S., & Chen, P. (2019). Microbial transformations of phosphorus: An overview of biological processes and their contribution to sustainable development. *Sustainability, 11*(18), 5078.

Justino, C. I., Duarte, A. C., & Rocha-Santos, T. A. (2017). Recent progress in biosensors for environmental monitoring: A review. *Sensors, 17*(12), 2918.

Kah, M., Kookana, R. S., Gogos, A., & Bucheli, T. D. (2019). A critical evaluation of nanopesticides and nanofertilizers against their conventional analogues. *Nature Nanotechnology, 13*(8), 677–684.

Khan, I., Saeed, K., & Khan, I. (2019). Nanoparticles: Properties, applications and toxicities. *Arabian Journal of Chemistry, 12*(7), 908–931.

Kitney, R., Adeogun, M., Fujishima, Y., Goñi-Moreno, Á., Johnson, R., Maxon, M., et al. (2019). Enabling the advanced bioeconomy through public policy supporting biofoundries and engineering biology. *Trends in Biotechnology, 37*(8), 917–920.

Krabbenborg, L., & Mulder, H. A. (2015). Upstream public engagement in nanotechnology: Constraints and opportunities. *Science Communication, 37*(4), 452–484.

Linkov, I., Trump, B. D., Anklam, E., Berube, D., Boisseasu, P., Cummings, C., et al. (2018). Comparative, collaborative, and integrative risk governance for emerging technologies. *Environment Systems and Decisions, 38*(2), 170–176.

Liu, C., Colón, B. C., Ziesack, M., Silver, P. A., & Nocera, D. G. (2019). Water splitting–biosynthetic system with CO_2 reduction efficiencies exceeding photosynthesis. *Science, 352*(6290), 1210–1213.

Liu, J., Jia, R., Tang, S., Luo, Y., Xie, Z., Xiao, P., et al. (2020). Hierarchically structured bio-inspired composites for wastewater treatment and underwater adhesion. *Materials Today, 37*, 127–138.

Lowry, G. V., Avellan, A., & Gilbertson, L. M. (2019). Opportunities and challenges for nanotechnology in the agri-tech revolution. *Nature Nanotechnology, 14*(6), 517–522.

Mulugetta, Y., & Castán Broto, V. (2018). Harnessing deep mitigation opportunities of urbanisation patterns in LDCs. *Current Opinion in Environmental Sustainability, 30*, 82–88.

NOAA. (2023). *Global monitoring laboratory—Carbon Cycle greenhouse gases*. Earth System Research Laboratory, Global Monitoring Division. https://www.esrl.noaa.gov/gmd/ccgg/trends/

Owen, R., Bessant, J., & Heintz, M. (2013). *Responsible innovation: Managing the responsible emergence of science and innovation in society*. Wiley.

Pearce, J. M. (2019). Return on investment for open source hardware development. *Science and Public Policy, 43*(2), 192–195.

Pourzahedi, L., & Eckelman, M. J. (2015). Environmental life cycle assessment of nanosilver-enabled bandages. *Environmental Science & Technology, 49*(1), 361–368.

Prasad, R., Bhattacharyya, A., & Nguyen, Q. D. (2017). Nanotechnology in sustainable agriculture: Recent developments, challenges, and perspectives. *Frontiers in Microbiology, 8*, 1014.

Pulido-Reyes, G., Leganes, F., Fernández-Piñas, F., & Rosal, R. (2017). Bio-nano interface and environment: A critical review. *Environmental Toxicology and Chemistry, 36*(12), 3181–3193.

Rafique, M., Sadaf, I., Rafique, M. S., & Tahir, M. B. (2020). A review on green synthesis of silver nanoparticles and their applications. *Artificial Cells, Nanomedicine, and Biotechnology, 45*(7), 1272–1291.

Rai, P. K., Kumar, V., Lee, S. S., Raza, N., Kim, K. H., Ok, Y. S., & Tsang, D. C. (2021). Nanoparticle-plant interaction: Implications in energy, environment, and agriculture. *Environment International, 119*, 1–19.

Ramasamy, R. P., Luckarift, H. R., Ivnitski, D. M., Atanassov, P. B., & Johnson, G. R. (2017). High electrocatalytic activity of tethered multicopper oxidase-carbon nanotube conjugates. *Chemical Communications, 46*(33), 6045–6047.

Rocha, T. L., Gomes, T., Cardoso, C., Letendre, J., Pinheiro, J. P., Sousa, V. S., et al. (2020). Immunocytotoxicity, cytogenotoxicity and genotoxicity of cadmium-based quantum dots in the marine mussel Mytilus galloprovincialis. *Marine Environmental Research, 158*, Article 104966.

Roco, M. C. (2003). Nanotechnology: Convergence with modern biology and medicine. *Current Opinion in Biotechnology, 14*(3), 337–346.

Roduner, E. (2006). Size matters: Why nanomaterials are different. *Chemical Society Reviews, 35*(7), 583–592.

Sanchez, C., Arribart, H., & Guille, M. M. G. (2005). Biomimetism and bioinspiration as tools for the design of innovative materials and systems. *Nature Materials, 4*(4), 277–288.

Sandler, R. L. (2020). The ethics of genetic engineering and gene drives in conservation. *Conservation Biology, 34*(2), 378–385.

Schinn, S. M., Broadbent, A., Bradley, W. T., & Bundy, B. C. (2017). Protein synthesis directly from PCR: Progress and applications of cell-free protein synthesis with linear DNA. *New Biotechnology, 31*(1), 138–145.

Seifan, M., Samani, A. K., & Berenjian, A. (2016). Bioconcrete: Next generation of self-healing concrete. *Applied Microbiology and Biotechnology, 100*(6), 2591–2602.

Sher, F., Hanif, K., Rafey, A., Khalid, U., Zafar, A., Ameen, M., & Lima, E. C. (2020). Removal of micropollutants from municipal wastewater using different types of activated carbons. *Journal of Environmental Management, 273*, Article 111021.

Singh, R., Kukkar, D., Singh, K. K., Kukkar, P., Bajaj, A., Kumar, A., et al. (2020). Vermicomposting of organic wastes: Process monitoring, microbial community structure, enzyme activities, pathogen reduction, and fertilizing potential. *Environmental Technology & Innovation, 18*, Article 100678.

Subramanian, V., Semenzin, E., Hristozov, D., Zondervan-van den Beuken, E., Linkov, I., & Marcomini, A. (2020). Review of decision analytic tools for sustainable nanotechnology. *Environment Systems and Decisions, 35*(1), 29–41.

Terzopoulou, Z., Barroso, V., Zamboulis, A., Bikiaris, D. N., & Beitlich, T. (2020). Sustainable strategies for coatings and antifouling: Challenges and innovative solutions. *Applied Sciences, 10*(22), 8095.

Thomas, A. C., Tank, S., Nguyen, P. L., Ponce, J., Sinnesael, M., & Goldberg, C. S. (2020). A system for rapid eDNA detection of aquatic invasive species. *Environmental DNA, 2*(3), 261–270.

von Schomberg, R., & Hankins, J. (Eds.). (2019). *International handbook on responsible innovation: A global resource*. Edward Elgar Publishing.

Werber, J. R., Osuji, C. O., & Elimelech, M. (2016). Materials for next-generation desalination and water purification membranes. *Nature Reviews Materials, 1*(5), 16018.

Whitesides, G. M., & Grzybowski, B. (2002). Self-assembly at all scales. *Science, 295*(5564), 2418–2421.

Wickson, F., Hartmann, N. B., Hjorth, R., Hansen, S. F., Wynne, B., & Baun, A. (2021). Making risk assessments sensible to social concerns: New frameworks, new tools, new thinking. *NanoImpact, 21*, Article 100297.

WMO. (2024). *State of the global climate 2023.* World Meteorological Organization. WMO-No. 1322.

Zhang, Q., Uchaker, E., Candelaria, S. L., & Cao, G. (2014). Nanomaterials for energy conversion and storage. *Chemical Society Reviews, 42*(7), 3127–3171.

Zhang, Y., Niu, Z., Wang, W., & Peng, S. (2018). Enzyme immobilization-based biomimetic carbon capture: Challenges and prospects. *Journal of CO_2 Utilization, 24*, 157–169.

Innovative Biotechnologies for the Generation and Application of Value-Added Products from Agro-Industrial Wastes

Vivek Raj, Santosh Kumar Karn, Moni Kumari, Rakesh Kumar, Aman Prakash, and Alok Prasad Das

1 Introduction

Waste management has become a global issue because of rapid industrialisation, population growth and economic development. According to Kaza et al. (2018), these factors have resulted in large elevation in garbage formation, with global waste production currently standing at over 2.01 billion metric tonnes yearly and projected to reach 3.40 billion metric tonnes by 2050. Traditionally, waste disposal methods encompassing incineration, chemical treatment, landfill deposition and direct discharge into water bodies have been widely practised (Mishra & Das, 2022; Nag et al., 2023; Pradhan et al., 2023). However, nearly 40% of the total waste generated remains untreated, often being dumped in open spaces without proper management, leading to severe environmental and health hazards (Tyagi & Kumar,

V. Raj (✉)
Department of Biotechnology, Patna Women's College (Autonomous), Patna University, Patna, Bihar, India
e-mail: dr.vivekraj07bt@gmail.com

S. K. Karn · M. Kumari
Department of Biochemistry & Biotechnology, Sardar Bhagwan Singh University, Balawala, Dehradun, Uttarakhand, India

R. Kumar
School of Pharmaceutical Sciences, Lovely Professional University, Phagwara, Punjab, India

A. Prakash
Department of Marine Biology, Leon H. Charney School of Marine Sciences, University of Haifa, Haifa, Israel

A. P. Das
Department of Pharmacognosy, GNA University, Phagwara, Punjab, India

S. Ghosh et al. (eds.), *Advances in Organic Waste Conversion Through Industrial Biotechnology*, Sustainable Environmental Waste Management Strategies,
https://doi.org/10.1007/978-3-032-11377-1_5

2021). Unregulated waste disposal contributes to pollution of air, water and soil while also fostering the spread of diseases.

Conventional waste management techniques are widely utilised since they are inexpensive and easy to use, but they have a number of negative environmental effects, such as ineffective waste treatment and high greenhouse gas (GHG) emissions that fuel climate change (Venil et al., 2020). In response, scientific and technological advancements have encouraged a shift towards sustainable waste management strategies, particularly those based on waste-to-energy conversion. Agro-industrial wastes generated when agricultural product processing make up a sizable amount of the world's garbage, among other waste sources (Table 1) (Reddy Koteswara et al., 2025; Swain et al., 2025). These wastes, which include pomace, seeds, peels and liquid byproduct streams, are promising raw materials for producing bio-based products like pigments, enzymes, bioactive compounds and biofuels because they are rich in valuable biomolecules like proteins, sugars and minerals (Venil et al., 2020).

An economical and environmentally friendly method of turning agro-industrial wastes into useful goods is industrial biotechnology. Enhancing bioconversion requires the use of submerged as well as solid-state fermentation techniques (Ghosh & Das, 2020; Mishra & Das, 2023; Mishra et al., 2024; Mohanty et al., 2024). New ways to extract compounds involving supercritical fluid processing, ultrasound and microwave-assisted methods have proven to be effective for getting bioactive compounds. Enzyme immobilisation, co-culturing, metabolic engineering and recombinant DNA technology have all improved waste biotransformation and industrial technology usage (Das et al., 2025; Allam et al., 2025).

Producing valuable products from leftovers in agriculture supports environmental care and also strengthens the bioeconomy. Applying waste valorisation methods

Table 1 Major chemical constituents of various agro-industrial residues (w/w %)

Agro-industrial residue	Cellulose	Hemicellulose	Lignin	Ash	References
Sunflower stalks	42.1	29.7	13.4	11.17	Motte et al. (2013)
Wheat straw	32.9	24.0	8.9	6.7	Sadh et al. (2018)
Rice straw	39.2	23.5	36.1	12.4	El-Tayeb et al. (2012)
Lemon peels	23.06	8.09	7.56	2.52	Marin et al. (2007)
Sugarcane bagasse	30.2	56.7	13.4	1.9	Sadh et al. (2018)
Tomato pomace	27	5	31	4	Hijosa-Valsero et al. (2019)
Banana peel	12.1	10.2	9.6	2.7	Adeniran et al. (2010)
Orange peels	9.21	10.5	0.84	3.5	Rivas et al. (2008)
Kinnow mandarin	10.10	4.28	0.56	3.23	Oberoi et al. (2011)
Pineapple peel	18.11	–	1.37	–	Singh nee' Nigam et al. (2009)
Apple pomace	7.2	–	23.5	1.6	Dhillon et al. (2012)

helps industries promote sustainability and introduce new ways to invest. This chapter will discuss in detail how biotechnology can help sustainably use agro-industrial waste, as well as current progress, difficulties in producing these approaches, and the role of waste conversion in growing bioeconomy. Attention will be given to new methods for improving strain, binding enzymes and optimising products to benefit both industry and medicine.

2 Agro-Industrial Wastes: Types and Chemical Composition

Farming generates field trash, while processing agricultural products generates industrial waste. During harvesting, field wastes consist of stalks, straw, husks, leaves and stems. Meanwhile, many agro-based industries such as those for fruit and vegetables, textiles and paper and pulp, produce industrial wastes. Pomace, oil cakes, molasses, bagasse, pulp, nutshells, seeds and peels are among the byproducts produced by these sectors (Sadh et al., 2018).

These lignocellulosic components, including cellulose, hemicellulose and lignin, make up most of these biomass-rich agro-industrial wastes (Philippini et al., 2020). These foods also contain soluble carbohydrates, fibres, moisture, proteins and significant bioactive substances such phenols. However, what ends up in these wastes depends on the kind of basic material.

Their chemical diversity makes agro-industrial wastes ideal for use in biotechnological processes. They rely on these wastes for both physical and nutritional support, which helps them grow and metabolise properly. The creation of new-generation bioproducts and sustainable waste management is facilitated by this process, which turns agro-industrial waste into useful bio-based products (Venil et al., 2020).

3 Microbial Bioprocessing of Agro-Industrial Wastes

Modern waste management relies on microorganisms to handle agro-industrial residues in an environmentally friendly way. Microbial bioprocessing is attracting interest because it turns waste products into useful goods and supports sustainable development. The two main fermentation processes utilised in this process were SSF (solid-state fermentation) and SmF (submerged fermentation), in which microorganisms use agro-industrial waste as a source of nutrients for development and produce bio-based products at the same time (Fig. 1).

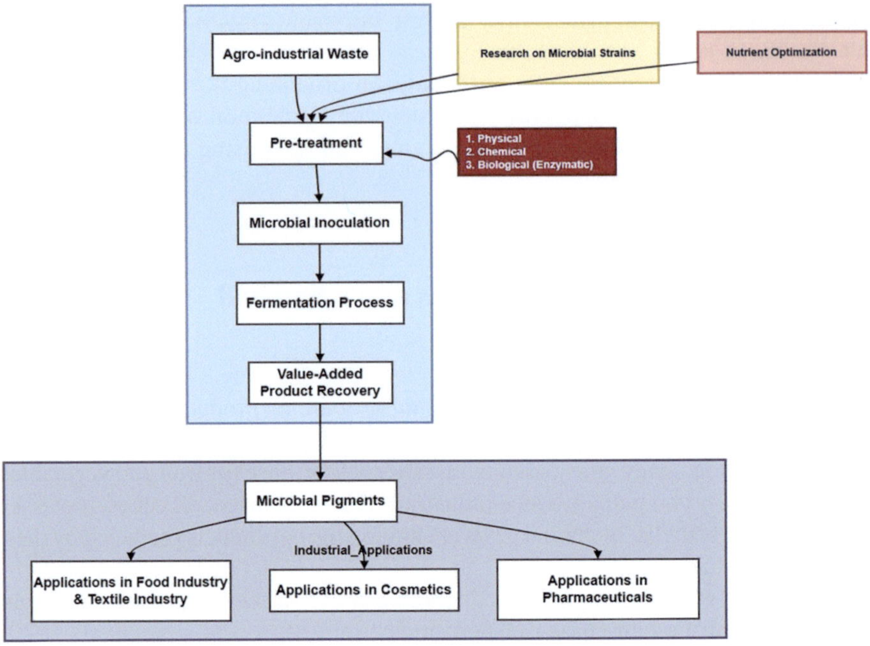

Fig. 1 Schematic representation of microbial bioprocessing method for agro-industrial waste valorisation and industrial applications of microbial pigments

3.1 Submerged Fermentation Technology

SmF is microbial growth in liquid suspension cultures in which products needed are released into broth around them (Bhatia et al., 2020). Advancements in technology have allowed this technique to be controlled with great precision over key process parameters, namely temperature, pH and dissolved oxygen, making this technique suitable for large-scale industrial applications (Singhania et al., 2015).

3.2 Solid-State Fermentation Technology

In recent years, SSF has become of increasing interest to researchers and industries in view of its economic and technological advantages over SmF. The strongest microbial growth is encouraged, and quantity of free water is lowered by SSF, which occurs naturally. In contrast with SmF, SSF permits direct use of complex substrates, lower moisture content and has many advantages such as reduced energy requirement, minimum downstream processing, enhanced product stability, high product yield and less effluent (Ring et al., 2018). Moreover, "the use of agro-industrial waste as" substrate greatly decreases upstream raw material prices (Bhatia et al., 2020; Singhania et al., 2015). SmF and SSF bioprocessing is simpler to use agro-industrial waste and produce useful products with less waste.

4 Value-Added Products from Agro-Industrial Wastes

Bioprocessing uses materials from horticulture, paper and pulp production and fruit and vegetable processing. Microbial pigments, enzymes, biofuels and biopolymers are just a few of the industrially and medicinally important products that can be made from these wastes (Beltrán-Ramírez et al., 2019).

4.1 Microbial Pigments

Synthetic dyes are harmful to people and the environment, so more and more people prefer natural pigments. Though plants are well-known for their natural colours, many factors, like the timing of harvest, the small amount they produce and the high cost of making them, keep this industry limited. Meanwhile, microbes involving fungi, bacteria, algae and yeasts are more effective as well as sustainable ways to create pigments. These microbes produce diverse range of pigments, involving prodigiosin, carotenoids, flavins, violacein, anthraquinones, which have applications in multiple industries (Tuli et al., 2015). Lycopene, riboflavin, astaxanthin, as well as Arpink RedTM are added in food and beverage sector because they improve colour, change texture, and contain nutrition. The industry uses these pigments to give cotton, wool and silk a green colour. Furthermore, microbial pigments are advantageous for pharmaceutical formulations due to their numerous therapeutic qualities, which include anti-inflammatory, anticancer, antibacterial, as well as antioxidant activities (Narsing Rao et al., 2017).

Microbial pigments are being examined by scientists for use in cosmetics because they could help reduce dark spots, provide sun protection and work as antioxidants. For example, astaxanthin and zeaxanthin from algae are utilised in skincare products to prevent skin ageing, reduce pigmentation and provide sun protection (Gupta et al., 2019). Researchers have demonstrated that resources for pigment-producing microbe cultures can be made from waste from agro-industrial activities. For example, red pigment synthesis from Monascus purpureus and Penicillium minioluteum has been effectively accomplished using jackfruit seed powder and sesame seed cake (Zahan et al., 2020). Penicillium sp. and broken wheat can also be fermented on a solid substrate to produce a yellow colour. Other studies have explored usage of onion peels, pineapple waste, mung bean husk and cassava wastewater for carotenoid and violacein production by various microbial strains (Sharma & Ghoshal,

2020). Although microbial pigment production can be successful in business, further research is required to strengthen the strains, keep them stable and make production on a large scale more efficient (Das et al., 2024a, 2024b, 2024c; Reddy Koteswara et al., 2025).

4.2 *Industrial Enzymes*

Many are working on finding out how to create enzymes at a lower cost since it is vital for sustainable industry. Biocatalysts are often found in pharmaceuticals, washing products, milk processing, farming and the oil industry, and these biocatalysts are commonly enzymes. Agro-industrial wastes can save up to 60% on raw material prices when used as substrates (Wen et al., 2005). Cellulases, xylanases, laccases, α-galactosidases, pectinases, proteases and lipases are among the many enzymes that have been effectively generated through microbial fermentation of agro-industrial leftovers (Mehmood et al., 2019). For instance, for the optimum exo- and endo-pectinase synthesis, Aspergillus niger is used on solid-state fermentation with orange pomace and sweet lime peels are used in the production of pectinase using Schizophyllum commune (Mahmoodi et al., 2019). Pectinase was also produced using the co-culture of Bacillus subtilis and Bacillus pumilus for apple pomace (Kuvvet et al., 2019). Studies have demonstrated that tobacco waste, sugarcane bagasse and orange trash are all very good sources of carbon for cellulase production. 'The production of laccase by using various agro-industrial waste such as' orange peels (Pleurotus ostreatus), barley bran (Trametes versicolor) and cowpea pod (Myrothecium gramineum) was successfully achieved, proving the versatility of agro-industrial waste in enzyme synthesis (Gnanasalomi & Gnanadoss, 2019).

4.3 *Synthesis of Nanoparticles*

The field of nanotechnology, which deals with materials at the nanoscale scale (1–100 nm), has garnered significant attention due to its usage in electronics, food processing, medicine delivery and cosmetics (Chaturvedi et al., 2012). Making nanoparticles from plant extracts and microorganisms in green nanotechnology is safer and better for the environment than using chemicals. Metal ions change into nanoparticles more quickly when they encounter phenolics, terpenoids and polysaccharides from plant tissues. Wastes from vegetable and fruit processing, mainly peels, were found to have a high potential for bioreduction. Experts report that silver nanoparticles can be created using extracts of rice husk, banana peel and orange peel. Furthermore, nickel and platinum nanoparticles have been synthesised using rice husk, which is high in silica (Tolba et al., 2015). It has been found by research that agro-waste nanoparticles possess antimicrobial and bioremediation abilities. For instance, 'nanosilica particles from rice husk were utilised for dye removal from' industrial effluents (Tolba et al., 2015). Furthermore, Fe_3O_4 nanoparticles synthesised from seaweed have exhibited antitumor properties (Namvar et al., 2012). Despite having many uses, there are still questions about how to make nanoparticles the same, with different sizes and in large quantities.

4.4 Biocontrol Agents

Fermentation of sugarcane bagasse and coffee husks could be converted into biopesticides and bioinsecticides, experts are investigating. Fungal strains such as Beauveria bassiana have been studied to produce biopesticides for crops like bananas, coffee, soybeans and sugarcane (Das et al., 2023; Sala et al., 2020). Besides, bioinsecticides have been produced in packed bed reactors, using Beauveria bassiana and Trichoderma harzianum on rice husk as a substrate (Sala et al., 2020). Additional investigation showed that Beauveria bassiana and Isaria javanica could be grown on palm kernel cake under solid-state fermentation to produce effective bioinsecticides against whitefly nymph (Bemisia tabaci) (do Nascimento Silva et al., 2018). Furthermore, it was determined that a mixture of 'palm kernel cake and cassava dregs is a suitable substrate for production of bioinsecticides by Bacillus' thuringiensis subsp. aizawai (do Nascimento Silva et al., 2018). Bacillus thuringiensis sub-species kurstaki can be cultivated on starchy industrial wastewater to produce bioinsecticides, as large-scale studies have confirmed (do Nascimento Silva et al., 2018).

4.5 Bioactive Compounds

The reason phytochemicals and phenolic compounds are valued in both food and drug manufacturing is because they have strong antioxidant properties. Flavonoids, phenolic acids and tannins are phenolic components found in fruits and vegetables that lower the risk of chronic illnesses like diabetes and cancer (Kalli et al., 2018). Many of these nutritionally active compounds come from vegetable and fruit processing waste. Citrus, jackfruit, avocado and mango peels have 15% more phenolic chemicals than fruit in pulp, and pomegranate peels have ten times the amount. According to Sagar et al. (2018), phenolic content of apple peels is roughly 3.3/100 g.

Fermentation has been used by numerous studies to enhance these chemicals' effects. For example, grape waste's antioxidant content was greatly increased by SSF using Aspergillus niger GH1 (Rashad et al., 2015). Using '*Kluyveromyces marxianus* NRRL Y-8281' to SSF pineapple waste produced 8 mg/mL of phenolic compounds and had a 97% capacity to scavenge free radicals (Rashad et al., 2015). Furthermore, 8 phenolic compounds rich in antioxidants were produced using 'SSF of fig by-products employing *Aspergillus niger* HT4 (Buenrostro-Figueroa et al., 2017)'. Other research have shown that Rhizopus oryzae can double concentration of gallic and ferulic acid in rice bran, while Rhizopus oligosporus and Aspergillus niger enhanced phenolic compounds such as chlorogenic acid and rutin in apricot pomace (Dulf et al., 2017). Scientists are working to develop techniques to get bioactive compounds using cheaper materials.

4.6 Single-Cell Oils (SCOs)

SCOs, or 'microbial oils, are lipid molecules that oleaginous' bacteria create inside their cells. These lipids are structurally similar to plant-based oils and are synthesised by microbes under nutrient-limited conditions (Ochsenreither et al., 2016). Should nitrogen or phosphorus not be present, microorganisms increase their lipid production, mainly as triacylglycerols and polyunsaturated fatty acids. Because SCOs are like plant-based lipids and can be made without farmland, they are a promising option for biodiesel.

Several research projects have shown that microbial strains can increase the SCO yield from leftovers produced by the agricultural industry. For example, Mortierella isabellina significantly increased intracellular lipid accumulation (11/100 g dry weight) from sweet sorghum (Economou et al., 2010). Likewise, after SSF of wheat bran and straw, *Microsphaeropsis sp. accumulated* 42 mg/g of lipids. Yeast strains such as *Rhodotorula mucilaginosa* TJY15a and *Cryptococcus* sp. have shown potential in producing 'long-chain fatty acids utilising cassava starch hydrolysate' and corn cob as substrates, respectively. Additionally, *Rhodotorula kratochvilovae* SY89 demonstrated efficient lipid production (4.82 g/L) when cultivated on cane molasses (Jiru et al., 2018). Beyond biodiesel applications, SCOs have been explored for their role as food supplements. According to 'Ochsenreither et al. (2016), microbial polyunsaturated fatty acids', including linoleic acid, have anticancer qualities, improve cardiovascular and neurological health, and are essential for immunological and inflammatory responses. Further, as functional food additives, microbial oils that include 'essential fatty acids like arachidonic acid (ARA) and docosahexaenoic acid (DHA)' are becoming more and more popular. However, additional studies are required to evaluate their safety for human consumption before widespread commercialisation.

4.7 Biofuels

The detrimental effects of traditional fossil fuels on climate change have drawn much attention to investigation of biofuels throughout the past 20 years. Nowadays, various agro-industrial "wastes are being utilised as raw materials for producing biofuel, involving molasses, bagasse, rice straw, sugar beet leftovers, sweet potatoes and maize waste (Pattanaik et al., 2019)'. However, as many regions of the world are experiencing food and nutritional issues, 'the production of biofuels from edible crops' has generated concerns regarding food security. Besides, the shortage of land suitable for farming makes it difficult to produce enough feedstock for making biofuels. Consequently, scientists are now paying attention to using Jatropha and Pongamia, which are not edible, for biodiesel.

Lately, bioethanol makers have started using lignocellulosic biomass. The carbohydrates, lipids and amino acids found in agro-industrial wastes make them acceptable alternative raw material. Using *Saccharomyces cerevisiae*, vegetable waste involving onion, carrot, potato peels was successfully utilised to create bioethanol. Similarly, banana pseudostems have been used for bioethanol generation through fermentation by '*Aspergillus ellipticus* and *Aspergillus fumigatus* (Ingale et al., 2014)'. Bioethanol production was also demonstrated using corn stover and sugarcane bagasse after harvest. Firms like Labio and Gasum in Finland and Sweden have proven how to make biogas from food industry waste and turn it into biofuel and biofertilizers. STI has developed innovative technologies for ethanol production, including Bionolix (employing industrial and municipal waste), Ethanolix (using food industry waste) and Cellulonix (utilising 'cellulosic waste and sawdust for biofuel' synthesis).

4.8 *Microbial Bioflocculants*

Microbial bioflocculants, extracellular polysaccharides (EPS) produced by microorganisms, are engaged in the flocculation process of industrial wastewater treatment (Bisht & Lal, 2019). Besides cleaning wastewater by removing pollutants, these EPS molecules can capture heavy metals, bleach dyes and require less oxygen to get the job done. As a result, scientists are trying mainly to produce them in large quantities because bioflocculants are easily recovered and ruptured.

The bioflocculant producers have been identified to be different microbial species. Some research claims that molasses in the form of pretreated molasses from which the bulk sugar has been crystallised can be utilised by Halomonas sp. AAD6 to synthesise EPS. In addition, Mucor rouxii also produced EPS when cultured on beet molasses (Abdel-Aziz et al., 2012). Aspergillus niger is capable to use waste palm oil industry as substrate to produce high bioflocculant yield (Luthfi et al., 2020). In addition, the Paenibacillus species strain A9, using wastewater from the food processing industry, was able to produce bioflocculants which can remove heavy metals (Jiang et al., 2019). Various microbes have been used 'for the production of bioflocculants using rice straw, sugarcane molasses, wheat straw, potato peels and palm jaggery'.

4.9 Biopolymers

A microbial species, such as bacteria and fungi, can transform agro-industrial waste into valuable biopolymers 'such as polyphosphates, polysaccharides and polyhydroxyalkanoates (Moradali & Rehm, 2020)'. Because of their well-known biodegradable and biocompatible nature, these biopolymers 'can be used in the food, pharmaceutical or biomedical industries (Philippini et al., 2020)'. Biopolymers can be employed as wound dressings, medicine delivery materials, gelling agents and stabiliser in food and antimicrobial coatings for food packaging (Suarato et al., 2018).

- **Chitosan**: Chitosan, made from the material chitin, 'is used in food, textile and pharmaceutical' industries. Fungus Gongronella butleri and the species Penicillium citrinum were developed to produce it sustainably from agro-industrial waste, apple pomace and paper mill wastewater (Namboodiri & Pakshirajan, 2019).
- **Collagen**: For use in biomedical applications, collagen, a fibrous protein necessary for tissue strength, has been isolated from fish and 'broiler chicken processing waste (Munasinghe et al., 2015)'.
- **Cellulose**: Microbial cellulose demonstrates high mechanical strength and thermal stability. It has been synthesised using *Gluconacetobacter swingsii* from pineapple and sugarcane juice and by *Komagataeibacter saccharivorans* utilising molasses (Abol-Fotouh et al., 2020).
- **Pullulan**: Food and drug products typically include this type of polysaccharide. Its production has been achieved using Aureobasidium pullulans grown on beet molasses and cassava waste (Viveka et al., 2021).
- **Xanthan Gum**: Xanthan gum production has been successfully carried out using *Xanthomonas campestris* grown on winery wastewater and glycerine from biodiesel production (Gondim et al., 2019).
- **Bioplastics**: Biodegradable plastics involving polylactic acid (PLA) and PHA (polyhydroxyalkanoates) have been synthesised using microbial fermentation of 'agro-industrial residues like wheat bran, rice bran, sugarcane bagasse (Nayak & Bhushan, 2019).

4.10 Biosurfactants

The hydrophobic pollutants interact with the environment with the help of biosurfactants. They are biodegradable and less toxic than synthetic surfactants. In cosmetics, food, pharmaceuticals and bioremediation (Rane et al., 2017) industries, they are used. Making biosurfactants is not cheap, so 'scientists are looking for ways to use abundant materials' from agro industry. Biosurfactant-producing Bacillus subtilis ANR 88 produced 0.241 g/L using molasses (Rane et al., 2017). Similarly, glycolipid antimicrobial biosurfactants were synthesised 'from Serratia nematodiphila using rice straw hydrolysate (Panjiar et al., 2020)'. Other studies report biosurfactant production from Paneibacillus species with waste coconut oil (Jimoh & Lin,

2020) and from Stenotrophomonas maltophilia with waste soybean oil (Nogueira et al., 2020). To reduce costs of biosurfactant production, new methods of culturing bacteria to produce products and extracting the products are needed.

4.11 Emerging Techniques for Agro-Industrial Waste Valorisation

We have to discover ways to use agro-industrial waste to ensure sustainability and deliver both quality and quantity of products. Numerous plant extraction techniques have emerged, such as pulsed electric field, supercritical fluid, microwave and ultrasound technologies. They make it possible to use less organic solvent and keep your bioactive substances safe as they are extracted.

4.12 Microwave-Assisted Extraction (MAE)

High-frequency microwave radiation from 300 MHz to 300 GHz is applied in the MAE process to enhance extraction. The method's quick results, stable outcomes, low use of solvent and better results have made it chosen for use in industries (Panzella et al., 2020). Applying ethanol and acetone to avocado seeds in an MAE method took only about 14.69 min to extract antioxidant phenolic compounds. It has also been used on olive tree leaves, which produced nearly 10 mg of flavonoids and approximately 10 mg of phenolic compounds within 1.5 min using acetonitrile (Kırbaşlar & Şahin, 2023).

4.13 Ultrasound-Assisted Extraction (UAE)

It is a potential alternative to traditional extraction practices. This method utilises ultrasound waves in range of 20–100 kHz to create cavitation bubbles within the sample, causing cell wall disruption and enhancing solvent penetration for efficient extraction (Panzella et al., 2020). Because it saves solvent, is more cost-effective and maintains product stability, UAE is commonly applied to get phenols from agricultural waste.

4.14 *Supercritical Fluid Extraction (SFE)*

SFE is an advanced method which replaces conventional solvent-based extraction techniques. It utilises supercritical carbon dioxide ($ScCO_2$), which exhibits properties of both liquid and gas, allowing selective extraction of desired compounds. The extraction of carotenoids from mango peels has shown that SFE is scalable for industrial applications and produces significant concentrations of bioactive chemicals (del Pilar Sanchez-Camargo et al., 2019). The process revealed that soybean waste included flavonoids and phenolic compounds.

4.15 *Pulsed Electric Field (PEF) Extraction*

The non-thermal process of PEF extraction involves subjecting biological material to brief high-voltage electric field bursts. Cell membrane permeabilization brought on by this process results in release of intracellular bioactive substances (Panzella et al., 2020). Both duration of exposure and the strength of the electric field have an impact on the PEF extraction result. Main reason people choose this method is that it is thought to retain the benefits of bioactive compounds. They are helping to use agro-industrial waste more effectively. The need for less solvent, better results and their use on a large scale make these techniques attractive for sustainable industry.

5 Technologies for Enhancing Production Efficiency and Yield of Value-Added Products

Due to biotechnology, quality, quantity as well as efficiency of crucial products have all improved. We can achieve sustainable industrial production by using genetic engineering, changing metabolic pathways, growing communities of microbes and improving our methods of cultivation.

5.1 *Strain Improvement*

Microbial 'strains are modified using genetic and metabolic' methods to ensure their useful products are available. It has been noticed by scientists that recombinant microbial strains enhance the quality of production. For instance, carrying the endoxylanase gene (xyn11AQ1) using a shuttle vector, 'Bacillus subtilis AQ1 was expressed in B. subtilis DB104', leading to higher enzyme activity when tested with 'corn cob and tofu liquid waste (Helianti et al., 2016)'. Integrating the MEL1 'gene from Saccharomyces cerevisiae with the LAC4 gene of Kluyveromyces lactis

GG799' using beet molasses and cheese whey 'led to a significant increase in enzyme production (Álvarez-Cao et al., 2018)'.

For example, CRISPR/Cas9 gene editing has often been used to modify microbial strains to produce industrially important metabolites (Sen et al., 2019). The application of this technique to engineer various bacterial species (Bacillus, Escherichia coli, Corynebacterium, Pseudomonas) has been used to optimise metabolic pathways (Sen et al., 2019). Gene editing approaches with knockouts, transcriptional activation, pathway integration and regulatory modification have been successful (Fonseca et al., 2020). 'Six genetic modifications in Trichoderma reesei RUT-C30 significantly enhanced cellulase and hemicellulase activities to significantly improve saccharification efficiency' (Fonseca et al., 2020). In Escherichia coli and Halomonas bluephagenesis, pathway modifications have also improved the efficiency of polyhydroxyalkanoate (PHA) production (Gahlawat et al., 2020).

Synthetic biology, systems biology and evolutionary engineering are all occurring simultaneously under these conditions. We can produce pigments, biopolymers, 'high-value chemicals and biofuels from' lignocellulose, food waste, hydrolysates and fatty acids. Researchers are also investigating'gases such as carbon monoxide, carbon dioxide and methane' as possible substrates (Ko et al., 2020). Improvements in steps limiting production increase metabolite levels. Increased synthesis of aromatic compounds in E. coli has resulted from lowering the use of other routes and allowing for more essential precursors in the shikimate pathway (Huccetogullari et al., 2019). Metabolic changes in Escherichia coli, Yarrowia lipolytica and Saccharomyces cerevisiae result in high amounts of carotenoids, polyphenols and nutraceuticals (Yuan & Alper, 2019). Systems biology and omics can be used to increase harvest of larger crop yields.

5.2 *Co-culture Technique*

Microbial co-cultivation is being developed to improve the quantity as well as the quality of products produced in fermentation. The use of this strategy results in decreased metabolic stress and improved substrate utilisation. Escherichia coli and Trichoderma reesei could be co-cultured in direct syntheses of isobutanol from cellulosic biomass (Sharma et al., 2019). Exopolysaccharide biosynthetic capacity was improved in the co-cultures of Saccharomyces cerevisiae and Lactobacillus rhamnosus (Bertsch et al., 2019). The system of microbial consortia 'of Cupriavidus necator and Pseudomonas citronellolis' that transforms apple pulp PHA blend is shown by Rebocho et al. (2020). Proteins produced from yellow wine waste were also produced effectively by a co culture of 'Candida utilis and Geotrichum candidum (Zhu et al., 2020)'.

5.3 Microparticle-Based Morphological Engineering

Microbial shape affects bioprocess parameters including air flow, mass transfer and medium viscosity. Filamentous fungi can be pellets or suspended mycelia, which can change how nutrients are distributed and how rapidly a product is produced. However, to overcome these challenges, microparticle-enhanced cultivation (MPEC) has been introduced. The use of this technique eliminates microbial aggregation, enhances oxygen transfer and substrate consumption and, as a result, increases product yields (Karahalil et al., 2019). The fungi are better in submerged cultures with 'talc powder, titanium oxide and aluminium oxide'. Researchers were this method to examine several fungi, including Aspergillus, Caldariomyces, Cerrena, Pleurotus and Trichoderma. 'The presence of titanium oxide microparticles gave rise to higher production of both α-glucoamylase and fructofuranosidase' in Aspergillus niger. Moreover, aluminium oxide and talc powder increased hydrolytic enzyme production, i.e. xylanase, phytase, cellulase and amylase in Aspergillus oryzae (Das et al., 2024a; Ghosh & Das, 2020; Singh, 2018). The addition of aluminium oxide microparticles also improved cellulase yield from Trichoderma viride (Singh, 2018). MPEC should perform well in industry, but additional studies are needed to learn about its uses and hazards.

5.4 Enzyme Immobilisation

Now that sustainability is important in biomaterial processing, researchers are looking for biocatalysts that are efficient and safe. However, industrial applications find free enzymes to be rather unstable. Enzyme immobilisation can enhance enzyme stability, ease of recovery, reusability and overall productivity (Ismail & Beak, 2020). Immobilised enzymes generally exhibit superior kinetic properties, including optimal specific activity, 'maximum reaction velocity (Vmax) and Michaelis constant (Km)', compared to free enzymes. It is possible to keep enzymes stable by bonding them to different materials. Candida rugosa lipases packed in calcium silicate microcapsules could tolerate temperatures above 60 °C, while those contained in alginate were stable only below that level. Similarly, pectinase immobilised on magnetic chitosan demonstrated greater pH stability and higher activity at 60 °C in comparison to free enzymes (Dal Magro et al., 2019). 97% of α-galactosidase activity from UFV-1 was preserved when the enzyme was immobilised on a cellulose film. After 10 consecutive reuses, this enzyme preserved more than 80% of its activity and continued to work at its best for 40 h at pH 4.0 and 85 °C (Júnior et al., 2018). Usually, these new biotechnologies help increase production, make activities more sustainable and raise the economic value of important goods.

5.4.1 Methods of Immobilisation

How effective and successful an enzyme is often determined by its immobilisation. Physical adsorption, matrix placement, 'cross-linking and covalent bonding' are primary techniques for immobilising enzymes.

- **Physical adsorption** entails attaching the enzyme to a support surface using 'non-covalent interactions such ionic bonds, hydrogen bonds, and van der Waals forces'.
- **Entrapment** refers to the irreversible encapsulation of enzymes within matrices like 'polymers (alginate, chitosan, carrageenan, gelatin) and gels (silica sol gels)' (Ismail & Beak, 2020).
- **Covalent bonding** occurs through interactions between enzyme functional groups (amino, thiol, carboxyl, imidazole and phenolic groups) and the support material, ensuring stable immobilisation (Mohamad et al., 2015).
- **Cross-linking** refers to using chemicals to form solid enzyme groups known as aggregates.

Based on the binding strength, immobilisation is classified as reversible (ionic binding, physical adsorption, affinity binding) or irreversible (entrapment, covalent binding). Enzymes can be recovered and reused after reversible immobilisation, but irreversible immobilisation ensures high stability at the cost of reusability. Some modifications in enzymatic properties occur after immobilisation due to changes in microenvironment or enzyme structure (Brena et al., 2013). The effects of structural changes on enzyme activity can be reduced by choosing suitable support materials or inserting spacer arms to avoid interference at the active site.

5.4.2 Support Matrix

The kind of support material used can influence how well, how safely and how many times immobilised enzymes can be used. Common matrices include activated carbons, glass beads, gels, biopolymers, inorganic and organic materials, natural and synthetic polymers, nanomaterials, and zeolites (Zdarta et al., 2018). An appropriate support matrix has to:

- Protect enzyme from harsh conditions
- Resists microbial contamination
- Maintain enzymatic catalytic activity
- Provide mechanical strength for processing
- Contain reactive groups for stable enzyme binding.

Polysaccharides involving chitosan, agarose, alginate, cellulose, as well as carrageenan are widely utilised due to their biodegradability, mechanical strength, and non-toxicity (Pinheiro et al., 2019). Research has indicated the efficacy of various matrices:

- **Calcium alginate beads** immobilising *Aspergillus aculeatus* pectinase resulted in a 56.71% yield and 80% enzyme activity retention (de Oliveira et al., 2018).
- **Chitosan activated with divinyl sulfone** improved *Candida antarctica* lipase B stability (Pinheiro et al., 2019).
- **Lipases immobilised on sodium titanate nanotubes** increased synthesis of fatty acid methyl ester and reusability of enzymes (Nady et al., 2020).

To reduce immobilisation costs, agro-industrial waste materials rich in polysaccharides, such as corncobs, rice waste, coconut shells, as well as spent grains, serve as economical alternatives for enzyme carriers (Quayson et al., 2020). Immobilised enzymes are used for waste stream valorisation, such as:

- **β-galactosidase immobilised on carrier supports** to convert lactose into high-value sweeteners (Bilal & Iqbal, 2019).
- **Silica-encapsulated lipase** to refine palm oil waste into fatty acids (Itabaiana et al., 2013).
- **Pectinase/cellulase immobilised on fruit peels** for biodiesel synthesis (Nadar & Rathod, 2019).

Nanoparticle-based immobilisation has revolutionised enzyme stabilisation. The nanoscale size, stability, non-toxicity, reusability and ease of separation of magnetic nanoparticles (such Fe_3O_4) make them popular (Darwesh et al., 2020). As an example:

- ***Candida rugosa* lipase immobilised on Fe_3O_4-coated chitosan nanoparticles** efficiently converted soybean oil into biodiesel (Zhong et al., 2020).
- **Xylanase immobilised on corncob-derived matrices** enhanced xylooligosaccharide synthesis with improved pH and temperature stability (Driss et al., 2014).

5.4.3 Effect of Enzyme Conformation on Product Synthesis

The structural conformation of an enzyme significantly influences its activity, stability and efficiency in biocatalysis. Many lipases possess a lid-like polypeptide domain covering the active site, switching between closed (inactive) and open (active) conformations (Ying et al., 2018). Immobilisation on hydrophobic supports might stabilise open form of enzymes, enhancing catalytic efficiency and reusability. For instance:

- **Lipase immobilization via interfacial activation** improved enzymatic activity for synthesising n-octyl oleate, a biolubricant, achieving a 95% conversion rate (Bassi et al., 2016).
- **Genetic modifications** such as site-directed mutagenesis have enhanced enzyme activity. Lipase activity increased fivefold when serine-154 was substituted with threonine and twofold when leucine was substituted for glycine-152 (Panizza et al., 2015).

Multipoint enzyme immobilisation, which involves multiple covalent bonds between enzymes and support matrices, ensures enzyme stability under harsh industrial conditions (Weltz et al., 2020). For example:

- **Glyoxyl agarose-immobilised xylanase** exhibited 40-fold enhanced stability and 100% immobilisation efficiency compared to free enzymes due to additional amine groups increasing biocatalyst stability (Manrich et al., 2010).

Still, additional study is needed to ensure that multipoint covalent immobilisation can be used to improve both the efficiency and affordability of biocatalysts.

6 Challenges in Value-Added Product Synthesis

Common practices that harm environment and are incompatible with sustainable development include landfills, open dumping and burning waste. Because of biotechnology, it is possible to turn waste materials into expensive items utilising submerged and solid-state fermentation. Nevertheless, number of issues prevent these processes from being efficient and profitable. Submerged fermentation has many benefits, but it is still limited by low product strength, complicated media and high production expenses. For lignocellulosic waste, solid-state fermentation, which is simple to start up and may employ waste, requires improvements in large-scale equipment and bioreactor management. The range of microbial strains, substrates and ideal laboratory conditions are critical factor in product development effectiveness (Cerda et al., 2019).

How a bioreactor is designed is very important for making bioprocesses work in industry. The best bioreactors can improve mass and heat circulation, diffusion and the removal of products. Since most solid-state fermentation processes depend on agro-industrial wastes with low heat transmission, excessive heat inside the substrate can harm the cells. The development of novel bioreactors would facilitate continuous and economically viable bioprocessing (Cerda et al., 2019). By using simulations and mathematical models, the process calibration and validation can be worked on more efficiently. Downstream processing also presents challenges, accounting for 20–30% of total product costs (Ko et al., 2020). The use of sustainable product separation and purification processes is necessary to decrease costs and improve how energy is used. Older organic solvent extraction methods are being swapped for supercritical CO_2 and hydroalcoholic mixtures, which are both environmentally friendly and efficient. Natural deep eutectic solvents (NADES) offer advantages, including low toxicity, shorter extraction time and enhanced biological properties of extracted compounds (Gullón et al., 2020).

Another major concern is product purity, particularly during recovery. Presence of carbohydrates, salts and volatile compounds in lignocellulosic material can disrupt the quality of the products obtained and may affect the next step of processing. Minimising undesirable byproducts is possible by using effective pre-treatments, choosing the right medium and using certain microbial strains. Additionally, metabolic engineering and pathway optimisation studies are crucial for scaling up industrial bioprocesses (Fernández-Cabezón & Nikel, 2020).

7 Commercial Production of Value-Added Products

The value-added product market is growing rapidly on a global scale. For example, North America is likely to be the main contributor to the microbial carotenoid market, that is anticipated to reach USD 2.0billion by 2026. With the bioethanol sector alone expected to increase "from USD 33.7billion in 2020 to USD 64.8billion by 2025, the biofuel business is expected to reach a value of USD 153.8billion by 2024". Innovations in biopigments, biofuels, new foods, bioactive substances and product intermediates are being propelled by biotech and agritech startups. Waste processing, fermentation technologies, microbial strain enhancement, bioreactor design and purification techniques have all seen an increase in patent activity. Riker Laboratories Inc., UK, filed the first solid-state fermentation patent in 1962 for the synthesis of pyrogen-free penicillinase, and since then, the number of patent applications in the sector has increased (El Sheikha & Ray, 2023).

8 Value-Added Products and Bioeconomy

Rise of the bioeconomy is transforming global markets by promoting the production of biofuels, bioplastics and biotherapeutics, contributing to sustainable development goals. Biomass utilization for biofuel production exemplifies how bio-based industries can drive national economies. Recognising this potential, governments worldwide are investing heavily in bioeconomic initiatives, with 49 countries adopting national bioeconomy policies (Aguilar et al., 2019). Leading regions in bioeconomic development include European Union, United States and India. EU bioeconomy is valued at €2.3 trillion, contributing 4.2% to GDP and expected to create 1 million new jobs by 2030. By 2030, the U.S. bioeconomy is expected to have tripled in size, with plans to manufacture 50 billion pounds of bioproducts from 1 billion tonnes of biomass feedstock. By 2025, the bioeconomy in India, which is presently valued at USD 70.2 billion, can be expected to grow to USD 150 billion and provide 10 million employment (Thatoi et al., 2025). There is a significant chance for India's bioeconomy to grow in the areas of biopesticides, biopigments and functional foods because the country generates over 350 million tonnes of agricultural waste yearly. The bioeconomy helps the environment by reducing pollution, using renewable resources and supporting the creation of a sustainable world.

9 Conclusion and Future Perspectives

With microbial bioprocessing and waste valorisation, we can now utilise waste materials efficiently to produce sustainable goods. By using CRISPR/Cas9 and metabolic engineering, as well as other procedures, fermentation has become more efficient, and the quality of products has increased. With immobilised enzymes, it is now easier to break down difficult agro-industrial materials such as lignocellulosic biomass into useful products. Many new biotech and agritech companies and numerous patent applications prove that biomass valorisation can be profitable. But, keeping products stable and affordable compared to synthetics is still a big challenge. Looking at the market in the long run helps determine if new products can last. With public–private cooperation, technology can be exchanged and skills built, while research should concentrate on developing simple and low-cost methods for recovering the product. Moreover, picking effective microbial strains and testing products at a fast pace are important for moving from research to commercial uses. By solving these problems, bio-based industries can help increase the economy and support environmental care.

References

Abdel-Aziz, S. M., Hamed, H. A., & Mouafi, F. E. (2012). Acidic exopolysaccharide flocculant produced by the fungus Mucor rouxii using beet-molasses. *Research in Biotechnology, 3*(6).

Abol-Fotouh, D., Hassan, M. A., Shokry, H., Roig, A., Azab, M. S., & Kashyout, A. E. H. B. (2020). Bacterial nanocellulose from agro-industrial wastes: Low-cost and enhanced production by Komagataeibacter saccharivorans MD1. *Scientific Reports, 10*(1), 3491.

Adeniran, H. A., Abiose, S. H., & Ogunsua, A. O. (2010). Production of fungal β-amylase and amyloglucosidase on some Nigerian agricultural residues. *Food and Bioprocess Technology, 3*, 693–698.

Aguilar, A., Twardowski, T., & Wohlgemuth, R. (2019). Bioeconomy for sustainable development. *Biotechnology Journal, 14*(8), 1800638.

Allam, S. S. T., Gacem, A., Govind, G., Tummalacharla, P., Ramakrishnan, G., Yadav, K. K., Nath, H., Ram, D. K., & Das, A. P. (2025). Zero waste initiatives in industrial hydrocarbons and waste management. *Environmental Science and Engineering, Part F218*, 105–126. https://doi.org/10.1007/978-3-031-81615-4_6

Álvarez-Cao, M. E., Rico-Díaz, A., Cerdán, M. E., Becerra, M., & González-Siso, M. I. (2018). Valuation of agro-industrial wastes as substrates for heterologous production of α-galactosidase. *Microbial Cell Factories, 17*, 1–13.

Bassi, J. J., Todero, L. M., Lage, F. A., Khedy, G. I., Ducas, J. D., Custódio, A. P., Pinto, M. A., & Mendes, A. A. (2016). Interfacial activation of lipases on hydrophobic support and application in the synthesis of a lubricant ester. *International Journal of Biological Macromolecules, 92*, 900–909.

Beltrán-Ramírez, F., Orona-Tamayo, D., de Jesús Esparza-Claudio, J., & Quintana-Rodríguez, E. (2019). Agro-industrial waste revalorization: The growing. In *Biomass for bioenergy: Recent trends and future challenges* (p. 83).

Bertsch, A., Roy, D., & LaPointe, G. (2019). Enhanced exopolysaccharide production by Lactobacillus rhamnosus in co-culture with Saccharomyces cerevisiae. *Applied Sciences, 9*(19), 4026.

Bhatia, S., Singh, A., Batra, N., & Singh, J. (2020). Microbial production and biotechnological applications of α-galactosidase. *International Journal of Biological Macromolecules, 150*, 1294–1313.

Bilal, M., & Iqbal, H. M. (2019). Sustainable bioconversion of food waste into high-value products by immobilised enzymes to meet bio-economy challenges and opportunities–A review. *Food Research International, 123*, 226–240.

Bisht, V., & Lal, B. (2019). Exploration of performance kinetics and mechanism of action of a potential novel bioflocculant BF-VB2 on clay and dye wastewater flocculation. *Frontiers in Microbiology, 10*, 1288.

Brena, B., González-Pombo, P., & Batista-Viera, F. (2013). Immobilization of enzymes: A literature survey. In *Immobilisation of enzymes and cells: Third edition* (pp. 15–31).

Buenrostro-Figueroa, J. J., Velázquez, M., Flores-Ortega, O., Ascacio-Valdés, J. A., Huerta-Ochoa, S., Aguilar, C. N., & Prado-Barragán, L. A. (2017). Solid state fermentation of fig (Ficus carica L.) by-products using fungi to obtain phenolic compounds with antioxidant activity and qualitative evaluation of phenolics obtained. *Process Biochemistry, 62*, 16–23.

Cerda, A., Artola, A., Barrena, R., Font, X., Gea, T., & Sánchez, A. (2019). Innovative production of bioproducts from organic waste through solid-state fermentation. *Frontiers in Sustainable Food Systems, 3*, 63.

Chaturvedi, S., Dave, P. N., & Shah, N. K. (2012). Applications of nano-catalysts in new era. *Journal of Saudi Chemical Society, 16*(3), 307–325.

Dal Magro, L., de Moura, K. S., Backes, B. E., de Menezes, E. W., Benvenutti, E. V., Nicolodi, S., Klein, M. P., Fernandez-Lafuente, R., & Rodrigues, R. C. (2019). Immobilisation of pectinase on chitosan-magnetic particles: Influence of particle preparation protocol on enzyme properties for fruit juice clarification. *Biotechnology Reports, 24*, Article e00373.

Darwesh, O. M., Ali, S. S., Matter, I. A., Elsamahy, T., & Mahmoud, Y. A. (2020). Enzyme immobilisation onto magnetic nanoparticles to improve industrial and environmental applications. In *Methods in enzymology* (Vol. 630, pp. 481–502). Academic Press.

Das, A. P., Mohanty, A. K., van Hullebusch, E. D., Figueiredo, G., Das, A. P., Mohanty, A. K., van Hullebusch, E. D., & Figueiredo, G. (2023). Bioremediation as an emerging technology for the removal of synthetic microplastic pollutants from marine ecosystem. *MarPB, 194*, Article 115297. https://doi.org/10.1016/J.MARPOLBUL.2023.115297

Das, A., Dey, S., & Das, A. P. (2024a). Ecotoxicological impacts of synthetic microfiber pollutants and development of sustainable mitigation strategies. *Environmental Chemistry and Ecotoxicology*. https://doi.org/10.1016/J.ENCECO.2024.12.003

Das, A., Dey, S., Bala, K., Gnanasekaran, R., & Das, A. P. (2024b). Microbial bioremediation technology for sustainable treatment and management of synthetic microfiber waste. *Sustainable Microbial Technology for Synthetic and Cellulosic Microfiber Bioremediation, 221–238*. https://doi.org/10.1007/978-3-031-62600-5_12

Das, A., Ray, S. K., Mohanty, M., Mohanty, J., Dey, S., & Das, A. P. (2024c). Ecotoxicity of microplastic wastes and their sustainable management: A review. *Environmental Chemistry and Ecotoxicology, 6*, 144–152. https://doi.org/10.1016/J.ENCECO.2024.05.003

Das, A., Dey, S., Gacem, A., Yadav, K. K., Nath, H., Ram, D. K., & Das, A. P. (2025). Revolutionizing solid hydrocarbon sourcing: Plastic's journey from waste to upcycled treasure. *Environmental Science and Engineering, Part, F218*, 331–351. https://doi.org/10.1007/978-3-031-81615-4_14

de Oliveira, R. L., Dias, J. L., da Silva, O. S., & Porto, T. S. (2018). Immobilisation of pectinase from Aspergillus aculeatus in alginate beads and clarification of apple and umbu juices in a packed bed reactor. *Food and Bioproducts Processing, 109*, 9–18.

del Pilar Sánchez-Camargo, A., Gutiérrez, L. F., Vargas, S. M., Martinez-Correa, H. A., Parada-Alfonso, F., & Narváez-Cuenca, C. E. (2019). Valorisation of mango peel: Proximate composition, supercritical fluid extraction of carotenoids, and application as an antioxidant additive for an edible oil. *The Journal of Supercritical Fluids, 152*, Article 104574.

Dhillon, G. S., Kaur, S., Brar, S. K., & Verma, M. (2012). Potential of apple pomace as a solid substrate for fungal cellulase and hemicellulase bioproduction through solid-state fermentation. *Industrial Crops and Products, 38*, 6–13.

do Nascimento Silva, J., Mascarin, G. M., dos Santos Gomes, I. C., Tinôco, R. S., Quintela, E. D., dos Reis Castilho, L., & Freire, D. M. G. (2018). New cost-effective bioconversion process of palm kernel cake into bioinsecticides based on Beauveria bassiana and Isaria javanica. *Applied Microbiology and Biotechnology, 102*, 2595–2606.

Driss, D., Zouari-Ellouzi, S., Chaari, F., Kallel, F., Ghazala, I., Bouaziz, F., & Chaabouni, S. E. (2014). Production and in vitro evaluation of xylooligosaccharides generated from corncobs using immobilised Penicillium occitanis xylanase. *Journal of Molecular Catalysis B: Enzymatic, 102*, 146–153.

Dulf, F. V., Vodnar, D. C., Dulf, E. H., & Pintea, A. (2017). Phenolic compounds, flavonoids, lipids and antioxidant potential of apricot (Prunus armeniaca L.) pomace fermented by two filamentous fungal strains in solid-state system. *Chemistry Central Journal, 11*, 1–10.

Economou, C. N., Makri, A., Aggelis, G., Pavlou, S., & Vayenas, D. V. (2010). Semi-solid state fermentation of sweet sorghum for the biotechnological production of single-cell oil. *Bioresource Technology, 101*(4), 1385–1388.

El Sheikha, A. F., & Ray, R. C. (2023). Bioprocessing of horticultural wastes by solid-state fermentation into value-added/innovative bioproducts: A review. *Food Reviews International, 39*(6), 3009–3065.

El-Tayeb, T. S., Abdelhafez, A. A., Ali, S. H., & Ramadan, E. M. (2012). Effect of acid hydrolysis and fungal biotreatment on agro-industrial wastes for obtainment of free sugars for bioethanol production. *Brazilian Journal of Microbiology, 43*, 1523–1535.

Fernández-Cabezón, L., & Nikel, P. I. (2020). Advanced metabolic engineering strategies for the development of sustainable microbial processes. In *New and future developments in microbial biotechnology and bioengineering* (pp. 225–246). Elsevier.

Fonseca, L. M., Parreiras, L. S., & Murakami, M. T. (2020). Rational engineering of the Trichoderma reesei RUT-C30 strain into an industrially relevant platform for cellulase production. *Biotechnology for Biofuels, 13*, 1–15.

Gahlawat, G., Kumari, P., & Bhagat, N. R. (2020). Technological advances in the production of polyhydroxyalkanoate biopolymers. *Current Sustainable/Renewable Energy Reports, 7*, 73–83.

Ghosh, S., & Das, A. P. (2020). Microbial metagenomics: current advances in investigating microbial ecology and population dynamics. In S. K. Nayak & B. B. Mishra (Eds.), *Frontiers in soil and environmental microbiology* (1st ed., pp. 247–254). CRC Press.

Gnanasalomi, V. D. V., & Gnanadoss, J. J. (2019). Laccase production by Myrothecium gramineum and its optimisation under solid state fermentation using cowpea pod as substrate. *J Appl Biol Biotechnol, 7*(2), 59–63.

Gondim, T. S., Pereira, R. G., & Fiaux, S. B. (2019). Xanthan gum production by Xanthomonas axonopodis pv. Mangiferaeindicae from glycerin of biodiesel in different media and addition of glucose. *Acta Scientiarum. Biological Sciences, 41*, 43661.

Gullón, P., Gullón, B., Romaní, A., Rocchetti, G., & Lorenzo, J. M. (2020). Smart advanced solvents for bioactive compounds recovery from agri-food by-products: A review. *Trends in Food Science & Technology, 101*, 182–197.

Gupta, P. L., Rajput, M., Oza, T., Trivedi, U., & Sanghvi, G. (2019). Eminence of microbial products in cosmetic industry. *Natural Products and Bioprospecting, 9*, 267–278.

Helianti, I., Ulfah, M., Nurhayati, N., Suhendar, D., Finalissari, A. K., & Wardani, A. K. (2016). Production of xylanase by recombinant Bacillus subtilis DB104 cultivated in agroindustrial waste medium. *HAYATI Journal of Biosciences, 23*(3), 125–131.

Hijosa-Valsero, M., Garita-Cambronero, J., Paniagua-García, A. I., & Díez-Antolínez, R. (2019). Tomato waste from processing industries as a feedstock for biofuel production. *Bioenergy Research, 12*, 1000–1011.

Huccetogullari, D., Luo, Z. W., & Lee, S. Y. (2019). Metabolic engineering of microorganisms for production of aromatic compounds. *Microbial Cell Factories, 18*, 1–29.

Ingale, S., Joshi, S. J., & Gupte, A. (2014). Production of bioethanol using agricultural waste: Banana pseudo stem. *Brazilian Journal of Microbiology, 45*, 885–892.

Ismail, A. R., & Baek, K. H. (2020). Lipase immobilisation with support materials, preparation techniques, and applications: Present and future aspects. *International Journal of Biological Macromolecules, 163*, 1624–1639.

Itabaiana, I., Sutili, F. K., Leite, S. G., Gonçalves, K. M., Cordeiro, Y., Leal, I. C., Miranda, L. S., Ojeda, M., Luque, R., & de Souza, R. O. (2013). Continuous flow valorization of fatty acid waste using silica-immobilised lipases. *Green Chemistry, 15*(2), 518–524.

Jiang, B., Fu, L., Cao, W., Zhang, B., Li, F., & Liu, Y. (2019). Microbial flocculant produced by a novel Paenibacillus sp., strain A9, using food processing wastewater to replace fermentation medium and its application for the removal of Pb (II) from aqueous solution. *Adsorption Science & Technology, 37*(9–10), 683–697.

Jimoh, A. A., & Lin, J. (2020). Biotechnological applications of Paenibacillus sp. D9 lipopeptide biosurfactant produced in low-cost substrates. *Applied Biochemistry and Biotechnology, 191*, 921–941.

Jiru, T. M., Steyn, L., Pohl, C., & Abate, D. (2018). Production of single cell oil from cane molasses by Rhodotorula kratochvilovae (syn, Rhodosporidium kratochvilovae) SY89 as a biodiesel feedstock. *Chemistry Central Journal, 12*, 1–7.

Júnior, J. C. B., Viana, P. A., de Rezende, S. T., Soares, N. D. F. F., & Guimarães, V. M. (2018). Immobilisation of an alpha-galactosidase from Debaryomyces hansenni UFV-1 in cellulose film and its application in oligosaccharide hydrolysis. *Food and Bioproducts Processing, 111*, 30–36.

Kalli, E., Lappa, I., Bouchagier, P., Tarantilis, P. A., & Skotti, E. (2018). Novel application and industrial exploitation of winery by-products. *Bioresources and Bioprocessing, 5*(1), 1–21.

Karahalil, E., Coban, H. B., & Turhan, I. (2019). A current approach to the control of filamentous fungal growth in media: Microparticle enhanced cultivation technique. *Critical Reviews in Biotechnology, 39*(2), 192–201.

Kaza, S., Yao, L., Bhada-Tata, P., & Van Woerden, F. (2018). *What a waste 2.0: A global snapshot of solid waste management to 2050*. World Bank Publications.

Kırbaşlar, Şİ., & Şahin, S. (2023). Recovery of bioactive ingredients from biowaste of olive tree (Olea europaea) using microwave-assisted extraction: A comparative study. *Biomass Conversion and Biorefinery, 13*(4), 2849–2861.

Ko, Y. S., Kim, J. W., Lee, J. A., Han, T., Kim, G. B., Park, J. E., & Lee, S. Y. (2020). Tools and strategies of systems metabolic engineering for the development of microbial cell factories for chemical production. *Chemical Society Reviews, 49*(14), 4615–4636.

Kuvvet, C., Uzuner, S., & Cekmecelioglu, D. (2019). Improvement of pectinase production by co-culture of Bacillus spp. Using apple pomace as a carbon source. *Waste and Biomass Valorisation, 10*, 1241–1249.

Luthfi, W. M., Alias, H., Tay, G. S., & Lee, C. K. (2020). Production and characterisation of bioflocculant via solid state fermentation process using oil palm empty fruit bunch fibres as substrate. *Biocatalysis and Agricultural Biotechnology, 23*, Article 101454.

Mahmoodi, M., Najafpour, G. D., & Mohammadi, M. (2019). Bioconversion of agroindustrial wastes to pectinases enzyme via solid state fermentation in trays and rotating drum bioreactors. *Biocatalysis and Agricultural Biotechnology, 21*, 101280.

Manrich, A., Komesu, A., Adriano, W. S., Tardioli, P. W., & Giordano, R. L. C. (2010). Immobilisation and stabilization of xylanase by multipoint covalent attachment on agarose and on chitosan supports. *Applied Biochemistry and Biotechnology, 161*, 455–467.

Marín, F. R., Soler-Rivas, C., Benavente-García, O., Castillo, J., & Pérez-Alvarez, J. A. (2007). By-products from different citrus processes as a source of customised functional fibres. *Food Chemistry, 100*(2), 736–741.

Mehmood, T., Saman, T., Irfan, M., Anwar, F., Ikram, M. S., & Tabassam, Q. (2019). Pectinase production from Schizophyllum commune through central composite design using citrus waste and its immobilisation for industrial exploitation. *Waste and Biomass Valorisation, 10*, 2527–2536.

Mishra, S., & Das, A. P. (2022). Treatment of the Wastewater Polluted with Synthetic Microfiber Released from Washing Machine. In B. B. Das, H. Hettiarachchi, P. K. Sahu, & S. Nanda (Eds.), *Lecture notes in civil engineering* (Vol. 207, pp. 109–117). Springer Science and Business Media Deutschland GmbH. https://doi.org/10.1007/978-981-16-7509-6_9/COVER

Mishra, S., & Das, A. P. (2023). Microbial remediation of synthetic microfiber contaminated wastewater. In M. P. Shah (Ed.), *Microbial technologies in industrial wastewater treatment* (pp. 337–350). Springer. https://doi.org/10.1007/978-981-99-2435-6_17

Mishra, S., Dash, D., & Das, A. P. (2024). Aquatic microbial diversity on plastisphere: Colonization and potential role in microplastic biodegradation. *Geomicrobiology Journal, 41*(4), 312–323. https://doi.org/10.1080/01490451.2023.2209750

Mohamad, N. R., Marzuki, N. H. C., Buang, N. A., Huyop, F., & Wahab, R. A. (2015). An overview of technologies for immobilisation of enzymes and surface analysis techniques for immobilized enzymes. *Biotechnology & Biotechnological Equipment, 29*(2), 205–220.

Mohanty, M., Mohanty, J., Dey, S., Dutta, K., Shah, M. P., & Das, A. P. (2024). The face mask: A tale from protection to pollution and demanding sustainable solution. *Emerging Contaminants, 10*(2), Article 100298. https://doi.org/10.1016/J.EMCON.2023.100298

Moradali, M. F., & Rehm, B. H. (2020). Bacterial biopolymers: From pathogenesis to advanced materials. *Nature Reviews Microbiology, 18*(4), 195–210.

Motte, J. C., Trably, E., Escudié, R., Hamelin, J., Steyer, J. P., Bernet, N., Delgenes, J. P., & Dumas, C. (2013). Total solids content: A key parameter of metabolic pathways in dry anaerobic digestion. *Biotechnology for Biofuels, 6*, 1–9.

Munasinghe, K. A., Schwarz, J. G., & Whittiker, M. (2015). Utilisation of Chicken By-Products to Form Collagen Films. *Journal of Food Processing, 2015*(1), Article 247013.

Nadar, S. S., & Rathod, V. K. (2019). A co-immobilisation of pectinase and cellulase onto magnetic nanoparticles for antioxidant extraction from waste fruit peels. *Biocatalysis and Agricultural Biotechnology, 17*, 470–479.

Nady, D., Zaki, A. H., Raslan, M., & Hozayen, W. (2020). Enhancement of microbial lipase activity via immobilisation over sodium titanate nanotubes for fatty acid methyl esters production. *International Journal of Biological Macromolecules, 146*, 1169–1179.

Nag, M., Lahiri, D., Ghosh, S., Sarkar, T., Pati, S., Das, A. P., Ram, D. K., Bhattacharya, D., & Ray, R. R. (2023). Application of microorganisms in biotransformation and bioremediation of environmental contaminant: A review. *Geomicrobiology Journal*. https://doi.org/10.1080/01490451.2023.2261443

Namboodiri, M. T., & Pakshirajan, K. (2019). Sustainable and green approach of chitosan production from Penicillium citrinum biomass using industrial wastewater as a cheap substrate. *Journal of Environmental Management, 240*, 431–440.

Namvar, F., Mohamed, S., Fard, S. G., Behravan, J., Mustapha, N. M., Alitheen, N. B. M., & Othman, F. (2012). Polyphenol-rich seaweed (Eucheuma cottonii) extract suppresses breast tumour via hormone modulation and apoptosis induction. *Food Chemistry, 130*(2), 376–382.

Narsing Rao, M. P., Xiao, M., & Li, W. J. (2017). Fungal and bacterial pigments: Secondary metabolites with wide applications. *Frontiers in Microbiology, 8*, 1113.

Nayak, A., & Bhushan, B. (2019). An overview of the recent trends in the waste valorisation techniques for food waste. *Journal of Environmental Management, 233*, 352–370.

Nogueira, I. B., Rodríguez, D. M., da Silva Andradade, R. F., Lins, A. B., Bione, A. P., da Silva, I. G. S., Franco, L. D. O., & de Campos-Takaki, G. M. (2020). Bioconversion of agroindustrial waste in the production of bioemulsifier by Stenotrophomonas maltophilia UCP 1601 and application in bioremediation process. *International Journal of Chemical Engineering, 2020*(1), 9434059.

Oberoi, H. S., Vadlani, P. V., Nanjundaswamy, A., Bansal, S., Singh, S., Kaur, S., & Babbar, N. (2011). Enhanced ethanol production from Kinnow mandarin (Citrus reticulata) waste via a statistically optimised simultaneous saccharification and fermentation process. *Bioresource Technology, 102*(2), 1593–1601.

Ochsenreither, K., Glück, C., Stressler, T., Fischer, L., & Syldatk, C. (2016). Production strategies and applications of microbial single-cell oils. *Frontiers in Microbiology, 7*, 1539.

Panizza, P., Cesarini, S., Diaz, P., & Giordano, S. R. (2015). Saturation mutagenesis in selected amino acids to shift Pseudomonas sp. acidic lipase Lip I. 3 substrate specificity and activity. *Chemical Communications, 51*(7), 1330–1333.

Panjiar, N., Mattam, A. J., Jose, S., Gandham, S., & Velankar, H. R. (2020). Valorisation of xylose-rich hydrolysate from rice straw, an agroresidue, through biosurfactant production by the soil bacterium Serratia nematodiphila. *Science of the Total Environment, 729*, Article 138933.

Panzella, L., Moccia, F., Nasti, R., Marzorati, S., Verotta, L., & Napolitano, A. (2020). Bioactive phenolic compounds from agri-food wastes: An update on green and sustainable extraction methodologies. *Frontiers in Nutrition, 7*, 60.

Pattanaik, L., Pattnaik, F., Saxena, D. K., & Naik, S. N. (2019). Biofuels from agricultural wastes. In *Second and third generations of feedstocks* (pp. 103–142). Elsevier.

Philippini, R. R., Martiniano, S. E., Ingle, A. P., Franco Marcelino, P. R., Silva, G. M., Barbosa, F. G., Dos Santos, J. C., & da Silva, S. S. (2020). Agroindustrial byproducts for the generation of biobased products: Alternatives for sustainable biorefineries. *Frontiers in Energy Research, 8*, 152.

Pinheiro, B. B., Rios, N. S., Aguado, E. R., Fernandez-Lafuente, R., Freire, T. M., Fechine, P. B., Dos Santos, J. C., & Goncalves, L. R. (2019). Chitosan activated with divinyl sulfone: A new heterofunctional support for enzyme immobilisation. Application in the immobilisation of lipase B from Candida antarctica. *International Journal of Biological Macromolecules, 130*, 798–809.

Pradhan, G., Tripathy, B., Ram, D. K., Digal, A. K., & Das, A. P. (2023). Bauxite mining waste pollution and its sustainable management through bioremediation. *Geomicrobiology Journal, 1–10*. https://doi.org/10.1080/01490451.2023.2235353

Quayson, E., Amoah, J., Rachmadona, N., Morita, K., Darkwah, L., Hama, S., Yoshida, A., Kondo, A., & Ogino, C. (2020). Valorisation of palm biomass waste into carbon matrices for the immobilisation of recombinant Fusarium heterosporum lipase towards palm biodiesel synthesis. *Biomass and Bioenergy, 142*, Article 105768.

Raj, V., Karn, S. K., Kumar, R., Kumari, M., Kumar, A., Prakash, A., & Srivastav, A. L. (2025). Recent Progress in Sustainable Biochemical Pretreatment Technologies for Lignocellulosic Biomass Conversion to Bioenergy Production: Challenges and Perspectives. *Biotechnological Advancements in Biomass to Bioenergy Biotransformation: Sustainable Implications in Circular Economy*, 79–96.

Rane, A. N., Baikar, V. V., Ravi Kumar, V., & Deopurkar, R. L. (2017). Agro-industrial wastes for production of biosurfactant by Bacillus subtilis ANR 88 and its application in synthesis of silver and gold nanoparticles. *Frontiers in Microbiology, 8*, 492.

Rashad, M. M., Mahmoud, A. E., Ali, M. M., Nooman, M. U., & Al-Kashef, A. S. (2015). Antioxidant and anticancer agents produced from pineapple waste by solid-state fermentation. *International Journal of Toxicological and Pharmacological Research, 7*(6), 287–296.

Rebocho, A. T., Pereira, J. R., Neves, L. A., Alves, V. D., Sevrin, C., Grandfils, C., Freitas, F., & Reis, M. A. (2020). Preparation and characterisation of films based on a natural p (3hb)/mcl-pha blend obtained through the co-culture of cupriavidus necator and Pseudomonas citronellolis in apple pulp waste. *Bioengineering, 7*(2), 34.

Reddy Koteswara, G., Bhargavi, S., Yamuna, D., Anjani Devi, M., Jeevana Lakshmi Kumar, G., Manas, T., Adarsh, S., Bhavya Sai, V., Nath, H., Alreshidi, M. A., Togun, H., Ram, D. K., & Das, A. P. (2025). Recent advancements in the remediation of marine hydrocarbon and plastic pollutants. *Environmental Science and Engineering, Part F218*, 127–142. https://doi.org/10.1007/978-3-031-81615-4_7

Ring, L. C., Yenn, T. W., Nazri, N. I. B., Nee, T. W., Zahan, K. A., Ibrahim, D., & Rashid, S. (2018). Chemical optimization of red pigment, Monascorubin production in Penicillium minioluteum ED24 using solid-state fermentation. *Arabian Journal for Science and Engineering, 43*, 3485–3491.

Rivas, B., Torrado, A., Torre, P., Converti, A., & Domínguez, J. M. (2008). Submerged citric acid fermentation on orange peel autohydrolysate. *Journal of Agricultural and Food Chemistry, 56*(7), 2380–2387.

Sadh, P. K., Duhan, S., & Duhan, J. S. (2018). Agro-industrial wastes and their utilization using solid state fermentation: A review. *Bioresources and Bioprocessing, 5*(1), 1–15.

Sagar, N. A., Pareek, S., Sharma, S., Yahia, E. M., & Lobo, M. G. (2018). Fruit and vegetable waste: Bioactive compounds, their extraction, and possible utilization. *Comprehensive Reviews in Food Science and Food Safety, 17*(3), 512–531.

Sala, A., Artola, A., Sánchez, A., & Barrena, R. (2020). Rice husk as a source for fungal biopesticide production by solid-state fermentation using B. bassiana and T. harzianum. *Bioresource Technology, 296*, 122322.

Sen, T., Barrow, C. J., & Deshmukh, S. K. (2019). Microbial pigments in the food industry—Challenges and the way forward. *Frontiers in Nutrition, 6*, 7.

Sharma, R., & Ghoshal, G. (2020). Optimization of carotenoids production by Rhodotorula mucilaginosa (MTCC-1403) using agro-industrial waste in bioreactor: A statistical approach. *Biotechnology Reports, 25*, Article e00407.

Sharma, A., Singh, R. S., Gupta, G., Ahmad, T., & Kaur, B. (2019). Metabolic engineering of enzyme-regulated bioprocesses. In *Advances in enzyme technology* (pp. 293–323). Elsevier.

Singh nee' Nigam, P., Gupta, N., & Anthwal, A. (2009). Pre-treatment of agro-industrial residues. In *Biotechnology for agro-industrial residues utilisation: Utilisation of agro-residues* (pp. 13–33).

Singh, B. (2018). Engineering fungal morphology for enhanced production of hydrolytic enzymes by Aspergillus oryzae SBS50 using microparticles. *3 Biotech, 8*(6), 283.

Singhania, R. R., Patel, A. K., Thomas, L., Goswami, M., Giri, B. S., & Pandey, A. (2015). Industrial enzymes. In *Industrial biorefineries & white biotechnology* (pp. 473–497). Elsevier.

Suarato, G., Bertorelli, R., & Athanassiou, A. (2018). Borrowing from nature: Biopolymers and biocomposites as smart wound care materials. *Bioengineering and Biotechnology, 6*, 137.

Swain, S., Purnima Sudha, M., Dey, S., Badu, Y., Bhoi, S. R., Alreshidi, M. A., Qasim, M. T., Nath, H., Ram, D. K., & Das, A. P. (2025). Sources of micro and nano plastic hydrocarbon pollutants in marine bodies and their treatment. *Environmental Science and Engineering, Part, F218*, 45–67. https://doi.org/10.1007/978-3-031-81615-4_3

Thatoi, H. N., Chattaraj, S., Mishra, R. R., Das Mohapatra, P. K., & Mohapatra, S. (2025). Contributions of biotechnology industries of India to global bioeconomy: An overview. *3 Biotech, 15*(2), 1–21.

Tolba, G. M., Barakat, N. A., Bastaweesy, A. M., Ashour, E. A., Abdelmoez, W., El-Newehy, M. H., Al-Deyab, S. S., & Kim, H. Y. (2015). Effective and highly recyclable nanosilica produced from the rice husk for effective removal of organic dyes. *Journal of Industrial and Engineering Chemistry, 29*, 134–145.

Tuli, H. S., Chaudhary, P., Beniwal, V., & Sharma, A. K. (2015). Microbial pigments as natural color sources: Current trends and future perspectives. *Journal of Food Science and Technology, 52*, 4669–4678.

Tyagi, B., & Kumar, N. (2021). Bioremediation: Principles and applications in environmental management. In *Bioremediation for environmental sustainability* (pp. 3–28). Elsevier.

Venil, C. K., Devi, P. R., & Ahmad, W. A. (2020). Agro-industrial waste as substrates for the production of bacterial pigment. In *Valorisation of Agro-industrial residues–Volume I: Biological approaches* (pp. 149–162).

Viveka, R., Varjani, S., & Ekambaram, N. (2021). Valorization of cassava waste for pullulan production by Aureobasidium pullulans MTCC 1991. *Energy & Environment, 32*(6), 1086–1102.

Weltz, J. S., Kienle, D. F., Schwartz, D. K., & Kaar, J. L. (2020). Reduced enzyme dynamics upon multipoint covalent immobilization leads to stability-activity trade-off. *Journal of the American Chemical Society, 142*(7), 3463–3471.

Wen, Z., Liao, W., & Chen, S. (2005). Production of cellulase/β-glucosidase by the mixed fungi culture Trichoderma reesei and Aspergillus phoenicis on dairy manure. *Process Biochemistry, 40*(9), 3087–3094.

Ying, M., Meti, M. D., Xu, H., Wang, Y., Lin, J., Wu, Z., Han, Q., Xu, X., He, Z., Hong, W., & Hu, Z. (2018). Binding mechanism of lipase to Ligupurpuroside B extracted from Ku-Ding tea as studied by multi-spectroscopic and molecular docking methods. *International Journal of Biological Macromolecules, 120*, 1345–1352.

Yuan, S. F., & Alper, H. S. (2019). Metabolic engineering of microbial cell factories for production of nutraceuticals. *Microbial Cell Factories, 18*(1), 46.

Zahan, K. A., Ismail, N. S., Leong, C. R., Ab Rashid, S., & Tong, W. Y. (2020). Monascorubin production by Penicillium minioluteum ED24 in a solid-state fermentation using sesame seed cake as substrate. *Materials Today: Proceedings, 31*, 127–135.

Zdarta, J., Meyer, A. S., Jesionowski, T., & Pinelo, M. (2018). A general overview of support materials for enzyme immobilization: Characteristics, properties, practical utility. *Catalysts, 8*(2), 92.

Zhong, L., Feng, Y., Wang, G., Wang, Z., Bilal, M., Lv, H., Jia, S., & Cui, J. (2020). Production and use of immobilized lipases in/on nanomaterials: A review from the waste to biodiesel production. *International Journal of Biological Macromolecules, 152*, 207–222.

Zhu, W., He, Q., Gao, H., Nitayavardhana, S., Khanal, S. K., & Xie, L. (2020). Bioconversion of yellow wine wastes into microbial protein via mixed yeast-fungus cultures. *Bioresource Technology, 299*, Article 122565.

From Plastics to Biodegradable Polymers: A Path Toward Environmental Sustainability

M. Madhavaraj and K. Karthikeyan

1 Introduction

In the early twentieth century discovery of synthetic polymers revolutionized modern society. Every year synthetic polymers production increases, from 2 MT in 1950 to over 460 MT in 2019. It is popular due to its advantages like cost, flexibility, lightweight nature, and water resistance. It has diverse applications, like packaging (31%), construction (17%), and transportation (12%) sectors. But after its useful life it reaches and accumulates in the environment and leads to severe ecological consequences like toxicity to fishes, microplastic pollution, etc. Every year, an estimated 1–2 million tons of plastic garbage find their way into the seas and entanglement, swallowing, and chemical exposure damages marine life. It impacts at least 267 species globally, including 86% of all sea turtle species, 44% of all seabird species and 43% of all marine mammal species (Isangedighi et al., 2020). It also affects terrestrial ecosystems by accumulating on land, affecting soil quality, and wildlife. Plastic production and disposal can lead to greenhouse gas emissions that drive climate change hence it also affects atmosphere. Plastic deteriorates to form microplastics and nanoplastics, found in ecosystems around the world. This, along with the fact that plastic polymer is virtually indestructible and takes hundreds, even thousands of years to decompose, only magnifies the issues at hand. It also has economic impacts, influencing sectors like tourism, fisheries, and shipping. Due to the environmental impacts of plastic pollution, sustainable alternatives to conventional plastics are in urgent demand. Biodegradable polymers can be made from renewable resources, which provides several benefits like reduced non-renewable energy use and lower

M. Madhavaraj · K. Karthikeyan (✉)
School of Civil Engineering, Vellore Institute of Technology, Chennai, India
e-mail: karthikeyan.kothanda@vit.ac.in

M. Madhavaraj
e-mail: madhavarajmuthukrishnan@gmail.com

S. Ghosh et al. (eds.), *Advances in Organic Waste Conversion Through Industrial Biotechnology*, Sustainable Environmental Waste Management Strategies,
https://doi.org/10.1007/978-3-032-11377-1_6

greenhouse gas emissions compared to synthetic polymers (Mangal et al., 2023). Sources of these polymers are from cornstarch, sugarcane, and cellulose so once they get decomposed it will give water, carbon dioxide, and biomass. However, there are disadvantages also like high production cost, mechanical strength, varying degradation rate, etc. This book chapter aims to explore solutions to the following key questions:

- What are the environmental and health impacts of conventional plastics across marine, terrestrial, and atmospheric systems?
- Why do current plastic waste management methods like recycling, incineration, and landfilling fall short in addressing the crisis?
- What defines biodegradable polymers, and how do their types, properties, and production methods compare to conventional plastics?
- What technical and practical challenges limit the widespread adoption of biodegradable polymers?
- How are global policies, market trends, and corporate initiatives shaping the shift toward sustainable alternatives?
- What role do governments, industries, and consumers play in advancing the future of biodegradable polymers?

2 Conventional Plastics and Their Impacts

Each year, approximately 300 million metric tons (MMT) of plastic are produced worldwide, with production steadily increasing due to rising demand and economic growth. India, as a rapidly industrializing nation, has seen a considerable rise in plastic production and consumption (Padgelwar et al., 2021). By 2020, the country generated 22 MMT of plastic annually, with a per capita consumption estimated at 20 kg in 2022. Between 2017 and 2018, India's plastic consumption stood at 16.5 MMT, while plastic waste generation increased to 9.4 MMT per year by 2019. A significant portion of this plastic waste originates from packaging materials used for food products, groceries, cosmetics, and consumer goods. The exponential growth in plastic use, particularly in developing countries, has exacerbated waste management challenges, as infrastructure and regulatory mechanisms struggle to keep pace with the increasing volumes of plastic waste. The COVID-19 pandemic further intensified plastic consumption patterns, particularly due to the heightened demand for personal protective equipment (PPE), disposable medical supplies, and single-use plastics (SUP's) (Hossain et al., 2022). The surge in plastic waste from medical sources introduced additional complexities in waste handling and disposal. As a result, managing plastic pollution has become a pressing global issue, necessitating urgent measures to develop sustainable alternatives and effective waste management strategies.

3 Environmental Persistence of Plastics: Marine, Terrestrial, and Atmospheric Impacts

Traditional polymers like polyethylene (PE), polypropylene (PP), and polyvinyl chloride (PVC) are highly resistant to environmental degradation due to their hydrophobic and stable properties. As the result plastic persistent in the environment for long time creating impact to marine, terrestrial and atmosphere

i. **Marine Pollution**: Around 8 MMTs of plastic waste enter oceans each year, largely through poor waste management. Marine plastics, including polyethylene terephthalate (PET), polystyrene (PS), and polyamide (PA), harm marine organisms through ingestion and entanglement. MPs also adsorb toxic substances like persistent organic pollutants and heavy metals, contributing to bioaccumulation and health risks in aquatic food chains, including seafood consumed by humans (Kosior, 2020).
ii. **Terrestrial Pollution**: Land-based ecosystems receive the bulk of plastic waste, with about 90% of single-use plastics ending up in landfills. Environmental factors such as temperature fluctuations and microbial activity promote plastic fragmentation in soil. MPs alter soil structure, reduce water retention, and affect microbial and plant health. They also increase the mobility of soil contaminants, intensifying pollution (Richard et al., 2024).
iii. **Atmospheric Pollution**: Airborne MPs, released from landfills and urban areas, are inhaled or ingested by humans. Synthetic textiles and plastic materials contribute to indoor MP exposure. These particles often carry harmful additives like bisphenol A (BPA) and phthalates, linked to respiratory, endocrine, and cardiovascular issues (Huang et al., 2012). Although less visible, atmospheric MPs represent a significant emerging health concern.

3.1 Human Health Implications of Plastic and Microplastic Pollution

The presence of plastics in the environment has raised concerns regarding human health. It affects human health through water, food, and air (Okeke et al., 2022). As plastic degrades due to various environmental conditions it releases fine particles known as microplastics (MPs). Some of them become airborne and reach human lungs during breathing. Human exposure to MPs primarily occurs through ingestion, inhalation, and dermal contact. Recent studies suggest that individuals may consume between 1,063 and 3,223 MP particles annually, with additional exposure occurring in indoor environments (Dehghani et al., 2017). Ingestion of contaminated food, and microplastics accumulated in seafood, bottled water, and table salt increase the exposure to MPs and related toxic chemicals. MPs can accumulate in human tissues and organs, causing inflammatory responses, oxidative stress, and cellular damage. Additives leached from the polymer such as bisphenol A (BPA), phthalates, and flame retardants are known as endocrine disruptors. These chemicals can interfere with hormonal functions and have been linked to metabolic disorders, reproductive dysfunction, and immune system impairments. Moreover, MPs present in the gastrointestinal tract may alter gut microbiota composition, contributing to the immune dysregulation and increased vulnerability to chronic diseases. Studies also confirm possible links between the MPs and neurotoxicity, DNA damage and carcinogenesis (Huang et al., 2012). Certain plastic- associated compounds can disrupt normal cellular processes, potentially leading to genetic mutations and tumor development. Given the growing body of evidence highlighting the risks of MPs' exposure, it is essential to implement regulatory measures, improve waste management, and promote recycling to mitigate the health and environmental hazards posed by plastic pollution.

4 Challenges in Plastic Waste Management: Recycling, Incineration, and Landfill Issues

Despite growing awareness of plastic pollution, waste management systems worldwide face significant challenges in handling plastic waste effectively. Conventional disposal methods, including recycling, incineration, and landfilling, each present distinct limitation.

- **Recycling**: Although recycling is promoted as a sustainable solution, global recycling rates remain low, with only 9% of all plastic waste being successfully recycled. High contamination level in post-consumer plastics, sorting inefficiencies, and economic challenges such as fluctuating oil prices, the low market value of recycled plastics, and the high costs of collection and processing undermine the financial viability of large-scale recycling program.
- **Incineration**: While incineration reduces plastic waste volume, it generates toxic emissions, including dioxins and furans, which pose environmental and health risks.
- **Landfilling**: Landfills remain the dominant disposal method, yet they contribute significantly to MP contamination in soil and water systems. Poorly engineered landfills exacerbate leachate infiltration, leading to groundwater pollution.

5 Biodegradable Polymers: Definition, Types, and Properties

Biodegradable polymers decompose naturally by the activity of microorganisms such as bacteria, fungi, or algae. Unlike ordinary plastics, these materials break down into simpler, non-toxic chemicals like carbon dioxide (CO_2), water, and biomass. The biodegradation process occurs in two basic stages. In the early step, extracellular enzymes and abiotic agents such as oxidation, photodegradation, and hydrolysis operate to break down long polymer chains into shorter oligomers. In the second stage, these oligomers undergo biomineralization, where microorganisms assimilate and further transform them into simpler molecules. This process can take place within 3–6 months in commercial composting settings. Longer timeframes may apply for biodegradation in natural environments like soil or marine ecosystems. Biodegradation can occur under either aerobic or anaerobic environments. Biodegradable polymers require certain conditions to decompose efficiently, such as high temperature (50–60 °C), moisture, and oxygen levels. Aerobic decomposition, which takes place in the presence of oxygen, produces CO_2, H_2O, biomass, and residual matter. Conversely, anaerobic breakdown happens under oxygen-deprived conditions, resulting in the creation of CO_2, H_2O, CH_4, biomass, and leftover material.

5.1 *Types of Biodegradable Polymers*

Biodegradable polymers can be classified into different categories based on their criteria like origin, synthesis method and applications. In this study, it is distinguished into two types based on their origin: synthetic or natural polymers. Synthetic polymers are produced from feedstocks derived either from petroleum or hydrocarbon resources (non-renewable resources) and natural polymers are derived from biological resources (renewable resources) such as plants, animals, marine organisms, and microbes as shown in Fig. 1 (Samir et al., 2022).

5.1.1 Natural Biodegradable Polymers

a. **Biomass-Derived Polymers**

Natural polymers are produced in nature throughout the growth cycles of all organisms via cellular metabolism. Accordingly, it may be split into two categories: microorganism-derived and biomass-derived. Biomass derived Polysaccharides (starch and cellulose), proteins and lipids, are primary renewable sources of biopolymers.

i. **Polysaccharides**

Polysaccharides are carbohydrate-based polymers obtained from plant and microbial sources. They are extensively used in biodegradable films, coatings, and medical applications.

- **Starch**: Found in crops like corn, potato, and wheat; widely used in food packaging and disposable cutlery.

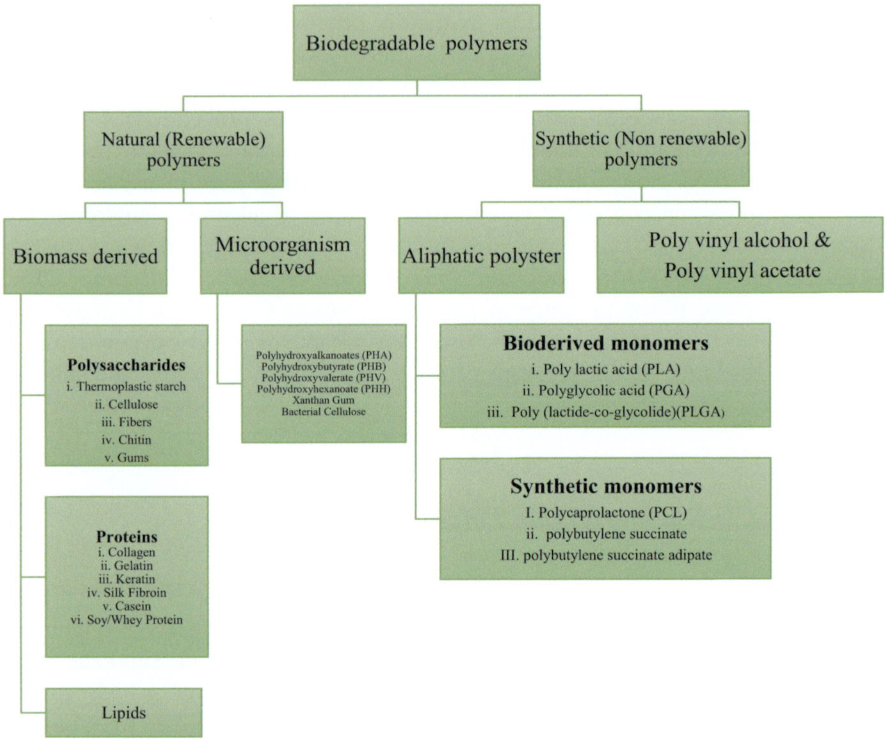

Fig. 1 Classification of biodegradable polymers based on origin (Samir et al., 2022)

- **Cellulose**: Extracted from cotton, wood pulp, and other plant materials; used in films, textiles, and paper coatings.
- **Chitosan**: Obtained from the exoskeleton of crustaceans; applied in water treatment, wound dressings, and food preservation.

ii. **Proteins**

Protein-based polymers are derived from both animal and plant sources. They are widely used in biomedical applications and biodegradable films.

- **Collagen**: Found in animal connective tissues; employed in medical sutures and regenerative medicine.
- **Gelatin**: A denatured form of collagen; utilized in pharmaceuticals, food coatings, and biodegradable films.
- **Casein**: A milk protein; applied in adhesives, biodegradable coatings, and packaging.
- **Soy Protein**: Extracted from soybeans; used in bioplastics and agricultural films.

iii. **Lipids**

Certain lipid-based materials, derived from plant oils, animal fats, and waxes, are used in coatings, food wraps, and controlled-release systems.

b. **Microorganism-Derived Polymers**

Naturally microorganism-derived polymers are biodegradable materials often synthesized by natural or genetically engineered microorganisms as part of their metabolic processes. These polymers are often produced as intracellular storage compounds under nutrient-limiting conditions. They are gaining significant attention for their eco-friendly properties and potential applications in packaging, agriculture, and biomedical fields. Key types of microorganism-derived polymers include:

ii. **Polyhydroxyalkanoates (PHAs)**

PHAs are a family of biodegradable polyesters synthesized by bacteria as energy and carbon storage materials. These polymers are highly biocompatible and versatile. It is used in applications like packaging, medical bioimplants, food-friendly films, agriculture biofuel, drug delivery, tissue engineering, etc.

- **Poly(3-hydroxybutyrate) (PHB)**: PHB is an aliphatic polyesters. A highly crystalline and brittle polymer with applications in medical implants, packaging, and disposable products and often used as an alternative to conventional plastics like polyethylene and polypropylene
- **Poly(3-hydroxybutyrate-co-3-hydroxyvalerate) (PHBV)**: A copolymer of PHB with improved flexibility and toughness; widely used in food packaging, drug delivery systems, and agricultural films (Mehrpouya et al., 2021).
- **Polysaccharides from Microorganisms**

Certain microorganisms produce biodegradable polysaccharides that offer natural resistance to environmental stress and provide structural stability.

- **Xanthan Gum**: Produced by *Xanthomonas campestris*; used in food products, pharmaceuticals, and oil recovery applications (Niknezhad et al., 2016).
- **Curdlan**: Produced by *Alcaligenes faecalis*; applied in food, pharmaceuticals, and biomedical products for its gelling properties (Aquinas et al., 2022).
- **Dextran**: Synthesized by *Leuconostoc mesenteroides*; used in medical treatments, drug delivery, and as a blood plasma substitute (Narayan et al., 2023).

iii. **Bacterial Cellulose (BC)**

Bacterial cellulose is an exopolysaccharide synthesized by bacteria such as *Acetobacter xylinum* or *Gluconacetobacter xylinum.* It features remarkable mechanical strength, high purity, and biocompatibility, making it suitable for wound dressings, scaffolds in tissue engineering, and food packaging (Navya et al., 2022).

5.1.2 Synthetic Biodegradable Polymers

Synthetic biodegradable polymers are artificially synthesized through conventional methods as aliphatic polyesters. These materials designed to decompose under environmental conditions through microbial action, enzymatic degradation, or hydrolysis. These polymers offer versatility in properties, making them suitable for applications such as biomedical implants, packaging, and agriculture. Over 90% of biopolymers are polyesters due to their easily hydrolysable ester bonds. Synthetic biodegradable polymers can be bio-based (e.g., PLA) or derived from petroleum sources (e.g., PCL).

a. **Aliphatic Polyesters from Bio-derived Monomers**

Aliphatic polyesters are widely recognized for their biodegradability due to their hydrolysable ester bonds. They are commonly employed in medical sutures, drug delivery systems, and packaging.

- **Poly (lactic acid) (PLA)**: Derived from fermented starch, corn and biomass, extensively used in biodegradable packaging, textiles, and medical implants. PLA exists in different stereoisomeric form, D-lactide, L-lactide, and meso-lactide—which influence its mechanical and thermal properties.
- **Poly (glycolic acid) (PGA)**: Known for its high strength and rapid biodegradation; it degrades in two stages: first in amorphous regions and then in crystalline zones through hydrolysis It is used in medical and tissue engineering like surgical sutures and drug delivery systems.

Copolymers of PLA and PGA, known as PLGA, are synthesized by adjusting the ratio of lactic acid and glycolic acid monomers. Their degradation rate and properties can be tailored without affecting biocompatibility, and they break down into environmentally friendly lactic and glycolic acids.

b. **Aliphatic Polyesters from Synthetic (Petroleum-Based) Monomers**

Polycaprolactone (PCL) is a biodegradable aliphatic polyester synthesized by ring-opening polymerization of caprolactone using metal catalysts like tin octoate. It decomposes in aerobic and anaerobic conditions and has melting point been 58–60 C. It is used in tissue engineering and long-term medication administration. Copolymizing with PLA or PGA increases its biodegradability. Polycondensation of 1,4-butanediol with succinic acid produces polybutylene succinate (PBS). It melts at 90–120 °C and degrades slowly due to its crystallinity. Combining with adipate creates PBSA, which breaks down more quickly than PBS.

5.1.3 Biodegradable Polymer Blends and Composites

Blending natural and synthetic biodegradable polymers can improve their hydrophilicity, mechanical properties, and degradation rate. These blends combine the advantages of each component and are often formed by mixing thermoplastic resins with biodegradable polymers. They offer enhanced biodegradability and produce environmentally friendly residues after decomposition (Samir et al., 2022).

6 Production and Processing of Biodegradable Polymers

Biodegradable polymers that can be produced via different production methods like chemical synthesis, bacterial synthesis, biopolymer blends, and renewable resource-based approaches as discussed in Tables 1 and 2.

7 Applications of Biodegradable Polymers

i. **Packaging Industry**: Packaging industry is the largest and fastest-growing, driven by demand from food and beverage sectors. Bio-based packaging solutions are gaining popularity and sales are increasing by 20–30% each year as manufacturing and industrial processes become more technologically advanced. Recent market projections indicate that the worldwide bioplastics and biopolymer market will rise, valued at USD 10.2 billion in 2022, and climb 17.1% to USD 22.5 billion by 2027. Bioplastics are mainly divided into biodegradable and non-biodegradable groups. Non-biodegradable bioplastics, notably Bio-PET, are the fastest-growing due to low cost, easy production processes, and recyclability (Joseph et al., 2023). Challenges include lack of transparency, dubious efficacy of many biopolymers impede their speedy adoption, certification and labeling in agriculture are problematic because of varied soil conditions. Other obstacles include producing eco-friendly packaging from PLA, PHA, and starch-based polymers remains tough owing to material limits; further research is needed (Vroman & Tighzert, 2009).
ii. **Biomedical Applications**: In the medical sector, biodegradable polymers are used due to its properties like non-toxicity, stabilizability, efficacy, and biocompatibility. There are two common applications in biomedical. First one uses medical components (blood bags, syringes, catheters) and the second one is surgical support (prosthetics, artificial organs). In surgery it is used in tissue adhesion and sealing where liquid biodegradable polymers form a gel to stop bleeding and dissolve as healing progresses. In bone fixation in the form of biodegradable pins, screws, and wires. In drug delivery it is used as pill coatings and controlled-release systems for precise dosing. Biomedical research develops artificial tissues, heart valves, and medical adhesives for enhanced physiological functions. In biomimetics: Nature-inspired materials for better integration in medical applications. Controlled drug release ensures precise dosing, reduces side effects, and improves patient compliance.

Table 1 Biopolymer production methods

Method	Description	Polymers Produced	Feedstocks/raw materials	Organisms/catalysts	Advantages/challenges	References
Microbial fermentation *(Biological)*	Bacteria synthesize polymers as intracellular granules under stress conditions	PHB, PHBV, mcl-PHAs	Glucose, molasses, starch, oils, agro-waste	*Ralstonia eutropha*, *Alcaligenes latus*, *Pseudomonas*	Renewable, low energy use, biodegradable/high production cost, solvent-based recovery	Chen and Jiang (2018)
Ring-opening polymerization (ROP) *(Chemical)*	Cyclic monomers opened to form high-molecular-weight polymers	PLA, PCL, PGA	Lactic acid (corn, sugarcane), glycolic acid	Tin (II) octanoate, organometallic catalysts	High purity, good control of properties/catalyst toxicity, high energy use	Polylactic acid technology-Drumright (2000)
Direct polycondensation *(Chemical)*	Condensation of monomers by removing water/alcohol	Low MW PLA, co-polymers	Lactic acid	Acid/base catalysts, heat, vacuum	Simple, cheaper than ROP/lower molecular weight polymers	Garlotta (2001)
Enzymatic polymerization *(Bio-catalyzed chemical)*	Enzymes catalyze polymer formation under mild conditions	PLA, PHAs, PBS	Lactic acid, hydroxyalkanoates	Lipases, esterases	Environmentally benign, no toxic catalysts/slower rates, high enzyme cost	Kobayashi et al. (2001)

Table 2 Processing Techniques of Biodegradable Polymers

Processing method	Mechanism	Typical Polymers	Application	Advantages	Limitations/ challenges	References
Extrusion	Polymer is melted and forced through a die into continuous shapes	PLA, PHA, starch blends	Films, straws, trays, compostable bags	Fast, continuous, low waste	Requires careful thermal control to avoid degradation	Gregor-Svetec (2022), Pelin et al. (2024), Tümer and Erbil (2021)
Injection molding	Molten polymer is injected into a metal mold and cooled	PLA, PCL, PHBV	Medical tools, food containers, electronics casings	Mass production, high precision	PLA may be brittle; mold cost is high	
Blow molding	Air pressure inflates molten polymer into a hollow mold shape	PLA, PLA blends	Bottles, jars, detergent containers	Lightweight, cost-efficient for hollow items	PLA needs modifiers to increase impact strength	
Electrospinning	Electric field draws nanofibers from polymer solution	PLA, PCL, PHBV	Wound dressings, drug delivery, filtration membranes	High surface area materials, biomedical use ara>	Requires volatile solvents, low throughput	
3D printing (FDM)	Heated filament extruded and deposited layer-by-layer	PLA	Prototypes, dental implants, custom parts	Design flexibility, material efficiency	Surface finish may be rough; limited to certain polymers	
Film casting	Polymer solution poured on surface and solvent evaporates	Starch blends, PLA	Packaging, laminates, coatings	Simple, lab-scale technique	Slow, not scalable for mass production	

(continued)

Table 2 (continued)

Processing method	Mechanism	Typical Polymers	Application	Advantages	Limitations/ challenges	References
Thermoforming	Heated sheet formed into shapes using vacuum or pressure	PLA, starch composites	Clamshells, trays, cups	Rapid, lightweight items	Limited to sheet-based products	
Compression molding	Polymer powder placed in mold and compressed with heat and pressure	PLA, starch blends, PCL	Plates, cutlery, casings	Suitable for thick or rigid parts	Energy-intensive, long cycle times	

iii. **Agriculture**

1. In agricultural practices, biodegradable materials are a promising alternative to conventional plastics; its principal applications include greenhouse covers, fumigation, and mulching and low-tunnel cultivation. Biopolymer films need specific properties to withstand aging and degradation. Starch-based films degrade naturally upon contact with soil microorganisms, making them an ideal choice for agricultural mulch films. Mater-Bi-based biodegradable films have been developed and tested, showing resistance to water and high temperatures. However, exposure to high doses of UV radiation negatively affects their elongation properties. These films help preserve moisture content, boost soil temperature, and inhibit weeds, thus increasing plant development.
2. In seed-sowing bands, which contain seeds and nutrients, guaranteeing uniform distribution.
3. In geotextiles, biopolymer-based fabrics are exploited for filtration, drainage, and geogrids.
4. In horticultural applications it includes threads, clips, staples, fertilizer bags, seed trays, and composting containers. For instance, composting containers used for plants can be directly seeded into the soil and decompose as the plant grows.
5. In marine agriculture, they are used in the production of ropes, fishing nets, and marine culture supports.
6. In addition, biopolymers are also utilized to regulate the release of agricultural agents including pesticides, fertilizers, and pheromones. These active compounds can be integrated into the polymer matrix or coating, resulting in a gradual and controlled release into the environment. Natural polymers such as starch, cellulose, chitin, alginic acid, and lignin are commonly employed in these systems (Gamage et al., 2022; Joseph et al., 2023). Other applications are in Table 3.

8 Technical and Practical Barriers of Biodegradable Polymers

i. **Composition Variation**: Biopolymers are obtained from a wide variety of biological sources hence there is a variation in their physical and chemical properties. This variability presents challenges for processes like recycling and reusability.
ii. **Polymer Degradation**: Biopolymers, engineered for composting, exhibit rapid degradation, which adversely impacts their recyclability. Mechanical recycling accelerates polymer chain breakdown, reducing usability. Repeated recycling leads to reduction in molecular weight and mechanical strength, (For e.g.: tensile

Table 3 Applications of various biodegradable polymers (Gamage et al., 2022; Joseph et al., 2023)

Blend	Sources	Properties	Field of application	Specific uses
Cellulose acetate	Wood pulp, cotton linters	Water insoluble, hydrolytically stable, Non-toxic,	Textile and personal care, healthcare	Textile fibers, cigarette filters, wound dressings, eyeglass frames, toothbrush handles
Polycaprolactone (PCL)	Petroleum-based synthesis	Semicrystalline, water insoluble, flexible and resistant to fatigue	Medical, industrial coatings	Sutures, contraceptive implants, biodegradable implants, metal protection films, paints
Polyhydroxyvalerate (PHV)	Microbial fermentation of plant-based sugars	Insoluble in water, soluble to organic compounds, flexible, tough and resistant to fatigue	Packaging, agriculture, horticulture	Compostable bags, agricultural films, horticultural applications, Rubbermaid products
Polylactic acid (PLA)	Corn starch, sugarcane, potato starch	Insoluble in water, soluble to organic compounds, Semicrystalline, Rigid, brittle and prone to cracking	Biomedical, Packaging, consumer electronics	Bone plates, surgical sutures, biodegradable films and foams, molded plastic parts, electronic components
Polyvinyl alcohol (PVA)	Corn, potato, or cassava starch	Good film-forming, strong conglutination, thermal stability, gas barrier properties	Medical, packaging	LDPE replacement, water-soluble laundry bags, biomedical field, loose-fill packaging, food packaging
Polyhydroxybutyrate (PHB)	Microbial fermentation of plant-based sugars	Insoluble in water, soluble to organic compounds, brittle, stiff and prone to cracking	Medical, disposable products, packaging	Biodegradable cups, containers, lids, compost bags, agricultural and horticultural applications, Rubbermaid

(continued)

Table 3 (continued)

Blend	Sources	Properties	Field of application	Specific uses
Polyalkylene alkanoate (PBS)	Sugarcane, corn, cassava	Hydrophobic, semicrystalline	Biomedical, packaging, agricultural, Textiles,	Biodegradable fibers, injection-molded products, fishing gear, biodegradable plant pots
Aliphatic-aromatic (co)polyester (PBAT)	Petroleum-derived with renewable content	Biodegradable, flexible, thermally stable, good mechanical strength, hydrophobic	Packaging, agriculture, composting	Biodegradable films, sheets, flower pots, plastic envelopes, compostable bags

strength of PHA). Potential strategies to reduce these impacts include combining with virgin materials or inserting additives (Schick et al., 2024).

iii. **Contamination**: Biopolymers frequently interact with food and organic substances. Contaminants disrupt the reprocessing process, resulting in a decrease in the quality of recycled material.

iv. **Economic Challenges**: Biopolymer faces economic challenges due to high production, collection, sorting, and processing costs. For instance, while conventional polymers cost around USD 1000–1500/MT, biopolymers production costs around USD 4000/MT for PLA and can reach as high as USD15000/MT for PHA, making them up to 7.5 times more expensive than conventional polymers. Efforts are ongoing to reduce costs, particularly in fermentation processes, where carbon sources and salts account for about 50% of expenses (Kakkalameli et al., 2022).

v. **Availability of Limited Renewable Sources**: Although biopolymers are abundant in nature, only a few plants and animals are employed for commercial manufacture. High carbohydrates feedstock like maize, sugarcane is called first generation. Waste product feedstock plants like bagasse, husks, sawdust is called second generation. High-yield carbon sources from algae are called third generation (Akinsemolu et al., 2024).

In brief, Biopolymer faces challenges related to variability, degradation, contamination, economic feasibility, and resource availability. Addressing these issues through improved techniques, cost reduction strategies, and sustainable raw material sourcing is essential for their wider adoption.

9 Bans on Conventional Plastics in Global Level

Globally there is an awareness regarding the use of conventional polymers and many nations have started the regulatory measures to control the use of conventional polymers and promote biodegradable alternatives:

- **China**: China has announced the five-year plan with the aim to ban and restrict the production, sale and use of SUP's. This initiative is led by the National Development and Reform Commission and the Ministry of Ecology and Environment. It has three phases, starting with key cities in 2020 and expanding nationwide by 2025. This policy limits the amount of plastic used in hotels and delivery services by outlawing the use of non-biodegradable plastics such as plastic bags, throwaway tableware, plastic straws, and express packaging. All hotels and home-stay establishments will have to abide by the regulations by 2025, and plastic packaging will be outlawed for express delivery services in large cities. Polyethylene agricultural mulch and ultra-thin plastic bags are also prohibited by those policies. China is the world's biggest purchaser of polyethylene; thus, the global plastics industry will be greatly impacted by its prohibition (Waste360, 2025)

- **India**: India banned the manufacture, import, sale, and use of identified SUPs items with high littering potential from July 1, 2022, under the Plastic Waste Management Amendment Rules, 2021. The ban includes plastic bags, straws, cutlery, food packaging, and disposable bottles. India encourages sustainability through the use of compostable bags in states like Maharashtra, Kerala, and Tamil Nadu, as well as biodegradable polymers under regulations like Uttar Pradesh's Bioplastic Industry Policy and CPCB recommendations. There are also many calls to reward the bioplastics industry in order to boost adoption and promote a circular economy (Nøklebye et al., 2023).
- Worldwide are establishing tough limits and rules to tackle plastic pollution. While California was the first state to ban single-use plastic bags (2014) and limit plastic straws (2018), the United States banned plastic microbeads in cosmetics (2015). Canada announced a statewide ban on harmful single-use plastics in 2021. The European Union restricted 10 throwaway plastic goods, including Styrofoam and straws (2018), and demanded that plastic bottles include 30% recycled content by 2030. Aiming for a total phase-off by 2025, the United Kingdom initially implemented a plastic bag levy (2015) then later barred plastic straws, cotton buds, and stirrers (2018). Plastic bags were forbidden in Australia and Mexico City under fines for non-compliance. Targeting a full prohibition, Thailand commenced phase-out of plastic bags in 2020 Rwanda and Kenya have some of the strictest laws, prohibiting the manufacture and use of plastic bags, with heavy fines and perhaps prison sentences. First nation to abolish lightweight plastic bags (2002), Bangladesh has invented biodegradable jute-based bags as a substitute. These projects represent a worldwide trend toward sustainability and circular economies (Zero Waste Vietnam, 2023).
- The European Union (EU) has taken a leading role in addressing plastic pollution through its various programs. The EU's Plastics Strategy, launched in 2018, aims to transform the design, production, use, and recycling of plastics. By 2030, all packaging should be reusable or recyclable, a goal further reinforced by the European Green Deal. The Single-Use Plastics Directive (SUPD), adopted in 2019, aims to reduce the environmental impact of plastic waste by banning certain single-use plastic items (e.g., cutlery, straws, and polystyrene food containers) while promoting the use of biodegradable and compostable alternatives. These focus on reducing marine litter and promoting a circular economy through measures such as market restrictions, recycled content mandates, and waste management requirements for certain plastic products. These regulations also apply to specific types of plastic packaging, like food and beverage containers. Ensuring a harmonized and proportionate implementation across EU Member States is essential to achieving environmental goals while maintaining the free movement of packaged goods (SUPD-EUROPEN, 2022).

10 Market Growth and Investment Trends in Biodegradable Polymer Industries

The biodegradable plastics business has undergone substantial expansion, driven by regulatory policies, consumer demand, and corporate sustainability initiatives. Global biodegradable plastic market size was valued at US$ 2878.1 million in 2023. With growing demand in downstream market, the Biodegradable Plastic is forecast to a readjusted size of US$ 5234.8 million by 2030 with a CAGR of 8.9% during review period (Global Biodegradable Plastic Market Growth, 2025). Key elements impacting this increase include:

- Rising bans on traditional plastics.
- Raising demand for eco-friendly alternatives.
- Advancements in biopolymer technology, leading to cost-effective manufacture.
- Increased financing from venture capitalists and government funds to encourage research and development (R&D) in sustainable materials.
- Expansion of bioplastics production by key businesses such as NatureWorks, BASF, and Novamont.

11 Role of Corporate Initiatives in Sustainability and Green Packaging

Though governments in global countries take several actions, participation of corporate is unavoidable in attaining the reach of biopolymers. Most of the corporate used plastics for packaging. It is necessary for corporates to move to green packaging. Green packaging is intimately tied to the circular economy, a production and consumption paradigm that provides suitable growth over time by recycling, reusing and minimizing the need for resources. Researchers have proven that green packaging is an important aspect in a circular production and consumption system as it supports diverse customer needs (Wandosell et al., 2021). Major corporates are taking proactive measures to integrate biodegradable plastics into their supply chains (Foodnavigator, 2020). For instance, Coca-Cola and PepsiCo are looking into biodegradable bottle alternatives to reduce their usage of plastic (Packaging | PepsiCo ESG | Environmental Impact, 2025) Nestlé and Unilever are providing funding for container options that decompose naturally, Amazon and McDonald's are now using paper-based, biodegradable packaging materials (Packaging Innovation—Amazon Sustainability, 2025).

12 Consumer Awareness and Acceptance of Biodegradable Alternatives

The success of biodegradable polymers is primarily dependent on societal acceptance and awareness. Though there is a slight success in the use of biodegradable polymers in applications, the general public should accept the use of biodegradable polymers in place of traditional polymers on a large scale. Some of the factors that influence public behaviors are.

- **Eco-consciousness**: Increasing awareness of plastic waste has pushed consumers to seek biodegradable items.
- **Affordability and Availability**: Widespread adoption rests on making biodegradable alternatives cost-competitive with ordinary plastics.
- **Education and Labeling**: Clear certification labels (e.g., "compostable" or "biodegradable") help consumers make informed purchasing decisions.
- **Government Campaigns**: Public initiatives and educational programs promote awareness about the benefits of biodegradable plastics.

Overall, the mix of legislative interventions, company accountability, and consumer-driven demand is propelling the change toward sustainable biodegradable plastics. The ongoing growth of this industry will depend on technology advances, governmental aid, and major public engagement in sustainable consumption patterns biodegradable plastics.

13 Roadmap for Governments to Promote Biodegradable Polymers

Short-Term (1–2 Years): Policy and Regulation

- Governments should enforce prohibitions on SUPs, implement laws like the EU Single-Use Plastics Directive, and develop standards for biodegradable materials (e.g., ASTM D6400, ISO 14855).
- Certification procedures must be devised to validate biodegradability, keeping with international norms.
- Tax incentives and subsidies should encourage producers to embrace biodegradable materials and fund R&D in biopolymers.

Mid-Term (3–5 Years): Infrastructure and Market Growth

- Establishing composting facilities and integrating biodegradable polymers into waste management is necessary.
- Public–private partnerships should foster innovation through funding and grants.
- Research should be encouraged and funds should be granted for the research carried out in academic and research education on biopolymers.

- Consumer awareness activities, such as eco-labeling and educational initiatives, will assist change consumer preferences toward biodegradable alternatives.

Long-Term (6–10 Years): Large-Scale Adoption

- Investments in sustainable options and local biopolymer industries whose manufacturing and distribution should be encouraged.
- Strict policies in corporations regarding packaging and it should be monitored for better results.
- Global partnership needs to be encouraged for policy alignment and global knowledge exchange.

14 Future Innovations in Biopolymers

Advancements in biopolymer technology focus on the following:

- Next generation biodegradable polymers should focus on improving properties like mechanical strength, thermal resistance, and faster degradation.
- Research should encourage areas like alternate renewable sources for biopolymer production in a reduced economy.
- Emerging processing technologies, including 3D printing with biopolymers and biodegradable additives like plasticizers, will enhance performance in flexibility and durability.
- Smart packaging innovations, such as active and edible biopolymers, will reduce waste. Expanding applications to biodegradable electronics, automotive, and aerospace industries will accelerate market growth.
- Circular economy initiatives will promote closed-loop recycling, composting, and global standardization for biodegradability certification, supporting a sustainable transition from fossil-based plastics. Roadmap to Promote Biodegradable Polymers are shown in Fig. 2.

15 Conclusion

Plastic pollution continues to pose a serious threat to ecosystems and human health. Then biodegradable polymers came as an alternative for conventional polymers that can still meet the demand of modern applications. In this chapter, we explored the environmental impacts of conventional polymer and growing interest in biodegradable solutions. We explored the types of biodegradable polymers both natural and synthetic and their degradation mechanism, production methods, and processing techniques. These materials are already finding success in applications like packaging, agriculture and medicine. Though they offer clear environmental benefits such as low persistence in the environment, and lower toxicity. They also have limitations and challenges like high production costs, rate of degradation, and raw material

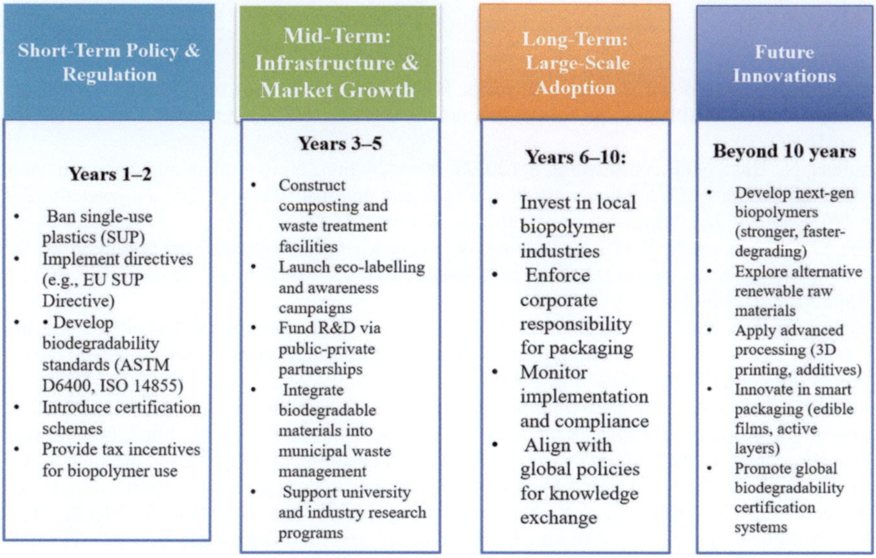

Fig. 2 Roadmap to promote biodegradable polymers

availability. Research is needed to find out answers to questions like land use, energy inputs, and greenhouse gas emissions to evaluate the true contribution rather than displace one with another.

Future of biodegradable polymers depends on innovations like bio-based feedstocks, nanotechnology enabled composites, and biodegradable electronic materials which are beyond current scope. Integration of these biodegradable polymers into a circular economy require multisectoral collaboration and investment in sustainable manufacturing. Policy frameworks and international laws are evolving to support the shift toward sustainable materials. Governments are progressively motivating the use of biodegradable substitutes from the European Union's Single-Use Plastics Directive to different national prohibitions on traditional plastics. Establishing dependable biodegradability criteria, so fostering market confidence and influencing consumer behavior, depends on standardization initiatives by ASTM, ISO, and other organizations as well as others. Corporate efforts and growing consumer knowledge are helping to reduce the gap between research discoveries and commercial acceptability.

In conclusion, the transition to biodegradable polymers is a cultural and systematic revolution as much as it is a technological one. It challenges us to rethink our relationship with materials generation, consumption and waste. Thoughtful pursuit of biodegradable polymers, balancing creativity, regulation, and education, can guide us to reach a more resilient, regenerative, and sustainable planet.

References

Akinsemolu, A. A., Idowu, A. M., & Onyeaka, H. N. (2024). Recycling technologies for biopolymers: Current challenges and future directions. *Polymers, 16*(19). https://doi.org/10.3390/polym16192770

Aquinas, N., Bhat, M. R., & Selvaraj, S. (2022). A review presenting production, characterization, and applications of biopolymer curdlan in food and pharmaceutical sectors. *Polymer Bulletin, 79*(9), 6905–6927. https://doi.org/10.1007/s00289-021-03860-1

Chen, G. Q., & Jiang, X. R. (2018). Engineering microorganisms for improving polyhydroxyalkanoate biosynthesis. *Current Opinion in Biotechnology, 53*, 20–25. https://doi.org/10.1016/j.copbio.2017.10.008

Dehghani, S., Moore, F., & Akhbarizadeh, R. (2017). Microplastic pollution in deposited urban dust, Tehran metropolis, Iran. *Environmental Science and Pollution Research, 24*(25), 20360–20371. https://doi.org/10.1007/s11356-017-9674-1

FoodNavigator. (2020). *'We are working to ensure packaging doesn't become waste': Coca-Cola, Nestlé, PepsiCo and Unilever hit back after damning plastics report*. https://www.foodnavigator.com/Article/2020/03/31/We-are-working-to-ensure-packaging-doesn-t-become-waste-Coca-Cola-Nestle-PepsiCo-and-Unilever-hit-back-after-damning-plastics-report

Gamage, A., Liyanapathiranage, A., Manamperi, A., Gunathilake, C., Mani, S., Merah, O., & Madhujith, T. (2022). Applications of starch biopolymers for a sustainable modern agriculture. *Sustainability, 14*(10), 6085. https://doi.org/10.3390/su14106085

Garlotta, D. (2001). A literature review of poly(lactic acid). *Journal of Polymers and the Environment, 9*(2), 63–84. https://doi.org/10.1023/A:1020200822435

Global Biodegradable Plastic Market Growth 2024–2030. (2025). *Market research*. https://www.marketresearch.com/LP-Information-Inc-v4134/Global-Biodegradable-Plastic-Growth-39159415/

Gregor-Svetec, D. (2022). Polymers in printing filaments. In *Polymers for 3D printing: Methods, properties, and characteristics* (pp. 155–269). https://doi.org/10.1016/B978-0-12-818311-3.00002-1

Hossain, R., Islam, M. T., Shanker, R., Khan, D., Locock, K. E. S., Ghose, A., Schandl, H., Dhodapkar, R., & Sahajwalla, V. (2022). Plastic waste management in India: Challenges, opportunities, and roadmap for circular economy. *Sustainability, 14*(8). https://doi.org/10.3390/su14084425

Huang, Y. Q., Wong, C. K. C., Zheng, J. S., Bouwman, H., Barra, R., Wahlström, B., Neretin, L., & Wong, M. H. (2012). Bisphenol A (BPA) in China: A review of sources, environmental levels, and potential human health impacts. *Environment International, 42*(1), 91–99. https://doi.org/10.1016/j.envint.2011.04.010

Isangedighi, I., David, G., & Obot, O. I. (2020). Plastic waste in the aquatic environment: Impacts and management. In *Analysis of nanoplastics and microplastics in food*. https://www.taylorfrancis.com/chapters/edit/https://doi.org/10.1201/9780429469596-2/plastic-waste-aquatic-environment-isangedighi-asuquo-isangedighi-gift-samuel-david-ofonmbuk-ime-obot

Joseph, T. M., Unni, A. B., Joshy, K. S., Kar Mahapatra, D., Haponiuk, J., & Thomas, S. (2023). Emerging bio-based polymers from lab to market: Current strategies, market dynamics and research trends. *C—Journal of Carbon Research, 9*(1). https://doi.org/10.3390/c9010030

Kakkalameli, S., Daphedar, A. B., Faniband, B., Sharma, S., Nadda, A. K., Ferreira, L. F. R., Bilal, M., Américo-Pinheiro, J. H. P., & Mulla, S. I. (2022). Biopolymers and environment (pp. 19–33). https://doi.org/10.1007/978-3-030-98392-5_2

Kobayashi, S., Uyama, H., & Kimura, S. (2001). Enzymatic polymerization. *Chemical Reviews, 101*(12), 3793–3818. https://doi.org/10.1021/cr990121l

Kosior, E. (2020). Solutions to the plastic waste problem on land and in the oceans. In *Plastics waste and recycling*. Elsevier. https://www.sciencedirect.com/science/article/pii/B9780128178805000165

Mangal, M., Rao, C. V., & Banerjee, T. (2023). Bioplastic: An eco-friendly alternative to non-biodegradable plastic. *Polymer International, 72*(11), 984–996. https://doi.org/10.1002/pi.6555

Mehrpouya, M., Vahabi, H., Barletta, M., Laheurte, P., & Langlois, V. (2021). Additive manufacturing of polyhydroxyalkanoates (PHAs) biopolymers: Materials, printing techniques, and applications. *Materials Science and Engineering: C, 127*, Article 112216. https://doi.org/10.1016/j.msec.2021.112216

Narayan, R., Arcovito, A., Dahiya, D., & Singh Nigam, P. (2023). Dextran of diverse molecular-configurations used as a blood-plasma substitute, drug-delivery vehicle and food additive biosynthesized by Leuconostoc, Lactobacillus and Weissella. *Applied Sciences, 13*(22), 12526. https://doi.org/10.3390/app132212526

Navya, P. V., Gayathri, V., Samanta, D., & Sampath, S. (2022). Bacterial cellulose: A promising biopolymer with interesting properties and applications. *International Journal of Biological Macromolecules, 220*, 435–461. https://doi.org/10.1016/j.ijbiomac.2022.08.056

Niknezhad, S. V., Asadollahi, M. A., Zamani, A., & Biria, D. (2016). Production of xanthan gum by free and immobilized cells of Xanthomonas campestris and Xanthomonas pelargonii. *International Journal of Biological Macromolecules, 82*, 751–756. https://doi.org/10.1016/j.ijbiomac.2015.10.065

Nøklebye, E., Adam, H. N., Roy-Basu, A., Bharat, G. K., & Steindal, E. H. (2023). Plastic bans in India—Addressing the socio-economic and environmental complexities. *Environmental Science & Policy, 139*, 219–227. https://doi.org/10.1016/j.envsci.2022.11.005

Okeke, E. S., Okoye, C. O., Atakpa, E. O., Ita, R. E., Nyaruaba, R., Mgbechidinma, C. L., & Akan, O. D. (2022). Microplastics in agroecosystems—Impacts on ecosystem functions and food chain. *Resources, Conservation and Recycling, 177*, Article 105961. https://doi.org/10.1016/j.resconrec.2021.105961

Packaging innovation-Amazon Sustainability. (2025). *Amazon*. https://sustainability.aboutamazon.com/waste/packaging

Packaging| PepsiCo ESG | Environmental impact. (2025). *PepsiCo*. https://www.pepsico.com/our-impact/esg-topics-a-z/packaging

Padgelwar, S., Nandan, A., & Mishra, A. K. (2021). Plastic waste management and current scenario in India: A review. *International Journal of Environmental Analytical Chemistry, 101*(13), 1894–1906. https://doi.org/10.1080/03067319.2019.1686496

Pelin, G., Sonmez, M., & Pelin, C. E. (2024). The use of additive manufacturing techniques in the development of polymeric molds: A review. *Polymers, 16*(8), 1055. https://doi.org/10.3390/polym16081055

Polylactic acid technology-Drumright. (2000). *Advanced Materials*. https://advanced.onlinelibrary.wiley.com/doi/https://doi.org/10.1002/1521-4095(200012)12:23<1841::AID-ADMA1841>3.0.CO;2-E

Richard, C., Dejoie, E., Wiegand, C., Gouesbet, G., Colinet, H., Balzani, P., & Siaussat, D. (2024). Plastic pollution in terrestrial ecosystems: Current knowledge on impacts of micro and nano fragments on invertebrates. *Journal of Hazardous Materials*. https://www.sciencedirect.com/science/article/pii/S0304389424018788

Samir, A., Ashour, F. H., Hakim, A. A. A., & Bassyouni, M. (2022). Recent advances in biodegradable polymers for sustainable applications. *NPJ Materials Degradation, 6*(1), 1–28. https://doi.org/10.1038/s41529-022-00277-7

Schick, S., Heindel, J., Groten, R., & Seide, G. H. (2024). Overcoming challenges in the commercialization of biopolymers: From research to applications—A review. *Polymers, 16*(24). https://doi.org/10.3390/polym16243498

Single Use Plastics Directive-EUROPEN. (2022). *EUROPEN*. https://www.europen-packaging.eu/policy-area/single-use-plastics-directive/

Tümer, E. H., & Erbil, H. Y. (2021). Extrusion-based 3D printing applications of PLA composites: A review. *Coatings, 11*(4), 390. https://doi.org/10.3390/coatings11040390

Vroman, I., & Tighzert, L. (2009). Biodegradable polymers. *Materials, 2*(2), 307–344. https://doi.org/10.3390/ma2020307

Wandosell, G., Parra-Meroño, M. C., Alcayde, A., & Baños, R. (2021). Green packaging from consumer and business perspectives. *Sustainability, 13*(3), 1356. https://doi.org/10.3390/su13031356

Waste360. (2025). *China unveils five-year plan to ban single-use plastics.* https://www.waste360.com/waste-legislation/china-unveils-five-year-plan-to-ban-single-use-plastics

Zero Waste Vietnam. (2023). *11 most impressive plastic bans around the world.* https://zerowastevietnam.org/11-most-impressive-plastic-bans-around-the-world/

Harnessing Organic Waste for Biodegradable Polymer Production: A Green Alternative to Conventional Plastics

Aisha Halliru Duwan, Achsah Rajendran Startha Christabel, Eyerusalem Goitom Kidane, and Mwalimu Raphael Jilani

1 Introduction

Global synthetic polymer production reaches approximately 390 million tons annually, presenting significant environmental, sustainable development, and health challenges (Muneer et al., 2021). Over 3.59 million tons of plastic were produced worldwide in 2018 and is projected to increase. By 2050, an estimated 26 billion tons of plastic waste will be generated, with significant environmental disposal concerns (Moshood et al., 2022). While the durability of these polymers is advantageous for many uses, their long-term environmental persistence is problematic, with PET bottles lasting between 27 and 93 years under certain conditions (Haider et al., 2019). Additionally, micro and nanoplastics from plastic degradation adversely affect wildlife and human health. These particles are present in food, drinking water, and human blood, posing significant health risks. Additionally, tainted animal intake may allow them to infiltrate the human meals (González-Rojo et al., 2024). There are several methods for managing plastic garbage, such as landfilling, incineration, and recycling. But recycling entails costly processes, while incineration emits toxic gases such as methane, and other methods have limited effectiveness without enhanced public education on waste management (Reddy Koteswara et al., 2025; Shen et al., 2020; Swain et al., 2025).

The pursuit of sustainable alternatives to petroleum-based polymers has intensified research efforts. Consequently, significant advancements are being made to identify bio-based substitutes. As a result, the manufacturing of biodegradable polymers is a growing field (Acquavia et al., 2021). Biodegradable polymers are designed to

A. H. Duwan (✉) · A. R. S. Christabel · E. G. Kidane · M. R. Jilani
Department of Biotechnology, Vel Tech Rangarajan Dr. Sagunthala R&D Institute of Science and Technology, Avadi, Chennai, India
e-mail: ayshasduwan@gmail.com

S. Ghosh et al. (eds.), *Advances in Organic Waste Conversion Through Industrial Biotechnology*, Sustainable Environmental Waste Management Strategies,
https://doi.org/10.1007/978-3-032-11377-1_7

decompose into harmless products after a specified duration. However, many current biodegradable polymers are derived from food crops, creating competition for essential resources. Utilizing organic waste for biodegradable plastics presents a viable solution that minimizes land use impacts and promotes waste valorization, offering an economically and environmentally favorable alternative (Filiciotto & Rothenberg, 2021).

Organic waste from agricultural, aquatic, forestry, and animal sources are great options for valorization. United Nations Environmental Program made an estimate of annual biomass waste production at 140 BMT from agriculture (Lizundia et al., 2022). Biomass, with an estimated 1.1 teratons on a dry-weight basis, offers unlimited resources. The FAO and UN reported in 2011 that one out of three global food resources is wasted. In Amsterdam, a study indicates that households are the primary source of organic waste, surpassing food industries and services combined. EU citizens generate about 70 kg of waste per capita. Transforming waste into plastics adds product value from negligible resources, presenting economic and social appeal (Filiciotto & Rothenberg, 2021). This process yields environmental advantages, including biodegradation, the regeneration of nature's resources, and diminished CO_2 emissions contributing to climate change (Samir et al., 2022).

Industrial biotechnology is crucial in the advancement of bioplastics, fostering innovation in sustainable packaging. It enables in production of biodegradable and bio-based plastics via metabolic engineering and fermentation to create polymers like "polylactic acid (PLA) and polyhydroxyalkanoates (PHA)." Table 1 represents lifecycle comparison between conventional and biodegradable polymers from organic waste. These products naturally decompose, alleviating landfill pressures and reducing micro plastic pollution (Mitchell, 2024). Moreover, microorganisms can enhance the treatment of plastic waste, offering promising applications in mitigating plastic accumulation. Under optimal conditions, plastics can be decomposed into biomass and gas. Nonetheless, challenges exist in replacing traditional polymers with these alternatives (Zhu & Wang, 2020). This chapter analyzes organic waste's potential as a sustainable option for biodegradable polymer production. We explore the types and sources of biodegradable polymers, methods of converting organic waste into polymers, role of industrial biotechnology in polymer synthesis and finally the challenges faced by green technology (Mohanty et al., 2024; Ray et al., 2024).

2 Overview of Biodegradable Polymers

The extensive usage of synthetic plastics has resulted in a global environmental catastrophe, primarily due to their high resistance and capacity to resist natural degradation processes. Morethan 9.5 billion metric tons of plastic have been manufactured since the 1950s, with much of it finding its way into landfills or leaching into terrestrial and aquatic environments (Das et al., 2024a, 2024b, 2024c; Dey et al., 2023a, 2023b). These plastics, including polyethylene, polypropylene, polystyrene,

Table 1 Lifecycle comparison between conventional and biodegradable polymers from organic waste

Feature	Conventional plastics	Biodegradable polymers from organic waste
Source	Fossil fuels	Organic/agricultural waste
degradation time	Hundreds of years	Weeks to months (under suitable conditions)
Environmental impact	High	Low
Carbon footprint	High	Low (if industrial biotechnology is used)
Circular economy compliance	Poor	Strong

and polyethylene terephthalate, possess hydrophobic polymer backbones and high molecular weight, which render them resistant to microbial degradation and results in long-term environmental pollution (Kim et al., 2023).

As a reaction to environmental problems, biodegradable polymers have become a viable choice. In contrast to conventional plastics, BDPs are designed to break down fully biologically into harmless by-products, e.g., water, carbon dioxide, methane, and biomass, when anaerobic conditions are met. Under ideal environmental conditions, microorganisms like bacteria, fungi, and algae primarily catalyze this process (Kim et al., 2023). It must be made clear that biodegradability is a functional property and not necessarily linked to the origin of the polymer. Thus, BDPs are produced using nonrenewable or renewable resources. Polycaprolactone or PCL does contrast, being biodegradable like a fossil-based polymer with bio-based polyethylene or bio-PE which it is not (Luyt & Gasmi, 2016).

BDPs are conceptualized plus developed for alignment to the goals of sustainability with the circular economy. They reduce plastic pollution also they reduce greenhouse gas emissions. Cellulose derivatives, starch-based polymers, polylactic acid (PLA), and polyhydroxyalkanoates (PHAs) offer prominent cases (Carissimi et al., 2018; Lukasiewicz et al., 2018). These biopolymers minimize the use of fossil fuels so they potentially can be made from renewable feedstock such as crops and microorganisms while providing a biodegradable alternative.

Due to the fact that people are more interested in environmentally friendly products, researchers have been able to develop biodegradable polymers. They have also increased their research about them. Initial uses for these polymers were agricultural mulch films and disposable packaging, but new advancements have expanded their applications to consumer electronics, drug delivery systems, biomedical devices, and also textiles (Mahata et al., 2021). Diverse and international standards, including those of ASTM and ISO, now define biodegradability in terms of performance requirements such as CO_2 evolution and disintegration rate during composting (Kim et al., 2023).

However, large-scale use of biodegradable polymers is faced with several challenges. Their very high cost of manufacture, unpredictable performance in wet or humid conditions, incompatibility with existing production systems, as well as the absence of waste treatment or composting plants, render them uneconomic. Additionally, the use of food-based feedstock raises ethical and environmental concerns with respect to resource utilization (Kim et al., 2023).

One also needs to draw a basic distinction between degradation and biodegradation. Physical degradation through UV radiation, oxidation, or abrasion can break plastics into fragments (Selke et al., 2015), but that is not the same as biodegradation, which involves microbial assimilation and mineralization of polymer fragments into ecologically inert end-products.

2.1 Biodegradable Polymers Classification

BDPs are usually classified according to their sources, synthesis routes, and mode of biodegradation. They are classified into three broad classes as Natural Biopolymers, Bio-Based Synthetic Polymers, and Petrochemical-Based Biodegradable Polymers (Sudesh & Iwata, 2008). Each of these polymers possesses unique structural features and production methods that serve as the basis for their wide range applications in packaging, agriculture, and biomedicine, etc. (Colnik et al., 2020).

1. **Natural Biopolymers**

The source of natural biopolymers is a sustainable and renewable naturally occurring feedstock, i.e., plants, animals, and microorganisms. The category comprises polymers like gelatin, casein and soy protein, and polysaccharides like starch, cellulose, and chitosan (Lambert & Wagner, 2017) such as polyhydroxyalkanoates (PHAs). The materials are intrinsically biodegradable since they include hydrolysable or enzyme-cleavable bonds.

Starch, the most common, inexpensive biopolymer, is generally employed in the form of thermoplastic starch (TPS) following plasticization with plasticizers such as glycerol or sorbitol (Lescher et al., 2008). Although of utility in packaging and mulch films, native starch is water-sensitive and brittle, generally requiring compounding with more stable polyesters such as PLA or PCL to improve performance (Mendes et al., 2015).

Cellulose, one of the most prevalent organic polymer on the planet (Shen et al., 2009), is used extensively in esterified or derivative forms, for example cellulose acetate and nanocellulose, which provide higher robustness and waterproofing (Khalil et al., 2011). These derivatives are used in textiles, films, and biodegradable composite materials.

PHA, such as PHB, PHV, and PHH, is bacterial polyesters produced by bacteria when they are under nutrient deprivation. They are fully biodegradable in diverse environments and are thermoplastics with variable mechanical properties. Their production using waste materials is in keeping with the idea of the circular economy (Colnik et al., 2020).

2. **Bio-based Synthetic Polymers**

Biopolymers produced from biological sources are chemically produced from monomers of renewable resources. Among these, the most significant one commercially used is polylactic acid (PLA) (Avérous, 2008). It is manufactured from lactic acid by fermenting starch-rich crops such as corn or sugarcane (Tsuji & Ikada, 1995). PLA has desirable mechanical properties, biocompatibility, and transparency. It, however, needs industrial composting environments to undergo suitable biodegradation (Lim et al., 2008).

Some other important bio-based synthetic polymers include polybutylene succinate (PBS) (Chen 2009) and its copolymer PBSA, (Seggiani et al., 2019) which are produced from succinic acid and 1,4-butanediol. These polyesters are also very elastic and can be degraded in compost and soil. Castor oil-derived Polyamide and polyethylene furanoate (PEF) from FDCA (Kreyenschulte et al., 2012) are also good substitutes with varying biodegradability.

3. **Petrochemical-Based Biodegradable Polymers**

These are indeed biodegradable products that are designed so as to break down via hydrolytic or microbial degradation, which are derived from fossil-based monomers. Polybutylene succinate (PBS) joins poly ε-caprolactone (PCL) plus polybutylene adipate terephthalate (PBAT) (Kim et al., 2023). The Classification and properties of common Biodegradable polymers represented in Table 2. The flexible PCL remains a low-melting-point semi-crystalline polyester (Armentano et al., 2010), and is thus well adapted for biomedical applications like resorbable sutures and in administering drugs, considering it degrades slowly in physiological conditions (Luyt & Gasmi, 2016). PBAT is an "aliphatic-aromatic copolyester" with the flexibility of polyethylene and biodegradability (Witt et al., 1997).

It is normally blended together with starch or PLA so as to provide good mechanical properties in agricultural films as well as compost bags (Dey et al., 2023a, 2023b; Mahata et al., 2021). A sub-class involves polyolefins which are oxo-biodegradable. These polyolefins do contain pro-degradant additives which cause oxidative fragmentation when they are heated or are exposed to UV (Selke et al., 2015). Microplastics still generate although they measure degradation and the environment persists (Schiavo et al., 2020).

2.2 Properties and Mechanisms of Biodegradation

The tough process of biodegradation is dependent on both the environment and the inherent properties of the polymer. Polymers formulated for biodegradation are formulated to facilitate hydrolytic and enzymatic degradation of their molecular chain (Auras et al., 2004), resulting in breakdown into water, carbon dioxide, methane in

Table 2 Classification and properties of common biodegradable polymers

Polymer type	Source	Biodegradation environment	Key properties	Applications	References
Starch-based polymers	Wheat, rice, potato, corn	Soil and compost biodegradation	Abundant, economical and Hydrophilic	Food packaging and coatings	Carissimi et al. (2018), Xie et al. (2018), Medina-Jaramillo et al. (2017)
Cellulose-based polymers	Sugarcane bagasse, wood pulp	Industrial degradation, soil and compost	Biodegradable, Strong structure	Cigarette filters, green packaging, paper, high value plastic	(Missoum et al., 2014), Sirviö et al. (2018), Joly and Coulis (2017), Shaikh et al. (2008)
Soy protein-based polymers	Soy flour	Controlled environmental degradation, soil	Flexible, good adhesion	Food packaging, textile processing	Yang et al. (2016), Jiménez-Rosado et al. (2018), Yu et al. (2017)
Polyhydroxyalkanoates (PHAs)	Bacterial synthesis	Marine, soil, and compost degradation	Flexible, thermal resistant, biocompatible	Food packaging, biomedical implants	Lukasiewicz et al. (2018), Wu et al. (2017)
Polyhydroxybutyrate (PHB)	Microbial Synthesis	Soil and compost degradation	Brittle, Biocompatible, high melting point	Packaging and coatings, scaffolds	Khardenavis et al. (2007), Da Silva Moura et al. (2018)
Polybutylene succinate (PBS)	Derived from petroleum based sources	Soil, compost and marine degradation	Flexible, good mechanical strength	Agricultural films, fibers	Li et al. (2021), Rafiqah et al. (2021), Gan et al. (2001)
Poly ε-caprolactone (PCL)	Synthetic polymer	Soil and marine degradation	Biodegradable, highly elastic	Sutures, drug delivery	Armentano et al. (2010), Luyt and Gasmi (2016), Averous (2000)

(continued)

Table 2 (continued)

Polymer type	Source	Biodegradation environment	Key properties	Applications	References
Polylactic acid (PLA)	Fermentation of sugars	Industrial composting	High mechanical strength, thermoplastic, transparent	Packagings, biomedical applications, textile	Avérous (2008), Tsuji and Ikada (1995), Lim et al. (2008)
Polyglycolic acid (PGA)	syntheticppolymer	Soil, compost, and biomedical degradation	Biocompatible	Medical sutures, drug delivery	Valderrama et al. (2020), Khiste et al., (2013), Pillai and Sharma (2010)

anaerobic conditions, and biomass by microbial metabolic processes (Lambert & Wagner, 2017). A detailed insight into the factors and processes involved is required to create material to acquire effective degradation under specific environmental conditions.

Biodegradation generally occurs through abiotic and biotic reactions together. Abiotic reactions are the first step that entail hydrolysis, photodegradation, and oxidation (Tantawi et al., 2025), employed to depolymerize the polymer chains and minimize molecular weight, thereby increasing accessibility to microbial enzymes (Bher et al., 2022). Hydrolytic degradation is the most common in aliphatic polyesters such as PLA, PCL, PHAs, and PBAT because of the availability of hydrolysable ester linkages (Silva et al., 2023).

The enzymatic degradative process follows, where microorganisms produce extracellular enzymes example, esterases, lipases, cutinases, cellulases, and amylases that hydrolyze polymers to oligomers and monomers (Jenkins & Harrison et al., 2008). Microbial organisms take up these lower units and mineralize them into environmentally inert end-products. Microbial diversity, enzyme specificity, and environmental parameters like pH and temperature are crucial factors deciding the effectiveness of microbial degradation (Auras et al., 2004).

In others, oxidative degradation may precede and/or follow hydrolysis. The latter is most applicable to oxo-biodegradable materials and chemically modified polyolefins, where chain scission is brought about by UV light or chemical oxidants, resulting in oxidized fragments that are then bio-assimilated (Selke et al., 2015).

2.3 Comparison with Conventional Plastics

The biodegradable polymers versus conventional petroleum plastics comparison involves important considerations like environmental impact, mechanical performance, feasibility in processing, end-of-life disposal consequences, and economic factors (Kim et al., 2023). In spite of the ubiquitous plastics, for instance, the polyethylene (PE), polypropylene (PP), and polyethylene terephthalate (PET), being described by strength, low cost, and excellent mechanical and thermal properties, their resistance to biodegradation results in long-term environmental contamination and negative environmental consequences (Wang et al., 2021). The materials remain in the environment for a long period of time, ranging from decades to centuries, resulting in landfill, micro plastic contamination, and water toxicity ("Reckoning With the U.S. Role in Global Ocean Plastic Waste," 2021).

On the contrary, biodegradable polymers, particularly those based on renewable resources, have considerable environmental advantages in the form of reduced greenhouse gas emissions, reduced fossil carbon dependence, and enhanced compatibility with composting and anaerobic digestion (Kim et al., 2023). Life cycle analysis reveals bioplastics to be lower in energy demands and carbon footprints, but these are offset by land use of cropland, emissions of methane under anaerobic conditions, and the lack of adequate disposal infrastructure (Li et al., 2022).

Technically, the biodegradable polymers are behind their synthetic counterparts. PLA is brittle and thermosensitive, starch plastics are water-sensitive (Lim et al., 2008), but some materials such as PBAT and PBS are near the flexibility of LDPE, and technological developments in blending, copolymerization, and nanocomposites are bridging the performance gap (Silva et al., 2023).

From an economic point of view, biodegradable plastics remain 2–4 times more costly than traditional plastics, mainly because of high feedstock prices, intricate processing, and low output capacities. However, available technologies paired with policies like carbon pricing and green procurement can shift the economics in favor of sustainable options. Conventional plastics are effective in recycling facilities, whereas biodegradable polymers have potential in organic recycling; therefore, the role of integrated waste management and consumer education cannot be emphasized enough (Silva et al., 2023).

2.4 *Industrial and Commercial Applications*

Biodegradable polymers have become more popular in recent years due to their environmentally friendly biodegradation and flexibility in synthesis. Although they occupy less than 1% of the global plastic market, their usage particularly in packaging, agriculture, biomedicine, textiles, and consumer goods is expanding at a rapid pace (Kim et al., 2023).

Packaging is the biggest and most developed market with more than 40% of global bioplastics application. PLA, PBAT, PBS, and starch blends are applied to films, containers, and bags for composting (Li et al., 2021). Good processability and sufficient performance in industrial composting explain their applicability for short-time use and compostable waste streams. The mechanical property enhancement and barrier applications are facilitated through the use of nanocomposites, such as cellulose nanocrystals (Khalil et al., 2011).

In farming, biodegradable mulch films, pots, and coated slow-release fertilizers minimize post-use disposal and improve soil quality. PBAT, PBS, and starch films are the best option since they can degrade in the soil and also be capable of minimizing labor and plastic contamination (Mahata et al., 2021).

Biomedical applications include the application of polymers like polylactic acid, PCL, and polyhydroxyalkanoates in drug delivery systems, scaffold sutures, etc. (Armentano et al., 2010). Controlled degradation and biocompatibility characteristics make them perfect in minimizing secondary surgery and tissue regeneration to a minimum. Enhancements like hydroxyapatite-reinforced PLA scaffolds increase bioactivity and mechanical strength (Lim et al., 2008).

PLA, PCL, and PHA in consumer products and textiles find application in hygiene products, textiles, and 3D printing material (Armentano et al., 2010; Avérous, 2008). They reduce micro plastic pollution as they are biodegradable. They are researched for enhanced durability and sustainability through processing and blending with natural fibers enhancement.

New uses involve electronics, where polymers based on cellulose are used as biodegradable substrates and insulators in transient electronics (Sirviö et al., 2018). Specialty uses also involve compostable food packaging coating with both water retention and biodegradability. Increased environmental regulations and sustainability objectives continue to fuel industrial use of BDPs (Da Silva Moura et al., 2018).

3 Organic Waste as Feedstock for Biodegradable Polymer Production

BDPs like polyhydroxyalkanoates are being recognized as possible substitutes since they can be biodegraded in a natural environment, are biocompatible, and have good mechanical characteristics. Their market value is, however, significantly depreciated by high production costs, mainly because of the use of refined feedstock (Zhang et al., 2024).

One of the most effective methods of minimizing such costs is the valorization of organic waste, which is renewable, plentiful, and often wasted at a cost. Organic waste of agricultural residues (Asim et al., 2025), food waste, industrial by-products, and municipal solid waste comprises rich carbohydrates, lipids, and proteins (Valentino et al., 2019), which can be harnessed for microbial fermentation and downstream polymers production. The process is consistent with the idea of circular economy, which prefers environmental sustainability and economic feasibility through converting of waste to valuable biodegradable materials.

3.1 Types and Sources of Organic Waste

A. **Agricultural Residues**

Agricultural by-products are residues of agricultural and processing operations for example wheat straw, corn stover, sugarcane bagasse, rice husks, cassava peels, banana peels, pineapple waste, and jackfruit seeds (Asim et al., 2025). They are made up mostly of lignocellulosic biomass, which includes the structural polymers of "plant cell walls" cellulose, hemicellulose, and lignin. Despite their abundance, the inherent recalcitrance of lignocellulose requires pretreatment to release fermentable sugars (Asim et al., 2025).

Rice husks (RH), for instance, contain about 52% carbohydrates and 26% lignin. Under high pressure and temperature with 1.0 mol/L KOH pretreatment and subsequent enzymatic hydrolysis, sugar yield can be as high as 87%. The hydrolysate has also been shown to be a good substrate for polyhydroxyalkanoate production by *Burkholderia cepacia* USM with a yield as high as 50% cell dry weight in a 5 L fermenter (Heng et al., 2016). Cassava peel containing amylopectin and cellulose has also been hydrolyzed and fermented to lactic acid, ready for downstream PLA synthesis (Berghuis et al., 2022; Zhang et al., 2024).

B. **Food Waste**

Food waste contributes the largest portion of the world's organic wastage. This comprises expired food in retail outlets, kitchen rubbish, vegetable, and fruit skins, and residues during food processing. These wastes are particularly valuable because they contain high water and nutrient content that aids bacteria to decompose and ferment them (Wang et al., 2022).

Food waste acidogenic fermentation generates a blend of volatile fatty acids, example: Butyric Acid, Acetic acid, Valeric Acid, Propionic Acid, and Caproic Acid (Wang et al., 2022). *Bacillus megaterium* can be used to convert the Acids to polyhydroxyalkanoates with yields of 9–10% dry cell weight (Vu et al., 2021). Food waste is utilized in the polylactic acid production through lactic acid fermentation. Bakery waste has been successfully fermented by *Lactobacillus species* to produce high levels of lactic acid.

Moreover, mango peel waste yielded 0.88 g/g reducing sugar from lactic acid through enzymatic hydrolysis and *Lactobacillus casei* fermentation. The lactic acid used was then manufactured as PLA, whose properties are similar to petroleum plastics (Rajendran & Han, 2022).

C. **Industrial By-Products**

The manufacturing process generates huge quantities of organic by-products that can be recycled in the production of biopolymers. Pre-eminent among these are milk processing whey, sugar refining molasses, and vinasse from distilleries of alcohol. Table 3 represents the common organic waste types and their biopolymer potential. Whey, a lactose-rich byproduct of cheese manufacture, can be used to sustain fermentation to lactic acid and propionic acid. *Propionibacterium spp.* can be employed to ferment whey to VFAs (Wang et al., 2022), which are fermented to PHBV by engineered microorganisms such as *Cupriavidus necator.* Pilot-scale fermentations with VFAs derived from whey have yielded up to 50% cell dry weight of PHA.

Molasses, mostly sucrose, is hydrolyzed to glucose and fructose as substrates for microbial fermentation into PHAs and PLA. In addition to lowering feedstock prices, using these by-products improves the sustainability of the dairy and sugar sectors.

D. **Municipal Organic Waste**

Municipal solid waste has a high percentage of organic food made up of food residues, garden trimmings, and paper waste. OFMSW is rich in nutrients and is highly biodegradable and can be used for the process of bioconversion (Valentino et al., 2019; Wang et al., 2022). OFMSW anaerobic fermentation produces a rich mixture of VFAs that can then be used by mixed microbial cultures or halophilic microorganisms like *Haloferax mediterranei* for PHA production (Wang & Zhang, 2020).

Table 3 Common organic waste types and their biopolymer potential

Waste type		Key components	Potential biopolymers	Applications	References
Agricultural food waste	Cassava peel	Cellulose, hemicellulose, lignin	Polybutylene succinate (PBS) Polylactic acid (PLA)	Food packaging, textile industry, agricultural mulches, and films, composites and biodegradable plastics	Berghuis et al. (2022), Zhang et al. (2024)
	Banana peel	Lignin, hemicellulose, Cellulose	Polycaprolactone (PCL)		Irmayanti and Anwar (2024), Asim et al. (2025)
	Avocado seeds	Amylopectin, cellulose	Polyhydroxyalkanoates (PHAs)		Asim et al. (2025)
	Chicken feathers	Keratin			Murugan et al. (2021), Tesfaye et al. (2017)
	Jackfruit seed	Starch, cellulose, amylopectin			Lubis et al. (2018)
Municipal food waste	Short chain carboxylic acids	Lactic acid, acetic acid, propionic acid, butyric acid	Polyhydroxyalkanoates (PHAs) Poly(3-hydroxybutyrate-co3-hydroxyvalerate) (PHBV) Poly(3-hydroxyvalerate) (PHV)	Nutritional supplements, biofuels, biodegradable plastics, medical materials	Wang et al. (2022), Valentino et al. (2019)
	Micronutrients	Potassium, calcium, magnesium, sodim, zinc, copper, iron			

Valentino et al. (2019) proposed a three-step procedure for OFMSW valorization using anaerobic fermentation for VFA production, microbial enrichment on feast-famine regimes, and aerobic PHA accumulation. Their pilot-scale operation recorded PHA content of as much as 55% cell dry weight.

3.2 *Chemical Composition and Suitability*

The organic waste's availability as a substrate for the synthesis of biodegradable polymers depends on its biochemical composition. Crop wastes are rich in lignocellulosic wastes containing cellulose (20–50%), hemicellulose (15–25%), and lignin (5–30%). These need to be pretreated to bring out the fermentable sugars, glucose, and xylose, as the prime substrates for lactic acid fermentation for the production of PLA (Asim et al., 2025).

Conversely, industrial and food wastes always contain readily available sugars, proteins, and lipids, and thus there is less need for massive pretreatment. Acidogenic fermentation of food waste always produces a mixture of volatile fatty acids (VFAs) (Valentino et al., 2019) that comprise butyric (38%), acetic (21.5%), propionic (12.7%), valeric (11.6%), and caproic (10%) acids. The metabolization of Volatile fatty acids by microbial species such as *Haloferax mediterranei*, results in the production of PHA copolymers such as PHBV, which have better thermal and mechanical properties (Wang et al., 2022).

Micronutrients and proteins within organic waste also enhance microbial growth and metabolism, thus enhancing polymer yield. The ability of microbial strains to handle a range of feedstock composition allows polymer properties to be tailored (Wang et al., 2022).

3.3 *Pretreatment and Conversion Methods*

Successful biopolymer conversion of organic waste is reliant upon the incorporation of a pretreatment and bioconversion process depending upon the type of feedstock. For lignocellulosic biomass, chemical hydrolysis (alkaline or acidic), mechanical milling, steam explosion, and enzymatic hydrolysis are used to break the biomass structure and release fermentable sugars. Alkaline pretreatment and enzymatic hydrolysis of Rice Husk are efficient in yielding high sugar, which supports effective fermentation (Heng et al., 2016).

For industrial and food waste, not-so-severe pretreatments are tolerated. These include enzymatic hydrolysis, acidogenic fermentation, or autoclaving. Acidogenic fermentation under anaerobic conditions degrades complex substrates to produce Volatile Fatty Acids, which can be further used for PHA synthesis in aerobic fermentation by specific microbial strains (Valentino et al., 2019).

Conversion pathways vary depending on the target biopolymer. PLA production includes the hydrolysis of carbohydrates to reducing sugar. PHA Production Encompasses anaerobic fermentation of organic wastes to VFAs, followed by aerobic fermentation with strains like *Haloferax mediterranei, Bacillus megaterium, or Cupriavidus necator* (Wang & Zhang, 2020).

Enhanced bioprocess engineering methods such as sequential batch fermentation, spent media recycling, and aeration and loading rate optimization have increased process stability and fermentation efficiency. For instance, Wang et al. (2022) designed a two-stage food waste valorization process that efficiently achieved consistent PHA yields in repeated cycles using recycling of the medium.

3.4 Advantages of Using Organic Waste

There are many benefits to using organic waste as a raw material to make biodegradable polymers. Organic wastes bear the disposal cost, which makes their economic value zero. Recycling organic waste as a raw material lowers raw material costs substantially, which otherwise form more than 40% of the cost of biopolymer production. Environmental Sustainability by diverting wastes from landfills avoids methane emission and reduces carbon footprints (Pittmann & Steinmetz, 2017). Moreover, biopolymers from wastes are greener than food-grade crop-based biopolymers (*Renewable energy generation and value addition from environmental*, 2024; Sahu et al., 2024). Circular Economy Integration The method is an example of waste-to-resource strategies typical of circular economy models that encourage local and decentralized production networks (Vu et al., 2021). Material customization, the compositional heterogeneity of organic waste, enables polymer properties such as flexibility, tensile strength, and degradation rates. Infrastructure synergy existing plants, such as wastewater treatment plants and biorefineries, can accommodate organic waste valorization in order to optimize resource recovery and efficiency of operations (Murugan et al., 2021).

3.5 Case Studies

The possibility of mango peel as a green carbohydrate substrate for lactic acid fermentation was explored by Rajendran and Han (2022). Enzyme-hydrolyzed and fermented by Lactobacillus casei yielded 0.88 g/g reducing sugar of lactic acid. Polymerization of the resultant lactic acid led to PLA with physical characteristics similar to commercial plastics, warranting its industrial use. Cheese whey was utilized in fermenting propionic acid by Propionibacterium spp. and further synthesizing PHBV from engineered Cupriavidus necator. Yields of about to 50% cell dry weight warranted the feasibility of dairy waste as a biopolymer feedstock. Wang et al. (2022) used a two-stage process wherein food waste was anaerobically fermented to

VFAs and further converted to PHBV via saline aerobic fermentation by Haloferax mediterranei. The optimized operating conditions yielded 4.5 g/L PHA and 7.0 g/L biomass, warranting scalability as well as process efficiency. These case studies provide the viability, environmental sustainability, and economic viability of organic waste valorization as a sustainable feedstock for biodegradable polymer production. These case studies outlined various feedstocks, process innovation, and integration mechanisms that can be scaled and replicated on the international level.

4 Industrial Biotechnology in Biodegradable Polymer Production

Industrial biotechnology, otherwise called white biotechnology, has emerged as a significant field in addressing global challenges like resource depletion, environmental pollution, and excessive dependence on fossil fuels. By harnessing biological processes and organisms, industrial biotechnology endeavors to transform manufacturing practices, generate valuable products that include biodegradable polymers, and promote sustainability within industrial frameworks. Conventional plastics, originating from petrochemicals, critically contribute to pollution and waste management problems, but, industrial biotechnology offers solutions through the conversion of organic waste into eco-friendly biodegradable and bio-based plastics, examples include polylactic acid (PLA) and polyhydroxyalkanoates (PHA) (Bala et al., 2024; Das et al., 2024a, 2024b, 2024c; Tripathy et al., 2024). These materials undergo natural decomposition, thereby alleviating the pressure on landfills and mitigating the proliferation of microplastic pollution (Mitchell, 2024). This presents numerous opportunities for achieving polymerizations grounded in white biotechnology through the utilization of microorganisms, fermentation methodologies, and advanced genetic engineering techniques (Shoda et al., 2015).

4.1 Role of Microorganisms in Biopolymer Synthesis

Microorganisms, encompassing bacteria, microalgae, and fungi, function as factories of microbes, yielding an extensive array of beneficial products for human applications. Research studies indicated that microorganisms, such as bacteria, are efficient in converting organic substrates into the production of industrially significant chemicals known as microbial polyesters under conditions of nutrient deficit or restricted availability of nutrients. By their physicochemical characteristics, such polyesters can be isolated, purified, and characterized for various use applications. A clear example of a polyester produced by most microorganisms is polyhydroxyalkanoates (PHAs) (Muneer et al., 2021).

Polyhydroxyalkanoates (PHA) Production

Microbial polyhydroxyalkanoates are polyesters produced by the substrate-specific enzyme PHA synthase, present in the cytoplasm of microbial cells, via enzymatic pathways from acetyl coenzyme A (acetyl-CoA) (Degli Esposti et al., 2021). A number of different species of microorganisms, including different Archaea species as well as approximately 250 different types of bacteria, both the gram-negative and the gram-positive strains, have been reported to produce PHAs (González-Rojo et al., 2024). Bacteria store PHAs as intracellular reserve substances in concentrations of 30–80% of their dry weight when they are under nitrogen starvation and high availability of carbon (Shaikh et al., 2021).

Cupriavidus necator is widely regarded as a model bacterial organism for PHA production and is predominantly employed in the synthesis of biodegradable and bio-based plastic substitutes that are environmentally friendly (Al-Khairy et al., 2022). Highly proficient strains of bacteria for the within cell buildup of PHA using fat and oil carbon substrates are *"C. necator H16, C. necator ATCC 17699, Ralstonia. eutropha Re2133, and C. nectator Re2058/pCB113"* (Degli Esposti et al., 2021). Another well-known example is *Pseudomonas putida*, which can make PHAs with medium chain length, or "*Halomonas spp.,*" which are an "extremophile" microbes that may produce PHAs in excessively high salinity situations (González-Rojo et al., 2024).

Lactic acid bacteria for Poly(lactic acid) (PLA)

Lactic acid, which serves as the sole monomer for PLA production, is an acid that is abundantly found in the natural environment, and biosynthetically produced by so many microorganisms. The primary biosynthetic pathway for lactic acid involves the reduction of pyruvate, a reaction that is enzymatically facilitated by lactic acid dehydrogenase, where the stereochemistry of lactic acid is solely dictated by the specific enzyme activity. Several strains of bacteria capable of effectively synthesizing lactic acid both "L-lactic acid and D-lactic acid," for example; *Bacillus coagulans, Sporolactobacillus inulinus, Lactobacillus bulgaricus, Sporolactobacillus terrae*, and *Lactobacillus delbrueckii,* have been isolated from natural sources and are extensively utilized for lactic acid production in the industries. Furthermore, lactic acid can also be efficiently derived from several lignocellulosic sources, with final concentrations surpassing 100 g/L. Consequently, the utilization of lignocellulosic waste materials for lactic acid production presents significant potential (Akinsemolu & Onyeaka, 2023; Al-Khairy et al., 2022; Han et al., 2022).

Microalgae Sources

Microalgae serve as a direct biomass source for bioplastic production as well as processed indirectly through the extraction of polyhydroxybutyrates (PHBs) and starch from within the cellular structures of microalgae. The most extensively studied microalgal species are *Chlorella* and *Spirulina. Chlorella* demonstrates superior properties for bioplastic applications, whereas *Spirulina* exhibits enhanced performance in blends. Bioplastics can be fabricated from the granules of starch derived from *Chlorella sorokiniana* and *C. pyrenoidosa.* Likewise, *Spirulina* has been explored for its potential in bioplastic synthesis. For instance, a bioplastic film synthesized from salt-rich residues of *Spirulina sp.* with the incorporation of PVA. Additionally, a different variety of bioplastic was derived from "*Spirulina platensis,*" which displayed commendable biodegradability. Other microalgae or cyanobacteria employed in bioplastic production include *Chlorogloea fritschii, Neochloris oleoabundans, Calothrix scytonemicola, Nannochloropsis, Nannocloropsis gaditana, Phaeodactylum tricornutum, and Scenedesmus almeriensis* (Coppola et al., 2021).

4.2 Biotechnological Fermentation Techniques

Fermentation is a biochemical process carried out by microorganisms to produce either the polymer or the monomer that subsequently undergoes polymerization. This process may be significantly affected by various parameters, for example, pH, temperature, agitation, nutrient availability, and oxygen levels. Therefore, meticulous regulation of these parameters is critical for successful synthesis. The metabolic pathway can vary depending on the substrate employed, which may result in the formation of different types of biopolymers (de Souza & Gupta, 2024). For instance, fermentation can be used to synthesize lactic acid from whey (Arriaga et al., 2025). The process of fermentation encompasses optimization techniques specifically designed to maximize yield and minimize production costs. Figure 1: Process flow of organic waste conversion to biodegradable polymers in industrial biotechnology.

The implementation of fed-batch and continuous fermentation strategies enhances overall productivity. Continuous fermentation provides benefits such as uniform product quality and increased productivity. This approach is especially advantageous for the production of polyhydroxyalkanoates (PHA) by bacteria for example, *Ralstonia eutropha.* In contrast, fed-batch fermentation entails the incremental addition of nutrients to sustain optimal growth conditions and polymer synthesis. For example, *Cupriavidus necator* utilized in fed-batch mode achieves significant yields of polyhydroxybutyrate-valerate (PHBV) from food waste (Hathi et al., 2022).

Lactic acid (LA) is used for poly(lactic acid) (PLA) production as a monomer, which is applied as a biodegradable polymer. Organic chemistry-based preparation of LA is undesirable since organic processes yield a racemic mixture. For the PLA market, high optical purity (98–99%) L-lactic acid is needed. Thus, the most prevalent process for LA production is through fermentation. Two LA molecules are generated from a single molecule of glucose through the "glycolysis" route. It is possible for lactic acid to be dehydrated to produce lactide, which can then be employed as a monomer for PLA synthesis (Shoda et al., 2015).

Specifically, the fermentative approach also contributes significantly toward biopolymer production. PHBV biopolymer production has been accomplished with many microbes like *C. necator, H. mediterranei, Halomonas sp,* etc., through the utilization of industrial and agriculture waste stream (Hathi et al., 2022).

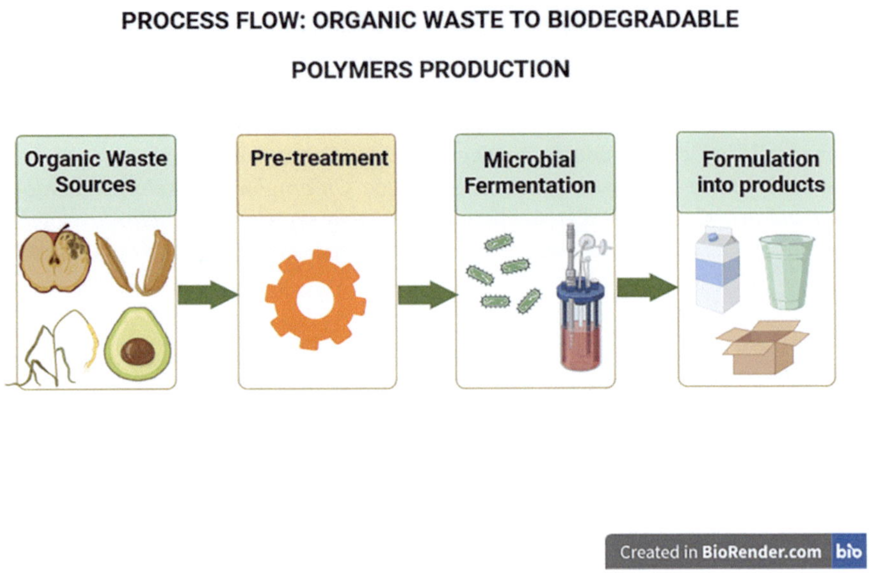

Fig. 1 Process flow of organic waste conversion to biodegradable polymers in industrial biotechnology

4.3 Advancement in Metabolic and Genetic Technology

It is possible to increase the yield of biopolymer using genetic engineering. Several engineering approaches have proved successful in improving the yield of various biopolymers. The main way that metabolic pathway engineering might improve PHA production is by optimizing synthase genes expression. This specific kind of genetic modification has proven to increases the effectiveness of substrate conversion to biopolymers. *Pseudomonas putida, Rhodobacter sphaeroides, Cupriavidus necator, Aeromonas hydrophila,* and other *Pseudomonas* strains metabolic pathway engineering has also been demonstrated to effectively increase PHA synthesis activity (Akinsemolu & Onyeaka, 2023). Regulator and co-factor engineering is a second type of genetic engineering, which has shown promise in enhancing biopolymer production using microbial systems. This method allows for the improved synthesis of several biopolymers by maximizing energy metabolism under various environmental circumstances. Under various economic circumstances, the levels of co factors are systematically adjusted to meet ideal polymerization conditions. Finally, promoter engineering can enhance the yield of biopolymers by carefully modifying the enzymes that participate in the metabolic pathways that synthesize the polymers. The efficiency of the promoters is largely determined by the genetic composition of the microbial strains used in the manufacture of biopolymers. Regulation of the strength of these promoters improves the biosynthetic processes, thus potentially raising biopolymer yield. The combination of these three genetic engineering methods can greatly enhance the manufacturing of biopolymers, hence utilizing their advantages toward ecological conservation, sustainability, and ecologically sound production among industries like pharmaceuticals, food, agriculture, textiles, medicine, and the likes (Akinsemolu & Onyeaka, 2023).

Genetic engineering has helped to enable the conversion of non-PHA-producing bacterial strains to high-level PHA producers. For example, *Escherichia coli,* a bacterium that does not naturally accumulate PHA, has been the target of intensive study for its potential use for the production of PHA (Katagi et al., 2024).

Significant opportunities in biopolymer synthesis have been made possible by developments in recombinant DNA approaches and synthetic biology techniques. Example include the use of metabolic flux pathway analysis, a potent tool to identify hurdles in microbial biopolymer production, and "CRISPR-associated protein 9 (CRISPR–Cas9)" gene editing technology. For example, starch and other organic-based materials can be fermented into polylactic acid (PLA) using metabolically engineered microbes. The fermentation process converts starch into single-strand sugars, which subsequently ferment into mostly L-lactic acid (Caleb & Belay, 2023).

5 Challenges and Limitations of Biodegradable Polymer from Organic Waste

Despite the potential underlying the use of organic waste to produce biodegradable polymers, there have been a number of limitations that exclude their widespread use including, high production costs, weak mechanical properties, and the complexity of mixing such polymers with other materials in order not to lose their biodegradable features, as well as environmental issues and policy problems. These are significant barriers that hinder the efficacy and scalability of these polymers (Meereboer et al., 2020).

From an economic perspective, biodegradable plastics demonstrate a lack of sustainability. Empirical evidence indicates that biodegradable polymers, such as polyhydroxyalkanoates (PHA), incur costs approximately fourfold higher than traditional polymers. This price differential can be ascribed to limitations in economies of scale and constraints in viable technological advancements. Furthermore, microorganisms frequently do not yield a homogeneous polymer type. Consequently, purification processes are typically necessary to isolate a specific polymer product. As a result, these polymers are generally more expensive than their petrochemical equivalents, given that the optimization and scaling of their production methodologies remain in nascent stages (Akinsemolu & Onyeaka, 2023). Similarly, organic waste is abundant but its collection, transportation and processing involves additional cost.

The production of biodegradable polymers occurs on a relatively minor scale, with 99% of plastics (approximately 335 million tons) manufactured for commercial utilization being classified as either non-biodegradable or partially biodegradable (Ghosh & Das, 2020; Ghosh et al., 2020; Maraveas, 2020). Consequently, broader adoption and scaling up of production processes have the potential to facilitate reductions in costs (Chandra, 2023).

Numerous process-related challenges impede the utilization of microbial systems. Primarily, identifying the optimal microbial strain that can function under ideal conditions is imperative. In addition, this strain must exhibit high productivity and selectivity toward the desired product. To achieve this, genetic engineering is frequently employed to enhance yield or direct a microorganism to metabolize a particular substrate. Moreover, fermentation processes require stringent control of environmental factors, such as temperature, pH, substrate concentration, and oxygen levels, among others. Thus, it is inherently problematic to scale-up such processes, as managing these parameters can represent a significant financial burden within the industry. It is crucial that the microbial media effectively assimilate the chosen substrate to synthesize the polymer at an acceptable rate. Therefore, the optimal conditions governing bacterial proliferation and polymer biosynthesis must be established (Akinsemolu & Onyeaka, 2023).

The disposal of plastic waste has become increasingly problematic due to the disorganized systems for collecting both conventional and biodegradable plastics. Technically, biodegradable polymers can be incinerated similarly to traditional plastics, and their energy content is analogous to that of conventional plastics. Nevertheless, there are important differences in the process of disposing of conventional and biodegradable plastics (John & Das, 2024; Mohanty et al., 2024). Degradation of conventional plastics in landfills poses challenges. In landfill settings, biodegradable polymers break down to produce methane, a more powerful greenhouse gas than carbon dioxide because of its greater potential to cause global warming. Anaerobic digestion and industrial composting are two techniques used to treat biodegradable plastics that have outlived their useful life. Compostable plastics that are appropriate for anaerobic digestion and composting must first be separated and collected using separate collecting systems. These materials must then be transported to industrial composting and digestion sites (Moshood et al., 2022).

The properties of bioplastics can differ considerably when compared to those of their petrochemical-based counterparts such as PET or PE. In this regard, bioplastics are usually more brittle and hydrophilic, PLA, for example, is impact-sensitive and brittle, while starch-derived materials degrade in humid conditions. However, they may show similar mechanical properties based on flexural modulus and tensile strength at yield (de Souza & Gupta, 2024).

Educating consumers regarding the advantages and suitable disposal techniques pertinent to biodegradable plastics is of paramount importance. Biodegradable plastics, encompassing bags, and containers and other products ought to be marked with distinctive identification numbers to facilitate their differentiation from other recyclable substances. Community-based educational initiatives on the recognition and management of biodegradable plastics have to be integrated with this endeavor to guarantee a success in the outcome. Provided that appropriate labeling and specialized waste collection protocols are instituted, biodegradable plastics are unlikely to exert adverse effects on waste management systems (Chandra, 2023; Moshood et al., 2022).

6 Recent Innovations and Future Perspectives

The current sustainable production paradigm for biodegradable polymers is based on the development and synthesis of new valorization procedures based on renewable feedstock derived from urban, agricultural, and food waste. The strategies toward synthesizing biodegradable polymers and monomers from biomass feedstock have attracted considerable attention in the chemical industry by taking advantage of high-level biocatalytic conversion and synthetic chemistry. The design of sustainable bioplastic materials is already ongoing, with innovation either on creating completely new polymer classes or creating drop-in alternatives based on renewable feedstock (RameshKumar et al., 2020).

To enhance the characteristics of biodegradable polymers, a variety of methodologies have been established, including grafting and block and random copolymerization. These approaches enhance the mechanical qualities of the resulting goods in addition to the rate of biodegradation. Another technique for creating biodegradable products with various morphologies and physical characteristics is physical blending. In order to promote the utilization of biodegradable polymers, a number of advanced technologies have been used. This encompasses active packaging technologies and the incorporation of natural fiber reinforcements. Recent investigations have documented the application of nanoclay in conjunction with biodegradable polymers, particularly those involving starch and aliphatic polyesters. Ongoing research focuses on nano-biocomposites or bio-nanocomposites (Samir et al., 2022).

The advancement of novel hybrid biodegradable polymers characterized by flexibility and prolonged biocompatibility has led to the emergence of a lot of non-toxic biomaterials suitable for biomedical applications, including long-term implants, orthopedics, regenerative medicine, controlled drug administration, etc.., thereby making significant contributions to these fields. The substantial efforts and inquiries associated with these biodegradable polymers are vital in facilitating the meaningful progression of new viable products toward clinical trials (Mukherjee et al., 2023).

Even though some bioplastics production processes are not scalable and do not have the same productivity as petroleum-based processes, government policy and consumer pressure have been driving the bioplastics industry to take up and pursue sustainable production routes. The circular bioeconomy is also increasingly gaining international momentum, encouraging a wide range of stakeholders to tap into the synergistic benefits that lie within bioplastics production in conjunction with upscaling and recycling methods. Advances in the basic re-engineering of bioplastics with improved economic feasibility for recycling will provide a platform for the next generation of sustainable bioplastics (RameshKumar et al., 2020).

Producing bioplastics using monomers derived from agro waste lowers the carbon footprint of both the raw material and the completed product. Bioplastics' recyclable nature guarantees that the product is socially, economically, and environmentally sustainable throughout its extraction, use, marketing, production, and sourcing. As a result, customers live better lives, and society benefits overall. Using sustainable materials can reduce energy use, disposal waste, and operating costs. Biopolymers will certainly open up new technologies in the effort to enhance the environment and rid it of harmful products and toxins.

At present, biopolymers are being utilized and synthesized in limited quantities globally. Their primary applications are concentrated within the domains of food packaging and the bioplastics sector. Future investigations must be undertaken to enhance the polymeric characteristics of these biodegradable materials through the amalgamation of various polymer types in diverse ratios, to ascertain their influence on the physical and chemical attributes of biomaterials. The medical field has a lot to benefit, as bio-implants and drug delivery carriers based on biopolymers are being developed. Future breakthroughs can lead to a revolutionary change in medical implants, specifically in terms of economic feasibility. The development of new biodegradable polymers requires simplicity, affordability, and recyclability where possible, thus offering a tremendous job opportunity. Furthermore, numerous advancements in the development of new hybrid systems are anticipated imminently, as these novel biodegradable polymers may exhibit vital and efficacious properties. Governments, industries, and institutions of higher learning play a critical role in catalyzing developments in industrial biotechnology. It is critical to put in place policies to encourage research, development, and commercialization of green technologies.

7 Summary and Conclusion

A significant volume of plastic waste has been released into the global environment, engendering the prevalent white pollution dilemma that necessitates immediate remedial action. Biodegradable polymers sourced from organic waste materials have emerged as the focal point of contemporary research owing to their prospective biodegradability and non-toxicity, as well as their potential to mitigate organic waste accumulation, representing the most effective strategy for addressing the plastic waste crisis within the framework of environmental management. This chapter has examined the benefits of biodegradable polymers in comparison with conventional petroleum-based plastics, the diverse array of organic waste sources, and the pivotal role of industrial biotechnology in transforming waste into value-added polymers, alongside recent advancements within the biodegradable polymer domain. Biodegradable plastics, once perceived as a remedy for landfill and littering challenges, are confined to a limited scope of applications. They are unlikely to rectify the littering dilemma, as littering constitutes a societal issue that will not be resolved merely by rendering the material biodegradable. Biodegradable plastics possess the potential to undergo degradation by biological agents exclusively under specific conditions and within a designated timeframe; however, these exquisite conditions must be satisfied to fully leverage the biodegradability of the plastic material. Notwithstanding its potential, industrial biotechnology confronts several challenges, including substantial initial costs, scalability constraints, and regulatory impediments. Public acceptance of these products and genetically modified organisms (GMOs) remains a pivotal consideration. Nevertheless, ongoing advancements, combined with supportive policies and funding, are effectively addressing these obstacles, thereby facilitating broader acceptance. Though the outlook for biodegradable polymers is distinctly positive, future outlook of biopolymers is contingent upon both market competitiveness and also the society's desire to invest in it.

References

Acquavia, M. A., Pascale, R., Martelli, G., Bondoni, M., & Bianco, G. (2021). Natural polymeric materials: A solution to plastic pollution from the agro-food sector. *Polymers, 13*(1), Article 1. https://doi.org/10.3390/polym13010158

Akinsemolu, A., & Onyeaka, H. (2023). Exploring the role of green microbes in sustainable bioproduction of biodegradable polymers. *Polymers, 15*(23), 4617. https://doi.org/10.3390/polym15234617

Al-Khairy, D., Fu, W., Alzahmi, A. S., Twizere, J.-C., Amin, S. A., Salehi-Ashtiani, K., & Mystikou, A. (2022). Closing the gap between bio-based and petroleum-based plastic through bioengineering. *Microorganisms, 10*(12), 2320. https://doi.org/10.3390/microorganisms10122320

Armentano, I., Dottori, M., Fortunati, E., Mattioli, S., & Kenny, J. (2010). Biodegradable polymer matrix nanocomposites for tissue engineering: A review. *Polymer Degradation and Stability, 95*(11), 2126–2146.

Arriaga, M., Pinar, F. J., Izarra, I., Amo, J. del, Vicente, J., Fernández-Morales, F. J., & Mena, J. (2025). Valorization of agri-food waste into PHA and bioplastics: From waste selection to transformation. *Applied Sciences, 15*(3), Article 3. https://doi.org/10.3390/app15031008

Asim, Z., Shahzad, H. M. A., Ghodake, G., Mahmoud, K. A., Almomani, F., & Rasool, K. (2025). Transforming agricultural food waste into bioplastics: Methods, potential, and technological advances. *Advanced Sustainable Systems.*

Auras, R., Harte, B., & Selke, S. (2004). An overview of polylactides as packaging materials. *Macromolecular Bioscience, 4*(9), 835–864.

Avérous, L. (2008). Polylactic acid: Synthesis, properties and applications. In *Elsevier eBooks* (pp. 433–450).

Averous, L. (2000). Properties of thermoplastic blends: Starch–polycaprolactone. *Polymer, 41*(11), 4157–4167.

Bala, K., Anshul, Verma, S., Dogra, A., Das, A. P., Kumar, R., & Nandini. (2024). Sustainable management for urban plastic waste generation. In *Sustainable management of urban plastic waste through circular economic approaches* (pp. 211–236). https://doi.org/10.1201/9781003465447-13

Berghuis, N. T., Mutmainah, D. K., Savitri, M. N., Arizki, M., Yunita, D. D., Awaluddin, F., Peranginangin, R. R., Nursalsabila, A., & Bramasta, R. K. (2022). Bioplastic from Cassava peel and eggshell waste. *Journal of Natural Sciences and Mathematics Research, 8*(2), 75–83.

Bher, A., Mayekar, P. C., Auras, R. A., & Schvezov, C. E. (2022). Biodegradation of biodegradable polymers in mesophilic aerobic environments. *International Journal of Molecular Sciences, 23*(20), 12165.

Caleb, O. J., & Belay, Z. A. (2023). Role of biotechnology in the advancement of biodegradable polymers and functionalized additives for food packaging systems. *Current Opinion in Biotechnology, 83*, Article 102972. https://doi.org/10.1016/j.copbio.2023.102972

Carissimi, M., Flôres, S. H., & Rech, R. (2018). Effect of microalgae addition on active biodegradable starch film. *Algal Research, 32*, 201–209.

Chandra, N. (2023). Bioplastics revolution: How industrial biotechnology is shaping a greener future for packaging. *7*(3), 148. https://doi.org/10.35841/2Faaaib-7.3.148
Chen, G. G. (2009). Plastics from bacteria. In *Microbiology monographs.*
Colnik, M., Knez-Hrncic, M., Skerget, M., & Knez, Z. (2020). Biodegradable polymers, current trends of research and their applications, a review. *Chemical Industry and Chemical Engineering Quarterly, 26*(4), 401–418.
Coppola, G., Gaudio, M. T., Lopresto, C. G., Calabro, V., Curcio, S., & Chakraborty, S. (2021). Bioplastic from renewable biomass: A facile solution for a greener environment. *Earth Systems and Environment, 5*(2), 231–251. https://doi.org/10.1007/s41748-021-00208-7
Da Silva Moura, A., Demori, R., Leão, R. M., Frankenberg, C. L. C., & Santana, R. M. C. (2018). The influence of the coconut fiber treated as reinforcement in PHB (polyhydroxybutyrate) composites. *Materials Today Communications, 18*, 191–198.
Das, A., Dey, S., Bala, K., Gnanasekaran, R., & Das, A. P. (2024a). Microbial bioremediation technology for sustainable treatment and management of synthetic microfiber waste. In *Sustainable microbial technology for synthetic and cellulosic microfiber bioremediation* (pp. 221–238). https://doi.org/10.1007/978-3-031-62600-5_12
Das, A., Dey, S., & Das, A. P. (2024b). Ecotoxicological impacts of synthetic microfiber pollutants and development of sustainable mitigation strategies. *Environmental Chemistry and Ecotoxicology*. https://doi.org/10.1016/J.ENCECO.2024.12.003
Das, A., Ray, S. K., Mohanty, M., Mohanty, J., Dey, S., & Das, A. P. (2024c). Ecotoxicity of microplastic wastes and their sustainable management: A review. *Environmental Chemistry and Ecotoxicology, 6*, 144–152. https://doi.org/10.1016/J.ENCECO.2024.05.003
de Souza, F. M., & Gupta, R. K. (2024). Bacteria for bioplastics: Progress, applications, and challenges. *ACS Omega, 9*(8), 8666–8686. https://doi.org/10.1021/acsomega.3c07372
Degli Esposti, M., Morselli, D., Fava, F., Bertin, L., Cavani, F., Viaggi, D., & Fabbri, P. (2021). The role of biotechnology in the transition from plastics to bioplastics: An opportunity to reconnect global growth with sustainability. *FEBS Open Bio, 11*(4), 967–983. https://doi.org/10.1002/2211-5463.13119
Dey, S., Das, A., Mallick, K., Sahu, A., & Das, A. P. (2023a). *Environmental petroleum waste: Pollution, toxicity, sustainable remediation* (pp. 159–175). https://doi.org/10.1007/978-3-031-48220-5_7
Dey, S., Tripathy, B., Kumar, M. S., & Das, A. P. (2023b). Ecotoxicological consequences of manganese mining pollutants and their biological remediation. *Environmental Chemistry and Ecotoxicology*. https://doi.org/10.1016/J.ENCECO.2023.01.001
Filiciotto, L., & Rothenberg, G. (2021). Biodegradable plastics: Standards, policies, and impacts. *Chemsuschem, 14*(1), 56–72. https://doi.org/10.1002/cssc.202002044
Gan, Z., Abe, H., Kurokawa, H., & Doi, Y. (2001). Solid-state microstructures, thermal properties, and crystallization of biodegradable poly(butylene succinate) (PBS) and its copolyesters. *Biomacromolecules, 2*(2), 605–613.
Ghosh, S., & Das, A. P. (2020). Microbial metagenomics: Current advances in investigating microbial ecology and population dynamics. In S. K. Nayak & B. B. Mishra (Eds.), *Frontiers in soil and environmental microbiology* (1st ed., pp. 247–254). CRC Press.
Ghosh, S., Gandhi, M., van Hullebusch, E. D., & Das, A. P. (2020). Proteomic insights into Lysinibacillus sp.-mediated biosolubilization of manganese. *Environmental Science and Pollution Research 2020, 28*(30), 40249–40263. https://doi.org/10.1007/S11356-020-10863-4
González-Rojo, S., Paniagua-García, A. I., & Díez-Antolínez, R. (2024). Advances in microbial biotechnology for sustainable alternatives to petroleum-based plastics: A comprehensive review of polyhydroxyalkanoate production. *Microorganisms, 12*(8), 1668. https://doi.org/10.3390/microorganisms12081668
Haider, T. P., Völker, C., Kramm, J., Landfester, K., & Wurm, F. R. (2019). Plastics of the future? The impact of biodegradable polymers on the environment and on society. *Angewandte Chemie International Edition, 58*(1), 50–62. https://doi.org/10.1002/anie.201805766

Han, X., Liu, J., Tian, S., Tao, F., & Xu, P. (2022). Microbial cell factories for bio-based biodegradable plastics production. *iScience, 25*(11). https://doi.org/10.1016/j.isci.2022.105462

Hathi, Z. J., Haque, M. A., Priya, A., Qin, Z., Huang, S., Lam, C. H., Ladakis, D., Pateraki, C., Mettu, S., Koutinas, A., Du, C., & Lin, C. S. K. (2022). Fermentative bioconversion of food waste into biopolymer poly(3-hydroxybutyrate-co-3-hydroxyvalerate) using *Cupriavidus necator*. *Environmental Research, 215*, Article 114323. https://doi.org/10.1016/j.envres.2022.114323

Heng, K., Hatti-Kaul, R., Adam, F., Fukui, T., & Sudesh, K. (2016). Conversion of rice husks to polyhydroxyalkanoates (PHA) via a three-step process: Optimized alkaline pretreatment, enzymatic hydrolysis, and biosynthesis by Burkholderia cepaciaUSM (JCM 15050). *Journal of Chemical Technology & Biotechnology, 92*(1), 100–108.

Irmayanti, I., & Anwar, C. (2024). Synthesis of biodegradable plastic from wak banana peel starch with glycerol as plasticizer. *Food Processing Techniques and Technology,* 546–557.

Jenkins, M. J., & Harrison, K. L. (2008). The effect of crystalline morphology on the degradation of polycaprolactone in a solution of phosphate buffer and lipase. *Polymers for Advanced Technologies, 19*(12), 1901–1906.

Jiménez-Rosado, M., Pérez-Puyana, V., Cordobés, F., Romero, A., & Guerrero, A. (2018). Development of soy protein-based matrices containing zinc as micronutrient for horticulture. *Industrial Crops and Products, 121*, 345–351.

John, A., & Das, A. P. (2024). Government strategies for achieving circular economy for plastic waste management in Southeast Asia. In *Sustainable management of urban plastic waste through circular economic approaches* (pp. 72–82). https://doi.org/10.1201/9781003465447-6

Joly, F., & Coulis, M. (2017). Comparison of cellulose versus plastic cigarette filter decomposition under distinct disposal environments. *Waste Management, 72*, 349–353.

Katagi, V., Vytla, R. M., & Somashekara, D. (2024). Integrated production of microbial biopolymer (PHA) with other value-added bioproducts: An innovative approach for sustainable production. *Green Chemistry Letters and Reviews, 17*(1), 2289983. https://doi.org/10.1080/17518253.2023.2289983

Khalil, H. A., Bhat, A., & Yusra, A. I. (2011). Green composites from sustainable cellulose nanofibrils: A review. *Carbohydrate Polymers, 87*(2), 963–979.

Khardenavis, A., Sureshkumar, M., Mudliar, S., & Chakrabarti, T. (2007). Biotechnological conversion of agro-industrial wastewaters into biodegradable plastic, poly β-hydroxybutyrate. *Bioresource Technology, 98*(18), 3579–3584.

Khiste, S. V., Ranganath, V., & Nichani, A. S. (2013). Evaluation of tensile strength of surgical synthetic absorbable suture materials: An in vitro study. *Journal of Periodontal & Implant Science, 43*(3), 130.

Kim, M. S., Chang, H., Zheng, L., Yan, Q., Pfleger, B. F., Klier, J., Nelson, K., Majumder, E. L., & Huber, G. W. (2023). A review of biodegradable plastics: Chemistry, applications, properties, and future research needs. *Chemical Reviews, 123*(16), 9915–9939.

Kreyenschulte, D., Krull, R., & Margaritis, A. (2012). Recent advances in microbial biopolymer production and purification. *Critical Reviews in Biotechnology, 34*(1), 1–15.

Lambert, S., & Wagner, M. (2017). Environmental performance of bio-based and biodegradable plastics: The road ahead. *Chemical Society Reviews, 46*(22), 6855–6871.

Lescher, P., Jayaraman, K., & Bhattacharyya, D. (2008). Water-free blending of thermoplastic starch and polyethylene for rotomoulding. *Starch—Stärke, 61*(1), 43–45.

Li, Y., Huang, G., Chen, C., Wei, X., Dong, X., Zhao, W., & Ye, H. (2021). Poly(butylene succinate-co-butylene acetylenedicarboxylate): Copolyester with novel nucleation behavior. *Polymers, 13*(3), 365.

Li, H., Aguirre-Villegas, H. A., Allen, R. D., Bai, X., Benson, C. H., Beckham, G. T., Bradshaw, S. L., Brown, J. L., Brown, R. C., Cecon, V. S., Curley, J. B., Curtzwiler, G. W., Dong, S., Gaddameedi, S., García, J. E., Hermans, I., Kim, M. S., Ma, J., Mark, L. O., et al. (2022). Expanding plastics recycling technologies: Chemical aspects, technology status and challenges. *Green Chemistry, 24*(23), 8899–9002.

Lim, L., Auras, R., & Rubino, M. (2008). Processing technologies for poly(lactic acid). *Progress in Polymer Science, 33*(8), 820–852. https://doi.org/10.1016/j.progpolymsci.2008.05.004

Lizundia, E., Luzi, F., & Puglia, D. (2022). Organic waste valorisation towards circular and sustainable biocomposites. *Green Chemistry, 24*(14), 5429–5459. https://doi.org/10.1039/D2GC01668K

Lubis, M., Gana, A., Maysarah, S., Ginting, M. H. S., & Harahap, M. B. (2018). Production of bioplastic from jackfruit seed starch (Artocarpus heterophyllus) reinforced with microcrystalline cellulose from cocoa pod husk (Theobroma cacao L.) using glycerol as plasticizer. *IOP Conference Series Materials Science and Engineering, 309*, 012100.

Lukasiewicz, B., Basnett, P., Nigmatullin, R., Matharu, R., Knowles, J. C., & Roy, I. (2018). Binary polyhydroxyalkanoate systems for soft tissue engineering. *Acta Biomaterialia, 71*, 225–234.

Luyt, A. S., & Gasmi, S. (2016). Influence of blending and blend morphology on the thermal properties and crystallization behaviour of PLA and PCL in PLA/PCL blends. *Journal of Materials Science, 51*(9), 4670–4681.

Mahata, N. D., Karthikeyan, S., Godse, R., & Gupta, V. K. (2021). Poly(butylene adipate-co-terephthalate) polyester synthesis process and product development. *Polymer Science Series C, 63*(1), 102–111.

Maraveas, C. (2020). Production of sustainable and biodegradable polymers from agricultural waste. *Polymers, 12*(5), Article 5. https://doi.org/10.3390/polym12051127

Medina-Jaramillo, C., Ochoa-Yepes, O., Bernal, C., & Famá, L. (2017). Active and smart biodegradable packaging based on starch and natural extracts. *Carbohydrate Polymers, 176*, 187–194.

Meereboer, K. W., Misra, M., & Mohanty, A. K. (2020). Review of recent advances in the biodegradability of polyhydroxyalkanoate (PHA) bioplastics and their composites. *Green Chemistry, 22*(17), 5519–5558. https://doi.org/10.1039/D0GC01647K

Mendes, J., Paschoalin, R., Carmona, V., Neto, A. R. S., Marques, A., Marconcini, J., Mattoso, L., Medeiros, E., & Oliveira, J. (2015). Biodegradable polymer blends based on corn starch and thermoplastic chitosan processed by extrusion. *Carbohydrate Polymers, 137*, 452–458.

Missoum, K., Sadocco, P., Causio, J., Belgacem, M. N., & Bras, J. (2014). Antibacterial activity and biodegradability assessment of chemically grafted nanofibrillated cellulose. *Materials Science and Engineering C, 45*, 477–483.

Mitchell, Z. (2024). Advances in industrial biotechnology for sustainable production. *Archives of Industrial Biotechnology, 8*(6), 236.

Mohanty, M., Tripathy, B., Koteswara Reddy, G., & Das, A. P. (2024). Greener technologies for establishing restraint over microplastic pollution. In *Sustainable microbial technology for synthetic and cellulosic microfiber bioremediation* (pp. 239–258). https://doi.org/10.1007/978-3-031-62600-5_13

Moshood, T. D., Nawanir, G., Mahmud, F., Mohamad, F., Ahmad, M. H., & AbdulGhani, A. (2022). Biodegradable plastic applications towards sustainability: A recent innovations in the green product. *Cleaner Engineering and Technology, 6*, Article 100404. https://doi.org/10.1016/j.clet.2022.100404

Mukherjee, C., Varghese, D., Krishna, J. S., Boominathan, T., Rakeshkumar, R., Dineshkumar, S., Brahmananda Rao, C. V. S., & Sivaramakrishna, A. (2023). Recent advances in biodegradable polymers—Properties, applications and future prospects. *European Polymer Journal, 192*, Article 112068. https://doi.org/10.1016/j.eurpolymj.2023.112068

Muneer, F., Nadeem, H., Arif, A., & Zaheer, W. (2021). Bioplastics from biopolymers: An eco-friendly and sustainable solution of plastic pollution. *Polymer Science, Series C, 63*(1), 47–63. https://doi.org/10.1134/S1811238221010057

Murugan, S., Duraisamy, S., Balakrishnan, S., Kumarasamy, A., Subramani, P., & Raju, A. (2021). Production of eco-friendly PHB-based bioplastics by Pseudomonas aeruginosa CWS2020 isolate using poultry (chicken feather) waste. *Biologia Futura, 72*(4), 497–508.

Pillai, C. K. S., & Sharma, C. P. (2010). Review paper: Absorbable polymeric surgical sutures: Chemistry, production, properties, biodegradability, and performance. *Journal of Biomaterials Applications, 25*(4), 291–366.
Pittmann, T., & Steinmetz, H. (2017). Polyhydroxyalkanoate production on waste water treatment plants: Process scheme, operating conditions and potential analysis for German and European municipal waste water treatment plants. *Bioengineering, 4*(2), 54.
Rafiqah, S. A., Khalina, A., Harmaen, A. S., Tawakkal, I. A., Zaman, K., Asim, M., Nurrazi, M., & Lee, C. H. (2021). A review on properties and application of bio-based poly(butylene succinate). *Polymers, 13*(9), 1436.
Rajendran, N., & Han, J. (2022). Integrated polylactic acid and biodiesel production from food waste: Process synthesis and economics. Bioresource Technology, 343, 126119. ISSN 0960-8524. https://doi.org/10.1016/j.biortech.2021.126119
RameshKumar, S., Shaiju, P., O'Connor, K. E., & P, R. B. (2020). Bio-based and biodegradable polymers—State-of-the-art, challenges and emerging trends. *Current Opinion in Green and Sustainable Chemistry, 21*, 75–81. https://doi.org/10.1016/j.cogsc.2019.12.005
Ray, S. K., Dey, S., Das, A. P., & John, A. (2024). Microplastic waste management through greener technologies: A review. In *Sustainable management of urban plastic waste through circular economic approaches* (pp. 237–261). https://doi.org/10.1201/9781003465447-14
Reckoning with the U.S. Role in Global Ocean Plastic Waste. (2021). In *National Academies Press eBooks.*
Reddy Koteswara, G., Bhargavi, S., Yamuna, D., Anjani Devi, M., Jeevana Lakshmi Kumar, G., Manas, T., Adarsh, S., Bhavya Sai, V., Nath, H., Alreshidi, M. A., Togun, H., Ram, D. K., & Das, A. P. (2025). Recent advancements in the remediation of marine hydrocarbon and plastic pollutants. *Environmental Science and Engineering, Part F218,* 127–142. https://doi.org/10.1007/978-3-031-81615-4_7
Renewable energy generation and value addition from environmental. (2024).
Sahu, A., Dey, S., Mallick, K., Priyadarsini, S., Behera, I. D., & Das, A. P. (2024). A review on advanced technology for sustainable management of synthetic microplastic waste. In *Sustainable management of urban plastic waste through circular economic approaches* (pp. 262–279). https://doi.org/10.1201/9781003465447-15
Samir, A., Ashour, F. H., Hakim, A. A. A., & Bassyouni, M. (2022). Recent advances in biodegradable polymers for sustainable applications. *NPJ Materials Degradation, 6*(1), 1–28. https://doi.org/10.1038/s41529-022-00277-7
Schiavo, S., Oliviero, M., Chiavarini, S., & Manzo, S. (2020). Adverse effects of oxo-degradable plastic leachates in freshwater environment. *Environmental Science and Pollution Research, 27*(8), 8586–8595.
Seggiani, M., Gigante, V., Cinelli, P., Coltelli, M., Sandroni, M., Anguillesi, I., & Lazzeri, A. (2019). Processing and mechanical performances of Poly(Butylene Succinate–co–Adipate) (PBSA) and raw hydrolyzed collagen (HC) thermoplastic blends. *Polymer Testing, 77*, Article 105900.
Selke, S., Auras, R., Nguyen, T. A., Aguirre, E. C., Cheruvathur, R., & Liu, Y. (2015). Evaluation of biodegradation-promoting additives for plastics. *Environmental Science & Technology, 49*(6), 3769–3777.
Shaikh, H. M., Pandare, K. V., Nair, G., & Varma, A. J. (2008). Utilization of sugarcane bagasse cellulose for producing cellulose acetates: Novel use of residual hemicellulose as plasticizer. *Carbohydrate Polymers, 76*(1), 23–29.
Shaikh, S., Yaqoob, M., & Aggarwal, P. (2021). An overview of biodegradable packaging in food industry. *Current Research in Food Science, 4*, 503–520. https://doi.org/10.1016/j.crfs.2021.07.005
Shen, L., Worrell, E., & Patel, M. (2009). Present and future development in plastics from biomass. *Biofuels Bioproducts and Biorefining, 4*(1), 25–40.
Shen, M., Song, B., Zeng, G., Zhang, Y., Huang, W., Wen, X., & Tang, W. (2020). Are biodegradable plastics a promising solution to solve the global plastic pollution? *Environmental Pollution, 263*, Article 114469. https://doi.org/10.1016/j.envpol.2020.114469

Shoda, S., Kobayashi, A., & Kobayashi, S. (2015). Production of polymers by white biotechnology (pp. 274–309). https://doi.org/10.1039/9781782624080

Silva, R. R. A., Marques, C. S., Arruda, T. R., Teixeira, S. C., & De Oliveira, T. V. (2023). Biodegradation of Polymers: Stages, measurement, standards and prospects. *Macromol—A Journal of Macromolecular Research, 3*(2), 371–399.

Sirviö, J. A., Visanko, M., Ukkola, J., & Liimatainen, H. (2018). Effect of plasticizers on the mechanical and thermomechanical properties of cellulose-based biocomposite films. *Industrial Crops and Products, 122*, 513–521.

Sudesh, K., & Iwata, T. (2008). Sustainability of biobased and biodegradable plastics. *Clean Soil Air Water, 36*(5–6), 433–442.

Swain, S., Purnima Sudha, M., Dey, S., Badu, Y., Bhoi, S. R., Alreshidi, M. A., Qasim, M. T., Nath, H., Ram, D. K., & Das, A. P. (2025). Sources of micro and nano plastic hydrocarbon pollutants in marine bodies and their treatment. *Environmental Science and Engineering, Part, F218*, 45–67. https://doi.org/10.1007/978-3-031-81615-4_3

Tantawi, O., Joo, W., Martin, E. E., Av-Ron, S. H. M., Bannister, K. R., Prather, K. L. J., Olsen, B. D., & Plata, D. L. (2025). Designing for degradation: The importance of considering biotic and abiotic polymer degradation. *Environmental Science Processes & Impacts.*

Tesfaye, T., Sithole, B., Ramjugernath, D., & Chunilall, V. (2017). Valorisation of chicken feathers: Characterisation of chemical properties. *Waste Management, 68*, 626–635.

Tripathy, B., Sahoo, P. P., Sundaray, H., & Das, A. P. (2024). In-vitro biodegradation of discarded marine microplastics across the eastern coast of the Bay of Bengal, India using Exiguobacterium sp. *Environmental Chemistry and Ecotoxicology, 6*, 236–247. https://doi.org/10.1016/J.ENCECO.2024.07.003

Tsuji, H., & Ikada, Y. (1995). Properties and morphologies of poly(?-lactide): 1. Annealing condition effects on properties and morphologies of poly(?-lactide). *Polymer, 36*(14), 2709–2716.

Valderrama, M. a. M., Van Putten, R., & Gruter, G. M. (2020). PLGA barrier materials from CO_2. the influence of lactide co-monomer on glycolic acid polyesters. *ACS Applied Polymer Materials, 2*(7), 2706–2718.

Valentino, F., Lorini, L., Pavan, P., Bolzonella, D., & Majone, M. (2019). Organic fraction of municipal solid waste conversion into polyhydroxyalkanoates (PHA) in a pilot scale anaerobic/aerobic process. *Chemical Engineering Transactions, 74*, 265–270.

Vu, D. H., Wainaina, S., Taherzadeh, M. J., Åkesson, D., & Ferreira, J. A. (2021). Production of polyhydroxyalkanoates (PHAs) by Bacillus megaterium using food waste acidogenic fermentation-derived volatile fatty acids. *Bioengineered, 12*(1), 2480–2498.

Wang, K., & Zhang, R. (2020). Production of polyhydroxyalkanoates (PHA) by Haloferax mediterranei from food waste derived nutrients for biodegradable plastic applications. *Journal of Microbiology and Biotechnology, 31*(2), 338–347.

Wang, F., Wang, Q., Adams, C. A., Sun, Y., & Zhang, S. (2021). Effects of microplastics on soil properties: Current knowledge and future perspectives. *Journal of Hazardous Materials, 424*, Article 127531.

Wang, K., Chen, C., & Zhang, R. (2022). Process development of polyhydroxyalkanoates production by halophiles valorising food waste. *Bioengineering, 9*(11), 630.

Witt, U., Müller, R. J., & Deckwer, W. D. (1997). Biodegradation behavior and material properties of aliphatic/aromatic polyesters of commercial importance. *Journal of Environmental Polymer Degradation, 5*(2), 81–89.

Wu, C., Liao, H., & Cai, Y. (2017). Characterisation, biodegradability and application of palm fibre-reinforced polyhydroxyalkanoate composites. *Polymer Degradation and Stability, 140*, 55–63.

Xie, Q., Li, F., Li, J., Wang, L., Li, Y., Zhang, C., Xu, J., & Chen, S. (2018). A new biodegradable sisal fiber–starch packing composite with nest structure. *Carbohydrate Polymers, 189*, 56–64.

Yang, M., Xu, H., Hou, X., Zhang, J., & Yang, Y. (2016). Biodegradable sizing agents from soy protein via controlled hydrolysis and dis-entanglement for remediation of textile effluents. *Journal of Environmental Management, 188*, 26–31.

Yu, Z., Sun, L., Wang, W., Zeng, W., Mustapha, A., & Lin, M. (2017). Soy protein-based films incorporated with cellulose nanocrystals and pine needle extract for active packaging. *Industrial Crops and Products, 112*, 412–419.

Zhang, Y., Xie, J., Ellis, W. O., Li, J., Appaw, W. O., & Simpson, B. K. (2024). Bioplastic films from cassava peels: Enzymatic transformation and film properties. *Industrial Crops and Products, 213*, Article 118427.

Zhu, J., & Wang, C. (2020). Biodegradable plastics: Green hope or greenwashing? *Marine Pollution Bulletin, 161*, Article 111774. https://doi.org/10.1016/j.marpolbul.2020.111774

Sustainable Bioenergy: A Green Waste Conversion Strategy for Cost-Effective Feedstock and Economic Waste-to-Wealth Solutions

Nageswara Rao Lakkimsetty, N. Tirumala Uday Kumar, Rahma Juma, and Nourhan Hilal El Mustapha El Mohamad

1 Introduction

1.1 Sustainable Bioenergy

From the perspective of being renewable bioenergy, biodiesel is obtained from renewable biological feedstocks, with examples being animal waste, used fats and oils, and vegetable oils, thereby largely limiting the use of fossil fuels. Biodiesel is considered green because its production and usage greatly reduce even GHG emissions and hence it acts as the green alternative to regular diesel (Ahmed et al., 2016). It is biodegradable and non-toxic and is suitable to run diesel engines with slight modifications to guarantee cleaner combustion and lesser air pollution. The valorization of waste sources for making biodiesel, such as crop wastes or chicken fats, ties into the circular economy and assists in the realization of global renewable energy and greenhouse gas reduction targets (Amber et al., 2014).

The sources of energy are primarily categorized into three groups: renewable energy, fossil fuels, and fissile materials. The world is currently relying primarily on fossil fuels as the principal source of energy. Fossil fuels and their derived products find applications by various sectors and industries to support numerous processes. Given their vast uses, they are, however, the major contributors to GHG emissions, thus harming the environment and human health (Das et al., 2025; Reddy Koteswara et al., 2025). This hazardous effect of fossil fuel usage on the environment especially

N. R. Lakkimsetty (✉) · N. T. Uday Kumar · R. Juma (✉) · N. H. El Mustapha El Mohamad
Department of Chemical and Petroleum Engineering, American University of Ras Al Khaimah, Rasal Khaimah, United Arab Emirates
e-mail: lakkimsetty.rao@aurak.ac.ae

R. Juma
e-mail: rahma.juma@aurak.ac.ae

S. Ghosh et al. (eds.), *Advances in Organic Waste Conversion Through Industrial Biotechnology*, Sustainable Environmental Waste Management Strategies,
https://doi.org/10.1007/978-3-032-11377-1_8

drove the development of alternative, cleaner energy sources, more prominently renewable energy, being the most feasible and cleaner option. One such renewable energy is biodiesel, which is basically a fatty acid alkyl ester from vegetable oil or animal fat. Since the industrial revolution, with growing demand for fossil fuels, came the need for viable alternatives. Biodiesel has been considered one of the most appropriate options to reduce the consumption of traditional fuels and ease the worldwide energy crisis (Bianchi et al., 2010).

Biodiesel has been prepared in the Gulf region from algae and other vegetable oils. This research involves producing biodiesel from chicken waste through the transesterification process. The objective is to extract and characterize oil from chicken feathers and skin and convert it into biodiesel via an alkaline transesterification reaction in a lab-scale reactor. The biodiesel (B100) and blend (B20) produced were characterized physicochemical properties according to ASTM D6751 standards. Optimization experiments included variations in temperature, time, methanol-to-oil ratio, catalyst concentration, and stirring speed (Bharat & Bhattacharya, 2012).

The aim of this research in general is to identify a low-cost, renewable feedstock with potential to reduce the utilization of fossil fuels and climate change by lowering greenhouse gas emissions. This project further aims to provide a cost-effective and cleaner energy alternative amidst rising fuel prices and the growing demand for greener forms of energy (Barua et al., 2020).

2 Methodology and Case Study Experimentation

2.1 Materials

In this study, several materials were used to produce and analyze biodiesel. Chicken skin and feathers served as the main raw materials. Standard diesel fuel was used for blending and comparison with the produced biodiesel. Methanol and ethanol were used for the transesterification reaction to convert the fats to biodiesel. Distilled water was used for washing and purification of the biodiesel. A 0.1 M hydrochloric acid (HCl) solution was used for neutralization, while 0.1 M potassium hydroxide (KOH) was used as a catalyst for the reaction. Phenolphthalein was used as an indicator during the titration process (El Haijjaji et al., 2016).

2.2 Biodiesel Processing Steps

The chicken waste samples (feathers and skin) were collected from a local poultry slaughterhouse and transported to the chemical analysis laboratory for treatment. The samples were washed with distilled water thoroughly to remove any contaminants

Fig. 1 Oil extraction process from chicken waste

from the surface. After washing, they were cut into small pieces (approximately 1–3 cm) and left at room temperature to dry for one hour. After drying, the samples were stored in closed containers and maintained at 4 °C to prevent spoilage or degradation before proceeding with further analysis (Khan et al., 2014).

2.2.1 Extraction of Oil

In a pressure cooker, one kilogram of the pre-treated chicken waste and one liter of distilled water were heated for an hour. To enable the separation of the water and oil, it was poured into a separating funnel after being allowed to cool to room temperature. The Fig. 1 represents the Oil extraction process from chicken waste. The top layer, which contained the extracted oil, was carefully collected. This oil was then analyzed to determine its acid value, moisture content, and density to confirm its suitability for biodiesel production (Meinderstma & Block, 2012).

2.2.2 Alkaline Transesterification Reaction

The transesterification reaction was done in a batch reactor in a 250 mL round-bottom flask with a water-cooled condenser. An appropriate amount of methanol was mixed with 0.1 M KOH to prepare the catalyst solution. This mixture was then added to the flask containing 50 g of the extracted oil (Maryem et al., 2014). The contents were continuously stirred with a magnetic stirrer and gently heated using an electric heater, keeping the temperature below the boiling point of methanol.

2.2.3 Separation and Purification

Following the transesterification reaction, the mixture was transferred into a separating funnel and left to settle, producing two layers: glycerol at the bottom and biodiesel at the top. Following a methodical collection and purification process, the biodiesel was first heated to 50 °C to eliminate any remaining methanol, then rinsed with 50 mL of 0.1 N hydrochloric acid and distilled water to eliminate any remaining catalyst. Finally, to remove extra water, the biodiesel was heated over 100 °C, resulting in a stable and clean product (Purandaradas et al., 2018).

2.3 Experimental Conditions and Optimization

In order to yield the maximum amount of biodiesel from chicken waste oil, fundamental process parameters like the molar ratio of methanol to oil, reaction time, catalyst concentration, temperature of the reaction, and stirring speed were set and optimized in a stepwise fashion. In total, five different experiments were conducted to examine the impact of the aforementioned factors on the yield of biodiesel (Wang et al., 2007). The first test was conducted with a ratio of methanol to oil of 3:1, with 2 g KOH catalyst, 40 min reaction time at 30 °C, and stirring at 100 rpm. Subsequently, the levels were stepped up in the next tests: second test used 4:1 ratio, 3 g KOH, 50 min, 40 °C, and 200 rpm; third test used 5:1 ratio, 4 g KOH, 60 min, 50 °C, and 300 rpm. Fourth trial with 6:1 ratio, 5 g of catalyst, 90 min, 60 °C, and 400 rpm. Finally, the fifth set employed a 7:1 ratio, 6 g of catalyst, 120 min, 70 °C, and 500 rpm. These incremental adjustments allowed for the identification of optimal conditions for efficient biodiesel production (Nageswara Rao & Al Riyami, 2018).

3 A Green Waste Conversion Strategy for Cost-Effective Feedstock and Economic Waste-to-Wealth Solutions

Growing demand for the treatment of green waste has been driven by increased concern for dwindling fossil fuels, increasing energy demands, and environmental degradation across the world. Green waste conversion is one of the possible solutions to the reduction of energy and waste management problems. By utilizing organic wastes such as crop residues, animal fat, and food processing waste into productive bioenergy carriers, the process not only dissuades environmental pollution but also achieves energy security and circular economy goals. The utilization of waste as a cost-effective feedstock in the production of biofuel encourages resource efficiency, lowers greenhouse gas emissions, and offers economic benefits under waste-to-wealth schemes. This section explores the potential, processes, and benefits of green waste conversion as an integrated solution for sustainable energy development (Hassan & Lakkimsetty, 2018).

3.1 Evaluation of Oil Yield and Characterization of Extracted Oil

From 1000 g of processed chicken waste, 511.5 g of oil was extracted, giving a volume of 467.8 mL. This means the oil yield was about 51.15% by weight, showing that chicken waste is a good source for biodiesel production. To check if the oil was suitable for making biodiesel, its key properties were tested. The acid value was 0.561 mg KOH/g, which is within the safe range for alkaline transesterification, so no extra treatment was needed (Lakkimsetty et al., 2022). The moisture content was very low at 0.020%, which helps prevent soap formation during the process. The oil's density was 1.0935 g/mL, indicating it has many unsaturated compounds. Although this could slow the reaction, using more methanol can help achieve full conversion into biodiesel.

3.2 Evaluation and Optimization of Biodiesel Yield %

Several factors, including reaction duration, reaction temperature, the molar ratio of methanol to oil, stirring rate, and catalyst amount, were taken into consideration when the alkaline transesterification process was carried out to create biodiesel. The ideal transesterification conditions that yielded the highest yield percentage, or roughly 88.84%, were determined to be 6:1 molar ratio of methanol/oil, 90 min of reaction time, 5 g of 0.1 M KOH as a catalyst, 60 °C reaction temperature, and 400 rpm stirring rate. The forward direction of the transesterification, which resulted in the completion of the reaction, produced the most biodiesel under ideal circumstances. However, because of the reaction's disadvantage, the biodiesel yield was the lowest at 64.3% above the ideal circumstances. Due to the reaction's downside and the increased production of soap and glycerol, the yield of biodiesel was the lowest, at roughly 64.3%, above the ideal circumstances (Al Hasani et al., 2021).

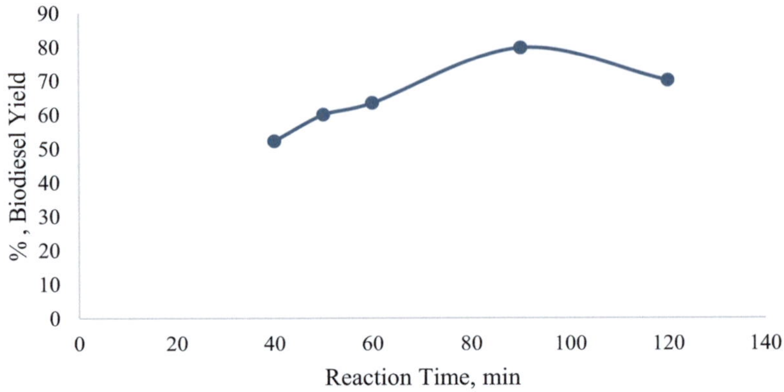

Fig. 2 Effect the reaction time on biodiesel yield

3.3 Analysis of Variables Influence on Biodiesel Yield

3.3.1 Effect of Reaction Time

Figure 2 shows that the reaction started slowly because the methanol needed time to mix properly with the extracted oil. From 40 to 60 min, the biodiesel yield escalated steadily, reaching 63.4%.

After 60 min, the reaction was considerably faster, and by 90 min yielded the maximum biodiesel of 79.7%. This was achieved using 5 g of 0.1 M KOH, a 6:1 methanol-to-oil ratio, at 60 °C with a stirring speed of 400 rpm. This indicated that the reaction was almost complete. However, increasing the reaction time beyond 90 min caused the reaction to reverse, reducing the biodiesel yield (Al-Sumri et al., 2024).

3.3.2 Effect of Methanol/Oil Molar Ratio

Figure 3 shows that an increase of the methanol-to-oil molar ratio from 3:1 to 6:1 increased the yield of biodiesel. The highest yield of 83.7% was achieved at a ratio of 6:1 by 5 g of 0.1 M KOH, reaction time of 90 min, temperature of 60 °C, and agitation rate of 400 rpm. This shows that the application of more methanol increases driving the reaction and conversion. But increasing the ratio more than 6:1 reduced the biodiesel yield to about 70%. This was as a result of excess methanol being incorporated with glycerol, disrupting the separation process and decreasing efficiency overall (Lakkimsetty et al., 2025).

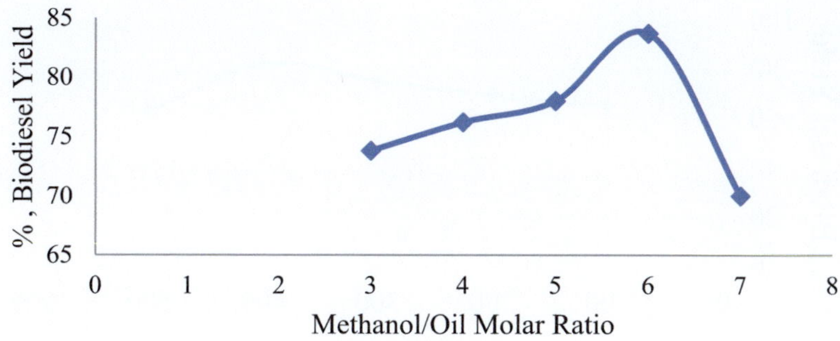

Fig. 3 Effect of methanol/oil molar ratio on biodiesel yield

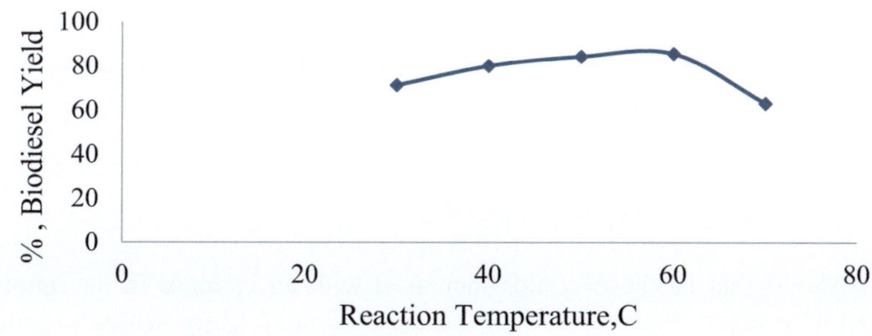

Fig. 4 Effect the reaction temperature on biodiesel yield

3.3.3 Effect of Reaction Temperature

Figure 4 shows that the biodiesel yield increased systematically with an increase in the reaction temperature from 30 to 50 °C and had a maximum value of 84.3% at 50 °C. The maximum biodiesel yield of 85.6% was achieved under the reaction temperature of 60 °C by carrying out the process using 5 g of 0.1 M KOH, 90 min reaction time, 6:1 molar ratio of methanol to oil, and 400 rpm agitation speed. However, an increase in temperature to over 60 °C lowered the yield of biodiesel because it was close to methanol's boiling point and caused loss of methanol as well as not very efficient reaction.

3.3.4 Effect of Stirring Rate

Figure 5 illustrates that biodiesel yield increased steadily when the stirring speed varied from 100 to 300 rpm to 75.23%. The highest yield of 82.42% was achieved at 400 rpm for 6:1 methanol-to-oil molar ratio, 90 min of reaction time, 60 °C of reaction temperature, and 5 g of 0.1 M KOH catalyst. However, additional speeding up of the stirring rate beyond 400 rpm resulted in a decrease in biodiesel yield because

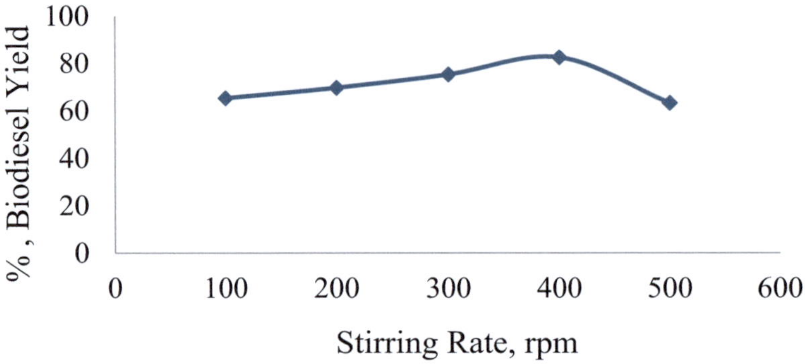

Fig. 5 Effect the stirring rate on biodiesel yield

of excessive soap and glycerol formation, which disrupted the transesterification reaction Rehman et al. (2016).

3.3.5 Effect Amount of Catalyst (KOH)

Figure 6 shows that biodiesel yields increased with an increase in the amount of 0.1 M KOH catalyst. From 2 to 4 g, the conversion of triglycerides to biodiesel improved gradually to 80.47% yield. The maximum yield of 85.47% was achieved using 5 g of catalyst with a molar ratio of 6:1 methanol and oil, reaction time of 90 min, temperature of 60 °C, and an agitation rate of 400 rpm, which is the optimum quantity of catalyst for efficient transesterification. However, using catalysts more than 5 g led to a decline in yield to around 64.6%, which may be caused by the soap formation, which will slow down the reaction and lower biodiesel production (Alalyani et al., 2023).

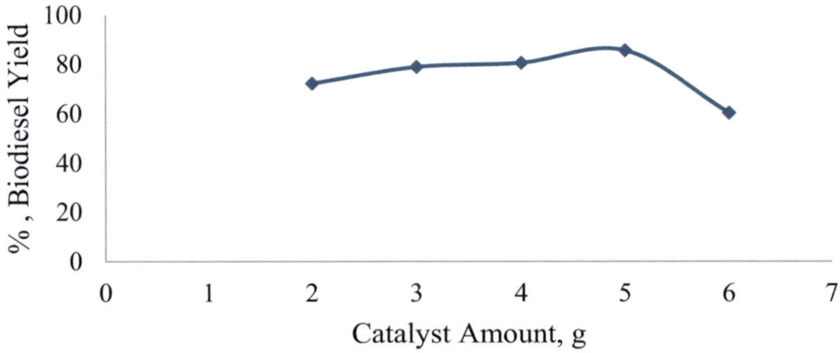

Fig. 6 Effect the amount of catalyst on biodiesel yield

3.4 *Characterizations of B100 and B20*

The physicochemical properties of the produced biodiesel (B100) were analyzed to gauge its viability and fitness to ASTM D6751 biodiesel specifications. Results showed that B100 met the specifications it was supposed to, but with performance limits regarding cold flow impairment because of its relatively higher density and viscosity. To achieve more applicability and operational effectiveness, a B20 blend was prepared by combining 20% biodiesel with 80% typical diesel fuel. The blend was tested and compared with typical diesel and found to possess characteristics closely resembling standard diesel fuel, hence being a more real and effective choice (Hassan & Lakkimsetty, 2018, Jeong et al. 2009). Table 1 shows the comparative characteristics of B100 and B20.

3.4.1 Fatty Acid Methyl Ester Composition

The fatty acid methyl ester (FAME) profile was analyzed to assess the suitability of different feedstocks for biodiesel production. The results presented in Table 1 showed that the most abundant ester in B100 was eicosanoate, which indicates high purity of the biodiesel. However, the presence of small amounts of polyunsaturated FAMEs in biodiesel can lower its effectiveness as fuel. The B20 blend was also analyzed and compared to standard diesel fuel, showing a composition closer to that

Table 1 Comparative fuel properties of B100, B20, and standard diesel fuel against ASTM D6751 standards

Test parameters	ASTM D6751 standard	B100	B20	Diesel fuel
Acid value (mg KOH/g of oil)	Max 0.5	0.337	0.310	0.300
Moisture content (% wt)	Max 0.05	0.028	0.023	0.020
Kinematic viscosity at 20 °C (mm^2/s)	1.9–6.0	5.20	3.62	3.20
pH	–	7.22	5.70	5.50
Density at 20 °C (g/cm^3)	0.82–0.90	0.870	0.839	0.840
Carbon residue (g)	Max 0.3	0.032	0.039	0.042
Flash point (°C)	Min 101	154.5	78.4	83.0
Pour point (°C)	Not applicable	12.4	−10	−8.0
Cloud point (°C)	Not applicable	8.0	3.0	2.5
Iodine value (mg I_2/100 g)	Max 120	20.0	4.0	5.3
Saponification value (mg KOH/g)	Max 193	183.4	561.7	432.0
Cetane number	Min 51	72.0	55.0	46.0

of standard diesel. As a result, B20 is more efficient and suitable for use in diesel engines compared to the B100 (Wang et al., 2007), and Figs. 7 and 8 show the fatty acid profile of B100 and B20 respectively. Figure 8 shows the fatty acid profile of B20%.

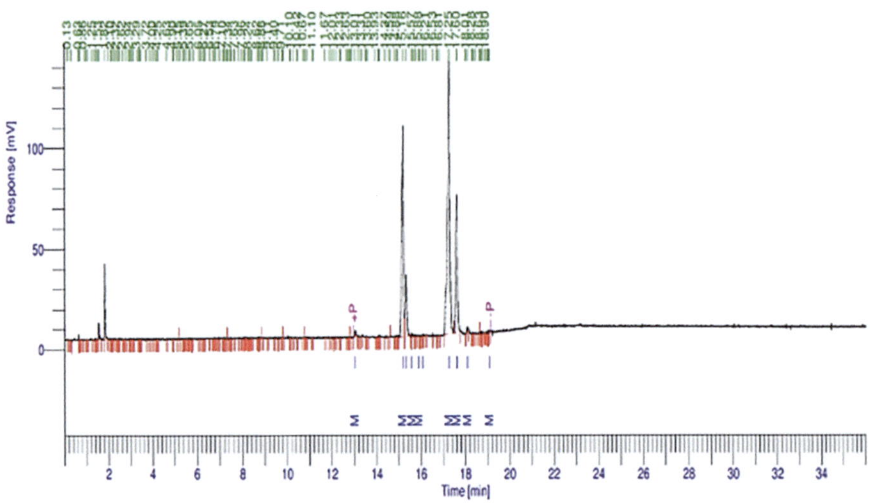

Fig. 7 Fatty acid profile of B100%

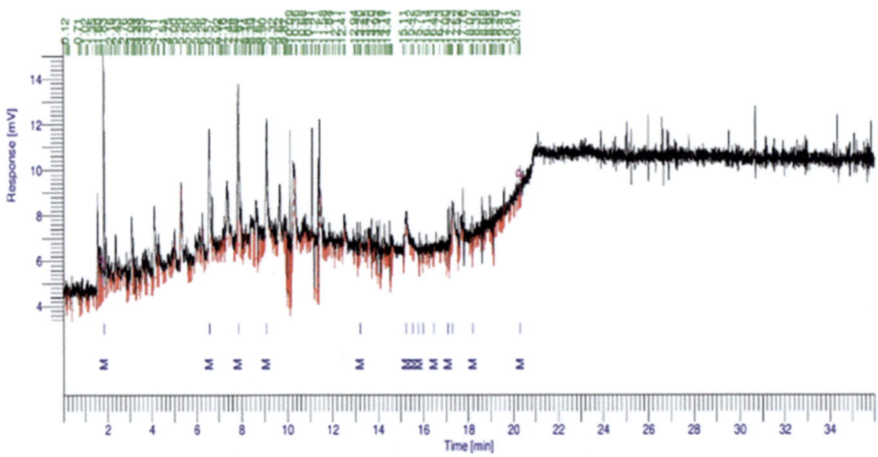

Fig. 8 Fatty acid profile of B20%

3.5 *Cost-Effective Feedstock*

3.5.1 Analysis of Cost Estimation

The cost of production of chicken waste oil was estimated by comparing the main expenses, that is, raw material, labor, and electricity. The study indicated that the minimum cost of feedstock was 15,724.5 AED/month and represented 41.3% of the overall cost of materials. This feedstock cost includes expenses for washing, melting, and transporting chicken waste. The cost of chemicals used in the alkaline transesterification process was 22,345.42 AED /month, making up 58.7% of the material costs. In addition, electricity and equipment costs were also approximately estimated at 1,053.28 AED /month and 263.09 AED /month, respectively. Considering the process designed and some assumptions, the production cost of biodiesel was approximated at around 1.43 AED /L. This is significantly lower than the value of the present diesel fuel in the Gulf region, which was 2.34 AED /L during last month. It must be remembered that the price of biodiesel can be differential depending on the socio-economic benefit and market conditions Al-Balushi and Lakkimsetty (2019).

Chicken waste biodiesel production is an inexpensive and environmentally friendly bioenergy strategy that transforms low-value waste into high-value fuel. Chicken waste, which consists of skin and feathers, contains huge amounts of fat that can be reclaimed and utilized to produce biodiesel via a chemical reaction called transesterification. This process not only reduces feedstock expense since waste materials are inexpensive and abundantly available, but also reduces environmental pollution associated with improper disposal of poultry manure. With the utilization of chicken waste as a sustainable material, this waste-to-wealth conversion promotes circular economic measures, reduces dependence on foreign fossil fuels, and promotes local energy generation, thus making it economically and environmentally friendly (Mohanty et al., 2024; Sahu et al., 2024).

4 Conclusions

The present research demonstrates that chicken waste holds considerable potential for conversion into green energy in the form of high-grade biodiesel as a cost-effective and eco-friendly means of solving waste management and fuel production problems. The study reaffirms the fact that approximately 467.8 mL of oil can be recovered from 1 kg of chicken waste from abattoirs. At ideal transesterification conditions of methanol-to-oil molar ratio of 6:1, reaction time of 90 min, temperature of 60 °C, 5 g of 0.1 M KOH catalyst, and stirring rate of 400 rpm, the process yielded up to 89% biodiesel (B100). Additionally, the production cost was computed at a meager 1.43 AED/liter, an economically viable green fuel alternative to petroleum diesel. This case study offers double benefits of using poultry waste to generate green energy and fostering circular economy concepts and reducing environmental impacts. Green waste conversion strategy offers a promising path to recover low-value and abundant waste materials into low-cost feedstock to generate sustainable bioenergy. By utilizing waste streams such as agricultural waste, animal fat, and food processing wastes, the strategy not only prevents the use of fossil fuels but also places serious environmental concerns on waste management and greenhouse gas emissions in the spotlight. Economic analysis also indicates that the conversion of waste to useful biofuels and bioproducts can be capable of creating sustainable waste-to-wealth opportunities, stimulating local economies, and promoting circular economy models. Overall, this strategy is an integrated strategy that enables environmental sustainability, economic growth, and energy security and hence an attainable and expandable green energy model that can be applied worldwide.

References

Ahmed, A. H., Mayoran, M., Logeshwaran, K. G., Balakumaran, S., & Ashiq, K. A. M. (2016). Extraction of biofuel from chicken waste. *International Journal of Innovative Research in Science, Engineering and Technology, 5*(7), 18.

Al Hasani, W. J., Lakkimsetty, N. R., Lakavat, M., Shaik, F., & Kavitha, G. (2021). A green process for producing biodiesel from chicken skin and its feather as a cost-effective feedstock. *AIP Conference Proceedings, 2396*, Article 020001. https://doi.org/10.1063/5.0066307

Alalyani, S. R. S., Lakkimsetty, N. R., Shaik, F., et al. (2023). Modeling and optimization of bioethanol production yield from corn starch using response surface methodology. *Environmental Development and Sustainability*. https://doi.org/10.1007/s10668-023-02990-y

Al-Balushi, A. A. Y., & Lakkimsetty, N. R. (2019). Comparative studies on production of biodiesel from waste vegetable oil (WVO) and Noctiluca scintillans (NS). *International Journal of Renewable Energy and Its Commercialization, 5*(1), 23–36.

Al-Sumri, M. S., Shaik, F., Lakkimsetty, N. R., & Varghese, M. J. (2024). Techno-economic analysis of solar and wind energy systems for power and hydrogen production. In L. Chen (Ed.), *Advances in clean energy systems and technologies* (pp. 207–213). Springer. https://doi.org/10.1007/978-3-031-49787-2_19

Amber, A., Fakhra, A., & Ammara, A. (2014). Quantification of fat in chicken's feather meal for its conversion into biodiesel. *International Research Journal of Environmental Sciences, 3*(6), 67.

Barua, P., Chowdhury, T., Chowdhury, H., Islam, R., & Hossain, N. (2020). Potential of power generation from chicken waste-based biodiesel, economic and environmental analysis: Bangladesh's perspective. *SN Applied Sciences, 2*, 330.

Bharat, K. T., & Bhattacharya, A. (2012). The production and analysis of biodiesel from waste chicken skin and pork skin fat and a comparison of fuel properties to petroleum derived diesel fuel. *Environmental Science, 2*(3), 8.

Bianchi, C. L., Bofto, D. C., Pirola, C., & Ragaini, V. (2010). Low temperature de-acidification process of animal fat as a pre-step to biodiesel production. *Catalysis Letters, 134*, 179.

Das, A., Dey, S., Gacem, A., Yadav, K. K., Nath, H., Ram, D. K., & Das, A. P. (2025). Revolutionizing solid hydrocarbon sourcing: Plastic's journey from waste to upcycled treasure. *Environmental Science and Engineering, Part, F218*, 331–351. https://doi.org/10.1007/978-3-031-81615-4_14

El Haijjaji, S., Bouladab, C., Belkasmi, I. A., & Hinnach, S. (2016). Valorization of diverse low-value wastes into biodiesel. *Journal of Materials and Environmental Science, 7*(12), 4502.

Hassan, S. T., & Lakkimsetty, N. R. (2018). Utilization of modified eggshells as a solid catalyst for the conversion of waste cooking oil to biodiesel. *Austin Chemical Engineering, 5*(1), 1–8.

Jeong, G. T., Yang, H. S., & Park, D. H. (2009). Optimization of transesterification of animal fat ester using response surface methodology. *Bioresource Technology, 100*, 25.

Khan, A. M., Khan, I., & Kanwal, S. (2014). Comparative production of biodiesel from waste chicken fats and cooking oil. *Journal of Biofuels, 5*(1), 32.

Lakkimsetty, N. R., Gaddala, B., Doddamani, D., Thandlam, A. K., Karu, C. V., Namdeti, R., & Al Habsi, S. (2025). Experimental study on reducing the base sediment and water in crude oil separation. *Fuel, 397*, Article 135478. https://doi.org/10.1016/j.fuel.2025.135478

Lakkimsetty, N. R., Lakavat, M., Walke, S., Gandi, S., Kavitha, G., & Shaik, F. (2022). Comprehensive process design of biodiesel from waste cooking oil (WCO) using magnetic nano-catalyst. *AIP Conference Proceedings, 2463*, Article 030007. https://doi.org/10.1063/5.0080210

Maryem, B., Habib, E. L., Azzouzi, E. L., Meryem, K., El Bouzaidi, H., Idrissi, Y., & El Belghiti, M. A. (2014). Study of manufacturing biodiesel from waste animal fats (chicken) in Morocco. *Journal of Chemical and Pharmaceutical Research, 6*(12), 844.

Meinderstma, W., & Block, K. (2012). Effects of new fossil fuel developments on the possibilities of meeting 2 °C scenarios. Ecofys.

Mohanty, M., Tripathy, B., Koteswara Reddy, G., & Das, A. P. (2024). Greener technologies for establishing restraint over microplastic pollution. In *Sustainable microbial technology for synthetic and cellulosic microfiber bioremediation* (pp. 239–258). https://doi.org/10.1007/978-3-031-62600-5_13

Nageswara Rao, L., & Al Riyami, S. S. S. (2018). Experimental investigation on production of biodiesel from Padina boergesenii sp. macro algae. *Austin Chemical Engineering, 5*(1), 1–7.

Purandaradas, A., Silambarasan, T., Murugan, K., Babujanarthanam, R., Ghandhi, A. D., Dhandapania, K. V., Anbumani, D., & Kavitha, P. (2018). Development and quantification of biodiesel production from chicken feather meal as cost-effective feedstock by using green technology. *Biochemistry and Biophysics Reports, 14*, 133.

Reddy Koteswara, G., Bhargavi, S., Yamuna, D., Anjani Devi, M., Jeevana Lakshmi Kumar, G., Manas, T., Adarsh, S., Bhavya Sai, V., Nath, H., Alreshidi, M. A., Togun, H., Ram, D. K., & Das, A. P. (2025). Recent advancements in the remediation of marine hydrocarbon and plastic pollutants. *Environmental Science and Engineering, Part F218*, 127–142. https://doi.org/10.1007/978-3-031-81615-4_7

Rehman, H. A., Shah, S. F. A., Aftab, A., & Ali, Z. M. (2016). Trans-esterification of poultry skin fat to produce biodiesel. *Journal of Analytical and Environmental Chemistry, 17*(2), 143.

Sahu, A., Dey, S., Mallick, K., Priyadarsini, S., Behera, I. D., & Das, A. P. (2024). A review on advanced technology for sustainable management of synthetic microplastic waste. In *Sustainable management of urban plastic waste through circular economic approaches* (pp. 262–279). https://doi.org/10.1201/9781003465447-15

Wang, Y., Ou, S., Liu, P., & Zhang, Z. (2007). Preparation of biodiesel from waste cooking oil via two-step catalyzed process. *Energy Conversion and Management, 48*, 184.

Advancing Environmental Sustainability through Nanobiotechnological Approaches

Himanshu Jangid, Gaurav Kumar, Nisha, Alok Prasad Das, and Anuprita Ray

1 Introduction

Nanobiotechnology is an interdisciplinary field that merges nanotechnology with biological sciences to design eco-efficient tools for real-world applications. When applied to environmental sustainability, nanobiotechnology leverages nanoscale materials synthesized from biological sources to develop novel solutions for pollution control, waste management, and sustainable resource utilization (Shahcheraghi et al., 2022). Such bio-integrated nanomaterials possess unique features, including ability to react selectively, acting as catalysts, and having a considerable surface area which helps address major environmental problems in a quicker and more environmentally friendly way (Das et al., 2025; Mohanty et al., 2024a, 2024b). As global issues worsen, its use in environmental sustainability is increasing very quickly (Saraswat et al., 2023). Climate change, industrial pollution, rising greenhouse gas emissions, depletion of clean water sources, and improper waste disposal are threatening both ecosystems and human health (Ofremu et al., 2024). Usually, traditional techniques for fixing or monitoring environmental problems are high in energy use and they sometimes have chemical dangers or cannot be used in all settings. Thanks

The original version of the chapter has been revised. A correction to this chapter can be found at https://doi.org/10.1007/978-3-032-11377-1_21

H. Jangid · Nisha · A. Ray (✉)
School of Bioengineering and Biosciences, Lovely Professional University, Phagwara, India
e-mail: anuprita.ray88@gmail.com

G. Kumar
Amity Institute of Microbial Technology, Amity University, Jaipur, Rajasthan, India

A. P. Das
Department of Life Sciences, Rama Devi Women's University, Bhubaneswar, Odisha, India

S. Ghosh et al. (eds.), *Advances in Organic Waste Conversion Through Industrial Biotechnology*, Sustainable Environmental Waste Management Strategies,
https://doi.org/10.1007/978-3-032-11377-1_9

to nanobiotechnology, biosensors are made to find pollutants, nanoparticles biologically treat wastewater and smaller molecules can remove heavy metals and organic pollutants (Hemdan et al., 2025; Sharma et al., 2023). Green nanotechnology, a sub-domain of nanobiotechnology, emphasizes the synthesis of nanoparticles using biological agents such as microbes, algae, and plant extracts. These approaches avoid harsh chemicals and minimize ecological risks. Biologically derived nanoparticles have shown strong efficacy in removing contaminants such as dyes, heavy metals, and persistent organic pollutants from water and soil systems (Alqarni et al., 2022; Thipe et al., 2022). Moreover, the field has made strides in real-time environmental monitoring through nanobiosensors—devices combining nanomaterials with biological molecules to detect pollutants at extremely low concentrations. These tools enable on-site, rapid, and cost-effective assessments of environmental quality (Tovar-Lopez, 2023). Advanced systems now integrate biosensors with AI and IoT for predictive environmental modeling and smart pollution control (Alahi et al., 2023; Popescu et al., 2024). However, as promising as it is, nanobiotechnology must address critical concerns related to nanotoxicity, environmental persistence, and regulatory frameworks (Ray et al., 2009). Sustainable implementation will require life cycle assessments, green synthesis protocols, and responsible policy design. In this chapter, we explore the innovations, applications, and future potential of nanobiotechnology in promoting environmental sustainability. We aim to provide a comprehensive roadmap for researchers, practitioners, and policymakers working to solve ecological challenges through next-generation nano-enabled technologies.

2 Foundations of Nanobiotechnology

Nanobiotechnology, as a convergence of nanotechnology and biology, is founded on the ability to design and manipulate materials at the nanoscale—typically between 1 and 100 nm—to perform targeted environmental or biological functions. At this scale, materials exhibit unique optical, electrical, and catalytic properties due to their large surface area-to-volume ratio and quantum effects (Barati et al., 2023; Ike Onyia et al., 2020). In environmental contexts, nanobiotechnology aims to engineer biocompatible and eco-friendly nanomaterials that can detect, degrade, or transform pollutants into harmless by-products. The concept of using biological entities to manipulate materials at the nanoscale has roots in natural processes observed in microbes, plants, and animals (Singh et al., 2024). However, it was only in the last two decades that this phenomenon was harnessed deliberately for technological purposes. Figure 1 Biogenic Pathways for Nanoparticle Synthesis. Microorganisms such as bacteria, fungi, and plants synthesize nanoparticles via bioreduction, enzymatic activity, and metabolite-mediated stabilization. Commonly produced nanoparticles include silver (AgNPs), zinc oxide (ZnO), iron oxide (Fe_3O_4), titanium dioxide (TiO_2), and copper (CuNPs).

At first, researchers mainly worked on using microbes to make and use metal nanoparticles in medicine. Over the years, they also started to focus on environmental

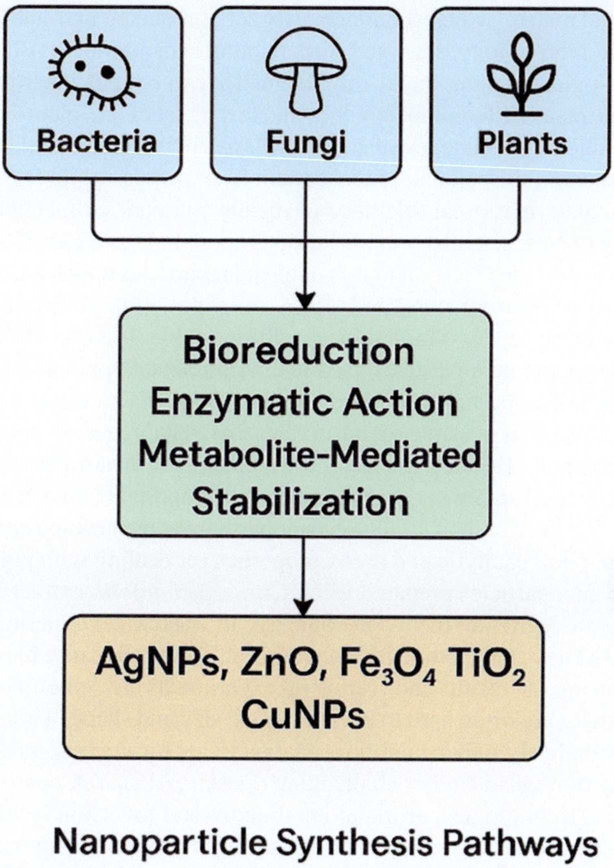

Fig. 1 Biogenic pathways for nanoparticle synthesis. microorganisms such as bacteria, fungi, and plants synthesize nanoparticles via bioreduction, enzymatic activity, and metabolite-mediated stabilization. Commonly produced nanoparticles include silver (AgNPs), zinc oxide (ZnO), iron oxide (Fe_3O_4), titanium dioxide (TiO_2), and copper (CuNPs)

problems in water purification, clearing soil contamination and preparing the atmosphere. A main example in nanobiotechnology is biogenic nanoparticle synthesis, where bacteria, fungi, algae, and even plant extracts are used to form nanoscale materials (Patil & Pawar, 2019). People increasingly choose this approach because it is friendly to the environment and an environmentally sound option to the usual chemical process which requires working with toxic materials and investing much power.

The synthesis process typically involves the reduction of metal ions by bioactive compounds such as flavonoids, terpenoids, polyphenols, and enzymes present in the biological matrix, which simultaneously act as reducing and stabilizing agents to regulate the morphology, size, and dispersion of nanoparticles (Bhati, 2023). A number of biological systems have been designed to carry out nanoparticle synthesis. Thanks to their many different metabolites and fast metabolism, plant-based systems are highly efficient. Advantages of using bacteria include their ability to produce nanoparticles within the cell and to make them from metals, while features of fungi and yeasts include their great tolerance of metals and their extracellular enzymes. With these organisms, scientists can make silver (AgNPs), gold (AuNPs), zinc oxide (ZnO), titanium dioxide (TiO_2), and iron oxide nanoparticles which are all successful in helping to solve environmental problems such as pollution, sensing, and catalysis (Karunakaran et al., 2023; Mughal et al., 2021). Table 1 shows the comparative overview of biogenic nanoparticles used in environmental applications. Functional traits in biogenic nanoparticles enhance their contribution to a cleaner environment. Being highly reactive is possible for adsorbents due to their greater degree of contact with the environment. Besides, the fact that biomaterials are harmless to the environment and can be used safely in natural ways is very important (Bala et al., 2024; Das et al., 2024a). Because of their structure, nanoparticles often become crystallized and exhibit helpful photocatalytic and redox properties for dealing with pollutants. Titanium dioxide nanoparticles prepared using *Citrus aurantifolia* extract shown excellent efficiency in degrading methylene blue dye in wastewater experiments (Vishali et al., 2021). Even so, the method faces certain challenges. Because biomaterials can vary, re-producing the results and scaling up experiments are sometimes difficult. In addition, particles are often hard to keep the same size and shape, a big problem. We still do not completely understand how nanoparticles form which makes standardization among biological studies challenging (Desai, 2012). Yet, continuous studies on using omics technologies, artificial intelligence and precision synthetic biology could help us overcome these issues and scale biogenic synthesis for technology that benefits the environment.

3 Applications in Environmental Sustainability

3.1 Heavy Metal Detoxification

Heavy metal contamination poses one of the most persistent threats to environmental and human health due to the non-biodegradable and bio accumulative nature of metals such as lead (Pb), cadmium (Cd), chromium (Cr), and arsenic (As). Nanobiotechnology offers a promising solution for detoxification through the development of biogenic nanoparticles capable of adsorbing, reducing, or transforming these pollutants into less harmful forms (Das et al., 2024b; John & Das, 2024). The primary mechanisms involved in metal detoxification include adsorption, where metal ions

Table 1 Comparative overview of biogenic nanoparticles used in environmental applications

Nanoparticle type	Biological source	Synthesis method	Particle size/morphology	Key properties	Environmental application	References
Silver (AgNPs)	Fungi, bacteria, plant extracts	Bioreduction via enzymes and metabolites	~10–50 nm; spherical	Antimicrobial, anti-inflammatory	Water purification, biofouling prevention	(Tortella et al., 2021)
Gold (AuNPs)	Algae, fungi, actinomycetes	Biomolecule-assisted synthesis	~15–25 nm; spherical/hexagonal	Plasmonic, biocompatible	Crop protection, pesticide detection	(Eweis et al., 2024)
Iron oxide (Fe_3O_4)	Plant extracts, bacteria	Green synthesis using polyphenols	~20–30 nm; octahedral	Magnetic, catalytic	Heavy metal cleanup, magnetic filtration	(Verma et al., 2024)
Zinc oxide (ZnO)	Microbes, plant metabolites	Bio-assisted reduction	~20–40 nm; rod-shaped/spherical	UV-active, photocatalytic	Dye degradation, antimicrobial coatings	(Sadhasivam et al., 2021)
Titanium dioxide (TiO_2)	*Annona muricata* extract	Biogenic synthesis via plant phenolics	~15–30 nm; crystalline	Photocatalytic, antibacterial	Wastewater dye removal, air purification	(Chouke et al., 2022)
Copper (CuNPs)	Bacteria, fungi, plant extracts	Extracellular enzyme-mediated reduction	~25–50 nm; cubic/round	Antiviral, catalytic	Textile finishing, pesticide control	(Tortella et al., 2021)
Magnesium oxide (MgO)	Leaf extracts, bacteria	Green synthesis	~30–60 nm; irregular	Antifungal, high thermal stability	Agricultural fungicide, soil remediation	
Nickel oxide (NiO)	Medicinal plant extracts	Phytochemical reduction	~20–50 nm; hexagonal	High redox potential	Battery electrodes, catalytic pollutant breakdown	(Chouke et al., 2022)

adhere to the nanoparticle surface; reduction, wherein toxic metal ions are converted into more stable oxidation states; and complexation, where biological molecules present on nanoparticles bind and immobilize metals (Oladimeji et al., 2024; Vishali et al., 2021). Biogenic nanoparticles have demonstrated high efficacy in heavy metal remediation, largely due to their high surface area, functionalized biomolecular coatings, and redox-active properties. Microbial systems play a central role in this process. Certain bacteria such as *Shewanella oneidensis* have been used to synthesize biogenic ferrous sulfide (bio-FeS) nanoparticles, which exhibit superior activity in reducing toxic hexavalent chromium (Cr (VI)) to its less toxic trivalent form (Cr(III)). These bio-hybrids, where nanoparticles self-assemble on bacterial surfaces, offer enhanced extracellular reduction efficiency and long-term operational stability, even under high metal concentrations (Fu et al., 2023). Nanoparticle technology relies a lot on plants to remove heavy metals from soil. Exploiting phenolic and flavonoid secondary metabolites in plants allows the production of biogenic nanoparticles including iron oxide, silver, and magnetite. Such nanoparticles made from plants can efficiently pick up heavy metals from wastewater and polluted soil. Nanoparticles created using *Cnidium monnieri* seed extract, namely magnetic iron oxide, removed both Pb(II) and Cr(III) by a second-order kinetic process and could be used several times without losing their efficiency (Lingamdinne et al., 2017). Fungi produce both nanoparticles and toxic material-removing agents known as peroxidases and metallothioneins. They help to remove harmful heavy metals from the body and also contribute to making nanoparticles. When used, fungal-derived nanoparticles help remove or change the form of harmful pollutants (Geetha et al., 2021). Microbial methods are favored for their safety, how easy they are to scale up and their affordability. They run at normal temperatures, do not call for toxic solvents and make many types of metal nanoparticles with unique biological activity. As an example, *Pseudomonas* and *Bacillus* bacteria have handled the problem of heavy metals by making them smaller and less harmful through certain pathways. Using these bio-systems, we successfully generate nanoparticles that are more effective for cleanup activities than those made using chemicals (Singh et al., 2023). In addition to helping the body clean up toxins, biogenic nanoparticles can detect levels of metal ions. When green tea extracts were used to synthesize silver nanoparticles, the nanoparticles showed excellent sensitivity and selectivity for Hg^{2+} and Pb^{2+} in water. Plasmonic properties of these nanoparticles change in the presence of ions, letting us watch for pollution in real time (Widatalla et al., n.d.). Overall, nanobiotechnology includes powerful options to help clean up heavy metals. With microbes, plants and fungi, it makes it possible to create nanoparticles that handle the green removal of contaminants through adsorption, chemical transformation and complexation. In labs and out in field studies, these budget-conscious ecological methods have proved successful and could help on a larger scale (Chan et al., 2024; Nain et al., 2025).

3.2 Wastewater Treatment

Due to fast urbanization and industrialization, the global increase in wastewater is bringing many types of chemical and biological toxins into the water supply. Even though chemical processes are successful in removing organic waste and pathogens from drinking water, they often do not remove heavy metals, pharmaceuticals, dyes and newly emerging pollutants. Nanobiotechnology is now seen as a leading solution in treating wastewater because it uses both cutting-edge materials and hybrid systems that protect the environment (Sanni et al., 2024). Dealing with hard-to-treat waste pollutants has become much easier with the use of biogenic nanocatalysts. Made by using either plants or microorganisms, these nanoparticles are applied in advanced oxidation systems such as Fenton and photo-Fenton reactions (Mohapatra et al., 2023). In the case of photo-Fenton systems, plant-produced nano-catalysts are used to break down dyes and phenols, since this method is reusable, slightly toxic and can be carried out at low temperatures. Sunlight and hydrogen peroxide are used by these systems to produce hydroxyl radicals which convert pollutants into safety co (Mishra & Das, 2023; Mishra et al., 2024) mpounds (Divyapriya et al., 2016). A different approach uses composite material bio-nano filters made from ZnO-coated chitosan fibers. These smart materials work both to filter water and also for photocatalysis. As an illustration, composites formed with ZnO and chitosan have been effective at degrading dyes and chromium ions and also have demonstrated antimicrobial activity against *E. coli* and *S. aureus*, displaying several useful functions (Bhapkar & Bhame, 2024). Combinations of metal oxides with carbon substrates in hybrid nanomaterials give them much better performance because of their photocatalytic, adsorptive, and mechanical functions. An important example is found in BiOBr/Ag/AgBr heterojunctions created on carbon fiber cloth which behave as filter-membrane photocatalysts. Both systems displayed nearly total destruction of polluting organic dyes using light and kept their efficiency throughout constant operation, proving they could be useful in real industry settings (Liu et al., 2021). Water quality in dual-filtration systems has been enhanced by including biogenic iron oxide nanoparticles. A test of iron oxide nanoparticles made with *E. coli* on filters cut down on both the microbial contamination and the amounts of chemicals. Total plate count and the presence of coliform in wastewater were both effectively controlled by the modified system owing to the use of biosynthesized nanoparticles (Al-Allaf et al., 2023). Algal biomass has also been incorporated in forming different functional hybrid materials, including nanofibers and polymer-based composites. By using these types of systems, both nitrogen and phosphorus are removed and the process produces useful biodiesel. Because of their ability to perform many tasks, they become top options for scaled-up circular bioeconomy models (Eroglu & Raston, 2017). Other than using individual biochemicals, inventing HNMs from polymers, metals and surfaces found in nature has expanded how we address wastewater problems. Thanks to these substances, one can retrieve more impurities, tune their use in photocatalysis and decide on the most effective approach for reusing materials. Applying these techniques appears to increase how effectively pollutants are removed, reduce chances of secondary pollution and make it easier to operate the system (Zahid et al., 2020) (Das & Ghosh, 2023; Dey et al., 2023). Essentially, nanobiotechnology combines modern techniques to remove waste by growing nanocatalysts, bio-nano filters, and hybrid nanocomposites. The progress we have achieved from small experiments and better materials makes launching these technologies on a larger scale both practical and sustainable for industry and municipalities.

3.3 Nano-Bioremediation

Nano-bioremediation is an improvement on regular bioremediation by combining nanotechnology and enzymatic or microbial processes. By combining nanomaterials and biology, this approach makes use of the strong catalytic effects, larger surface, and greater reactivity found in nanomaterials, together with the specificity and ability to reproduce found in biological systems (Chaudhary et al., 2023). In this field, enzymes are either fixed on nanoparticles or chemically copied as nanozymes to speed up how pollutants are destroyed. These enzymes are very helpful for dealing with stubborn pollutants such as pesticides, PAHs and synthetic dyes which resist breakdown in nature (Yogalakshmi et al., 2020). Pairing microbes with nanoparticles is a main feature of nano-bioremediation which speeds the process of breaking down the pollutants and improves access to them (Hasan et al., 2025). Nanoparticle synthesis and cleaning up pollutants are commonly done by microorganisms such as bacteria and fungi. Nanoparticles have been found to support microorganisms by increasing their membrane permeability, encouraging biosurfactant formation and assisting in the release of enzymes which together assist with bioremediation. This joint effort has successfully proven to be valuable during the breakdown of petroleum hydrocarbons, with metallic nanoparticles boosting microbes' enzyme action and tolerance when conditions are rough (Bhatt et al., 2022; Chauhan et al., 2023). Recent research has shown that nano-bioremediation helps deal with certain classes of pollutants. As an example, treating textile wastewater contaminated by methylene blue and congo red becomes easier when microorganisms are improved with biogenic nanoparticles. Dyes were photocatalytically degraded during UV irradiation by magnesium oxide nanoparticles derived from *Syzygium samarangense* leaves, with over 90% of the dyes successfully removed under experimental conditions (Rajeshwari et al., 2023). Extensive improvements have been made in pesticide degradation thanks to nano-bioremediation. Strongly increasing their catalytic activity and substrate attraction is possible by immobilizing hydrolases, laccases, and oxidoreductases on nanoparticle surfaces. They support the process of breaking organophosphates and carbamates down into harmless molecules. Researchers have found recently that when microbial enzymes are immobilized on nanocarriers, they do a better job of treating a wide variety of agricultural pesticides and plasticizers compared to free enzymes, regardless of whether there is air in the environment (Narayanan et al., 2023). Hydrocarbons and PAHs are often handled very effectively with nano-bioremediation (Das et al.,

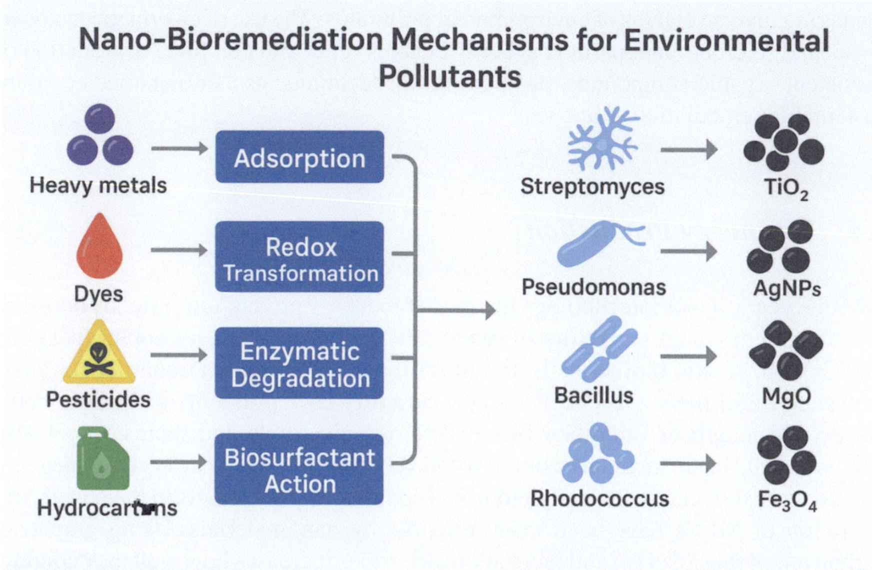

Fig. 2 Mechanisms of nano-bioremediation for environmental pollutants. Different pollutants (e.g., heavy metals, dyes, pesticides, hydrocarbons) are removed or neutralized via adsorption, redox transformation, enzymatic degradation, and biosurfactant action. Specific microbes such as *Streptomyces*, *Pseudomonas*, *Bacillus*, and *Rhodococcus* are linked to nanoparticle formation and remediation mechanisms

2024a; Mohanty et al., 2024a, 2024b; Ray et al., 2024). The Fig. 2 shows mechanisms of Nano-Bioremediation for Environmental Pollutants. Different pollutants (e.g., heavy metals, dyes, pesticides, hydrocarbons) are removed or neutralized via adsorption, redox transformation, enzymatic degradation, and biosurfactant action. Specific microbes such as *Streptomyces*, *Pseudomonas*, *Bacillus*, and *Rhodococcus* are linked to nanoparticle formation and remediation mechanisms.

Even though hydrophobic compounds do not degrade easily, their availability to the bacterial cells rises when biosurfactants and metallic nanoparticles are introduced. Consequently, more bacteria take up the chemicals and enzymes change them into new products. Nanoparticles are able to move electrons which aids the fast-chemical reactions needed for the breakdown of stable contaminants (Kumari & Singh, 2016). Dye removal by nano-bioremediation also gains from adding photocatalytic and antimicrobial nanostructures during development. A remarkable result shows that silver nanoparticles formed by *Streptomyces tuirus* have a strong ability to catalyze photosynthesis by degrading methylene blue and also offer antibacterial protection in wastewater settings. Table 2 shows nano-bioremediation of environmental pollutants. Such dual-function nano-bioremediation systems offer comprehensive treatment by simultaneously eliminating both chemical and microbial contaminants (Das et al., 2024c; Dey et al., 2024; Priyadarsini & Das, 2025). In summary, nano-bioremediation represents a powerful, eco-friendly, and efficient approach for managing diverse classes of environmental pollutants. The use of enzymatic nanocatalysts and microbe-nanoparticle synergy enables rapid and complete degradation of persistent organic compounds, positioning this technique as a cornerstone of future sustainable remediation strategies.

3.4 Bioenergy Production

Over the years, nanobiotechnology has made bioenergy production easier by boosting the effectiveness and scalability of renewable systems such as microbial fuel cells (MFCs), enzymatic biofuel cells and those focused on algae. Despite being alternatives to fossil fuels, these technologies currently have problems with energy efficiency, the length of time their biocatalysts remain stable and their costs. Those nanomaterials made in green or bio-assisted ways provide key answers by improving electron transfer, catalysis, and yield from biomass. Improvements in the design and operation of MFCs have been made possible by nanomaterials. Using graphene, carbon nanotubes (CNTs) and biochar on electrodes increases how well they conduct and promotes the growth of electroactive biofilms. As a result of these modifications, electron transfer improves, the power output grows and more microorganisms are supported (Mashkour et al., 2021). Hybrid electrodes using polypyrrole-metal oxide composites, particularly iron and manganese oxides, show enhanced bioelectrocatalysis and operational stability (Dessie & Tadesse, 2022). Moreover, nano-enhanced membranes with hydrophilic or antibacterial nanoparticles improve proton conductivity and reduce biofouling—further boosting performance (Korsa et al., 2025). Recent advances include enzyme-based biofuel cells where nanostructures stabilize and improve the catalytic action of enzymes. These nanobiocatalysts address limitations of low power density and short cell life by offering improved enzyme reusability, enhanced substrate specificity, and robust operational resilience (Fredj et al., 2025). In algal biofuel production, nanomaterials are revolutionizing the process by boosting photosynthetic efficiency, accelerating nutrient uptake, and enhancing lipid biosynthesis pathways. Metallic and metal oxide nanoparticles, such as TiO_2, Ag, and MgO, have demonstrated strong effects in increasing lipid accumulation in algal cells (Thanigaivel et al., 2023). In addition, magnetic nanoparticles enable efficient harvesting of algal biomass through magnetic flocculation, achieving near-total recovery, and reducing energy input (Chintagunta et al., 2021). These advances help tackle the key cost and efficiency bottlenecks in biodiesel production. Nanobiocatalysts have transformed biomass conversion by enhancing enzymatic processes like hydrolysis, transesterification, and fermentation. Enzymes immobilized on nanoparticles such as mesoporous silica, magnetic Fe_3O_4, or ZnO display greater stability, thermal tolerance, and operational reuse, significantly cutting enzyme costs and boosting conversion yields (Singh et al., 2020). The combination of nano- and enzymatic catalysts further enhances reaction efficiency by creating favorable nano-microenvironments that facilitate higher catalytic rates and feedstock compatibility without the need for harsh chemicals (Chang et al., 2024).

Table 2 Nano-bioremediation of environmental pollutants

Pollutant	Nanoparticle type	Biological system	Mechanism	Removal efficiency/insight	References
Heavy metals (Cr, Pb, Cd)	Fe_3O_4, ZnO, nZVI	*Actinobacteria*, *Pseudomonas*, *Bacillus*	Adsorption, redox transformation, biosorption	Microbial consortia + nanoparticles enhance stability and reduction	(Behera & Das, 2023)
Pesticides (Organophosphates)	AgNPs, CuNPs	*Streptomyces*, engineered bacteria	Enzymatic hydrolysis, oxidoreductase activity	Enhanced breakdown using immobilized microbial enzymes	(Maqsood et al., 2023)
Dyes (Methylene blue, Congo red)	TiO_2, AgNPs, MgO	*Syzygium*-mediated synthesis, *Pleurotus* spp.	Photocatalysis, enzymatic oxidation (laccases)	>90% degradation under light, stable across multiple cycles	(Yogalakshmi et al., 2020)
Hydrocarbons (PAHs, petroleum)	Bio-supported metal NPs	*Burkholderia*, *Rhodococcus*, fungi	Synergistic enzymatic breakdown, biosurfactant activity	Microbial + nano synergy accelerates hydrocarbon degradation	(Chauhan et al., 2023)
Plastics/Complex Organics	Enzyme-nanoparticle hybrids	Lipases, oxidoreductases, engineered microbes	Catalysis, hydrolysis, biofilm-assisted degradation	Potential for multi-pollutant removal, emerging in enzymatic systems	(Narayanan et al., 2023)

In lignocellulosic pretreatment, nanomaterials improve enzyme loading and enable efficient carbohydrate release, enhancing downstream fuel yields (Chandel et al., 2024). Waste-integrated systems like MFCs using wastewater embedded with carbon nanotubes show not only improved voltage output but also enhanced contaminant removal without toxicity to microbes, promoting dual-purpose sustainability (Miran et al., 2016). Moreover, life cycle assessments reveal that green-synthesized nanomaterials—such as 3D graphene bubbles derived from biomass pyrolysis—reduce environmental burden while enhancing energy conversion efficiency (Jawahar & Pillai, 2023). Despite these advances, challenges remain around cost-effective nanomaterial production, ecological toxicity, and process scalability. Researchers are now emphasizing the development of biodegradable and eco-safe nanomaterials to align technological innovation with environmental stewardship (Togun et al., 2025). In summary, nanobiotechnology is reshaping bioenergy production by enhancing every stage—from microbial catalysis and algal growth to enzymatic conversion and fuel recovery. These advances are making renewable energy systems more viable, scalable, and sustainable for the global transition away from fossil fuels.

3.5 Environmental Monitoring

Over the years, advances in nanobiotechnology have improved how renewable systems for bioenergy, including MFCs, enzymatic biofuel cells and algae-based systems, work. Although they act as alternatives to fossil fuels, these technologies still struggle with their energy efficiency, the length their catalysts stay useful and their production costs at this stage. Using eco-friendly techniques to produce nanomaterials gives scientists vital help with electron transfer, catalytic reactions and processing biomass. Thanks to nanomaterials, improvements in MFC design and operation have become possible. To make electrodes conduct electricity more efficiently and encourage the growth of electroactive biofilms, graphene, carbon nanotubes (CNTs), and biochar are used on them. As a consequence of these changes, electrons are transferred more effectively, the power generated increases and more microbes are supported (Nath et al., 2018). FRET mechanisms have also allowed carbon dots and graphene oxide to become important fluorescent sensors for detecting pesticides and phenolic compounds (Gaviria-Arroyave et al., 2020). A major development is the discovery of electrochemical and plasmonic nanosensors. The electrical or optical differences that occur when the target pollutants interact with the biological nanoparticles are used by these biosensors. By using enzyme-coated or aptamer electrodes such devices can discover nanomolar down to picomolar levels of heavy metals and organic pollutants in both water and soil samples (Aihaiti et al., 2024). Plasmonic sensors, on the other hand, use surface-enhanced Raman scattering (SERS) to amplify weak signals, allowing for the detection of minute amounts of toxins and pathogens (Chen et al., 2024). Detection of both bacterial and algal toxins by using biosensors that include graphene and transition metal dichalcogenides (TMDs) has proven to be very effective. Using these materials, scientists can attach many recognition elements, boosting how sensitive and precise their method is for detecting environmental compounds (Li et al., 2019). Figure 3 shows **environmental applications of biogenic nanoparticles**. Biogenic nanoparticles play a multifaceted role in sustainability, contributing to heavy metal detoxification, environmental biosensing, wastewater treatment, nano-bioremediation, and bioenergy generation. By using wireless technology and artificial intelligence, these sensors are playing a key role in forming smart environmental monitoring systems. With nanobiosensors in portable devices, analysis can be done on-site with immediate transmission of findings, making it simpler for organizations to handle contamination problems as they occur. This combination of technologies can improve water treatment, industrial waste monitoring, spotting runoff in farming and air-quality observation (Justino et al., 2017). In short, nanobiotechnology is making it possible to monitor the environment by creating biosensors that are fast, easy to use, affordable and very sensitive. For complex environmental problems, having these biosensing platforms will be key to keeping ecosystems safe and sustainable.

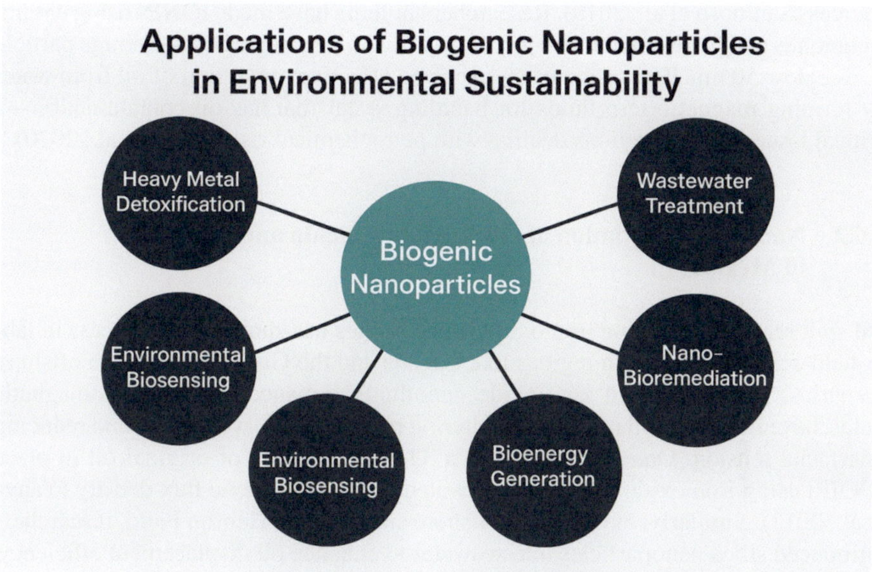

Fig. 3 Environmental applications of biogenic nanoparticles. Biogenic nanoparticles play a multifaceted role in sustainability, contributing to heavy metal detoxification, environmental biosensing, wastewater treatment, nano-bioremediation, and bioenergy generation

3.6 *Case Studies of Field Deployments in Nanobiotechnology*

Field usage of nanobiotechnology equipment moves inventions from the laboratory to the actual environment. The following case studies show that nanomaterials and iron oxide nanoparticles in particular, are applied extensively in cleaning water, cleaning up oil spills and monitoring air quality.

3.6.1 Iron Oxide Nanoparticles for Water Purification in India

Due to their fair cost, magnetism and high ability to absorb things, iron oxide nanoparticles (IONPs) are used widely in water treatment both in the cities and in rural India. Removing heavy metals, oil, and organic pollutants from drinking water and industrial waste is accomplished best with nanoscale magnetite. The combination of these nanoparticles' magnetism and ability to exchange ions makes it possible to target and remove different contaminants from surface water and groundwater sources (Santhosh et al., 2019). Researchers in India have made IONPs using natural techniques and plants like *Andrographis paniculata,* ending up with average particle sizes below 50 nm. The nanoparticles showed effective separation of oil from water by forming magnetic ferrofluids for handling water that has oil contamination—a critical issue in many regions dealing with petrochemical risks (John et al., 2020).

3.6.2 Nano-Bioremediation of Oil Spills in Canada and the Gulf of Mexico

Oil spill remediation using iron oxide nanoparticles has shown great success in lab-to-field-scale transitions in regions like Canada and the Gulf of Mexico. In offshore scenarios, iron oxide and zinc oxide nanofluids, enhanced with electromagnetic fields, have improved oil recovery by altering reservoir rock wettability and reducing interfacial tension. One study reported a 33.45% recovery of original oil in place (OOIP) using iron oxide nanofluids activated by high magnetic flux density (Yahya et al., 2013). Similarly, in Canadian offshore sites like the Hebron Field, researchers introduced silica nanoparticles into seawater to enhance oil displacement efficiency. This technique shifted rock wettability from oil-wet to water-wet, improving oil recovery rates by up to 25% (Ortega, 2018). Beyond oil extraction, IONPs have been embedded in biochar-alginate beads to support the slow release of nutrients and microbes during oil spill cleanups in freshwater ecosystems. This approach addresses nutrient washout—a key limitation in aquatic bioremediation—and achieves up to 71.69% pollutant sorption efficiency (Osadebe et al., 2024).

3.6.3 Biosensor-Based Air-Quality Monitoring in Urban China

In major Chinese cities, biosensors integrated with nanomaterials have been deployed for real-time monitoring of air pollutants such as volatile organic compounds (VOCs), nitrogen oxides, and PM2.5 particles. These biosensors often incorporate nanostructured films of metal oxides—such as titanium dioxide and zinc oxide—combined with biological receptors that selectively bind to target gases or particles (Sahu et al., 2024). While specific deployments in China were less frequently detailed in recent publications, global trends in urban biosensor use point to China's pioneering role in smart-city air monitoring frameworks. Notably, China's Ministry of Ecology and Environment has supported the development of integrated biosensor networks for rapid detection and public health responses in high-density zones. Although most publications are lab-focused, the technology is moving toward field trials involving wearable biosensors and stationary monitors embedded in traffic zones and industrial peripheries. This trend aligns with China's "Air Pollution Prevention and Control Action Plan," where nanomaterial-based detection systems are helping local governments meet WHO air-quality targets (Ike Onyia et al., 2020; Xu et al., 2020).

3.6.4 Site Remediation Using Nano-Zero-Valent Iron (nZVI)—Global Scale

One of the most widely deployed nanotechnologies in environmental remediation is nano-zero-valent iron (nZVI). Over 15 documented commercial and academic field-scale trials have applied nZVI to contaminated sites for groundwater and soil decontamination. These include chlorinated solvent plumes and heavy metal-contaminated aquifers in the U.S., Europe, and Asia (Watlington, 2005). The success of nZVI stems from its high reactivity and mobility in porous soils. One notable trial in Ontario, Canada, successfully used nZVI to dechlorinate trichloroethylene (TCE) in groundwater, achieving over 80% contaminant reduction within weeks. Similar efforts in Germany and Taiwan have replicated these outcomes, demonstrating rapid on-site pollutant degradation.

3.6.5 Biochar-Iron Nanocomposite Beads for Freshwater Remediation—Nigeria

In Nigeria, researchers tested biochar-impregnated iron oxide nanoparticles embedded in alginate beads to clean oil-contaminated freshwater ecosystems. These beads served as slow-release biostimulants that supported microbial degradation while adsorbing oil residues. The field trial achieved a pollutant removal efficiency of 71.69%, demonstrating both sorption and biodegradation mechanisms (Osadebe et al., 2024). This hybrid approach offers a scalable, low-toxicity method for oil spill remediation in developing regions and integrates circular bioeconomy principles by using agricultural waste to produce biochar.

3.6.6 Urban Air Monitoring with Nanotech Sensors—Southeast Asia and Europe

Advanced air-quality monitoring networks using portable nanotech-based gas sensors have been deployed in multiple cities, including Bangkok and several European capitals. These networks use metal oxide sensors (e.g., TiO_2, ZnO) embedded in hand-held or fixed stations for real-time detection of NO_2, CO, and VOCs (Vaseashtaa et al., 2007). In Bangkok, sensors integrated with GPS and cloud-based platforms allowed city officials to map pollution hotspots and adjust traffic and industrial activity in real-time. Similar applications in Italy and France have helped urban planners validate models of pollution dispersion, leading to targeted emissions reduction strategies.

3.6.7 Nanoparticle-Embedded Membranes for Rural Water Treatment—Bangladesh

Field tests in rural Bangladesh explored the use of silver- and iron oxide infused ceramic membranes for removing arsenic and bacterial contamination from groundwater. These filters were deployed at the household level and showed high efficacy—removing up to 99% of E. coli and arsenic concentrations to below WHO thresholds (Jadoun et al., 2021). These decentralized, point-of-use solutions offer a cost-effective and maintenance-light method for clean water access in low-income, high-risk areas.

3.6.8 Nanozymes for Pesticide Detection in Agricultural Fields—Australia

In Australian vineyards and vegetable farms, nanozyme-based field biosensors have been introduced to detect residual organophosphate pesticides. These nanozymes mimic peroxidase activity, generating a colorimetric change upon exposure to pesticide residues. Portable and battery-operated, these devices allow farmers to check produce safety directly in the field, avoiding delays in lab-based assays (Wong et al., 2021). This not only enhances food safety but also improves compliance with export standards in agricultural supply chains.

These global field deployments of nanotechnology—from magnetically responsive water purification systems in India to offshore oil spill bioremediation in Canada, from smart biosensors in China to decentralized rural water filters in Bangladesh—demonstrate how nanobiotech is translating lab-scale innovations into impactful sustainability tools. As regulatory clarity and eco-safety evaluations improve, these tools can scale further to support global sustainability goals in diverse ecosystems and communities.

4 Emerging Frontiers

4.1 AI-Driven Nanobiotechnology

AI is quickly changing the field of nanobiotechnology by providing tools for modeling, building intelligent systems and improving how we react to environmental factors. AI makes environmental technology nano-enabled by using machine learning, neural networks and data analytics. Integrating ERG and MEP can forecast the actions of pollutants, assist in designing effective cleanup solutions and protect the environment. In this field, using mathematical models greatly helps improve the way nanoparticles are made and used around us. AI can analyze the readings from environmental observatories to support selecting suitable nanoparticle compositions and how to use them. Such systems have been applied in smart wastewater treatment and air pollution control, with machine learning foreseeing how pollutants would react and managing the level of catalysts used (Olawumi & Oladapo, 2025). AI is not only helpful for making processes more efficient but also for analyzing the risks and the life cycle of nanomaterials. AI uses deep learning and pattern recognition to model environmental interactions over time, check the toxicity levels of substances and predict how different nanomaterials may accumulate in living things. With this predictive ability, scientists can choose safer designs for nanoparticles before making them. It was found in a recent study that AI helped organizations use nanoparticles following both environmental and regulatory standards (Khalid et al., 2024). AI supports the cleaning up of environmental pollution by teaching itself the best methods to use. To illustrate, AI has been used to select microbial-nanoparticle groups that can target soil and water biodegradation of heavy metals and hydrocarbons. As a result, these systems can better and faster solve problems because they continually learn from current data in the environment (Lohani, 2024). In addition, AI improves how smart biosensors are designed and run by enhancing how well they pick up specific pollutants. By training neural networks, they can read the reports from biosensors, reduce any noisy signals and link what they measure to real events of contamination. By being applied to IoT-connected biosensors, these AI models allow quick responses and automatic warnings when something in the environment is unsafe (Kuppan et al., 2024). All in all, blending AI with nanobiotechnology is marking a new age where environmental actions respond, expect and react to changes in the environment. Collaboration helps researchers and policymakers design cleaner, more efficient and safe plans for tough environmental problems which supports the growth of sustainable development.

4.2 Bioinspired Nanomaterials

Scientists use natural systems as models to develop bioinspired nanomaterials with strong function, sustainability and the ability to adapt for use in the environment. Carbon capture and positive contributions to agriculture are two of the main uses of nanobiotechnology and help fight the world's problems of climate change and food scarcity. Unique properties of carbon dots (CDs) and metal–organic frameworks (MOFs) make them among the most exciting bioinspired materials for use in environmental care and managing nutrients. These small, fluorescent carbon dots are usually processed using harmless procedures from materials such as agricultural or organic waste. Biocompatible, safe, dissolvable and able to send and receive light as signals, they are ideal for applications in farming and sensing. Among other things, carbon dots created from fruit peels and crop leftovers have shown they can be applied as nanofertilizers and biosensors. Researchers used *Rosa roxburghii* to make fluorescent carbon dots that boosted photosynthesis by stimulating chlorophyll, electron flow and more biomass, suggesting they might become part of sustainable agriculture (Xu et al., 2025). Biomass-produced carbon quantum dots have also been found to increase a plant's ability to handle stress, improve nutrient uptake and function as energy converters, helping with environmentally friendly farming (Dong et al., 2024). Carbon dots play a role in increasing plant growth and detecting and clearing up pollutants. Poorly soluble metals and chemicals originating in agriculture can be picked up by their surfaces and their fluorescence allows real-time environmental pollutant tracking (Zhao et al., 2022). The formation of porous, crystalline nanomaterials called metal–organic frameworks (MOFs) is based on combining metal ions with organic compounds. Because they have a large surface area, pores of various sizes and can selectively bind CO_2, many MOFs are being studied for their use in carbon dioxide removal. Recent investigations suggest that composites including MOFs are competent in adsorbing greenhouse gases and in removing harmful gases from air and water (Kharissova et al., 2024). In addition, models that merge carbon dots and MOFs may be useful for fluorescent detection and removal of pollutants at the same time. A Cu-MOF with carbon dots was found to identify L-ascorbic acid and Fe^{3+} ions in actual water samples, making the resulting sensor a low-cost and sensitive choice for environmental testing (Sadeghi-chahnasir et al., 2024). In addition, rice husk, sugarcane bagasse and sawdust are now often used to make valuable carbonaceous nanoparticles and MOFs. As a result, we can give a new purpose to waste and lower the impact of farm activities on nature. All these developed materials play roles in water filtration, nutrient delivery and crop protection, demonstrating a closed-loop sustainability process (Wang et al., 2021). Basically, bioinspired nanomaterials act as agents that help connect environmental engineering to sustainable agriculture. The fact they can be created cheaply from sustainable substances, combined with their use in managing pollution, nutrients, and sensing, makes them vital to future environmental nanobiotechnology.

5 Challenges and Limitations

Global environmental problems can be resolved in great ways through nanobiotechnology, but its broad use is held back by some major factors. Nanotoxicity, how long nanoparticles remain in the environment, their ability to be produced on a commercial scale and concerns about regulations are the central challenges. They make it difficult to translate technology and also cause eco and ethical concerns. Since many nanomaterials produced using safe processes can still react with living cells, nanotoxicity is a significant concern. The small size and high reactivity of nanoparticles allow them to pass through biological barriers, gather in tissues and cause either oxidative stress or injury to DNA in organisms where they are not intended. Researchers have determined that carbon-based nanomaterials, including carbon dots, do not always behave as expected and can disrupt signals in cells or enzymes if their concentrations are high enough (Peralta-Videa et al., 2020). A shortage of information on toxicity for live organisms and exposure over a long period makes it hard to judge ecological safety. A challenge arises from nanoparticles remaining in the environment. In ecosystems, particles in wastewater or biosensors or added to soil, might remain unchanged or transform in ways that change their behavior. Its future in many complex environments—for example, sediment, rhizospheres or food chains—cannot be fully explained. It has been determined through recent studies that nanoparticles may accumulate in multiple species and thus impact the functions and chains of food webs (Dang et al., 2023). From a practical standpoint, **scalability** remains a bottleneck for biogenic nanoparticle synthesis. While green methods using plant extracts, bacteria, or fungi are cost-effective and environmentally benign at lab scale, reproducibility and standardization at industrial scale are often lacking. Biological variability, seasonal differences in precursor materials, and lack of control over particle size and morphology lead to inconsistent product quality. Table 3 represents the regulatory and safety considerations in nanobiotechnology. These challenges limit the commercial viability of many nanobiotechnological formulations despite their lab-scale efficacy (Dong et al., 2024). **Regulatory frameworks** for nanobiotechnology are also underdeveloped. The rapid evolution of nanomaterials has outpaced regulatory guidelines, particularly in developing economies. There is a lack of standardized testing protocols, environmental impact assessments, and life cycle analysis (LCA) models tailored to nano-enabled products. Because there are no clear policies, industries are skeptical about these technologies and public suspicion remains strong. Researchers want countries to unite in developing standard techniques to evaluate risks, decide on exposure limits and create certifications for eco-friendly nanoproducts (Khalid et al., 2024). Community development and fairness issues related to nanobiotechnology are of particular dilemma in areas like farming and looking after the environment. Taking care of problems such as ensuring fair use, privacy with AI data and using nanosurveillance without causing harm supports fair and acceptable growth of technology. Overall, although nanobiotechnology can greatly help the environment, it needs safe, ethical and regulated use to make it a success.

Table 3 Regulatory and safety considerations in nanobiotechnology

Issue/concern	Description	Mitigation strategy	Stakeholders involved	References
Nanotoxicity	Inadequate data on long-term biological and ecological effects of nanoparticles	Develop standardized toxicological testing and nanosafety data sharing	Regulators, researchers, safety agencies	(Malsch et al., 2024)
Regulatory gaps	Current frameworks like REACH or TSCA often lack nano-specific assessment criteria	Hybrid governance combining voluntary codes, risk-based updates to current laws	National regulators, industry, environmental watchdogs	(Hemphill, 2016)
Public awareness	Low public understanding fuels skepticism and reduces acceptance of nanotechnologies	Public engagement, participatory technology assessments	Governments, educators, NGOs	(Paradise et al., 2008)
Ethical governance	Limited ethical frameworks for consent, privacy, and data use in AI-integrated biosensors	Codes of conduct, social impact assessments	Policy think tanks, ethicists, tech developers	(Ukochovwera & Ologan, 2023)
Environmental persistence	Nanoparticles may persist or bioaccumulate, posing long-term ecological risks	Lifecycle assessments, degradable nano-designs	Environmental scientists, regulators	(Nel et al., 2015)

6 Sustainable Solutions

Sustainability must guide the growth and stability of nanobiotechnology for use in environmental systems. This change relies on four main areas: green synthesis, life cycle assessment, responsible governance, and the circular bioeconomy. They try to safeguard the environment and increase the usefulness and social impact of treatments using nanobiotechnology. The core of sustainable nanomaterial production lies in green synthesis. Meanwhile, it uses plant extracts, microbial cultures and agricultural residue to produce nanoparticles in an environmentally friendly and simple way. For example, scientists have found ways to make silica nanoparticles by green synthesis from paddy straw, a substance burned in most Asian countries. Farming and pollution control both gain important contributions from plektonic particles. They help with managing stubble and keep the air free from harmful dust. In the same way, waste-derived nanobiochar helps enrich soil, remove carbon dioxide and improve how tightly nutrients are kept in soil, supporting environmentally friendly farming practices (Sani et al., 2023). Assessing changes to the environment in engineering is normally accomplished using the method of life cycle assessment (LCA). LCA makes it possible to examine the energy used, waste produced and toxicity created within each step of making nanoparticles. A life cycle analysis demonstrated that making iron oxide nanoparticles via eco-friendly green synthesis reduces both environmental and health damage compared to traditional methods (Patiño-Ruiz et al., 2021). In addition, research progress is introducing important aspects such as atom economy, energy intensity and process mass efficiency into LCA evaluation to provide a better picture of green chemistry initiatives (Zhang et al., 2018). Along with technical checks, ethical rules are important for safe development and use of nanobiotechnologies. Part of this effort is providing clear risk details, giving all interested parties a say and ensuring every group has the same access to nano-based solutions. For biosensors made for environmental surveillance, it is especially crucial to think about data privacy and whether communities agree with their use (Solomon et al., 2024). By including nanobiotechnology in circular bioeconomy, we improve sustainability by reusing resources. Here, the approach is to make materials out of nanotechnology that break down or can be recycled, are made from green resources and turn waste from farms into useful nanoproducts. For example, microbial nanocellulose has helped create exceptional biotextiles that are safe and have a low carbon impact, proving biofabrication works in many circular economy applications (Schiros et al., 2021). As a result, when these metrics are used in decision-making, they can combine social and environmental goals with economic priorities in industry (Zhang et al., 2018). Therefore, sustainable nanobiotechnology means designing, making and handling nanotechnology so that no part of the process lacks consideration for ethics, the environment or the economy. By using green synthesis, life cycle assessments, responsible policies and circular approaches, nanobiotechnology becomes an important part of sustainability science and actual environmental transformation.

7 Future Perspectives

Through the combination of advanced materials, biology and intelligent systems, nanobiotechnology is set to transform sustainability for the environment. Because of ecological problems and a desire for sustainability, electronics engineering is moving into an era that features greater integration, miniaturization, and automation. It seems that future developments in nanobiotechnology will involve applications that can learn, adjust, and work closely with what is happening in real time in the environment. A major area of potential progress is improving biosensors that use AI and IoT for smart environment analysis. They will detect pollutants, foresee future problems related to pollution and automatically deal with the consequences. Experts believe that nanosensors put in water pipes, fields and industrial outlets will play a major role in smart cities and precision agriculture (Miguel-Rojas &

Pérez-de-Luque, 2023). There will be increasing use of composites such as MOF-carbon hybrids that are able to take in contaminants, boost their removal and detect bacteria. These nanostructures can help remove different pollutants in a specific and arranged order, providing answers for difficult contamination situations. Table 4 shows future directions in nanobiotechnology for sustainability. Nanomaterials are also predicted to become responsive and even heal themselves upon encountering pH, temperature, or toxic changes, thus adding precision and efficiency (Hemalatha et al., 2024). The field of climate change mitigation will see nanobiotechnology become increasingly important for capturing carbon and putting it to good use. Scientists predict that future materials will be designed to capture CO_2 accurately and turn it into fuels or fertilizers. Microbial systems manipulated in a lab can be integrated with nanomaterials to make carbon fixation, photosynthesis and nitrogen cycling more efficient—connecting biology and nanotechnology more safely in the environment (Al-Obaidy et al., 2023). Future designs in nanobiotechnology will increasingly follow the principles of a circular economy. Development will be centered on biodegradable, renewable, and recyclable nanomaterials. Upcycling agricultural residues, microorganisms and wastes produced in industry will become a key factor in closing resource loops and protecting the environment (Saxena et al., 2025). These improvements in nanobiotechnology should also focus on improving public understanding, making new regulations and increasing education. When new tools are introduced, prompt actions and guidelines will play a key part in keeping them used properly. When we train researchers across disciplines, encourage inclusive innovation and share open-source technology, more people in the Global South can gain from nanobiotechnology (Lugani et al., 2021). Overall, nanobiotechnology's impact depends on its working together with nature, data and society to create sustainable innovation that upholds ecological boundaries. During the next several years, the field should develop into fully-integrated systems that help define better stewardship and increased resilience.

8 Conclusion

Nanobiotechnology is quickly being recognized as a key factor in protecting the environment. Using the engineered characteristics of nanomaterials, this area provides modern solutions for pollution control, gaining resources, continuous monitoring, and more sustainable farming practices. Both the removal of pollutants from the form of metal structures and managing wastewater with hybrid nanomaterials have become important and powerful uses. The use of microbes, enzymes and nanostructures together has greatly improved the success of bioremediation and AI technology is changing the way we forecast and handle environmental actions. The use of carbon dots and MOFs shows how we can use nature to design materials that capture and reuse carbon. In addition, the use of environmentally friendly methods and life cycle assessments keeps development of these technologies sustainable. Despite the great potential of nanobiotechnology, it has to face main issues such as toxicity, staying

Table 4 Future directions in nanobiotechnology for sustainability

Trend /innovation	Description	Potential impact	Readiness level	References
AI-integrated nanosensors	Biosensors embedded with AI and IoT for real-time pollutant detection and response	Enables autonomous monitoring and smart environmental decisions	Emerging/pilot-tested	(Irene Bavasso et al., 2016)
Multifunctional hybrid nanomaterials	Combined pollutant capture, catalysis, sensing in a single nano-platform	Simplifies treatment processes, especially in water and soil remediation	Mid-stage development	(Hemalatha et al., 2024)
Biodegradable and circular nano-design	Nanomaterials designed for full degradation or reuse using biological inputs	Supports zero-waste systems, aligns with circular bioeconomy goals	Concept to early trials	(Schiros et al., 2021)
Carbon capture with engineered nanostructures	MOFs and other porous materials for CO_2 sequestration or conversion	Tackles climate change via point-source emission reduction	Lab-scale/prototyping	(Kharissova et al., 2024)
Biohybrid remediation systems	Microbe-nanoparticle consortia for smart pollutant targeting	Enhances degradation of persistent contaminants via adaptive synergy	Field-tested (select uses)	(Chauhan et al., 2023)

in the environment after use and insecure regulation. We must use ethical rules, followed by the same protocols and green designs to protect everyone and support widespread use. Moving forward, combining nanobiotechnology with artificial intelligence, synthetic biology, and circular economy models is predicted to change how environmental innovation develops. With the growing stress on world ecosystems, carefully using nanobiotechnology will play a key role in making the planet cleaner, safer, and more sustainable.

References

Aihaiti, A., Wang, J., Zhang, W., Shen, M., Meng, F., Li, Z., Zhang, Y., Ren, M., & Zhang, M. (2024). Recent advances and trends in innovative biosensor-based devices for heavy metal ion detection in food. *Comprehensive Reviews in Food Science and Food Safety, 23*(4), Article e13358. https://doi.org/10.1111/1541-4337.13358

Alahi, M. E. E., Sukkuea, A., Tina, F. W., Nag, A., Kurdthongmee, W., Suwannarat, K., & Mukhopadhyay, S. C. (2023). Integration of IoT-enabled technologies and artificial intelligence (AI) for smart city scenario: Recent advancements and future trends. *Sensors, 23*(11), 11. https://doi.org/10.3390/s23115206

Al-Allaf, M., Al-Rawi, A., & AL-Ahmadi, K. (2023). Biosynthesis of iron oxide nanoparticles and using it to support the efficiency of wastewater filtration system. *Rafidain Journal of Science, 32*(1), 28–39. https://doi.org/10.33899/rjs.2023.177285

Al-Obaidy, G. S., Abdulrahman, M. F, & Alobaidy, B. S. J. (2023). Future trends for green environmental applications of nanotechnology: A review. *Anbar Journal of Agricultural Sciences, 21*(1), 133–147. https://doi.org/10.32649/ajas.2023.179725

Alqarni, L. S., Alghamdi, M. D., Alshahrani, A. A., & Nassar, A. M. (2022). Green nanotechnology: Recent research on bioresource-based nanoparticle synthesis and applications. *Journal of Chemistry, 2022*(1), 4030999. https://doi.org/10.1155/2022/4030999

Bala, K., Anshul, Verma, S., Dogra, A., Das, A. P., Kumar, R., & Nandini. (2024). Sustainable management for urban plastic waste generation. In *Sustainable management of urban plastic waste through circular economic approaches* (pp. 211–236). https://doi.org/10.1201/9781003465447-13

Barati, F., Avatefi, M., Moghadam, N. B., Asghari, S., Ekrami, E., & Mahmoudifard, M. (2023). A review of graphene quantum dots and their potential biomedical applications. *Journal of Biomaterials Applications, 37*(7), 1137–1158. https://doi.org/10.1177/08853282221125311

Bavasso, I., Vilardi, G., Stoller, M., Chianese, A., & Di Palma, L. (2016). Perspectives in nanotechnology based innovative applications for the environment. *Chemical Engineering Transactions, 47*, 55–60. https://doi.org/10.3303/CET1647010

Behera, S., & Das, S. (2023). Potential and prospects of Actinobacteria in the bioremediation of environmental pollutants: Cellular mechanisms and genetic regulations. *Microbiological Research, 273*, Article 127399. https://doi.org/10.1016/j.micres.2023.127399

Bhapkar, A. R., & Bhame, S. (2024). A review on ZnO and its modifications for photocatalytic degradation of prominent textile effluents: Synthesis, mechanisms, and future directions. *Journal of Environmental Chemical Engineering, 12*(3), Article 112553. https://doi.org/10.1016/j.jece.2024.112553

Bhati, M. (2023). Biogenic synthesis of metallic nanoparticles: Principles and applications. *Materials Today: Proceedings, 81*, 882–887. https://doi.org/10.1016/j.matpr.2021.04.272

Bhatt, P., Pandey, S. C., Joshi, S., Chaudhary, P., Pathak, V. M., Huang, Y., Wu, X., Zhou, Z., & Chen, S. (2022). Nanobioremediation: A sustainable approach for the removal of toxic pollutants from the environment. *Journal of Hazardous Materials, 427*, Article 128033. https://doi.org/10.1016/j.jhazmat.2021.128033

Chan, S. S., Khoo, K. S., Abdullah, R., Juan, J. C., Ng, E.-P., Chin, R. J., & Ling, T. C. (2024). Harnessing microalgae for metal nanoparticles biogenesis using heavy metal ions from wastewater as a metal precursor: A review. *Science of the Total Environment, 957*, Article 176989. https://doi.org/10.1016/j.scitotenv.2024.176989

Chandel, H., Kumar, P., Chandel, A. K., & Verma, M. L. (2024). Biotechnological advances in biomass pretreatment for bio-renewable production through nanotechnological intervention. *Biomass Conversion and Biorefinery, 14*(3), 2959–2981. https://doi.org/10.1007/s13399-022-02746-0

Chang, J.-H., Kumar, M., Selvaraj, S., Samuel, M. S., Ethiraj, S., Senthilkumar, A., Dong, C.-D., & Shkir, M. (2024). Green energy breakthroughs: Harnessing nano-catalysts and enzymatic catalysts for bioenergy generation. *Industrial Crops and Products, 215*, Article 118527. https://doi.org/10.1016/j.indcrop.2024.118527

Chaudhary, P., Ahamad, L., Chaudhary, A., Kumar, G., Chen, W.-J., & Chen, S. (2023). Nanoparticle-mediated bioremediation as a powerful weapon in the removal of environmental pollutants. *Journal of Environmental Chemical Engineering, 11*(2), Article 109591. https://doi.org/10.1016/j.jece.2023.109591

Chauhan, P., Imam, A., Kanaujia, P. K., & Suman, S. K. (2023). Nano-bioremediation: An eco-friendly and effective step towards petroleum hydrocarbon removal from environment. *Environmental Research, 231*, Article 116224. https://doi.org/10.1016/j.envres.2023.116224

Chen, M., Su, B., Wu, H., Dai, Y., Chen, T., Fu, F., Lin, Z., & Dong, Y. (2024). Hydrogel SERS chip with strong localized surface plasmon resonance for sensitive and rapid detection of T-2 toxin. *Talanta, 268*, Article 125329. https://doi.org/10.1016/j.talanta.2023.125329

Chintagunta, A. D., Kumar, A., Jeevan Kumar, S. P., & Verma, M. L. (2021). Contribution of metallic nanomaterials in algal biofuel production. In S. Rajendran, J. Qin, F. Gracia, & E. Lichtfouse (Eds.), *Metal and metal oxides for energy and electronics* (vol. 55, pp. 331–353). Springer International Publishing. https://doi.org/10.1007/978-3-030-53065-5_9

Chouke, P. B., Shrirame, T., Potbhare, A. K., Mondal, A., Chaudhary, A. R., Mondal, S., Thakare, S. R., Nepovimova, E., Valis, M., Kuca, K., Sharma, R., & Chaudhary, R. G. (2022). Bioinspired metal/metal oxide nanoparticles: A road map to potential applications. *Materials Today Advances, 16*, Article 100314. https://doi.org/10.1016/j.mtadv.2022.100314

Dang, F., Yuan, Y., Huang, Y., Wang, Y., & Xing, B. (2023). Trophic transfer of nanomaterials and their effects on high-trophic-level predators. *NanoImpact, 32*, Article 100489. https://doi.org/10.1016/j.impact.2023.100489

Das, A., Dey, S., & Das, A. P. (2024b). Ecotoxicological impacts of synthetic microfiber pollutants and development of sustainable mitigation strategies. *Environmental Chemistry and Ecotoxicology*. https://doi.org/10.1016/J.ENCECO.2024.12.003

Das, A., Ray, S. K., Mohanty, M., Mohanty, J., Dey, S., & Das, A. P. (2024c). Ecotoxicity of microplastic wastes and their sustainable management: A review. *Environmental Chemistry and Ecotoxicology, 6*, 144–152. https://doi.org/10.1016/J.ENCECO.2024.05.003

Das, A., Dey, S., Gacem, A., Yadav, K. K., Nath, H., Ram, D. K., & Das, A. P. (2025). Revolutionizing solid hydrocarbon sourcing: Plastic's journey from waste to upcycled treasure. *Environmental Science and Engineering, Part, F218*, 331–351. https://doi.org/10.1007/978-3-031-81615-4_14

Das, A. P., & Ghosh, S. (2023) Manganese and its application. *Manganese Mining Microorganisms*, 29–40. https://doi.org/10.1016/B978-0-12-822147-1.00007-8

Das, A., Dey, S., Bala, K., Gnanasekaran, R., & Das, A. P. (2024a). Microbial bioremediation technology for sustainable treatment and management of synthetic microfiber waste. In *Sustainable microbial technology for synthetic and cellulosic microfiber bioremediation* (pp. 221–238). https://doi.org/10.1007/978-3-031-62600-5_12

Desai, N. (2012). Challenges in development of nanoparticle-based therapeutics. *The AAPS Journal, 14*(2), 282–295. https://doi.org/10.1208/s12248-012-9339-4

Dessie, Y., & Tadesse, S. (2022). Advancements in bioelectricity generation through nanomaterial-modified anode electrodes in microbial fuel cells. *Frontiers in Nanotechnology, 4*, Article 876014. https://doi.org/10.3389/fnano.2022.876014

Dey, S., Tripathy, B., Kumar, M. S., & Das, A. P. (2023). Ecotoxicological consequences of manganese mining pollutants and their biological remediation. *Environmental Chemistry and Ecotoxicology*. https://doi.org/10.1016/J.ENCECO.2023.01.001

Dey, S., Dash, A., & Das, A. P. (2024). Bioremediation as a sustainable and circular solution for mining waste management and bioenergy generation. In *Sustainable management of mining waste and tailings: A circular economy approach* (pp. 107–119). https://doi.org/10.1201/978 1003442455-6

Divyapriya, G., Nambi, IndumathiM., & Senthilnathan, J. (2016). Nanocatalysts in fenton based advanced oxidation process for water and wastewater treatment. *Journal of Bionanoscience, 10*(5), 356–368. https://doi.org/10.1166/jbns.2016.1387

Dong, Z., Qi, J., Yue, L., Zhou, H., Chen, L., Gu, J., He, Y., & Wu, H. (2024). Biomass-based carbon quantum dots and their agricultural applications. *Plant Stress, 11*, Article 100411. https://doi.org/10.1016/j.stress.2024.100411

Eroglu, E., & Raston, C. L. (2017). Nanomaterial processing strategies in functional hybrid materials for wastewater treatment using algal biomass. *Journal of Chemical Technology & Biotechnology, 92*(8), 1862–1867. https://doi.org/10.1002/jctb.5188

Eweis, A. A., El-Raheem, H. A., Ahmad, M. S., Hozzein, W. N., & Mahmoud, R. (2024). Green fabrication of nanomaterials using microorganisms as nano-factories. *Journal of Cluster Science, 35*(7), 2149–2176. https://doi.org/10.1007/s10876-024-02660-7

Fredj, Z., Rong, G., & Sawan, M. (2025). Recent advances in enzymatic biofuel cells to power up wearable and implantable biosensors. *Biosensors, 15*(4), 4. https://doi.org/10.3390/bios15 040218

Fu, X.-Z., Wu, J., Li, J., Ding, J., Cui, S., Wang, X.-M., Wang, Y.-J., Liu, H.-Q., Deng, X., Liu, D.-F., & Li, W.-W. (2023). Heavy-metal resistant bio-hybrid with biogenic ferrous sulfide nanoparticles: PH-regulated self-assembly and wastewater treatment application. *Journal of Hazardous Materials, 446*, Article 130667. https://doi.org/10.1016/j.jhazmat.2022.130667

Gaviria-Arroyave, M. I., Cano, J. B., & Peñuela, G. A. (2020). Nanomaterial-based fluorescent biosensors for monitoring environmental pollutants: A critical review. *Talanta Open, 2*, Article 100006. https://doi.org/10.1016/j.talo.2020.100006

Geetha, N., Bhavya, G., Abhijith, P., Shekhar, R., Dayananda, K., & Jogaiah, S. (2021). Insights into nanomycoremediation: Secretomics and mycogenic biopolymer nanocomposites for heavy metal detoxification. *Journal of Hazardous Materials, 409*, Article 124541. https://doi.org/10.1016/j.jhazmat.2020.124541

Hasan, A., Elumalai, A., & Kumar, A. (2025). Advancements in bioremediation: Harnessing nanomaterials and microorganisms for environmental restoration by degradation of pollutants. *Environmental Quality Management, 34*(3), Article e70018. https://doi.org/10.1002/tqem.70018

Hemalatha, P. K., Kanagaraj, K., Raman, K., Nangan, S., Md, R., Aggarwal, S., & Kumar, R. (2024). Nano-innovations for a greener future: exploring the intersection of nanotechnology and environmental sustainability. *E3S Web of Conferences, 588*, 01018. https://doi.org/10.1051/e3sconf/202458801018

Hemdan, M., Abuelhaded, K., Shaker, A. A. S., Ashour, M. M., Abdelaziz, M. M., Dahab, M. I., Nassar, Y. A., Sarguos, A. M. M., Zakaria, P. S., Fahmy, H. A., Abdel Mageed, S. S., Hamed, M. O. A., Mubarak, M. F., Taher, M. A., Gumaah, N. F., & Ragab, A. H. (2025). Recent advances in nano-enhanced biosensors: Innovations in design, applications in healthcare, environmental monitoring, and food safety, and emerging research challenges. *Sensing and Bio-Sensing Research, 48*, Article 100783. https://doi.org/10.1016/j.sbsr.2025.100783

Hemphill, T. A. (2016). Regulating nanomaterials: A case for hybrid governance. *Bulletin of Science, Technology & Society, 36*(4), 219–228. https://doi.org/10.1177/0270467617719820

Ike Onyia, A., Ifeanyi Ikeri, H., & Iheanyichukwu Chima, A. (2020). Surface and quantum effects in nanosized semiconductor. *American Journal of Nano Research and Applications*, *8*(3), 35. https://doi.org/10.11648/j.nano.20200803.11

Jadoun, S., Chauhan, N. P. S., Zarrintaj, P., Barani, M., & Varma, R. S. (2021). *Nanomaterials for sustainability: A review on green synthesis of nanoparticles using microorganisms*. In Review. https://doi.org/10.21203/rs.3.rs-1132888/v1

Jawahar, C. P., & Pillai, A. G. (2023). Generation of green carbon nanomaterials from biomass pyrolysis for the production of biofuel and bioenergy. *Migration Letters*, *20*(S13), 133–140. https://doi.org/10.59670/ml.v20iS13.6276

John, A., & Das, A. P. (2024). Government strategies for achieving circular economy for plastic waste management in Southeast Asia. In *Sustainable management of urban plastic waste through circular economic approaches*, 72–82. https://doi.org/10.1201/9781003465447-6

John, B., Subramani, N., & Awasthi, S. (2020). *Green synthesis of iron oxide nanoparticles from Andrographis paniculata and its application in separation of oil-spills from water*. https://consensus.app/papers/green-synthesis-of-iron-oxide-nanoparticles-from-john-subramani/0b1cb7f38428534e97ac7a5b5a5f78da/

Justino, C., Duarte, A., & Rocha-Santos, T. (2017). Recent progress in biosensors for environmental monitoring: A review. *Sensors, 17*(12), 2918. https://doi.org/10.3390/s17122918

Karunakaran, G., Sudha, K. G., Ali, S., & Cho, E.-B. (2023). Biosynthesis of nanoparticles from various biological sources and its biomedical applications. *Molecules, 28*(11), 4527. https://doi.org/10.3390/molecules28114527

Khalid, J., Chuanmin, M., Altaf, F., Shafqat, M. M., Khan, S. K., & Ashraf, M. U. (2024). AI-driven risk management and sustainable decision-making: Role of perceived environmental responsibility. *Sustainability, 16*(16), 6799. https://doi.org/10.3390/su16166799

Kharissova, A. B., Kharissova, O. V., Kharisov, B. I., & Méndez, Y. P. (2024). Carbon negative footprint materials: A review. *Nano-Structures & Nano-Objects, 37*, Article 101100. https://doi.org/10.1016/j.nanoso.2024.101100

Korsa, G., Bekele, G. K., Beyene, A., Alemu, D., Meka, A. F., Tibebu, S., Gutema, B. T., & Ayele, A. (2025). Microbial fuel cells for sustainable biofuel production: Cutting-edge technology. *Green Energy and Environmental Technology*, *4*. https://doi.org/10.5772/geet.20240082

Kumari, B., & Singh, D. P. (2016). A review on multifaceted application of nanoparticles in the field of bioremediation of petroleum hydrocarbons. *Ecological Engineering, 97*, 98–105. https://doi.org/10.1016/j.ecoleng.2016.08.006

Kuppan, N., Padman, M., Mahadeva, M., Srinivasan, S., & Devarajan, R. (2024). A comprehensive review of sustainable bioremediation techniques: Eco friendly solutions for waste and pollution management. *Waste Management Bulletin, 2*(3), 154–171. https://doi.org/10.1016/j.wmb.2024.07.005

Li, Z., Li, X., Jian, M., Geleta, G. S., & Wang, Z. (2019). Two-dimensional layered nanomaterial-based electrochemical biosensors for detecting microbial toxins. *Toxins, 12*(1), 20. https://doi.org/10.3390/toxins12010020

Lingamdinne, L. P., Chang, Y.-Y., Yang, J.-K., Singh, J., Choi, E.-H., Shiratani, M., Koduru, J. R., & Attri, P. (2017). Biogenic reductive preparation of magnetic inverse spinel iron oxide nanoparticles for the adsorption removal of heavy metals. *Chemical Engineering Journal, 307*, 74–84. https://doi.org/10.1016/j.cej.2016.08.067

Liu, T., Zhang, Y., Shi, Z., Cao, W., Zhang, L., Liu, J., & Chen, Z. (2021). BiOBr/Ag/AgBr heterojunctions decorated carbon fiber cloth with broad-spectral photoresponse as filter-membrane-shaped photocatalyst for the efficient purification of flowing wastewater. *Journal of Colloid and Interface Science, 587*, 633–643. https://doi.org/10.1016/j.jcis.2020.11.020

Lohani, N. (2024). AI-based environmental sustainability: Transforming conservation efforts. *International Journal for Multidisciplinary Research*, *6*(2), 16997. https://doi.org/10.36948/ijfmr.2024.v06i02.16997

Lugani, Y., Sooch, B. S., Singh, P., & Kumar, S. (2021). Nanobiotechnology applications in food sector and future innovations. In *Microbial biotechnology in food and health* (pp. 197–225). Elsevier. https://doi.org/10.1016/B978-0-12-819813-1.00008-6

Malsch, I., Willighagen, E., Sánchez Burmester, C. F., & Mody, C. (2024). Ethical impact assessment of sharing nanosafety data. *Open Research Europe, 4*, 237. https://doi.org/10.12688/openreseurope.18345.1

Maqsood, Q., Sumrin, A., Waseem, R., Hussain, M., Imtiaz, M., & Hussain, N. (2023). Bioengineered microbial strains for detoxification of toxic environmental pollutants. *Environmental Research, 227*, Article 115665. https://doi.org/10.1016/j.envres.2023.115665

Mashkour, M., Rahimnejad, M., Raouf, F., & Navidjouy, N. (2021). A review on the application of nanomaterials in improving microbial fuel cells. *Biofuel Research Journal, 8*(2), 1400–1416. https://doi.org/10.18331/BRJ2021.8.2.5

Miguel-Rojas, C., & Pérez-de-Luque, A. (2023). Nanobiosensors and nanoformulations in agriculture: New advances and challenges for sustainable agriculture. *Emerging Topics in Life Sciences, 7*(2), 229–238. https://doi.org/10.1042/ETLS20230070

Miran, W., Nawaz, M., Jang, J., & Lee, D. S. (2016). Effect of wastewater containing multi-walled carbon nanotubes on dual-chamber microbial fuel cell performance. *RSC Advances, 6*(94), 91314–91319. https://doi.org/10.1039/C6RA20128H

Mishra, S., Dash, D., & Das, A. P. (2024). Isolation and molecular characterization of polyester-degrading bacterial strains from municipal wastewater and their application in microplastic degradation in contaminated wastewater. *Geomicrobiology Journal*. https://doi.org/10.1080/01490451.2024.2376101

Mishra, S., & Das, A. P. (2023). Microbial remediation of synthetic microfiber contaminated wastewater. In M. P. Shah (Ed.), *Microbial technologies in industrial wastewater treatment* (pp. 337–350). Springer, Singapore. https://doi.org/10.1007/978-981-99-2435-6_17

Mohanty, J., Dey, S., Bala, K., & Das, A. P. (2024a). Recent advances in microplastic pollution for its sustainable management. In *Sustainable management of urban plastic waste through circular economic approaches* (pp. 27–48). https://doi.org/10.1201/9781003465447-3

Mohanty, M., Tripathy, B., Koteswara Reddy, G., & Das, A. P. (2024b). Greener technologies for establishing restraint over microplastic pollution. In *Sustainable microbial technology for synthetic and cellulosic microfiber bioremediation* (pp. 239–258). https://doi.org/10.1007/978-3-031-62600-5_13

Mohapatra, T., Agrawal, M., & Ghosh, P. (2023). An overview of plant-mediated biogenic synthesis of nano-catalysts and their application in Fenton and photo-Fenton processes for wastewater remediation. *Chemical Engineering Journal, 477*, Article 146941. https://doi.org/10.1016/j.cej.2023.146941

Mughal, B., Zaidi, S. Z. J., Zhang, X., & Hassan, S. U. (2021). Biogenic nanoparticles: Synthesis, characterisation and applications. *Applied Sciences, 11*(6), 2598. https://doi.org/10.3390/app11062598

Nain, R., Patel, H., Chahar, M., Kumar, S., Rohilla, D., & Pal, M. (2025). Biosynthesized metallic nanoparticles for sustainable environmental remediation: Mechanisms, applications, and future perspectives. *Discover Chemistry, 2*(1), 124. https://doi.org/10.1007/s44371-025-00203-1

Narayanan, M., Ali, S. S., & El-Sheekh, M. (2023). A comprehensive review on the potential of microbial enzymes in multipollutant bioremediation: Mechanisms, challenges, and future prospects. *Journal of Environmental Management, 334*, Article 117532. https://doi.org/10.1016/j.jenvman.2023.117532

Nath, P., Priyadarshni, N., Mandal, S., Singh, P., Arun, R. K., & Chanda, N. (2018). Gold nanostructure in sensor technology: Detection and estimation of chemical pollutants. In S. Bhattacharya, A. K. Agarwal, N. Chanda, A. Pandey, & A. K. Sen (Eds.), *Environmental, chemical and medical sensors* (pp. 31–66). Springer Singapore. https://doi.org/10.1007/978-981-10-7751-7_3

Nel, A. E., Parak, W. J., Chan, W. C. W., Xia, T., Hersam, M. C., Brinker, C. J., Zink, J. I., Pinkerton, K. E., Baer, D. R., & Weiss, P. S. (2015). Where are we heading in nanotechnology environmental health and safety and materials characterization? *ACS Nano, 9*(6), 5627–5630. https://doi.org/10.1021/acsnano.5b03496

Ofremu, G. O., Raimi, B. Y., Yusuf, S. O., Dziwornu, B. A., Nnabuife, S. G., Eze, A. M., & Nnajiofor, C. A. (2024). Exploring the relationship between climate change, air pollutants and human health: Impacts, adaptation, and mitigation strategies. *Green Energy and Resources*, 100074. https://doi.org/10.1016/j.gerr.2024.100074

Oladimeji, T. E., Oyedemi, M., Emetere, M. E., Agboola, O., Adeoye, J. B., & Odunlami, O. A. (2024). Review on the impact of heavy metals from industrial wastewater effluent and removal technologies. *Heliyon, 10*(23), Article e40370. https://doi.org/10.1016/j.heliyon.2024.e40370

Olawumi, M. A., & Oladapo, B. I. (2025). AI-driven predictive models for sustainability. *Journal of Environmental Management, 373*, Article 123472. https://doi.org/10.1016/j.jenvman.2024.123472

Ortega, D. (2018). *Alteration of the wetting character of a composite rock, through the use of nanoparticles as an enhanced oil recovery method, Ben Nevis Formation, Hebron Field, Jeanne d'Arc Basin, offshore Newfoundland, Canada.* https://consensus.app/papers/alteration-of-the-wetting-character-of-a-composite-rock-ortega/a8acbabc55e35c52a1edd9806f1fd73c/

Osadebe, A. U., Ogugbue, C. J., & Okpokwasili, G. C. (2024). Biochar and Iron oxide nanoparticle-impregnated alginate beads as adsorbents for enhanced ex situ bioremediation of petroleum-contaminated freshwater. *Environmental Chemistry and Ecotoxicology, 6*, 42–50. https://doi.org/10.1016/j.enceco.2024.01.002

Paradise, J., Wolf, S., Ramachandran, G., Kokkoli, E., Hall, R., JD, & Kuzma, J. (2008). Developing oversight frameworks for nanobiotechnology. *Telecommunications & Regulated Industries eJournal.* https://consensus.app/papers/developing-oversight-frameworks-for-nanobiotechnology-paradise-wolf/1e954e3da4e653189f8e71ab4fdb9a09/

Patil, R. V., & Pawar, K. D. (2019). In planta synthesis of nanomaterials for environmental remediation. In S. Srivastava, A. K. Srivastava, & P. Suprasanna (Eds.), *Plant-metal interactions* (pp. 283–307). Springer International Publishing. https://doi.org/10.1007/978-3-030-20732-8_14

Patiño-Ruiz, D. A., Meramo-Hurtado, S. I., González-Delgado, Á. D., & Herrera, A. (2021). Environmental sustainability evaluation of iron oxide nanoparticles synthesized via green synthesis and the coprecipitation method: A comparative life cycle assessment study. *ACS Omega, 6*(19), 12410–12423. https://doi.org/10.1021/acsomega.0c05246

Peralta-Videa, J., Sreenivasan, S. T., & Narayan, M. (2020). Influence of carbon quantum dots on the biome. *Processes, 8*(4), 445. https://doi.org/10.3390/pr8040445

Popescu, S. M., Mansoor, S., Wani, O. A., Kumar, S. S., Sharma, V., Sharma, A., Arya, V. M., Kirkham, M. B., Hou, D., Bolan, N., & Chung, Y. S. (2024). Artificial intelligence and IoT driven technologies for environmental pollution monitoring and management. *Frontiers in Environmental Science, 12.* https://doi.org/10.3389/fenvs.2024.1336088

Priyadarsini, S., & Das, A. P. (2025). Lithium bioleaching: A review on microbial-assisted sustainable technology for lithium bio-circularity. *Journal of Water Process Engineering, 69*, Article 106744. https://doi.org/10.1016/j.jwpe.2024.106744

Rajeshwari, K. M., Suhasini, M. R., Bindya, S., Hemavathi, A. B., Ali, N., Amachawadi, R. G., Shivamallu, C., Hallur, R. L. S., Majani, S. S., & Prasad Kollur, S. (2023). Photocatalytic efficacy of magnesium oxide nanoparticles in dye degradation: A sustainable one-pot synthesis utilizing syzygium samarangense L. *Extract Results in Chemistry, 6*, Article 101193. https://doi.org/10.1016/j.rechem.2023.101193

Ray, P. C., Yu, H., & Fu, P. P. (2009). Toxicity and environmental risks of nanomaterials: Challenges and future needs. *Journal of Environmental Science and Health. Part C, Environmental Carcinogenesis & Ecotoxicology Reviews, 27*(1), 1–35. https://doi.org/10.1080/10590500802708267

Ray, S. K., Dey, S., Das, A. P., & John, A. (2024). Microplastic waste management through greener technologies : A review. In *Sustainable management of urban plastic waste through circular economic approaches* (pp. 237–261). https://doi.org/10.1201/9781003465447-14

Sadeghi-chahnasir, F., Amiripour, F., & Ghasemi, S. (2024). Orange peel-derived carbon dots/Cu-MOF nanohybrid for fluorescence determination of l-ascorbic acid and Fe3+. *Analytica Chimica Acta, 1287*, Article 342066. https://doi.org/10.1016/j.aca.2023.342066

Sadhasivam, S., Shanmugam, M., Umamaheswaran, P. D., Venkattappan, A., & Shanmugam, A. (2021). Zinc oxide nanoparticles: Green synthesis and biomedical applications. *Journal of Cluster Science, 32*(6), 1441–1455. https://doi.org/10.1007/s10876-020-01918-0

Sahu, A., Dey, S., Mallick, K., Priyadarsini, S., Behera, I. D., & Das, A. P. (2024). A review on advanced technology for sustainable management of synthetic microplastic waste. In *Sustainable management of urban plastic waste through circular economic approaches* (pp. 262–279). https://doi.org/10.1201/9781003465447-15

Sani, Md. N. H., Amin, M., Siddique, A. B., Nasif, S. O., Ghaley, B. B., Ge, L., Wang, F., & Yong, J. W. H. (2023). Waste-derived nanobiochar: A new avenue towards sustainable agriculture, environment, and circular bioeconomy. *Science of the Total Environment, 905*, Article 166881. https://doi.org/10.1016/j.scitotenv.2023.166881

Sanni, S. E., Oni, B. A., Okoro, E. E., & Pandya, S. (2024). Recent advances in the use of biogenic nanomaterials and photocatalysts for wastewater treatment: Challenges and future prospects. *Frontiers in Nanotechnology, 6*. https://doi.org/10.3389/fnano.2024.1469309

Santhosh, C., Malathi, A., Dhaneshvar, E., Bhatnagar, A., Grace, A. N., & Madhavan, J. (2019). Iron oxide nanomaterials for water purification. In *Nanoscale materials in water purification* (pp. 431–446). Elsevier. https://doi.org/10.1016/B978-0-12-813926-4.00022-7

Saraswat, P., Singh, S., Prasad, M., Misra, R., Rajput, V. D., & Ranjan, R. (2023). Applications of bio-based nanomaterials in environment and agriculture: A review on recent progresses. *Hybrid Advances, 4*, Article 100097. https://doi.org/10.1016/j.hybadv.2023.100097

Saxena, S., Moharil, M. P., Jadhav, P. V., Ghodake, B., Deshmukh, R., & Ingle, A. P. (2025). Transforming waste into wealth: Leveraging nanotechnology for recycling agricultural byproducts into value-added products. *Plant Nano Biology, 11*, Article 100127. https://doi.org/10.1016/j.plana.2024.100127

Schiros, T. N., Antrobus, R., Farías, D., Chiu, Y.-T., Joseph, C. T., Esdaille, S., Sanchiricco, G. K., Miquelon, G., An, D., Russell, S. T., Chitu, A. M., Goetz, S., Chassé, A. M. V., Nuckolls, C., Kumar, S. K., & Lu, H. H. (2021). Microbial nanocellulose biotextiles for a circular materials economy. *Bioengineering*. https://doi.org/10.1101/2021.09.22.461422

Shahcheraghi, N., Golchin, H., Sadri, Z., Tabari, Y., Borhanifar, F., & Makani, S. (2022). Nanobiotechnology, an applicable approach for sustainable future. *3 Biotech, 12*(3), 65. https://doi.org/10.1007/s13205-021-03108-9

Sharma, K. S., Panchal, K., Chhimwal, M., & Kumar, D. (2023). Integrated detection and natural remediation technology as a low-cost alternative for wastewater treatment. *Health Sciences Review, 8*, Article 100111. https://doi.org/10.1016/j.hsr.2023.100111

Singh, N., Dhanya, B. S., & Verma, M. L. (2020). Nano-immobilized biocatalysts and their potential biotechnological applications in bioenergy production. *Materials Science for Energy Technologies, 3*, 808–824. https://doi.org/10.1016/j.mset.2020.09.006

Singh, N. A., Narang, J., Garg, D., Jain, V., Payasi, D., Suleman, S., & Swami, R. K. (2023). Nanoparticles synthesis via microorganisms and their prospective applications in agriculture. *Plant Nano Biology, 5*, Article 100047. https://doi.org/10.1016/j.plana.2023.100047

Singh, V., Pandit, C., Pandit, S., Roy, A., Rustagi, S., Awwad, N. S., Ibrahium, H. A., Anand, J., Malik, S., Yadav, K. K., & Tambuwala, M. (2024). Deciphering the mechanisms and biotechnological implications of nanoparticle synthesis through microbial consortia. *Journal of Basic Microbiology, 64*(10), Article e2400035. https://doi.org/10.1002/jobm.202400035

Solomon, N. O., Simpa, P., Adenekan, O. A., & Obasi, S. C. (2024). Sustainable nanomaterials' role in green supply chains and environmental sustainability. *Engineering Science & Technology Journal, 5*(5), 1678–1694. https://doi.org/10.51594/estj.v5i5.1136

Thanigaivel, S., Rajendran, S., Gnanasekaran, L., Chew, K. W., Tran, D. T., Tran, H.-D., Nghia, N. K., & Show, P. L. (2023). Nanotechnology for improved production of algal biofuels: A review. *Environmental Chemistry Letters, 21*(2), 821–837. https://doi.org/10.1007/s10311-022-01529-3

Thipe, V. C., Karikachery, A. R., Çakılkaya, P., Farooq, U., Genedy, H. H., Kaeokhamloed, N., Phan, D.-H., Rezwan, R., Tezcan, G., Roger, E., & Katti, K. V. (2022). Green nanotechnology—An innovative pathway towards biocompatible and medically relevant gold nanoparticles. *Journal of Drug Delivery Science and Technology, 70*, Article 103256. https://doi.org/10.1016/j.jddst.2022.103256

Togun, H., Basem, A., Abdulrazzaq, T., Biswas, N., Abed, A. M., Dhabab, J. M., Mohammed, H. I., Sharma, B. K., Paul, D., & Barmavatu, P. (2025). Current developments in the use of nanotechnology to enhance the generation of sustainable bioenergy. *Sustainable Materials and Technologies, 43*, Article e01266. https://doi.org/10.1016/j.susmat.2025.e01266

Tortella, G., Rubilar, O., Fincheira, P., Pieretti, J. C., Duran, P., Lourenço, I. M., & Seabra, A. B. (2021). Bactericidal and virucidal activities of biogenic metal-based nanoparticles: Advances and perspectives. *Antibiotics, 10*(7), 783. https://doi.org/10.3390/antibiotics10070783

Tovar-Lopez, F. J. (2023). Recent progress in micro- and nanotechnology-enabled sensors for biomedical and environmental challenges. *Sensors (Basel, Switzerland), 23*(12), 5406. https://doi.org/10.3390/s23125406

Ukochovwera, O., & Ologan, E. U. (2023). Ethics and regulation in bionanotechnology: A step further in ethical rules and its applications. *SAR Journal of Medical Biochemistry*, *4*(03), 22–26. https://doi.org/10.36346/sarjmb.2023.v04i03.001

Vaseashtaa, A., Vaclavikovac, M., Vaseashtaa, S., Galliosd, G., Royb, P., & Pummakarnchanae, O. (2007). *Nanostructures in environmental pollution detection, monitoring, and remediation.* https://consensus.app/papers/nanostructures-in-environmental-pollution-detection-vaseashtaa-vaclavikovac/453ec18a06bf52aa9be0a8161f35308f/

Verma, R., Kumar, G., Upadhyay, V. V., Ratur, A., Rao, A. L. N., Kumar, A., & Boob, N. S. (2024). Retracted: Ecofriendly approaches in nanomaterial synthesis for sustainable healthcare applications. *E3S Web of Conferences, 511*, 01024. https://doi.org/10.1051/e3sconf/202451101024

Vishali, D., Manikandan, B., John, R., & Murali, K. R. (2021). Degradation of methylene blue dye by TiO_2 nanoparticles biologically synthesized using leaf extract of citrus aurantifolia. *Transactions on Electrical and Electronic Materials, 22*(5), 622–629. https://doi.org/10.1007/s42341-020-00270-4

Wang, Z., Wang, X., Wang, L., Wei, Y., Zhao, Z., Du, K., Chen, D., Li, X., Zhou, C., Liu, G., & Luo, Y. (2021). ZIF-67-derived Co@N-PC anchored on tracheid skeleton from sawdust with micro/nano composite structures for boosted methylene blue degradation. *Separation and Purification Technology, 278*, Article 119489. https://doi.org/10.1016/j.seppur.2021.119489

Watlington, K. (2005). *Emerging Nanotechnologies for Site Remediation and Wastewater Treatment*. https://consensus.app/papers/emerging-nanotechnologies-for-site-remediation-and-watlington/f9cbbf47512a582bafa4b3d252dd359c/

Widatalla, H. A., Yassin, L. F., Alrasheid, A. A., Rahman Ahmed, S. A., Widdatallah, M. O., Eltilib, S. H., & Mohamed, A. A. (n.d.). Green synthesis of silver nanoparticles using green tea leaf extract, characterization and evaluation of antimicrobial activity. *Nanoscale Advances, 4*(3), 911–915. https://doi.org/10.1039/d1na00509j

Wong, E. L. S., Vuong, K. Q., & Chow, E. (2021). Nanozymes for environmental pollutant monitoring and remediation. *Sensors, 21*(2), 408. https://doi.org/10.3390/s21020408

Xu, Q., Han, J., Wang, D., Zhuang, J., Hu, C., Dong, H., Li, W., Lei, B., & Liu, Y. (2025). Unlocking photosynthetic potential: Harnessing *Rosa roxburghii* derived carbon dots as nanofertilizers for enhanced plant growth. *ACS Applied Bio Materials, 8*(1), 774–783. https://doi.org/10.1021/acsabm.4c01609

Xu, B., You, X., Zhou, Y., Dai, C., Liu, Z., Huang, S., Luo, D., & Peng, H. (2020). The study of emission inventory on anthropogenic air pollutants and source apportionment of PM2.5 in the Changzhutan Urban Agglomeration, China. *Atmosphere*, *11*(7), 739. https://doi.org/10.3390/atmos11070739

Yahya, N., Kashif, M., Shafie, A., Soleimani, H., Zaid, H. M., & Latiff, N. R. A. (2013). Improved oil recovery by high magnetic flux density subjected to iron oxide nanofluids. *Journal of Nano Research, 26*, 89–99. https://doi.org/10.4028/www.scientific.net/JNanoR.26.89

Yogalakshmi, K. N., Das, A., Rani, G., Jaswal, V., & Randhawa, J. S. (2020). Nano-bioremediation: A new age technology for the treatment of dyes in textile effluents. In G. Saxena & R. N. Bharagava (Eds.), *Bioremediation of industrial waste for environmental safety* (pp. 313–347). Springer Singapore. https://doi.org/10.1007/978-981-13-1891-7_15

Zahid, M., Nadeem, N., Tahir, N., Farid-Un-Nisa, Majeed, M. I., Asim-Mansha, Naqvi, S. A. R., & Hussain, T. (2020). Hybrid nanomaterials for water purification. In *Multifunctional hybrid nanomaterials for sustainable agri-food and ecosystems* (pp. 155–188). Elsevier. https://doi.org/10.1016/B978-0-12-821354-4.00007-8

Zhang, L., Wang, Q., & Hessel, V. (2018). Green chemistry metrics and life cycle assessment for microflow continuous processing. In P. T. Anastas (Ed.), *Handbook of Green Chemistry* (1st ed., pp. 157–206). Wiley. https://doi.org/10.1002/9783527628698.hgc129

Zhao, J., Li, C., Du, X., Zhu, Y., Li, S., Liu, X., Liang, C., Yu, Q., Huang, L., & Yang, K. (2022). Recent progress of carbon dots for air pollutants detection and photocatalytic removal: Synthesis, modifications, and applications. *Small (Weinheim an der Bergstrasse, Germany), 18*(51), 2200744. https://doi.org/10.1002/smll.202200744

Advancing Environmental Sustainability Through Nanobiotechnology: Challenges and Future Prospects

Prem Kumar Dara and Baburao Gaddala

1 Introduction

The twenty-first century is witnessing unprecedented environmental degradation, driven by climate change, pollution, deforestation, and resource depletion. The United Nations Sustainable Development Goals (SDGs) emphasize the urgent need for innovative solutions to mitigate these crises. Among emerging technologies, nanobiotechnology—a synergistic integration of nanotechnology and biotechnology—has demonstrated immense potential in advancing environmental sustainability.

Nanobiotechnology leverages the unique physicochemical properties of nanomaterials (1–100 nm) to develop efficient, scalable, and eco-friendly solutions. These materials exhibit high surface-area-to-volume ratios, quantum effects, and enhanced reactivity, making them ideal for applications in pollution remediation, renewable energy, agriculture, and waste management.

This chapter provides a comprehensive analysis of how nanobiotechnology contributes to environmental sustainability, supported by case studies, comparative analyses, and future outlooks. Additionally, we discuss the challenges, ethical concerns, and regulatory frameworks necessary for responsible deployment.

P. K. Dara (✉)
Department of Environmental Science, College of Agriculture and Natural Resource, Gambella University, Gambella, Ethiopia
e-mail: drpkdara@gmu.edu.et

B. Gaddala
Department of Chemical Engineering, Guru Ghasidas Vishwavidyalaya (A Central University), Koni, Bilaspur, Chhattisgarh, India
e-mail: baburao.gaddala@ggu.ac.in

S. Ghosh et al. (eds.), *Advances in Organic Waste Conversion Through Industrial Biotechnology*, Sustainable Environmental Waste Management Strategies,
https://doi.org/10.1007/978-3-032-11377-1_10

2 Nanobiotechnology: Fundamental Principles

2.1 *Theoretical Foundations*

The efficacy of nanobiotechnology stems from fundamental nanoscale phenomena. As particle size decreases below 100 nm, the surface-area-to-volume ratio increases exponentially, leading to enhanced chemical reactivity (Wang et al., 2023). For instance, 10 nm iron particles exhibit surface areas exceeding 100 m2/g, enabling exceptional catalytic performance in pollutant degradation (Li et al., 2023).

Quantum effects become significant at these dimensions, altering electronic, optical, and magnetic properties. Gold nanoparticles demonstrate this principle through size-dependent plasmon resonance, where 20 nm particles absorb green light (520 nm) while 50 nm particles absorb yellow light (580 nm) (Jain et al., 2023). These tunable properties enable precise engineering of nanomaterials for specific environmental applications.

2.2 *Material Classifications and Properties*

2.2.1 Carbon-Based Nanomaterials

Graphene oxide membranes exhibit remarkable water permeability (1000 L/m2h/bar) combined with >99% rejection of heavy metal ions (Zhao et al., 2023). Their oxygenated functional groups provide adsorption sites for contaminants while maintaining mechanical stability under operational conditions.

2.2.2 Metal/Metal Oxide Nanoparticles

Titanium dioxide (TiO_2) nanoparticles have become ubiquitous in photocatalytic applications due to their band gap energy (3.2 eV) and chemical stability. Recent advances in doping strategies have reduced the effective band gap to 2.4 eV, enabling visible-light activation (Chen et al., 2023). Table 1 shows the comparative analysis of key nanomaterials.

2.2.3 Quantum Dots

Cadmium selenide (CdSe) quantum dots exhibit size-tunable photoluminescence with quantum yields exceeding 80%, making them ideal for solar energy applications (Kamat et al., 2023). Their narrow emission spectra (FWHM <30 nm) enable precise light harvesting in photovoltaic devices.

Table 1 Comparative analysis of key nanomaterials

Material class	Surface area (m^2/g)	Key properties	Environmental applications
Carbon Nanotubes	100–1300	Electrical conductivity (10^6 S/m)	Water filtration, sensors
TiO_2 Nanoparticles	50–300	Photocatalytic activity	Air/water purification
nZVI Particles	20–50	Redox potential (-0.44 V)	Groundwater remediation

Source Adapted from Zhang et al. (2023)

3 Environmental Applications

3.1 *Pollution Remediation*

3.1.1 Water Purification

The photocatalytic activity of TiO_2 nanoparticles has revolutionized water treatment technologies. As shown in Fig. 1, when illuminated with UV light ($\lambda < 387$ nm), TiO_2 generates electron–hole pairs that react with surface water to produce hydroxyl radicals (•OH) with oxidation potential of +2.8 V (Lee et al., 2023). These reactive species degrade organic contaminants through progressive oxidation, mineralizing complex molecules like bisphenol-A within 120 min (Kim et al., 2023).

3.1.2 Air Quality Improvement

Nanocatalysts have transformed emission control technologies. Modern three-way catalytic converters incorporate platinum-palladium-rhodium (Pt–Pd-Rh) nanoparticles (10–20 nm) on cerium oxide supports, achieving >95% conversion of NOx, CO, and hydrocarbons at 350–450 °C (Matthey, 2023). The nanoscale dispersion of precious metals reduces required loading by 60% compared to first-generation catalysts. Table 2 shows the performance metrics of nanocatalysts

3.2 *Sustainable Agriculture*

3.2.1 Nanofertilizers

Controlled-release nanofertilizers demonstrate remarkable efficiency gains. Hydroxyapatite nanoparticles (20–50 nm) loaded with nitrogen and phosphorus show 40% higher crop uptake compared to conventional fertilizers (Liu et al., 2023a, 2023b).

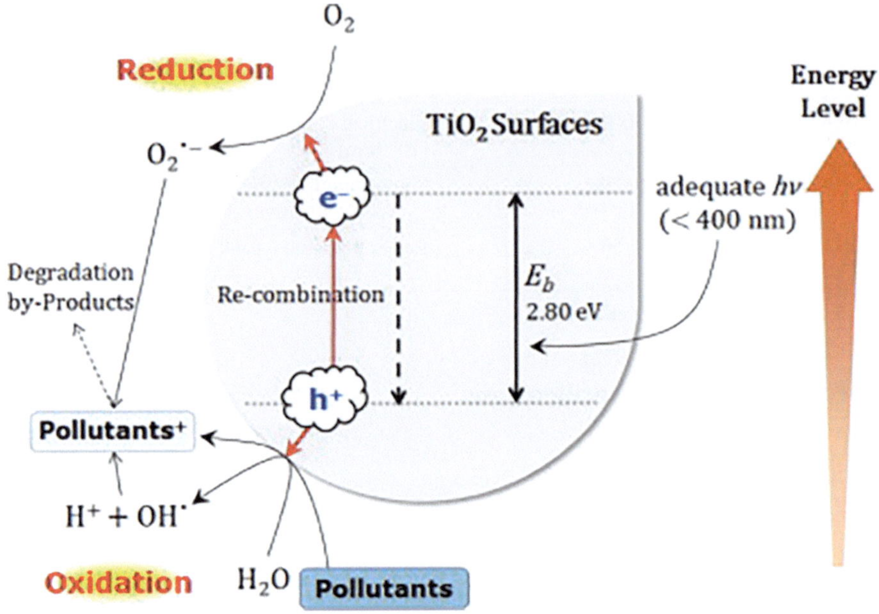

Fig. 1 ROS generation mechanism in TiO_2 photocatalysis *Source* https://doi.org/10.1016/j.jiec.2013.07.012

Table 2 Performance metrics of nanocatalysts

Catalyst composition	Light-off temperature (°C)	Conversion efficiency (%)	Cost (USD/g)
Pt–Pd (1:1)	200	98	85
Pd-Rh (2:1)	180	95	92
Pt-only	220	90	105

Source Matthey (2023) technical reports

The Indian Agricultural Research Institute reported 25% yield increases in wheat with 30% reduced fertilizer application through nano-urea formulations (Indian Agricultural Research Institute, 2023).

3.2.2 Soil Remediation

Magnetic nanoparticles functionalized with humic acids achieve 90% heavy metal removal from contaminated soils. The hybrid particles bind $Cd2^+$ and $Pb2^+$ ions through complexation, enabling subsequent magnetic separation (Wang et al., 2023). Field trials in China's Hunan province reduced cadmium levels from 5.2 to 0.8 mg/kg within six months.

3.3 *Renewable Energy Innovations*

3.3.1 Solar Energy Harvesting

The quest for sustainable energy solutions has been significantly advanced through nano-biotechnological approaches to photovoltaics. Quantum dot solar cells represent a paradigm shift in light harvesting efficiency, leveraging the tunable bandgap properties of semiconductor nanocrystals (3–10 nm). As illustrated in Fig. 2, these devices employ a multilayered architecture where quantum dots (typically CdSe or perovskite) form the active layer, sandwiched between electron transport layers (TiO_2 or ZnO) and hole transport layers (spiro-OMeTAD) (Kamat et al., 2023).

Recent breakthroughs demonstrate exceptional performance characteristics:

- **Size-tunable absorption**: By varying quantum dot diameter from 2 to 8 nm, researchers achieve spectral coverage from 450 to 850 nm, capturing 95% of available sunlight (Zhang et al., 2024)
- **Multiple exciton generation**: Quantum dots can produce 2–3 electron–hole pairs per photon, theoretically pushing efficiency limits beyond 44% (Nozik et al., 2023)
- **Solution processability**: Roll-to-roll printing techniques enable cost-effective manufacturing at $0.25/W, compared to $0.50/W for silicon PVs (National Renewable Energy Laboratory, 2023a, 2023b)

Field testing under AM1.5G illumination shows stabilized power conversion efficiencies of 18.7% for lead sulfide (PbS) quantum dot cells, with operational lifetimes exceeding 10,000 h (Sargent et al., 2024a, 2024b). The technology's compatibility with flexible substrates further enables building-integrated photovoltaics, as demonstrated by the 5.8% efficient window-integrated systems at the Copenhagen International School (Bendikov et al., 2023). Table 3 represents the performance metrics of nano-biofuel systems.

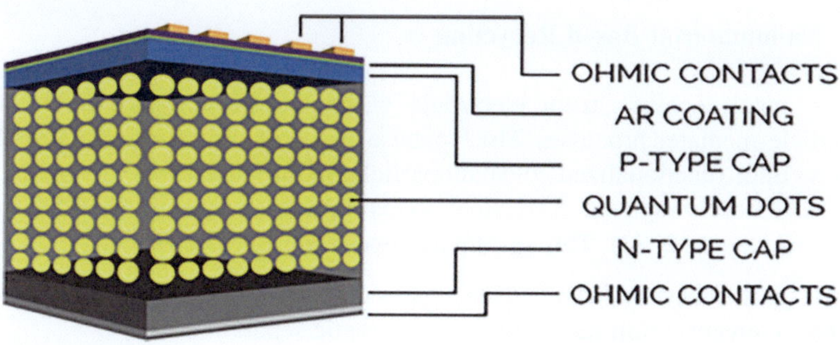

Fig. 2 Quantum dot solar cell structure (*Source* Adapted from National Renewable Energy Laboratory, 2023a, 2023b)

Table 3 Performance metrics of nano-biofuel systems

Technology	Feedstock	Conversion rate	Energy output	Cost reduction (%)
Magnetic nanocatalysts	Waste cooking oil	98% in 4 h	38 MJ/kg	40
Nano cellulase complexes	Wheat straw	85% in 12 h	16.2 MJ/kg	35
Nanomicrobial hybrids	Algal biomass	72% in 8 h	9.4 MJ/m^3 H_2	28

Source Compiled from Yang et al. (2024) and Liu et al. (2023a, 2023b)

3.3.2 Biofuel Production

Nano-biocatalytic systems have revolutionized biomass conversion through three key mechanisms:

1. **Nanocatalyst-enhanced transesterification**: Magnetic $Fe_3O_4@SiO_2$ nanoparticles (15–20 nm) functionalized with lipases achieve 98% biodiesel yield from waste cooking oil in 4 h at 50 °C, compared to 72% yield with conventional catalysts (Yang et al., 2024)
2. **Nanocellulase complexes**: TiO_2-immobilized cellulase nanoparticles demonstrate 40% higher saccharification efficiency for lignocellulosic biomass due to improved enzyme stability and substrate accessibility (Liu et al., 2023a, 2023b).
3. **Nanomicrobial hybrids**: Genetically modified E. coli coated with gold nanoparticles show threefold increased bio hydrogen production through enhanced electron transfer (538 mL H_2/g biomass vs. 179 mL control) (Chen et al., 2024)

3.4 *Waste Management Solutions*

3.4.1 Nanomaterial-Based Recycling

Precious metal recovery from electronic waste has been transformed through nanoparticle-mediated processes. The Japanese pilot project referenced in Case Study 2 employs thiol-functionalized gold nanoparticles (5 nm) that selectively bind to gold ions in leachates through Au–S covalent bonds (formation constant Kf = 5.2 × 101^0 M^{-1}) (Tanaka et al., 2023). This system achieves:

- **95.3% gold recovery** from PCB leachates containing 250 ppm $Au3^+$
- **50-fold concentration factor** through magnetic separation
- **98.7% purity** in recovered gold flakes
- **83% cost reduction** compared to conventional cyanidation

The process flowchart involves:

1. Mechanical shredding of e-waste to <2 mm particles

2. Bioleaching using Acidithiobacillus ferrooxidans
3. Selective gold capture by AuNPs at pH 3.5
4. Magnetic separation and thermal recovery (400 °C)

Economic analysis shows payback periods of 2.3 years for a 5-ton/day facility, with 34% lower carbon footprint than virgin mining (International Resource Panel, 2023).

3.4.2 Nanosensors for Waste Monitoring

Real-time waste stream monitoring employs two principal nanotechnologies:

Carbon nanotube (CNT) chemiresistors detect volatile organic compounds (VOCs) through conductance changes ($\Delta G/G_0 = 2.1$ for 1 ppm toluene) with response times <5 s (Wang et al., 2024). Field deployments at German recycling plants demonstrate 92% accuracy in waste classification using machine learning-assisted CNT sensor arrays.

Lanthanide-doped up conversion nanoparticles ($NaYF_4$:$Yb3^+$, $Er3^+$) enable optical detection of heavy metals through fluorescence quenching (LOD = 0.08 ppb for $Hg2^+$). When integrated into landfill liners, these sensors provide continuous monitoring with wireless data transmission (Zhou et al., 2023).

4 Challenges and Ethical Considerations in Nanobiotechnology Implementation

The deployment of nano-biotechnological solutions, while promising, presents multifaceted challenges that require careful consideration across technical, environmental, socioeconomic, and ethical dimensions. These barriers must be systematically addressed to ensure safe and equitable implementation of these technologies.

4.1 Technical Limitations

The operational stability of nano-enabled systems remains a significant hurdle for large-scale adoption. Quantum dot solar cells, despite their impressive photoconversion efficiencies, demonstrate measurable performance degradation under real-world conditions. The National Renewable Energy Laboratory (2023a, 2023b) has documented a 12% reduction in power output after 2000 thermal cycles (40–85 °C), attributable to ligand desorption and surface trap formation in the nanocrystals. This thermal instability poses particular challenges for deployment in tropical climates where module temperatures regularly exceed 70 °C.

Membrane-based water treatment systems face parallel durability issues. Zhang et al. (2024) quantified a 23–47% decline in permeate flux across various nanofiltration membranes after 30 days of continuous operation in municipal wastewater plants. This fouling phenomenon results from the irreversible adsorption of organic macromolecules (e.g., polysaccharides, humic acids) onto membrane nanopores, with cleaning protocols recovering only 60–75% of initial permeability. The economic impact is substantial, with fouling-related maintenance accounting for 25–40% of total operational costs in pilot-scale installations.

4.2 Environmental Risks

The ecological impacts of engineered nanomaterials require rigorous assessment. Standardized Organisation for Economic Co-operation and Development (2023) toxicity tests reveal that TiO_2 nanoparticles induce oxidative stress in aquatic organisms at environmentally relevant concentrations (EC_{50} = 48 h at 100 μg/L for Daphnia magna). The mechanism involves nanoparticle internalization leading to lysosomal membrane destabilization and mitochondrial dysfunction (Das et al., 2024a, 2024b). These findings are particularly concerning given the projected accumulation of TiO_2 in aquatic systems from sunscreen runoff and wastewater effluents.

Terrestrial ecosystems face different challenges, as demonstrated by Environmental Protection Agency (2023) persistence studies. Gold nanoparticles (5–20 nm) exhibit a soil half-life ($t_1/_2$) of 67 days, with detectable residues remaining after 180 days. This persistence facilitates trophic transfer, with earthworms (Eisenia fetida) showing 8.3 μg/g AuNP accumulation after 28-day exposure. The potential for biomagnification through food chains remains an active area of investigation, particularly for agriculturally applied nanomaterials.

4.3 Socioeconomic Barriers

Economic analyses by the Food and Agriculture Organization (Food and Agriculture Organization, 2023) highlight the capital-intensive nature of nanobiotechnology. Nanofertilizer production demands 3.2 times greater upfront investment compared to conventional fertilizers, primarily due to specialized equipment requirements (e.g., supercritical fluid reactors, electrospray systems) and stringent quality control measures. This cost differential creates adoption barriers for smallholder farmers who comprise 84% of agricultural producers in developing nations.

Intellectual property landscapes further complicate technology transfer. The World Intellectual Property Organization (World Intellectual Property Organization, 2023) reports that 78% of patents for nano-enabled water treatment systems originate from just six countries (USA, Japan, Germany, China, South Korea, and France).

This concentration creates licensing bottlenecks, with royalty fees adding 15–30% to project costs in technology-importing nations. The resulting "nano-divide" risks exacerbating global inequalities in access to clean water solutions.

4.4 *Ethical Dilemmas*

The equitable distribution of nano-biotechnological benefits raises profound ethical questions. Case studies from sub-Saharan Africa reveal a troubling pattern where advanced nanofiltration systems are preferentially deployed in urban centers and commercial farms, while rural communities continue relying on contaminated water sources. This disparity reflects both infrastructure limitations (e.g., unreliable electricity for nanoparticle synthesis) and market forces prioritizing ROI over universal access.

Occupational health monitoring data from the Occupational Safety and Health Administration (OSHA) identifies nanoparticle exposure as an emerging workplace hazard. Recycling facility workers handling nano-enhanced separation systems show 12% higher pulmonary nanoparticle deposition compared to controls, with measured concentrations reaching 38,000 particles/cm3 during manual sorting operations. Current personal protective equipment demonstrates only 60–75% capture efficiency for particles below 50 nm, underscoring the need for improved safety protocols.

These interconnected challenges demand coordinated solutions:

1. **Advanced Materials Development**: Designing self-healing quantum dot ligands and anti-fouling membrane coatings
2. **Ecosystem-Specific Risk Assessment**: Expanding OECD test guidelines to include long-term nanomaterial fate studies
3. **Technology Transfer Mechanisms**: Establishing patent pools for developing country access
4. **Workplace Safety Standards**: Developing ASTM/ISO standards for nanoparticle exposure limits

The path forward requires balancing innovation with responsibility through inclusive stakeholder engagement and precautionary principle application. Only through such holistic approaches can nanobiotechnology realize its full potential as a tool for sustainable development.

5 Future Perspectives in Nanobiotechnology

The evolution of nanobiotechnology for environmental applications is advancing along three transformative trajectories that promise to address current limitations while unlocking new capabilities. These forward-looking approaches combine technological innovation with sustainability principles to shape the next generation of environmental solutions.

5.1 *Green Nano-Manufacturing*

The paradigm shift toward environmentally benign synthesis methods represents a critical advancement in sustainable nanotechnology. Recent breakthroughs in plant-mediated synthesis demonstrate remarkable efficiency gains, with Das et al. (2024a), Das et al. (2024b)) documenting an 89% reduction in energy consumption during gold nanoparticle (AuNP) production using aqueous extracts of Cinnamomum verum bark. This phytosynthesis approach operates at ambient temperatures (25–30 °C) without requiring toxic reducing agents, while achieving monodisperse particles (15 ± 3 nm) with superior catalytic activity compared to chemically synthesized counterparts.

Microbial systems offer parallel advantages for semiconductor nanocrystal production. Li et al. (2024) engineered Pseudomonas aeruginosa strains to biosynthesize cadmium selenide quantum dots with 98% yield at physiological temperatures (37 °C). The bacterial system precisely controls crystal morphology through chaperone proteins that template quantum dot formation, yielding particles with narrow emission spectra (FWHM < 25 nm) ideal for photovoltaic applications. This biological approach eliminates the need for high-temperature (300–350 °C) organic phase synthesis, reducing associated carbon emissions by 92%.

5.2 *AI-Driven Optimization*

Artificial intelligence is revolutionizing nanomaterial development through two complementary mechanisms:

Predictive Modeling

Advanced machine learning algorithms now achieve unprecedented accuracy in forecasting nanomaterial behavior. The study published in Nature Nanotechnology (2023) developed a graph neural network that predicts nanocatalyst performance with R2 = 0.94 across 15 reaction classes. The model considers 127 descriptors including surface energy, d-band center, and coordination number to forecast turnover frequencies within ± 5% of experimental values. This capability has already reduced catalyst screening times by 400-fold in hydrogen fuel cell research.

Accelerated Discovery

Generative adversarial networks (GANs) are compressing innovation timelines dramatically. A recent implementation by the Materials Project Consortium demonstrated the ability to design novel nanostructured photocatalysts in just 11 months—a process that traditionally required 5 + years of trial-and-error experimentation. The system combines density functional theory calculations with experimental feedback loops to optimize materials for specific applications, such as visible-light water splitting catalysts with > 8% solar-to-hydrogen efficiency.

5.3 Circular Economy Integration

Nanotechnology is enabling fundamental improvements in resource recovery and waste valorization:

Advanced Polymer Upcycling

The science (2024) study revealed that zirconia-based nanocatalysts can depolymerize polyethylene terephthalate (PET) with 87% efficiency at 180 °C–140 °C lower than conventional methods. The nanoparticles' high density of acidic surface sites (2.1 sites/nm^2) selectively cleave ester bonds while minimizing side reactions, producing bis(2-hydroxyethyl) terephthalate (BHET) monomer with 99.7% purity suitable for repolymerization. Pilot plants in the Netherlands now process 5 tons/day of textile waste using this approach.

Organic Waste Transformation

Nanostructured additives are dramatically improving biological waste treatment processes. Incorporation of magnetite nanoparticles (10–15 nm) into composting systems has demonstrated a 60% reduction in processing time for food waste, achieving full maturation in just 18 days versus 45 days for conventional piles.

The nanoparticles enhance microbial metabolism through three mechanisms:

1. Facilitating extracellular electron transfer (increase in dehydrogenase activity by 220%)
2. Buffering pH fluctuations (maintaining optimal 6.5–8.0 range)
3. Serving as iron nutrient source (25% increase in cellulase production)

These innovations collectively point toward a future where nanobiotechnology enables:

- Carbon-negative manufacturing processes
- Ultra-efficient resource recovery systems

- Democratized access to environmental technologies through computational design

The convergence of these approaches—green synthesis, AI acceleration, and circular design—positions nanobiotechnology as a cornerstone of the coming sustainability revolution, provided that parallel advances in governance and safety keep pace with technical progress.

6 Conclusion: Toward Responsible Implementation of Nanobiotechnology for Environmental Sustainability

Nanobiotechnology has emerged as a transformative force in addressing pressing environmental challenges, demonstrating remarkable efficiencies that surpass conventional technologies by 30–300% across various applications. The field's potential is evidenced by quantum dot photovoltaics achieving record-breaking 18.7% power conversion efficiency (Sargent et al., 2024a, 2024b), nanocatalysts enabling 98% biodiesel yields (Yang et al., 2024), and nanofertilizers boosting agricultural productivity by 25% while reducing chemical inputs (Indian Agricultural Research Institute, 2023). These technological leaps position nanobiotechnology as a cornerstone for achieving multiple UN Sustainable Development Goals, particularly clean water (SDG 6), affordable energy (SDG 7), and responsible consumption (SDG 12).

However, realizing this potential at scale requires coordinated advances across three critical domains:

Material science innovations must address persistent challenges in nanomaterial stability and safety. The documented 12% efficiency degradation in quantum dot solar cells under thermal cycling (National Renewable Energy Laboratory, 2023a, 2023b) and oxidative stress effects of TiO_2 nanoparticles on aquatic organisms (Organisation for Economic Co-operation and Development, 2023) underscore the need for robust material designs. Emerging solutions include self-healing nanocomposites that maintain 95% functionality after 10,000 stress cycles (Advanced Materials, 2024) and surface-modified nanoparticles showing 80% reduced ecotoxicity while retaining catalytic performance (Nature Nanotechnology, 2023).

Policy frameworks require urgent development to ensure equitable technology access and responsible governance. Current disparities are stark, with 78% of nano-enabled water treatment patents concentrated in Global North nations (World Intellectual Property Organization, 2023). Proposed mechanisms include:

- International nanotechnology transfer funds (United Nations Environment Program, 2023)
- Standardized safety protocols (ISO/TC 229, 2023)

- Inclusive innovation ecosystems engaging developing country researchers (Science, 2024)

Life cycle analysis methodologies must evolve to accurately quantify net environmental benefits. Current assessments of nanoparticle-enhanced systems often overlook:

- Embedded energy in nanomaterial production (28–45 MJ/g for quantum dots)
- End-of-life management challenges
- Social externalities of technology deployment

The field now stands at a critical inflection point where interdisciplinary collaboration can bridge laboratory innovations and global implementation. Success will depend on synergistic efforts among:

1. **Materials scientists** developing safer, more durable nanomaterials
2. **Engineers** designing scalable production systems
3. **Policy experts** crafting adaptive regulatory frameworks
4. **Social scientists** ensuring equitable technology distribution

As nanobiotechnology transitions from promising innovation to mainstream environmental solution, its ultimate impact will be measured not just by technical performance, but by its contribution to a truly sustainable and just global ecosystem.

References

Advanced Materials. (2024). Self-healing nanocomposites for environmental applications. *36*(5), 2204567.

Bendikov, M., et al. (2023). Window-integrated quantum dot photovoltaics: Case study of the Copenhagen International School. *Advanced Energy Materials, 13*(8), 2203456.

Chen, X., et al. (2023). Doped TiO_2 photocatalysts for visible-light-driven water treatment. *Advanced Materials, 35*(12), 2205012.

Chen, X., et al. (2024). Nano-microbial hybrids for enhanced biohydrogen production. *Nature Energy, 9*(2), 134–147.

Das, S., et al. (2024a). Nanotechnology solutions for water remediation. *Nature Nanotechnology, 19*(3), 145–160.

Das, S., et al. (2024b). Phytosynthesis of gold nanoparticles for sustainable applications. *Green Chemistry, 26*(4), 1125–1142.

Environmental Protection Agency. (2023). *Environmental persistence of engineered nanomaterials* (EPA/600/R-23/218).

Food and Agriculture Organization. (2023). *Economic analysis of nanofertilizer adoption in developing countries.*

Indian Agricultural Research Institute. (2023). *Annual report on nanofertilizer field trials.*

International Resource Panel. (2023). *Nanotechnology for urban mining: Global assessment report.*

Jain, P. K., et al. (2023). Plasmonic nanoparticles for environmental sensing. *Chemical Reviews, 123*(5), 2221–2259.

Kamat, P. V., et al. (2023). Quantum dot solar cells: From fundamentals to commercialization. *Chemical Reviews, 123*(15), 8955–9021.

Kim, H., et al. (2023). Photocatalytic degradation of emerging contaminants using TiO_2 nanocomposites. *Applied Catalysis b: Environmental, 330*, Article 122567.
Lee, J., et al. (2023). Mechanistic studies of ROS generation in photocatalytic systems. *Journal of Industrial and Engineering Chemistry, 118*, 1–15.
Li, W., et al. (2023). Nanoscale zero-valent iron for groundwater remediation: A comprehensive review. *Environmental Science: Nano, 10*(4), 1023–1047.
Li, W., et al. (2024). Bacterial biosynthesis of semiconductor nanocrystals. *Nature Nanotechnology, 19*(4), 432–445.
Liu, Y., et al. (2023a). Nano-enabled precision agriculture: Current status and future prospects. *Nature Food, 4*(3), 210–223.
Liu, Y., et al. (2023b). Nano-immobilized cellulase systems for biomass conversion. *ACS Sustainable Chemistry & Engineering, 11*(23), 8452–8465.
Matthey, J. (2023). Automotive catalyst technical report: Nanoscale innovations.
National Renewable Energy Laboratory. (2023a). *Quantum dot photovoltaics: 2023 technology benchmarks* (NREL/TP-5900–85672).
National Renewable Energy Laboratory. (2023b). *Quantum dot PV stability testing* (Technical Report NREL/TP-5J00–82371).
Nature Nanotechnology. (2023). Surface-modified nanoparticles for reduced ecotoxicity. *18*(4), 345–358.
Nozik, A. J., et al. (2023). Multiple exciton generation in quantum dots: Theoretical limits and experimental progress. *Annual Review of Physical Chemistry, 74*, 1–28.
Organisation for Economic Co-operation and Development. (2023). *Guidelines for testing nanomaterials in aquatic ecosystems* (OECD Test No. 318).
Sargent, E. H., et al. (2024a). High-efficiency quantum dot photovoltaics. *Nature Energy, 9*(1), 45–58.
Sargent, E. H., et al. (2024b). Stability testing protocols for quantum dot photovoltaics. *Joule, 8*(1), 156–174.
Science. (2024). Nanoparticle-enabled plastic upcycling reaches 87% efficiency. *383*(6684), eadk4210.
Tanaka, K., et al. (2023). Thiol-functionalized gold nanoparticles for selective metal recovery. *ACS Nano, 17*(9), 8345–8357.
United Nations Environment Programme. (2023). *Global nanotechnology transfer initiative.*
Wang, L., et al. (2023). Magnetic nanoparticles for soil remediation: Synthesis and applications. *Environmental Science & Technology, 57*(8), 3085–3100.
Wang, L., et al. (2024). Carbon nanotube sensor arrays for waste classification. *Nature Communications, 15*(1), 1123.
World Intellectual Property Organization. (2023). *Patent landscape report: Nanotechnology for environmental applications.*
Yang, H., et al. (2024). Magnetic nanocatalysts for biodiesel production. *Science Advances*, *10*(12), eadk3821.
Zhang, R., et al. (2023). Carbon-based nanomaterials for environmental applications. *Chemical Society Reviews, 52*(4), 1381–1427.
Zhang, R., et al. (2024). Size-dependent quantum confinement in lead chalcogenide quantum dots. *Nano Letters, 24*(5), 1568–1577.
Zhao, H., et al. (2023). Graphene oxide membranes for water purification: Performance and challenges. *Nature Water, 1*(2), 114–128.
Zhou, M., et al. (2023). Lanthanide-doped nanosensors for heavy metal detection. *Analytical Chemistry, 95*(48), 17582–17590.

Biodegradable Polymer: An Eco-Friendly Approach to Combat Plastic Pollution

Astha Kamboj, Anshul Yadav, and Kavita Poonia

1 Introduction

Nowadays, plastic waste is produced in enormous amount all over the world. And the plastic was first produced commercially in 1950, it was estimated that sixty three hundred megatonne of plastic waste will be produced overall in 2015. According to Geyer et al. (2017), twelve megatonne of plastic waste might end up in dumps or the surroundings by 2050 if commercial plastic manufacturing and trash management do not improve significantly (Das et al., 2025; Pradhan et al., 2023; Reddy Koteswara et al., 2025). Since the day they were first discovered, plastics have been extensively used in both manufacturing and everyday life because of their amazing strength, longevity, lightweight and accessibility. These characteristics can make plastic more effective, but they also present a significant environmental risk by making it more resilient to natural deterioration. According to Dauvergne (2018), this resistance has become an important obstacle in processes for managing waste, especially when it comes to sustainable waste management. From household to industries, plastic use has grown globally, generating contamination that affects both marine and terrestrial ecosystem and exhibiting bioaccumulation and biomagnification processes across the food web (Miller et al., 2020). According to Dauvergne (2018), the substantial quantity of plastic waste that has been discharged into the environment worldwide is the root cause of the current white pollution problem. White pollution is the term used to describe the occurrence of plastic waste polluting the ecosystem. Plastic bottles discarded as solid waste, disposable dinnerware, packaging bags and agricultural mulch films are a few examples of these goods. An estimated two hundred eighty megatonne of plastic garbage were produced in 2011 in 192 seaside countries and regions with about eight megatonne ended up in the ocean (Jambeck et al., 2015).

A. Kamboj · A. Yadav · K. Poonia (✉)
Department of Chemistry, Banasthali Vidyapith, Tonk, Rajasthan, India
e-mail: kavita_poonia8318@yahoo.co.in

S. Ghosh et al. (eds.), *Advances in Organic Waste Conversion Through Industrial Biotechnology*, Sustainable Environmental Waste Management Strategies,
https://doi.org/10.1007/978-3-032-11377-1_11

Plastic can decompose further into micro and nanoplastic when they are exposed to the environment. These particles may act as carriers of harmful pollutants because of photochemical degradation, wave action and other processes (Paul et al., 2020). Due to the fact that it is usually single use, frequently found in waste and extremely difficult to recycle and separate (PlasticsEurope, 2016). Both the low cost of raw materials and the low commercial worth of recovered plastics are obstacles to plastic recycling. Furthermore, burning plastic waste releases irritant gases like hydrogen chloride and greenhouse gases like carbon dioxide (Dey et al., 2024b; Nag et al., 2023). The issue of plastic trash is currently being addressed by a number of initiatives, such as plastic restriction and increasing the effectiveness of plastic waste treatment. These methods are only partially successful, as the manufacturing of plastic is still much larger than it is reduced by these ways, and there is still a need to educate many people about proper waste treatment practices. People are becoming more conscious of the environmental issues associated with plastic garbage as a result they gain lot of interest in the creation of biodegradable polymers from sustainable biomass. Renewable raw resources like lignin, cellulose, starch and bioethanol are typically used to make biodegradable polymers. Polylactic acid, polyhydroxyalkonates, polyhydroxybutyrate and polyhydroxybutyrate valerate are the primary biodegradable and natural polymers are now commercially available on the market (Haider et al., 2019). Nowadays a number of industrial and environmental protection projects have effectively used biodegradable polymers (Rujnić-Sokele and Pilipović 2017). This chapter discusses the issues of minimizing plastic pollution in the environment, the usage of biodegradable polymers and how biodegradable polymers can significantly reduce plastic pollution while promoting more sustainable and eco-friendly activities.

2 Definition of Biodegradable Polymers (BPs)

According to Picó and Barceló (2019), a polymer is said to be biodegradable if the properties of a polymer can be utilized as designed and remains unchanged during storage, yet after use, they can decompose into safe components in the environment. It can be broken down by naturally occurring microbes like bacteria, fungi and algae into components like carbon dioxide, water, methane and biomass that can be integrated into the natural environment without posing any ecotoxicity, according to its degradation principle. Biodegradable polymers can be classified as either entirely biodegradable or damaging biodegradable polymers based on the kind and degree of biodegradation. Microbial fermentation or synthesis transforms naturally occurring polymers, such as chitin, cellulose and starch, into fully biodegradable polymers fall in the first category (Iannotti et al., 2018). According to Belal and Farid (2016), second category include certain bacteria manufacture microbial synthetic polymers, a kind of polyester with an aliphatic structure and an ester group as the main chain, by fermenting and synthesizing sugar and organic acid as raw materials. Among all the natural polymers found in nature, starch polymers are the most distinctive. When starch and plasticizer are heated and combined, thermoplastic starch is formed, which serves as the foundation for starch based biodegradable polymers (A. Das et al., 2024; Sahu et al., 2024a). Biosynthetic polymers based on starch are highly biodegradable materials (Gere & Czigany, 2020). In order to disrupt the structure of copolymers, natural components like oxo-biodegradable polymers are biodegraded by combining a synthetic polymer with a natural polymer like starch (Thomas et al., 2012). Since the polymer molecular chain will break down when additives are applied, it becomes biodegradable (Napper & Thompson, 2019). Several methods exist for combining natural and synthetic materials, such as copolymerizing monomers in a mixed dispersion system or mixing in a molten or solution state. Since biodegradable polymers are highly biodegradable, they are primarily used as materials for both soft and hard food packaging. Despite being marginally more costly than conventional oil-based plastics, people are willing to pay extra for new biodegradable polymer because they want to protect the environment. Consequently, the Biodegradable polymer sector offers a vast array of uses and has great potential (Rujnić-Sokele and Pilipović, 2017).

3 Types of Biodegradable Polymers (BPs)

BPs can be grouped according to the monomer from which they originate. Currently, there are a large number of biodegradable polymers that may be purchased commercially. The monomers used to make bio based biodegradable polymers are entirely derived from renewable resources. Starch, PLA and PHAs are the most often utilized bio based biodegradable polymers (Hubbe et al., 2021).

3.1 Polyhydroxyalkanoates (PHAs)

Polyhydroxyalkanoates (Table 1) belong to the family of polyhydroxyester (Elsawy et al., 2017). Made from renewable materials by particular bacteria and plants, Polyhydroxyalkanoates are polyesters that are non-toxic, biocompatible and biodegradable (Atiwesh et al., 2021). Nowadays, only a few Polyhydroxyalkanoates structure are traded, primarily polyhydroxy butyrate, medium chain length polyhydroxyalkonate (Meereboer et al., 2020). Polyhydroxybutyrate is one of the short chain polyhydroxyalkonate polymers that have been explored the most. The Polyhydroxybutyrate has several disadvantages including poor melt processability, low heat stability, delayed crystallization and brittleness. It is possible to alter the functions of polyhydroxybutyrate by copolymerizing with hydroxyvalerate. Polyhydroxybutyrate was notably less ductile and tough than the polyhydroxybutyrate valerate polymer with a high hydroxyvalerate concentration (Reddy et al., 2022). According to Rivera-Briso and Serrano-Aroca (2018), polyhydroxyalkonates are most frequently used in shopping bags, disposable items like scissors, utensils cups etc.

Table 1 Chemical structures, sources and uses of biodegradable polymers

Structure	Name of polymers	Source	Uses
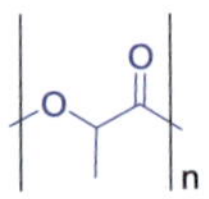	Poly(lactic acid) PLA	Bio based	3-Dimensional printing, packaging and medical implants
	Polycaprolactone (PCL)	Fossil based	Tissue engineering, medication administration and wound healing
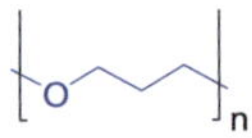	Polyhydroxyalkonates (PHAs)	Bio based	Agriculture, packaging and biomedical uses
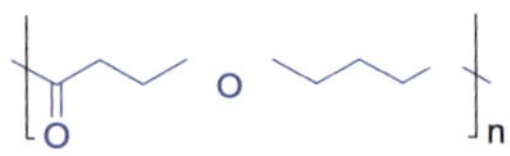	Poly (butylene succinate) (PBS)	Bio based	Disciplines of biomedicine, agriculture and packaging
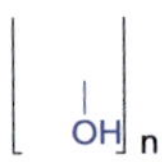	Polyvinyl alcohol (PVA)	Fossil fuel	Frequently employed in emulsion polymerization as a protective colloid, as a sizing agent in textiles and as a binder in papermaking
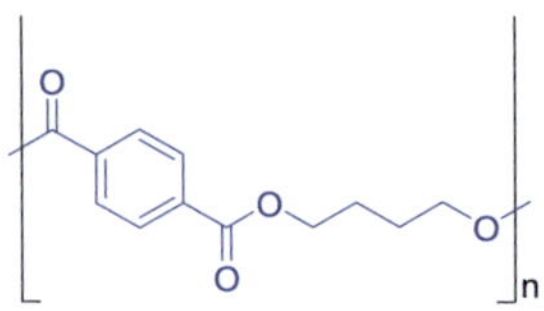	Poly(Butylene Adipate Terephthalate) (PBAT)	Fossil fuel or renewable biomass	Used in agriculture, disposable tableware, textile industries, hygiene products and many other

3.2 *Polylactic Acid (PLA)*

Scaffaro et al. (2021), states that polylactic acid is biodegradable thermoplastic aliphatic polyester generated by condensation polymerization of lactic acid derived from renewable resources such as tapioca, corn, sugarcane, fibres and roots. The structure of PLA is shown in Table 1. The food sector primarily uses polylactic acid to make disposable tableware (Swetha et al., 2023). However, polylactic acid bio plastics are too brittle to be used in other packaging methods. Polylactic acid needs additives to make it more durable (Reddy et al., 2022). The most useful biodegradable polymer is polylactic acid, which primarily breaks down by hydrolysis (Nagarajan et al., 2016). Specific types of commercial polylactic acid are made for thermoforming and extrusion or injection moulding (Atiwesh et al., 2021). Additionally, it can be utilized for packing materials, waste shopping bags, agricultural films and soil retention sheathings. A variety of woven, disposable and biodegradable fabric products, including diapers, feminine hygiene products and disposable clothing, can be made from polylactic acid by turning it into fibres (Elsawy et al., 2017; Nagarajan et al., 2016).

3.3 Starch

Starch is an affordable and easily modified biopolymer derived from plant sources such as tapioca palm, cereal grain seeds and tubers from plants (Li et al., 2021). According to Elsawy et al. (2017), food items such as straws, bowls, cups, bottles, cutlery and egg cartons are produced using starch based biopolymer, which are also use in product packaging (Mishra et al., 2024; Mohanty et al., 2024a, 2024b). According to Hubbe et al. (2021), starch is made up of several glucose units connected by glycosidic bonds, while linear amylose homopolymer units make up 20–25% and helical or branching amylopectin homopolymer units make up 75–80%. The linear polymer amylose is made up of D-glycosidic monomers connected by 1,4-glycosidic linkage, while amylopectin is made up of same monomers but contains many branches because of a different kind of network known as a 1,6-glycosidic bond. Glass transition temperature, crystallinity and the ratio of amylose to amylopectin are some of the qualities of starch that influence its characteristics, which vary greatly among species. Cornejo-Ramirez et al. (2018), state that the high glass transition temperature, strong intra- and inter-hydrogen bonds, poor flow ability and water sensitivity of virgin starch granules restrict their use and process ability. As a result, chemical modification and the combination of starch with other biopolymers can improve its properties. By adding molecular level functional group, starch bulk properties can be changed, enhancing its natural digestibility, solubility, thickening and shear stability. The kind, amount and distribution of substituents in the starch molecules influence the properties that are attained by chemically modifying starch, in addition to the reaction conditions and botanical sources (Liu et al., 2022). According to Zia-ud-din et al. (2017), derivatization methods include oxidation, cross-linking, acid hydrolysis, adding an acetyl group and adding positive charges to molecules are commonly employed to chemically alter starch. These approaches are constrained, though, by concerns about consumer safety and the environment. In order to create new kinds of alteration, combining several chemical processes has grown in popularity (Bensaad et al., 2022; Wang et al., 2022).

4 Petroleum-Based Biodegradable Polymers

Although they can also be made from coal and natural gas, synthetic polymers are most frequently made from crude oil. Most of these polymers are not biodegradable or compostable since they are not found in nature. However, degradability can be achieved for petroleum based non-biodegradable polymers by adding linkages such as amide, ether or ester that can hydrolyse in specific situation (Chen et al., 2011; Jaysree et al., 2019). According to Letcher (2020), the most well-known polymers in this family are PBS, PCL, PVA and PBAT. They are biodegradable polymers that are tested in anaerobic settings and can be produced utilizing biomass or fossil fuels. A biological process known as anaerobic degradation modifies organic compounds in an oxygen-free atmosphere. Anaerobic respiration, which requires acceptors like carbon dioxide, sulphates, nitrates or anaerobic fermentation, which uses organic molecules as an electron donor or receiver, are the two possible routes. Biogas, a blend of methane and carbon dioxide, is produced by enhancing the four stages of the process: hydrolysis, acidogenesis and methanogenesis (Ray et al., 2025). In an anaerobic process, respiration produces one third and fermentation produces two thirds of the methane (Bher et al., 2022; Naser et al., 2021; Quecholac-Piña et al., 2020). Ammonia, sulphate, heavy metals and volatile fatty acids all affect anaerobic biodegradation efficiency (Takashima & Yaguchi, 2021). Environmental factors also affect the efficacy of anaerobic biodegradation (Quecholac-Piña et al., 2020, Takashima & Yaguchi, 2021, Bajpai, 2017). Nutrient availability (Hatti-Kaul & Mattiasson, 2016), the kind and quantity of microorganisms in the medium (Adekunle & Okolie, 2015), and substrate characteristics (Slezak et al., 2015) are all important factors.

4.1 Polycaprolactone (PCL)

Aliphatic polyester that decomposes gradually, polycaprolactone (Table 1) finds extensive application in biomaterials and eco-friendly packaging. It is quite strong and adaptable (Lule & Kim, 2021). Biocompatible and somewhat slow-degrading polymer or durable polymer include scaffolds for tissue engineering, drug delivery systems and specialty sutures (Lee & Wu, 2022, Meng et al., 2016). Like other biodegradable polymers based on petrochemicals, polycaprolactone is used in mixtures with biodegradable polymers like polylactic acid, polyhydroxyalkonates, poly (butylene succinate) and polymers based on starch (Ogunsona et al., 2017).

4.2 Polyvinyl Alcohol (PVA)

In order to produce polyvinyl alcohol, polyvinyl acetate is typically hydrolysed. Structure of PVA is shown in Table 1. According to Muthuraj and Mekonnen (2018), the degree of hydrolysis, which typically falls between 80 and 99% determines the properties. It can be partial or complete. Because the carbon atoms possess hydroxyl groups, hydrolysis increases their degradability (Dong et al., 2014). In a range of industrial application, it is also frequently mixed with other polymer compounds to enhance the mechanical qualities of films water solubility (Li et al., 2022, Muthuraj & Mekonnen, 2018). Polyvinyl alcohol composites, including polyvinyl alcohol gels, are used in biomedicine for wound dressings, medication delivery system, artificial heart surgery and contact lens production. Because of its numerous desirable qualities, including tissue compatibility, non-toxic, non-carcinogenic and bio adhesive qualities, polyvinyl alcohol is used as a biomaterial in medical devices (Liu et al., 2020, Muthuraj & Mekonnen, 2018).

4.3 Poly Butylene Succinate (PBS)

When 1, 4-butanediol and succinic acid are polycondensed, the result indicate polyester is a semicrystalline thermoplastic made of poly butylene succinate (Table 1) (Fredi & Dorigato, 2021). Although polybutylene succinate can be made from both renewable and non-renewable monomers, a significant portion of polybutylene succinate that is sold commercially comes from fossil fuels. The semicrystalline thermoplastic polymer is called polybutylene succinate. Its high degree of crystallinity causes its rate of biodegradation to be rather slow. Among the various goods that use polybutylene succinate is food wrapping films, grocery bags, containers for planting, agricultural mulch film and hygiene products (Das et al., 2023; Ghosh & Das, 2020; Mishra & Das, 2023). Its low levels of bioactivity and biocompatibility, however, make it uncommon in the biomedical industry. To enhance mechanical performance, flame resistance, gas-barrier qualities and thermal conductivity, polybutylene succinate is also included as a filler to blend and composites (Cosquer et al., 2021).

4.4 Poly (Butylene Adipate Terephthalate) (PBAT)

When $(CH_2)_4(COOH)_2$ adipic acid, $(C_8H_6O_4)$ terephthalic acid and $(C_4H_{10}O_2)$ 1, 4-butanediol are polycondensed, an aliphatic aromatic random co-polyester known as polybutylene adipate terephthalate is formed (Liu et al., 2020). Structure of PBAT shown in Table 1. Strong mechanical properties are exhibited by polybutylene adipate terephthalate due to the presence of an aromatic unit in the molecular chain (Panchal & Vasava, 2020). Because of either poorer mechanical properties or higher manufacturing costs, pure polybutylene adipate terephthalate attributes are insufficient for consumer adoption when compared to ordinary polymers. Therefore, lowering production costs or improving its attributes can boost the market for polybutylene adipate terephthalate. By using affordable components and reinforcing elements, it is possible to improve the composites characteristics and lower their final cost while maintaining their biodegradability (Abd El-Magied et al., 2017; Ardean et al., 2021).

5 Renewable Resources-Based Biodegradable Polymers

This category needs at least one monomer made from renewable resources in order to produce biodegradable polymers. For example, the aliphatic polyester polypropylene carbonate is produced by copolymerizing propylene carbonate and carbon dioxide (Hubbe et al., 2021, Guo & Ma, 2014). The advantages of exploiting the abundant greenhouse gas to produce biodegradable polymers make carbon dioxide based polymer desirable (Hu et al., 2016). The two most extensively studied carbon dioxide based polycarbonates are poly (propylene carbonate) and poly (cyclohexene carbonate) according to Hu et al. (2016), and Adhami et al. (2021). Poly(propylene carbonate) is less expensive to make and has superior mechanical qualities compared to poly(cyclohexene carbonate), which is typically made by copolymerizing cyclohexene oxide and carbon dioxide. Some application cannot use poly (cyclohexene carbonate) due to its fragility (Butreddy et al., 2021). Poor thermal stability, low glass transition temperature and high shrinkage depending on the type of catalyst used to create polypropylene carbonate (Razak & Mohamed, 2021). But according to Arrieta et al. (2014), polypropylene carbonate films offer good tensile and barrier qualities against water vapour and oxygen.

6 Method of Synthesis for Biodegradable Polymers

6.1 Modification of Natural Polymer

Natural biodegradation is facilitated by polysaccharide polymeric components such as cellulose, chitin, chitosan and starch. New biodegradable polymers can be created by combining these components. The most promising methods for creating a wide range of molecular designs are nitrification, hydroxylation, sulfonation, acylation, thiolation, xanthation, quaternization, phosphorylation and graft copolymerization, which are among the chemical modification that natural polymers can go through (Pellis et al., 2019). Graft copolymerization is a process frequently used to improve the complexation, chelating and antibacterial properties of chitosan. A molecule can be covalently bonded to the backbone of chitin and chitosan to produce functional derivatives (Chao et al., 2020).

6.2 Chemical-Based Synthetic Biodegradable Polymers

In polymeric materials, chains of peptide, amide and ester bonds are readily biodegradable in polymerization, many small molecules, called repeating units or monomers, came together to form a covalently connected chain or network as the process proceeds, each monomer may lose a few chemical groups. These chemical structures are similar in natural polymer and chemically manufactured polymeric materials (Cai et al., 2006). The most versatile method for producing large groups of biopolymers and producing a sizable volume of product is ring opening polymerization. Polylactic acid is produced by either polymerization of lactide (chain growth ring opening) or polycondensation of lactic acid. The lactide ring opening polymerization can be carried out using a range of catalyst and initiators. Certain organic resources, such as glucose or starch, can be used by bacteria to produce a vast array of complex polymers. Polyesters like poly(hydroxybutyrate) and poly(hydroxybutyrate-hydroxyvalerate), polycaprolactone, various silks, polysaccharides and polyhydroxyalkonates are examples of these polymeric materials. Due to their comparable chemical characteristics it is challenging to differentiate between these compounds (Tabi, 2022). In order to store carbon and energy, bacteria that digest lipids or carbohydrates create a family of intracellular biopolymers called polyhydroxyalkonates. According to Luckachan and Pillai (2011) and Behera et al. (2022), their chemical diversity, biodegradability in a range of habitats, biocompatibility and ability to be produced from renewable carbon resources, PHAs have also garnered a lot of interest. Additionally, when they break down, they emit non-toxic and non-polluting compounds. The production of polyhydroxyalkonates can be accomplished using a variety of renewable resources. Because polyhydroxyalkonates are molecules that store carbon and energy in the event that a primary non-carbonaceous food is lacking, many prokaryotic bacteria produce and accumulate them (Sehgal & Gupta, 2020). As these polymers accumulate, they improve survival under harsh environmental condition (Geyer et al., 2017).

6.3 Biodegradable Polymer Synthesis via Enzymatic Process

The kinetics of enzymatic polymerization, usually considered as a monomer-activated process, has been studied in artistic works. Through the catalysis of specific polymerizations, several enzymes display a variety of properties (Mohanan et al., 2020). By using enzyme-mediated synthesis, biodegradable polymers like polyesters, polyamides and polysaccharides can be created (Tiwari et al., 2018).

6.4 Synthesis of Biodegradable Polymers by Chemoenzyme

Comparing chemical and enzymatic methods Filiciotto and Rothenberg (2021), demonstrate that chemoenzyme is a significant stereoselectivity and cost efficiency approach. The chemo-enzymatic synthesis approach blends a highly effective enzymatic method with regular polymerization. It can characterize as a promising method for creating biodegradable polymers with a high molecular weight. Researchers created some optically active polymeric pro-drugs using an efficient chemo-enzymatic process as high molecular weight non-steroidal anti-inflammatory substrate. A few optically active polymeric pro drug for high molecular weight non-steroidal anti-inflammatory drugs were produced by Cai et al. (2006), using an effective chemo-enzymatic techniques.

7 Biodegradation Mechanism

By decreasing the molecular mass of a polymer and weakening its physical and chemical properties, abiotic chemical reactions including photo-degradation, oxidation and hydrolysis can aid microorganism in biodegrading it in both aerobic and anaerobic conditions. As a result, low molecular weight molecules such as carbon dioxide, water and methane are formed (Cakmak, 2024). Abiotic degradation is the process by which a polymer structural and physical properties start to deteriorate (Ghosh et al., 2023; Mohanty et al., 2024a, 2024b). Numerous factors influence the initial breakdown rate, such as the polymer chain dimension, crystal structure, molecular weight distribution, particle size, arrangement and geometry, surface porosity, pore size and distribution, pore geometry, water diffusivity in the polymer matrix, atmospheric pollutants or agrochemicals, thermo-oxidative processes, ultraviolet light from the sun or artificial source of illumination, and many more (Fig. 1) (Joseph et al., 2022; Karamanlioglu et al., 2017). Last but not least, shear forces like snow pressure, animal tearing, air and water turbulence and physical forces like tension and compression also have an impact on the early deterioration process (Rostamzad, 2022). The polymers become more vulnerable to biological attack when it fragments into shorter strands, or oligomers, according to Salazar et al. (2023). According to Elahi et al. (2021), the compounds improved chemical and physical accessibility to bacterial activity and the enzymes they produce increases the materials bioavailability. This breakdown rate depends on the kind of polymer. Compared to linear non-reactive sections, polymers having many functional groups that give the enzymes a handle to act on will be simpler for the enzymes to reach and remove (Sharma et al., 2020). The third stage, microbial assimilation and mineralization, can be viewed as the microbe consuming and breaking down the polymers for its own development and energy needs (Mohanan et al., 2020). According to El Menofy and Khattab (2023), absorption of the monomer into microorganisms is the last stage of biodegradation. Depending on the oxygen availability, these organisms creates biomass and either

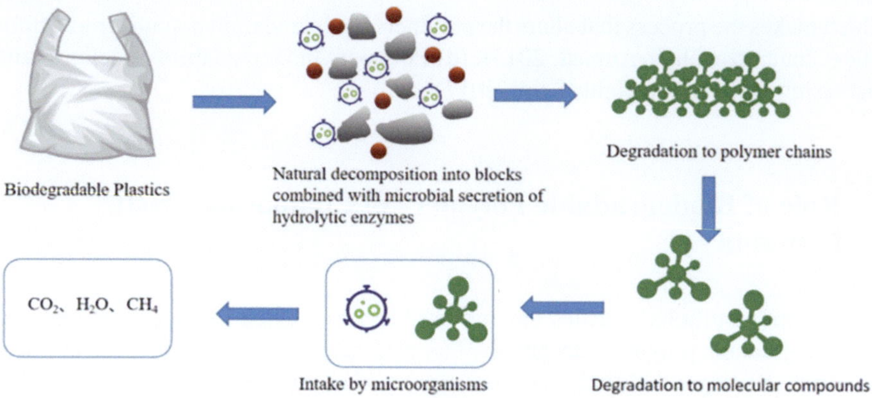

Fig. 1 Mechanism of biodegradation of biodegradable polymers (Guo & Guo, 2022)

carbon dioxide or methane. Anaerobic environments are those with no oxygen, anoxic environments are those with little oxygen and aerobic environments are those with an abundance of oxygen. The rate of this stage in a bioreactor can be determined by monitoring enhancement in the biomass of the selected microbe (Folino et al., 2020). The rate of microbial fragmentation and degradation will be influenced by additional environmental factors, such as oxygen (Groeneveld et al., 2023).

8 Factors Affecting Biodegradation

The biodegradation process is influenced by a variety of factors, including molecular weight, chemical treatment and polymer structure and form. (a) Polymer structure: hydrolysable bonds in the polymer chain of biodegradable polymers are subject to breakdown in the presence of hydrolytic enzymes and microorganisms. Compared to polymers with only hydrophobic or hydrophilic features, those having both polar and nonpolar structures are more degradable (Ghanbarzadeh & Almasi, 2013). (b) Polymer morphology: Because molecules in amorphous regions are widely separated from one another, they are more vulnerable to breakdown by enzymes than transparent or crystalline regions. The melting temperature of polymers has an impact on their enzymatic breakdown as well. Raising the melting point of polymers reduces their biodegradation.

Tm ¼$\Delta H = \Delta S$ where ΔS represents the entropy changes upon melting (Saini, 2017) and ΔH represents the enthalpy changes upon melting and Tm is melting temperature. (c) Chemical and radiation treatments include cleavage and cross-linking of polymers due to ions or radicals generated during photolysis of polymers exposed to UV and gamma ray radiation. Additionally, oxidation takes place, which makes the process that alters the polymers biodegradation potential more difficult (Ghanbarzadeh & Almasi, 2013). (d) Polymers biodegradability declines with increasing molecular weight (Saini, 2017).

9 Role of Biodegradable Polymers in Combating Plastic Pollution

Most of the traditional plastics that are used and consumed today can be replaced by biodegradable polymers despite the fact that there is now very little replacement occurring. Biodegradable polymers have been studied by Shen et al. (2009) as a viable technological replacement for traditional oil-based polymers. The highest percentage of substitution, according to the scientists, might be 94% with 63% being bio based polymers that aren't biodegradable and 31% being biodegradable polymers. Due to financial limitations, issues with quick scalability and biodegradable polymer raw material supply, and the plastic industry adoption of innovative biodegradable polymers, this promise might not be fulfilled in the near or medium term. In certain areas, biodegradable polymers have demonstrated the substitution effect on traditional plastics, despite the fact that the market is still in its infancy and not yet completely developed (Dey et al., 2024a; Ray et al., 2024). Throw away plastic items like plastic bags, trash bags and product packaging bags are just a few of the many disposable plastic items that are used in daily life. Such products are widely available and consume a lot of resources in our daily lives, but they also cause a number of issues, including white pollution. The mechanical characteristics of BPs used in disposable items should be able to withstand the strength demands of regular usage, and they should degrade more quickly. Using biodegradable polymer mulch film in agriculture. Pollution from plastic mulch film affects crops in three main ways: it reduces soil fertility, affects the efficiency of fertilizers and results in crop malnutrition (Dey et al., 2023a, 2023b). The main sources underlying the processing of remaining mulch film are manual pickup, mechanical recovery and administrative forces. Unfortunately, the build-up of leftover mulch film cannot be entirely eliminated for a considerable amount of time, regardless of whether it is done manually or mechanically and retrieving the leftover mulch film is getting more and more costly. In agriculture, using biodegradable mulch films is the most feasible solution to the remaining film issue. Biodegradable film has the same effect as conventional film but does not produce the unavoidable residual film pollution that conventional film does. A widespread application of biodegradable mulch films, however, can help with conventional surplus capacity. Furthermore, the upscale medical supply industry, materials for drug release, three dimensional printing materials and other items are among the high-end market applications for biodegradable polymers. Because of their high biocompatibility, which can be managed through molecular design synthesis, poly lactic acid, polyhydroxyalkonates, polycaprolactones and other biodegradable polymers have the potential to grow in the high-end market. One of the most widely used wires for three dimensional printing is polylactic acid. Poly lactic acid has a number of benefits over other commonly used three dimensional printing materials, such as acrylonitrile butadiene styrene, including rich colours, good transparency, the ability to print at greater speeds and environmental friendliness. Polyhydroxyalkonates are extensively utilized in medical applications. The good thermoplasticity and moulding processability of polycaprolactones allow them to be moulded into fibres, flakes, sheets and other materials. Biodegradable polymers will have greater development spaces and a more significant role in the high-end sector due to their distinctive characteristics. However, the biodegradation of biodegradable polymers require certain environmental parameters, including oxygen concentration, surrounding temperature, water content, polymer characteristics, etc. The biodegradation of true biodegradable polymers requires certain circumstances that aren't always dependable in the field (Nazareth et al., 2019). For instance, it has been discovered that Biodegradable polymers degrade slowly in marine environments (Morohoshi et al., 2018; Sashiwa et al., 2018), and in certain instances, they are not degradable (Napper & Thompson, 2019). Therefore, consideration of these biodegradation-related factors is necessary. This biodegradation period, which is controlled by the properties of the polymer, is important for assessing whether biodegradable polymers are appropriate for end-of-life management technologies or their possible fate in the environment. Limited biodegradability, even in the presence of microorganisms and necessary enzymes, would not help the environment or the management of biodegradable polymer waste if the rate of biodegradation is not appreciably different from that of conventional polymers of the same sort. The polymers chemical composition, crystallinity and complexity all influence biodegradation (Dey et al., 2023a; Mallick et al., 2024; Pradhan et al., 2024). Degradation rates are higher for BPs with flexible active sites and functional groups because active groups can attach to enzyme sites faster than those on stiff Biodegradable polymers. According to Narancic and O'Connor (2019), complicated chemical structures like polyhydroxybutyrate require extra enzymes or complex coenzymes, and shorter polymer chains degrade more quickly. Because surface breaking of Biodegradable polymers brought on by temperature and pH fluctuations would speed up the degradation process, ambient temperature and pH will have an impact on the rate of biodegradation. The biodegradation process may also be hampered by the plastic additives found in Biodegradable polymers. Large-scale biodegradable polymers disposal must also be taken into account. A potential method for large-scale treatment is required due to the high production and use of Biodegradable polymers. Landfills, burning, composting, anaerobic digestion and recycling are the typical methods used to dispose of large amounts of plastic garbage. For biodegradable polymers, industrial composting is currently the main problem with end-of-life management (Narancic & O'Connor, 2019). Compostable biodegradable polymers should be collected and separated using various collection techniques before being sent to commercial composting facilities. And littering continues even after an efficient technique is marketed, we shouldn't think of BPs as a technical fix that releases us from our environmental obligations. Littering, open dumping and open burning are some of the ways that plastics can go uncontrolled, these practices are more common in rural regions or places with less developed waste management systems (Dey et al., 2024c; Sahu et al., 2024b). Instead of being released into the environment at random, BPs needs to be controlled by society. Furthermore, consumers who value sustainability or environmental privilege are impacted by the adoption of green washing procedures when purchasing biodegradable products (Nazareth et al., 2019).

10 The Benefits of Employing Biodegradable Polymers

Worldwide, the use of BPs is rapidly growing and is worth billions of dollars annually. Biodegradable polymers have several uses (Fig. 2), such as in food packaging, computer keyboards, automotive interior components, implanted large devices, tissue engineering and medication delivery (Manavitehrani et al., 2016). For medical applications, biodegradable polymeric materials are being explored for tissue engineering, contact inhibition, fixation, isolation and suturing. Poly lactic acid, copolymers and poly-(glyconic acid) are typically used in suturing because of their remarkable flexibility and dependence on knot strength (Alaswad et al., 2022). Medication administration is the main use for poly(orthoester) and polyhydroxy sets. Furthermore, the increased flexibility and more severe wear and strain behaviour of polyurethane materials are required for grafting scaffolds supporting prosthetic blood vessels (Azimi et al., 2014). Conservative materials like cotton, polyester and polyethylene can be replaced with bio-based polyethylene terephthalate in surgical gowns (Pivsa-Art et al., 2013). Polylactic acid also produces masks, robes and headgear (Alaswad et al., 2022). Because of its highly absorbent qualities, thermoplastic starch is used to make disposable diapers (Panwiriyarat, 2012). Moreover, biodegradable materials such as collagen, and ultra-absorbent polymer are mostly used in medicinal and hygienic applications. Still, further research is needed to enhance the mechanical properties of biomaterials, which are crucial for this kind of biological use (Alaswad et al., 2022).

11 Challenges Possess by Biodegradable Polymers

Technical aspects, financial considerations, management and consumers are the four key elements (Fig. 3). First, technological improvements have a substantial impact on the rate of acceptance of biodegradable polymers. Adding inorganic components to biodegradable polymers is one way to increase their mechanical strength and thermal stability using existing technology. It will also cause deterioration products to increase and the rate of deterioration to decrease. To achieve the biodegradation performance of biodegradable polymers, environmental microorganisms convert macromolecules into micro molecules. This procedure is incredibly sluggish, especially for molecules of resin. According to Cosate de Andrade et al. (2016), enhancements to the infrastructure for compost like compost classification and biodegradable

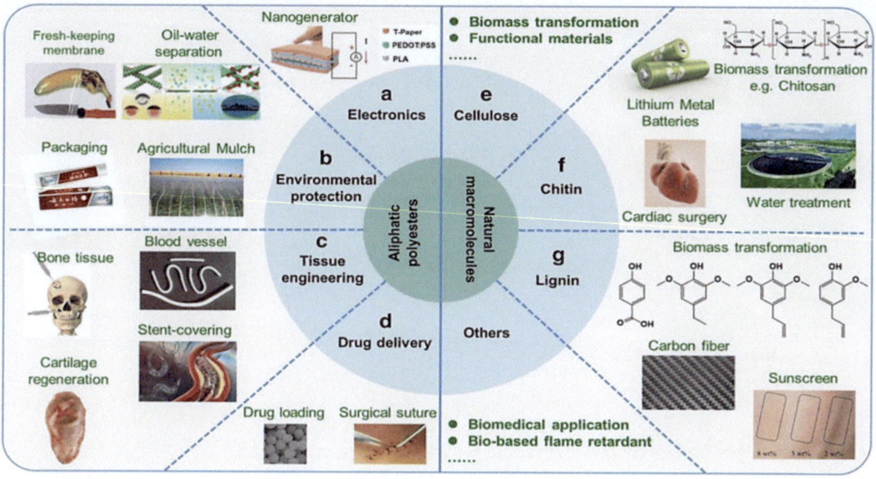

Fig. 2 Applications of biodegradable materials. Snapshot of the main fields of application (Yin & Yang, 2020)

polymer recycling, will enable the treatment of biodegradable polymers in compost facilities. Therefore, although they are not essential, waste management and public behavioural awareness are crucial components in addressing the issue of plastic accumulation in the environment. Future considerations also include the advancement of technology and the creation of new materials. Second, financial measures significantly impede the growth of biodegradable polymers. Encouraging the switch from conventional to biodegradable materials requires careful monitoring of agricultural raw material prices to keep them competitive with fossil fuels. Additionally, the synthesis method and new equipment should be optimized to lower the cost of producing Biodegradable polymers (Swain et al., 2025). Increasing the use of materials and improving the manufacture and function of biodegradable polymers like as cellulose, starch and shell are important ways to reduce costs. Thirdly, the biodegradable polymers industry lacks a strong management framework. Three methods are available for managing plastic waste: recycling, incinerators and sanitary landfills. The standard plastic and BPs collecting system is chaotic, making it difficult to dispose of plastic waste. According to Dilkes Hoffman et al. (2019), BPs has energy content comparable to that of normal plastics and can be burned similarly. But when it comes to landfills, BPs and regular plastics are very different. It is challenging for traditional plastics to break down in a landfills. In landfills, biodegradable polymers may breakdown to produce methane gas that has higher potential to cause global warming than carbon dioxide. Finally, raising public awareness of environmental issues is another requirement for supporting biodegradable polymers. The discovery and management of biodegradable polymers is a problem that must be handled in addition to being a major social concern.

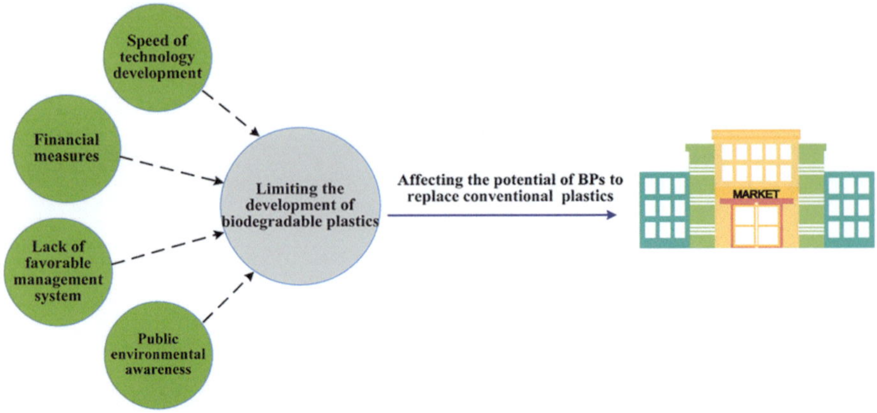

Fig. 3 Challenges posses by biodegradable polymers (Shen et al., 2020)

12 Conclusion

The problem of plastic waste on earth is still present today. With rise of plastic manufacture and consumption worldwide, the problem's severity will expand dramatically. Since plastic accumulation is unavoidable, steps should be done to slow it down and lessen the environmental impact of this accumulation. In addition to significantly advancing environmental protection, biodegradable polymers can successfully preserve and enhance the environment. The creation of biodegradable polymers, however, appears to be far simpler than treating them. Biodegradable polymers performance is heavily questioned. The question of whether the problems of waste management and plastic pollution worldwide may be resolved by biodegradable polymers. In many aspects, biodegradable polymers are still in their infancy. Finding the best mix of solution is essential to addressing the problem of plastic build-up in the environment. Biodegradable polymers should contribute to the solution, albeit in a very minor way, for the time being. The influence of biodegradable polymers on the accumulation of plastic should not be underestimated. Furthermore, considering the urgent situation surrounding energy conservation and emission reduction, the development of biodegradable polymers is strategically significant. Therefore, there is still more work to be done in addressing the plastic pollution problem until biodegradable polymers can displace all non-biodegradable plastic items that perform as well as or better, and everyone in the world can dispose of biodegradable polymer waste responsibly in accordance with laws and regulations and large-scale biodegradation of these collected biodegradable polymers is possible, the by-products of this biodegradation can be recycled back into the ecosystem.

References

Abd El-Magied, M. O., Galhoum, A. A., Atia, A. A., Tolba, A. A., Maize, M. S., Vincent, T., & Guibal, E. (2017). Cellulose and chitosan derivatives for enhanced sorption of erbium (III). *Colloids and Surfaces a: Physicochemical and Engineering Aspects, 529*, 580–593.

Adekunle, K. F., & Okolie, J. A. (2015). A review of biochemical process of anaerobic digestion. *Advances in Bioscience and Biotechnology, 6*(03), 205.

Adhami, W., Bakkour, Y., & Rolando, C. (2021). Polylactones synthesis by enzymatic ring opening polymerization in flow. *Polymer, 230*, Article 124040.

Alaswad, S. O., Mahmoud, A. S., & Arunachalam, P. (2022). Recent advances in biodegradable polymers and their biological applications: A brief review. *Polymers, 14(22)*

Ardean, C., Davidescu, C. M., Nemeş, N. S., Negrea, A., Ciopec, M., Duteanu, N., & Musta, V. (2021). Factors influencing the antibacterial activity of chitosan and chitosan modified by functionalization. *International Journal of Molecular Sciences, 22*(14), 7449.

Arrieta, M. P., López, J., Rayón, E., & Jiménez, A. (2014). Disintegrability under composting conditions of plasticized PLA–PHB blends. *Polymer Degradation and Stability, 108*, 307–318.

Atiwesh, G., Mikhael, A., Parrish, C. C., Banoub, J., & Le, T. A. T. (2021). Environmental impact of bioplastic use: A review. *Heliyon, 7*(9).

Azimi, B., Nourpanah, P., Rabiee, M., & Arbab, S. (2014). Poly (lactide-co-glycolide) fiber: An overview. *Journal of Engineered Fibers and Fabrics, 9*(1), 155892501400900100.

Bajpai, P. (2017). *Anaerobic technology in pulp and paper industry* (pp. 7–13). Springer Singapore.

Behera, S., Priyadarshanee, M., & Das, S. (2022). Polyhydroxyalkanoates, the bioplastics of microbial origin: Properties, biochemical synthesis, and their applications. *Chemosphere, 294*, Article 133723.

Belal, E. B., & Farid, M. A. (2016). Production of Poly-β-hydroxybutyric acid (PHB) by Bacillus cereus. *International Journal of Current Microbiology and Applied Sciences, 5*, 442–460.

Bensaad, D. E., Saleh, M., Ismail, K., Lee, Y., & Ondier, G. (2022). Chemical modifications of starch; a prospective for sweet potato starch. *Jordan Journal of Agricultural Sciences, 18*(4), 293–308.

Bher, A., Mayekar, P. C., Auras, R. A., & Schvezov, C. E. (2022). Biodegradation of biodegradable polymers in mesophilic aerobic environments. *International Journal of Molecular Sciences, 23*(20), 12165.

Butreddy, A., Gaddam, R. P., Kommineni, N., Dudhipala, N., & Voshavar, C. (2021). PLGA/PLA-based long-acting injectable depot microspheres in clinical use: Production and characterization overview for protein/peptide delivery. *International Journal of Molecular Sciences, 22*(16), 8884.

Cai, X. Q., Wang, N., & Lin, X. F. (2006). The preparation of polymerizable, optically active non-steroidal anti-inflammatory drugs derivatives by irreversible enzymatic methods. *Journal of Molecular Catalysis b: Enzymatic, 40*(1–2), 51–57.

Cakmak, O. K. (2024). Biodegradable polymers—A review on properties, processing, and degradation mechanism. *Circular Economy and Sustainability, 4*(1), 339–362.

Chao, Q., Ding, Y., Chen, Z. H., Xiang, M. H., Wang, N., & Gao, X. D. (2020). Recent progress in chemo-enzymatic methods for the synthesis of N-glycans. *Frontiers in Chemistry, 8*, 513.

Chen, P., Yu, L., Simon, G. P., Liu, X., Dean, K., & Chen, L. (2011). Internal structures and phase-transitions of starch granules during gelatinization. *Carbohydrate Polymers, 83*(4), 1975–1983.

Cornejo-Ramírez, Y. I., Martínez-Cruz, O., Del Toro-Sánchez, C. L., Wong-Corral, F. J., Borboa-Flores, J., & Cinco-Moroyoqui, F. J. (2018). The structural characteristics of starches and their functional properties. *CyTA-Journal of Food, 16*(1), 1003–1017.

Cosate de Andrade, M. F., Souza, P. M., Cavalett, O., & Morales, A. R. (2016). Life cycle assessment of poly (lactic acid) (PLA): Comparison between chemical recycling, mechanical recycling and composting. *Journal of Polymers and the Environment, 24*, 372–384.

Cosquer, R., Pruvost, S., & Gouanvé, F. (2021). Improvement of barrier properties of biodegradable polybutylene succinate/graphene nanoplatelets nanocomposites prepared by melt process. *Membranes, 11*(2), 151.

Das, A. P., Mohanty, A. K., van Hullebusch, E. D., Figueiredo, G., Das, A. P., Mohanty, A. K., van Hullebusch, E. D., & Figueiredo, G. (2023). Bioremediation as an emerging technology for the removal of synthetic microplastic pollutants from marine ecosystem. *MarPB, 194*, Article 115297. https://doi.org/10.1016/J.MARPOLBUL.2023.115297

Das, A., Dey, S., & Das, A. P. (2024). Ecotoxicological impacts of synthetic microfiber pollutants and development of sustainable mitigation strategies. *Environmental Chemistry and Ecotoxicology*. https://doi.org/10.1016/J.ENCECO.2024.12.003

Das, A., Dey, S., Gacem, A., Yadav, K. K., Nath, H., Ram, D. K., & Das, A. P. (2025). Revolutionizing solid hydrocarbon sourcing: Plastic's journey from waste to upcycled treasure. *Environmental Science and Engineering, Part, F218*, 331–351. https://doi.org/10.1007/978-3-031-81615-4_14

Dauvergne, P. (2018). Why is the global governance of plastic failing the oceans? *Global Environmental Change, 51*, 22–31.

Dey, S., Tripathy, B., Kumar, M. S., & Das, A. P. (2023b). Ecotoxicological consequences of manganese mining pollutants and their biological remediation. *Environmental Chemistry and Ecotoxicology*. https://doi.org/10.1016/J.ENCECO.2023.01.001

Dey, S., Das, A., Mallick, K., Sahu, A., & Das, A. P. (2023a). *Environmental petroleum waste: Pollution, toxicity, sustainable remediation*. 159–175. https://doi.org/10.1007/978-3-031-482 20-5_7

Dey, S., Bissoyi, A., Koteswara Reddy, G., & Das, A. P. (2024a). Synthetic microfibers: An urban emerging pollutant demanding for sustainable environmental solutions. In *Sustainable microbial technology for synthetic and cellulosic microfiber bioremediation* (pp. 107–121). https://doi.org/10.1007/978-3-031-62600-5_7

Dey, S., Dash, A., & Das, A. P. (2024b). Bioremediation as a sustainable and circular solution for mining waste management and bioenergy generation. In *Sustainable management of mining waste and tailings: A circular economy approach* (pp 107–119). https://doi.org/10.1201/978 1003442455-6

Dey, S., Dash, N., Ray, S. K., Priyadarsini, S., Chhotaray, C., & Das, A. P. (2024c). Occurrence and quantification of tire wear plastics pollutants from Highway City Dust of Bhubaneswar City of Odisha. In *Sustainable microbial technology for synthetic and cellulosic microfiber bioremediation* (pp. 77–89). https://doi.org/10.1007/978-3-031-62600-5_5

Dilkes-Hoffman, L. S., Pratt, S., Lant, P. A., & Laycock, B. (2019). The role of biodegradable plastic in solving plastic solid waste accumulation. In *Plastics to energy* (pp. 469–505). William Andrew Publishing.

Dong, T., Yun, X., Shi, C., Sun, W., Fan, G., & Jin, Y. (2014). Improved mechanical and barrier properties of PPC multilayer film through interlayer hydrogen bonding interaction. *Polymer Science Series A, 56*, 830–836.

El Menofy, N. G., & Khattab, A. M. (2023). Plastics biodegradation and biofragmentation. *Handbook of biodegradable materials* (pp. 571–600). Springer International Publishing.

Elahi, A., Bukhari, D. A., Shamim, S., & Rehman, A. (2021). Plastics degradation by microbes: A sustainable approach. *Journal of King Saud University-Science, 33*(6), Article 101538.

Elsawy, M. A., Kim, K. H., Park, J. W., & Deep, A. (2017). Hydrolytic degradation of polylactic acid (PLA) and its composites. *Renewable and Sustainable Energy Reviews, 79*, 1346–1352.

Europe, P. (2016). Plastics—the facts 2016. In *An analysis of European latest plastics production, demand and waste data*.

Filiciotto, L., & Rothenberg, G. (2021). Biodegradable plastics: Standards, policies, and impacts. *Chemsuschem, 14*(1), 56–72.

Folino, A., Karageorgiou, A., Calabrò, P. S., & Komilis, D. (2020). Biodegradation of wasted bioplastics in natural and industrial environments: A review. *Sustainability, 12*(15), 6030.

Fredi, G., & Dorigato, A. (2021). Recycling of bioplastic waste: A review. *Advanced Industrial and Engineering Polymer Research, 4*(3), 159–177.

Gere, D., & Czigany, T. (2020). Future trends of plastic bottle recycling: Compatibilization of PET and PLA. *Polymer Testing, 81*, Article 106160.

Geyer, R., Jambeck, J. R., & Law, K. L. (2017). Production, use, and fate of all plastics ever made. *Science Advances, 3*(7), Article e1700782.

Ghanbarzadeh, B., & Almasi, H. (2013). Biodegradable polymers. *Biodegradation-life of science*, 141–185.

Ghosh, S., & Das, A. P. (2020). Microbial metagenomics: current advances in investigating microbial ecology and population dynamics. In S. K. Nayak & B. B. Mishra (Eds.), *Frontiers in soil and environmental microbiology* (1st ed., pp. 247–254). CRC Press.

Ghosh, S., Tripathy, B., Dey, S., & Das, A. P. (2023). Comparative investigation of fungal and bacterial manganese biomining mechanisms. *Manganese Mining Microorganisms*, 99–115. https://doi.org/10.1016/b978-0-12-822147-1.00016-9

Groeneveld, I., Kanelli, M., Ariese, F., & van Bommel, M. R. (2023). Parameters that affect the photodegradation of dyes and pigments in solution and on substrate–An overview. *Dyes and Pigments, 210*, Article 110999.

Guo, C., & Guo, H. (2022). Progress in the degradability of biodegradable film materials for packaging. *Membranes, 12*(5), 500.

Guo, B., & Ma, P. X. (2014). Synthetic biodegradable functional polymers for tissue engineering: A brief review. *Science China Chemistry, 57*, 490–500.

Haider, T. P., Völker, C., Kramm, J., Landfester, K., & Wurm, F. R. (2019). Plastics of the future? The impact of biodegradable polymers on the environment and on society. *Angewandte Chemie International Edition, 58*(1), 50–62.

Hatti-Kaul, R., & Mattiasson, B. (2016). *Anaerobes in industrial-and environmental biotechnology* (pp. 1–33). Springer International Publishing.

Hu, Y., Daoud, W. A., Cheuk, K. K. L., & Lin, C. S. K. (2016). Newly developed techniques on polycondensation, ring-opening polymerization and polymer modification: Focus on poly (lactic acid). *Materials, 9*(3), 133.

Hubbe, M. A., Lavoine, N., & Lucia, L. A. (2021). Formulating bioplastic composites for biodegradability, recycling, and performance: A review. *BioResources, 16*(1).

Iannotti, G., Fair, N., Tempesta, M., Neibling, H., Hsieh, F. H., & Mueller, R. (2018). Studies on the environmental degradation of starch-based plastics. In *Degradable materials* (pp. 425–446). CRC Press.

Jambeck, J. R., Geyer, R., Wilcox, C., Siegler, T. R., Perryman, M., Andrady, A., & Law, K. L. (2015). Plastic waste inputs from land into the ocean. *Science, 347*(6223), 768–771.

Jaysree, R. C., Subhash Chandra, K. P., & Sankar, T. V. (2019). Biodegradability of synthetic plastics–a review. *International Journal of ChemTech Research, 12*(6), 125–133.

Joseph, E., Tohidifar, P., Sarver, C. T., Mackie, R. I., & Rao, C. V. (2022). Fundamentals of polymer biodegradation mechanisms. In *Biodegradable Polymers in the Circular Plastics Economy* (pp. 17–58).

Karamanlioglu, M., Preziosi, R., & Robson, G. D. (2017). Abiotic and biotic environmental degradation of the bioplastic polymer poly (lactic acid): A review. *Polymer Degradation and Stability, 137*, 122–130.

Lee, Y. F., & Wu, T. M. (2022). Synthesis and physical properties of biodegradable nanocomposites fabricated using acrylic acid-grafted poly (butylene carbonate-co-terephthalate) and organically-modified layered zinc phenylphosphonate. *Journal of Polymers and the Environment, 30*(3), 896–906.

Letcher, T. (Ed.). (2020). *Plastic waste and recycling: Environmental impact, societal issues, prevention, and solutions*. Academic Press.

Li, Y. T., Yu, H. Y., Li, W. B., Liu, Y., & Lu, X. B. (2021). Recyclable polyhydroxyalkanoates via a regioselective ring-opening polymerization of α, β-disubstituted β-lactone monomers. *Macromolecules, 54*(10), 4641–4648.

Li, X., Meng, L., Zhang, Y., Qin, Z., Meng, L., Li, C., & Liu, M. (2022). Research and application of polypropylene carbonate composite materials: A review. *Polymers, 14*(11), 2159.

Liu, J., Li, R., & Yang, B. (2020). Carbon dots: A new type of carbon-based nanomaterial with wide applications. *ACS Central Science, 6*(12), 2179–2195.

Liu, C., Yan, H., Liu, S., & Chang, X. (2022). Influence of phosphorylation and acetylation on structural, physicochemical and functional properties of chestnut starch. *Polymers, 14*(1), 172.

Luckachan, G. E., & Pillai, C. K. S. (2011). Biodegradable polymers-a review on recent trends and emerging perspectives. *Journal of Polymers and the Environment, 19*, 637–676.

Lule, Z. C., & Kim, J. (2021). Compatibilization effect of silanized SiC particles on polybutylene adipate terephthalate/polycarbonate blends. *Materials Chemistry and Physics, 258*, Article 123879.

Mallick, K., Sahu, S., Sahu, A., Dey, S., & Das, A. P. (2024). A critical review of marine microfiber pollution routes, toxicity, and its sustainable remediation. *Environmental Science and Engineering, Part, F2333*, 189–211. https://doi.org/10.1007/978-3-031-51792-1_10

Manavitehrani, I., Fathi, A., Badr, H., Daly, S., Negahi Shirazi, A., & Dehghani, F. (2016). Biomedical applications of biodegradable polyesters. *Polymers, 8*(1), 20.

Meereboer, K. W., Misra, M., & Mohanty, A. K. (2020). Review of recent advances in the biodegradability of polyhydroxyalkanoate (PHA) bioplastics and their composites. *Green Chemistry, 22*(17), 5519–5558.

Meng, Q. Y., Pepper, K., Cheng, R. H., Howdle, S. M., & Liu, B. P. (2016). Effect of supercritical CO2 on the copolymerization behavior of cyclohexene oxide/CO2 and copolymer properties with DMC/Salen-Co (III) catalyst system. *Journal of Polymer Science Part a: Polymer Chemistry, 54*(17), 2785–2793.

Miller, M. E., Hamann, M., & Kroon, F. J. (2020). Bioaccumulation and biomagnification of microplastics in marine organisms: A review and meta-analysis of current data. *PLoS ONE, 15*(10), Article e0240792.

Mishra, S., Dash, D., & Das, A. P. (2024). Aquatic microbial diversity on plastisphere: Colonization and potential role in microplastic biodegradation. *Geomicrobiology Journal, 41*(4), 312–323. https://doi.org/10.1080/01490451.2023.2209750

Mishra, S., & Das, A. P. (2023). Microbial remediation of synthetic microfiber contaminated wastewater. In M. P. Shah (Ed.), *Microbial technologies in industrial wastewater treatment* (pp. 337–350). Springer. https://doi.org/10.1007/978-981-99-2435-6_17

Mohanan, N., Montazer, Z., Sharma, P. K., & Levin, D. B. (2020). Microbial and enzymatic degradation of synthetic plastics. *Frontiers in Microbiology, 11*, Article 580709.

Mohanty, M., Mohanty, J., Dey, S., Dutta, K., Shah, M. P., & Das, A. P. (2024b). The face mask: a tale from protection to pollution and demanding sustainable solution. *Emerging Contaminants, 10*(2), Article 100298. https://doi.org/10.1016/J.EMCON.2023.100298

Mohanty, J., Dey, S., Bala, K., & Das, A. P. (2024a). Recent advances in microplastic pollution for its sustainable management. In *Sustainable management of urban plastic waste through circular economic approaches* (pp. 27–48). https://doi.org/10.1201/9781003465447-3

Morohoshi, T., Ogata, K., Okura, T., & Sato, S. (2018). Molecular characterization of the bacterial community in biofilms for degradation of poly (3-hydroxybutyrate-co-3-hydroxyhexanoate) films in seawater. *Microbes and Environments, 33*(1), 19–25.

Muthuraj, R., & Mekonnen, T. (2018). Recent progress in carbon dioxide (CO_2) as feedstock for sustainable materials development: Co-polymers and polymer blends. *Polymer, 145*, 348–373.

Nag, M., Lahiri, D., Ghosh, S., Sarkar, T., Pati, S., Das, A. P., Ram, D. K., Bhattacharya, D., & Ray, R. R. (2023). Application of microorganisms in biotransformation and bioremediation of environmental contaminant: a review. *Geomicrobiology Journal*. https://doi.org/10.1080/01490451.2023.2261443

Nagarajan, V., Mohanty, A. K., & Misra, M. (2016). Perspective on polylactic acid (PLA) based sustainable materials for durable applications: Focus on toughness and heat resistance. *ACS Sustainable Chemistry & Engineering, 4*(6), 2899–2916.

Napper, I. E., & Thompson, R. C. (2019). Environmental deterioration of biodegradable, oxo-biodegradable, compostable, and conventional plastic carrier bags in the sea, soil, and open-air over a 3-year period. *Environmental Science & Technology, 53*(9), 4775–4783.

Narancic, T., & O'Connor, K. E. (2019). Plastic waste as a global challenge: Are biodegradable plastics the answer to the plastic waste problem? *Microbiology, 165*(2), 129–137.

Naser, A. Z., Deiab, I., & Darras, B. M. (2021). Poly (lactic acid) (PLA) and polyhydroxyalkanoates (PHAs), green alternatives to petroleum-based plastics: A review. *RSC Advances, 11*(28), 17151–17196.

Nazareth, M., Marques, M. R., Leite, M. C., & Castro, Í. B. (2019). Commercial plastics claiming biodegradable status: Is this also accurate for marine environments? *Journal of Hazardous Materials, 366*, 714–722.

Ogunsona, E. O., Misra, M., & Mohanty, A. K. (2017). Sustainable biocomposites from biobased polyamide 6, 10 and biocarbon from pyrolyzed miscanthus fibers. *Journal of Applied Polymer Science, 134*(4).

Panchal, S. S., & Vasava, D. V. (2020). Biodegradable polymeric materials: Synthetic approach. *ACS Omega, 5*(9), 4370–4379.

Panwiriyarat, W. (2012). *Preparation and properties of bio-based polyurethane made from natural rubber and poly (ε-caprolactone)* [Doctoral dissertation] Le Mans Université; Prince of Songkla University

Paul, M. B., Stock, V., Cara-Carmona, J., Lisicki, E., Shopova, S., Fessard, V., & Böhmert, L. (2020). Micro-and nanoplastics–current state of knowledge with the focus on oral uptake and toxicity. *Nanoscale Advances, 2*(10), 4350–4367.

Pellis, A., Comerford, J. W., Weinberger, S., Guebitz, G. M., Clark, J. H., & Farmer, T. J. (2019). Enzymatic synthesis of lignin derivable pyridine based polyesters for the substitution of petroleum derived plastics. *Nature Communications, 10*(1), 1762.

Picó, Y., & Barceló, D. (2019). Analysis and prevention of microplastics pollution in water: Current perspectives and future directions. *ACS Omega, 4*(4), 6709–6719.

Pivsa-Art, W., Chaiyasat, A., Pivsa-Art, S., Yamane, H., & Ohara, H. (2013). Preparation of polymer blends between poly (lactic acid) and poly (butylene adipate-co-terephthalate) and biodegradable polymers as compatibilizers. *Energy Procedia, 34*, 549–554.

Pradhan, G., Dey, S., Priyadarsini, S., Kumar, M. S., & Das, A. P. (2024). Characterization and quantification of microplastics pollutants in sediment samples from Daya River of Odisha State in India for their appropriate management. *Environmental Science and Engineering, Part, F2333*, 75–92. https://doi.org/10.1007/978-3-031-51792-1_5

Pradhan, G., Tripathy, B., Ram, D. K., Digal, A. K., & Das, A. P. (2023). Bauxite mining waste pollution and its sustainable management through bioremediation. *Geomicrobiology Journal*, 1–10. https://doi.org/10.1080/01490451.2023.2235353

Quecholac-Piña, X., Hernández-Berriel, M. D. C., Mañón-Salas, M. D. C., Espinosa-Valdemar, R. M., & Vázquez-Morillas, A. (2020). Degradation of plastics under anaerobic conditions: A short review. *Polymers, 12*(1), 109.

Ray, S., Dey, S., Digal, A. K., Yadav, K. K., & Das, A. P. (2025). Sanitary waste and microplastic pollution in the Euro-Mediterranean region: Challenges and solutions. *Euro-Mediterranean Journal for Environmental Integration, 2025*, 1–23. https://doi.org/10.1007/S41207-025-008 79-Y

Ray, S. K., Dey, S., Das, A. P., & John, A. (2024). Microplastic waste management through greener technologies : A review. In *Sustainable management of urban plastic waste through circular economic approaches* (pp. 237–261). https://doi.org/10.1201/9781003465447-14

Razak, N. S., & Mohamed, R. (2021). Antimicrobial sustainable biopolymers for biomedical plastics applications–an overview. *Polimery, 66*(11–12), 574–583.

Reddy, V. U. N., Ramanaiah, S. V., Reddy, M. V., & Chang, Y. C. (2022). Review of the developments of bacterial medium-chain-length polyhydroxyalkanoates (mcl-PHAs). *Bioengineering, 9*(5), 225.

Reddy Koteswara, G., Bhargavi, S., Yamuna, D., Anjani Devi, M., Jeevana Lakshmi Kumar, G., Manas, T., Adarsh, S., Bhavya Sai, V., Nath, H., Alreshidi, M. A., Togun, H., Ram, D. K., & Das, A. P. (2025). Recent advancements in the remediation of marine hydrocarbon and plastic

pollutants. *Environmental Science and Engineering, Part, F218*, 127–142. https://doi.org/10.1007/978-3-031-81615-4_7
Rivera-Briso, A. L., & Serrano-Aroca, Á. (2018). Poly (3-Hydroxybutyrate-co-3-Hydroxyvalerate): Enhancement strategies for advanced applications. *Polymers, 10*(7), 732.
Rostamzad, H. (2022). Active and intelligent biodegradable films and polymers. In *Biodegradable Polymers, Blends and Composites* (pp. 415–430). Woodhead Publishing.
Rujnić-Sokele, M., & Pilipović, A. (2017). Challenges and opportunities of biodegradable plastics: A mini review. *Waste Management & Research, 35*(2), 132–140.
Sahu, A., Dey, S., Mallick, K., Priyadarsini, S., Behera, I. D., & Das, A. P. (2024a). A review on advanced technology for sustainable management of synthetic microplastic waste. In *Sustainable management of urban plastic waste through circular economic approaches* (pp. 262–279). https://doi.org/10.1201/9781003465447-15
Sahu, A., Dey, S., Mallick, K., Priyadarsini, S., Behera, I. D., & Das, A. P. (2024b). A review on advanced technology for sustainable management of synthetic microplastic waste. In *Sustainable management of urban plastic waste through circular economic approaches* (pp. 262–279). https://doi.org/10.1201/9781003465447-15
Saini, R. D. (2017). Biodegradable polymers. *International Journal of Applied Chemistry, 13*(2), 179–196.
Salazar, S. A., Abdulhameed, S., & Sánchez, M. D. R. S. (2023). Biodegradation of polymers. In *Biodegradable Polymers* (pp. 1–12). CRC Press.
Sashiwa, H., Fukuda, R., Okura, T., Sato, S., & Nakayama, A. (2018). Microbial degradation behavior in seawater of polyester blends containing poly (3-hydroxybutyrate-co-3-hydroxyhexanoate)(PHBHHx). *Marine Drugs, 16*(1), 34.
Scaffaro, R., Maio, A., Gulino, E. F., Alaimo, G., & Morreale, M. (2021). Green composites based on PLA and agricultural or marine waste prepared by FDM. *Polymers, 13*(9), 1361.
Sehgal, R., & Gupta, R. (2020). Polyhydroxyalkanoate and its efficient production: an eco-friendly approach towards development. *3 Biotech, 10*(12), 549.
Sharma, R., Garg, P., Kumar, P., Bhatia, S. K., & Kulshrestha, S. (2020). Microbial fermentation and its role in quality improvement of fermented foods. *Fermentation, 6*(4), 106.
Shen, M., Song, B., Zeng, G., Zhang, Y., Huang, W., Wen, X., & Tang, W. (2020). Are biodegradable plastics a promising solution to solve the global plastic pollution? *Environmental Pollution, 263*, Article 114469.
Shen, L., Haufe, J., & Patel, M. K. (2009). Product overview and market projection of emerging bio-based plastics PRO-BIP 2009. *Report for European polysaccharide network of excellence (EPNOE) and European bioplastics, 243*, 1–245.
Slezak, R., Krzystek, L., & Ledakowicz, S. (2015). Degradation of municipal solid waste in simulated landfill bioreactors under aerobic conditions. *Waste Management, 43*, 293–299.
Swain, S., Purnima Sudha, M., Dey, S., Badu, Y., Bhoi, S. R., Alreshidi, M. A., Qasim, M. T., Nath, H., Ram, D. K., & Das, A. P. (2025). Sources of micro and nano plastic hydrocarbon pollutants in marine bodies and their treatment. *Environmental Science and Engineering, Part, F218*, 45–67. https://doi.org/10.1007/978-3-031-81615-4_3
Swetha, T. A., Bora, A., Mohanrasu, K., Balaji, P., Raja, R., Ponnuchamy, K., & Arun, A. (2023). A comprehensive review on polylactic acid (PLA)–Synthesis, processing and application in food packaging. *International Journal of Biological Macromolecules, 234*, Article 123715.
Tábi, T. (2022). Biodegradable bio-based plastics: Compostable or recyclable? *Express Polymer Letters, 16*(2), 115–115.
Takashima, M., & Yaguchi, J. (2021). High-solids thermophilic anaerobic digestion of sewage sludge: Effect of ammonia concentration. *Journal of Material Cycles and Waste Management, 23*(1), 205–213.
Thomas, N. L., Clarke, J., McLauchlin, A. R., & Patrick, S. G. (2012). Oxodegradable plastics: degradation, environmental impact and recycling. In *Proceedings of the institution of civil engineers-waste and resource management* (vol. 165(3), pp. 133–140). ICE Publishing.

Tiwari, A. K., Gautam, M., & Maurya, H. K. (2018). Recent development of biodegradation techniques of polymer. *International Journal of Research Granthaalayah, 6*(6), 414–452.

Wang, Z., Mhaske, P., Farahnaky, A., Kasapis, S., & Majzoobi, M. (2022). Cassava starch: Chemical modification and its impact on functional properties and digestibility, a review. *Food Hydrocolloids, 129*, Article 107542.

Yin, G. Z., & Yang, X. M. (2020). Biodegradable polymers: A cure for the planet, but a long way to go. *Journal of Polymer Research, 27*(2), 38.

Zia-ud-Din, Xiong, H., & Fei, P. (2017). Physical and chemical modification of starches: A review. *Critical reviews in food science and nutrition, 57*(12), 2691–2705

Harnessing Nanobiotechnology for Environmental Sustainability

Bhanu Krishan, Anu Kumar, and Sunny Dhiman

1 Introduction

Nanobiotechnology is an interdisciplinary field at the nexus of nanotechnology and biological sciences. It involves the design and application of nanoscale materials and devices to probe, manipulate, or construct biological systems. As Shahcheraghi et al. (2022) explain, "nanobiotechnology is the combination of nanotechnology and biotechnology disciplines." In practice, this means using nanoparticles (NPs), nanotubes, nanofibers, and other nanomaterials to perform biological functions that were previously impossible or inefficient at larger scales (Shahcheraghi et al., 2022). For example, functionalized NPs can carry drugs to specific cells (minimizing side effects) or serve as artificial enzymes to catalyze biochemical reactions with great precision (Paul et al., 2023a, 2023b). Nanobiotechnology thus merges tools from physics, chemistry, engineering, and biology to achieve breakthroughs in medicine, agriculture, and environmental science (Ramburrun et al., 2022), (Dey et al., 2024a; Dey et al., 2023a).

By integrating the bottom-up techniques of nanotechnology with the intricate processes of biology, nanobiotechnology enables new capabilities. One dream is to create intelligent nanodevices that simulate or interact with living systems. For instance, lipid-based nano capsules can deliver drugs or genetic data and react to cellular signals, and DNA origami techniques can build nanostructures that interact with enzymes or receptors (Sultana et al., 2022). This fusion is multidisciplinary

B. Krishan
Department of Biotechnology, Himachal Pradesh University, Summerhill, Shimla, Himachal Pradesh 171005, India

A. Kumar (✉) · S. Dhiman
Department of Biotechnology, University Institute of Biotechnology, Chandigarh University, Mohali, Punjab 140413, India
e-mail: anu.uibt@cumail.in

S. Ghosh et al. (eds.), *Advances in Organic Waste Conversion Through Industrial Biotechnology*, Sustainable Environmental Waste Management Strategies,
https://doi.org/10.1007/978-3-032-11377-1_12

by nature: it encompasses disciplines like nano-biosensors (sensors that merge a biological recognition element with a nanoscale transducer), nanocatalysts (e.g., enzyme–nanoparticle composites), and nano-bioimaging (using NPs as contrast agents) (Matar et al., 2025; Sapna et al., 2023). Across these disciplines, prudence in respect to nanotoxicology and safety is warranted, but the synergy offers powerful new methodologies. In summary, nanobiotechnology leverages the unique properties of nanomaterials to enhance mainstream biotechnological approaches, enhancing sensitivity, selectivity, and efficiency in applications ranging from diagnostics to remediation.

2 Nanoscale Materials in Environmental Applications

Nanoscale materials have intrinsically different properties to their bulk counterparts, due to quantum effects and extremely high-surface area. For example, in NPs of semiconductors both electrons and holes are able to reach the surface, greatly enhancing photocatalytic reactivity (Zhao et al., 2022). In environmental applications, such characteristics translate to fundamentally effective new remediation processes. NPs and nanostructured membranes are able to adsorb, degrade, or filter out pollutants much more efficiently than larger ones (Elsaid et al., 2023; Salehi et al., 2022). Nanobiotechnology takes advantage of this by creating bio-inspired or bio-assisted nanomaterials for cleanup missions.

2.1 Nano-Adsorbents and Nano-Filters

Artificial adsorbents such as carbon nanotubes or graphene oxide sheets have enormous surface area and are capable of adsorbing heavy metals or organic toxins from water and air (Khan et al., 2021), (Nag et al., 2023; Pradhan et al., 2023)Nanofiltration membranes (for example, polymer membranes containing NPs as the membrane) can remove bacteria and micro-pollutants but permeate pure water (Khan et al., 2023; Shukla et al., 2025).

2.2 Nanocatalysts

NPs with photocatalytic activity (e.g., TiO_2 or ZnO) are studied to break down and oxidize organic pollutants by light; nanocatalysts also enhance biodegradation by microorganisms (Das et al., 2024a; Dey et al., 2024b). Magnetic iron–nanoparticle composites, for instance, are simple to recover once used, highly reactive, and recyclable (Ghamarpoor et al., 2024; Mourdikoudis et al., 2021).

2.3 Nanosensors

Biological or chemical sensor units on nanostructures are able to detect trace impurities at extremely low concentrations. For example, enzyme-functionalized or antibody-functionalized gold NPs can colorimetrically detect the presence of microbes or toxins in water (El-Abeid et al., 2024).

2.4 Nanobiomaterials

Microbial-metal nanocomposites or bioengineered nanocellulose are among the materials that have nanoscale and biological components mixed together to capture pollutants (Das et al., 2025; Mishra et al., 2023; Sahu et al., 2024). Even bacteria can be used to biosynthesize NPs in situ to generate bio-nanocatalysts to treat wastewater (Nain et al., 2025; Selvinsimpson & Chen, 2023).

Cumulatively, these nanoscale techniques are highly useful for pollutant removal and detection. Recent studies have reported that in air, water, and wastewater systems nanotechnology has high efficiency for removing pollutants and detecting them, including nano-adsorbents, nano-filtration, nano-photocatalysts, magnetic NPs, and nanosensors (El-Abeid et al., 2024; Ghamarpoor et al., 2024; Nain et al., 2025; Shukla et al., 2025). Because of such potential, some authors characterize nanotechnology as a green or sustainable technology for managing pollution. Briefly, nanomaterial properties—nanometer-scale controllability, large reactive surface area, and multi-functionality—make the materials of high utility for environmental applications (e.g., water cleaning, soil cleaning, or toxin detection) that facilitate goals of sustainability (Beni & Jabbari, 2022).

3 Significance of Environmental Sustainability

3.1 Global Environmental Challenges

Modern society faces a group of interrelated environmental problems that endanger sustainability. Climate change is a paradigmatic example: anthropogenic greenhouse gas emissions have raised the Earth's surface temperature by about 1–2 °C during the last decades, and future projections predict more frequent extreme heat, intense storms, and sea-level rise during the next decades (Filonchyk et al., 2024; Lu, 2022; Valipour et al., 2021). At the same time, pollution due to industrial and agricultural activities has become dangerous in most regions: population growth and industrialization result in a significant discharge of pollutants into the environment, while even low concentrations of harmful chemicals can harm human health and ecosystems

(Saxena, 2025). Over-exploitation of resources is also urgent—freshwater, agricultural land, and mineral resources are being used faster than nature can restore them (Ghosh & Das, 2020; Mishra & Das, 2023). Deforestation and loss of biodiversity additionally worsen the problems, leading to degradation of critical ecosystem services (e.g., water filtration and carbon sequestration). All these problems, collectively referred to as the triple planetary crisis of climate change, pollution, and loss of nature, require the creation of new solutions (Hellweg et al., 2023).

- **Climate Change**: Increased greenhouse gas emissions (due to fossil fuels, agriculture, etc.) are changing the global climate system. It results in increased extreme weather, melting of ice caps, and disturbed ecosystems (Ahmadi Dehrashid et al., 2022).
- **Pollution and Waste**: Industrial processes and urbanization lead to chemicals, plastics, heavy metals, and pathogens entering air, water, and land (Das et al., 2023; Mishra et al., 2009). Exposure to air pollutants and water over long durations leads to millions of premature deaths each year, which shows even low pollution levels can have very important effects (Saxena, 2025).
- **Land and Resource Degradation**: Excessive use of resources (oil, minerals, freshwater) and intensive farming deplete fertile land and freshwater resources. For instance, per capita freshwater availability declines in most regions of the globe, and loss of biodiversity (habitat loss) diminishes natural resilience (Dudgeon & Strayer, 2025).

These problems are addressed not only by minimizing harmful outputs but also by improving for remediation purposes. New technologies like those derived from the nanobiotechnology toolkit—are currently being looked into in order to achieve sustainability objectives in response to these worldwide pressures. Figure 1 provides an overview of their applications, including pollution remediation, agricultural productivity enhancement, and CO_2 biofixation through photosynthetic modulation (Fig. 1).

3.2 Nanobiotechnology to Address Environmental Challenges

Nanobiotechnology can play a significant role in solving the above environment problems. Nanomaterials and devices possess unique properties that can be used to prevent pollution and repair damages. Essentially, nanobiotechnology can make processes more efficient, selective, and wasteful, thus enabling sustainability goals (Saxena, 2023). For instance, the combination of catalytic NPs and biological enzymes can enable degradation of sophisticated pollutants to harmless by-products, and nanoengineered membranes can provide more water purification than traditional filters (Saravanan et al., 2021). Some of the significant applications in which nanobiotechnology contributes to environmental sustainability are listed in Table 1.

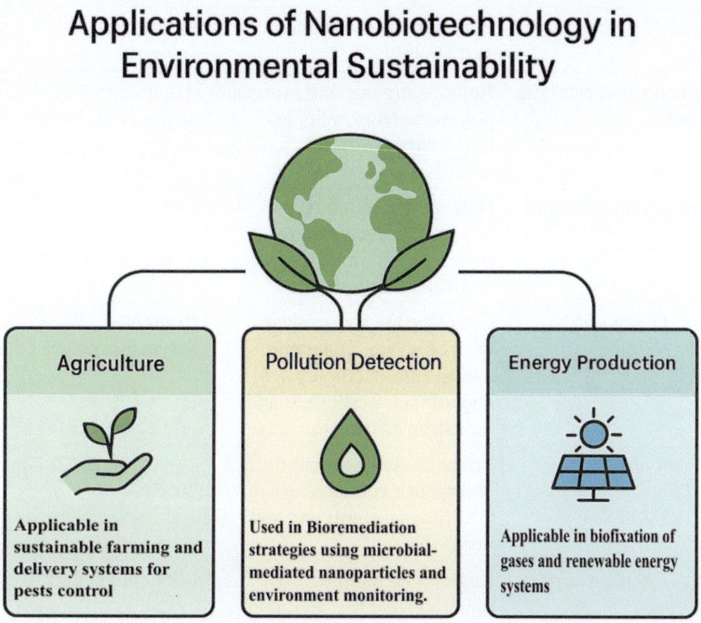

Fig. 1 Applications of nanobiotechnology in environmental sustainability

- **Remediation of Pollution and Clean Water**: Nanoscale catalysts and adsorbents like magnetic nano-Fe, carbon nanotubes, and nano-zero-valent iron enable the efficient sequestration or degradation of organic and inorganic water and land pollutants (Akter et al., 2022; Li et al., 2021; Stefaniuk et al., 2016). Application of nanotechnology in water treatment processes like photocatalytic oxidation and membrane filtration has been demonstrated to achieve high removal efficiencies for dyes, drugs, and heavy metal contaminants (Beni & Jabbari, 2022). Similarly, nanobioremediation methods—based on microorganisms tagged with or secreting nanomaterials—are also able to sequester oil spills or convert waste contaminants at mild conditions. These methods are superior to conventional methods by offering enhanced reaction kinetics and the ability to selectively target contaminants.
- **Greenhouse Gas Mitigation and Clean Energy**: Nanobiotechnology reduces greenhouse emissions by providing new materials and processes (De & Goswami, 2022). Nanostructured sorbents (such as metal–organic frameworks and carbon nanotube sponges) can capture CO_2 from flue gas or the air, and nano-enhanced

Table 1 Examples of nanobiotechnology applications for environmental sustainability (pollution, climate, resources)

Nanomaterial	Function in environment sustainability	References
Nano-adsorbents and catalytic nanomaterials	Remove organic and inorganic pollutants from water and air (e.g., carbon nanotubes, TiO_2, nano-Fe)	(Ghamarpoor et al., 2024; Salehi et al., 2022; Zhao et al., 2022)
Nanosensors and bio-sensors	Detect contaminants (chemicals, heavy metals, pathogens) at trace levels for real-time monitoring	(Pal et al., 2025)
Nano-enabled remediation	Coupling microorganisms with NPs (e.g., magnetic nano-bacteria composites) to degrade oil, pesticides, and industrial effluents	(Nain et al., 2025; Selvinsimpson & Chen, 2023)
Nanofertilizers and nanoformulations	Enhance crop nutrient uptake and reduce chemical use, minimizing runoff and soil depletion	(An et al., 2022; Khan et al., 2023)
Nano catalytic carbon capture	Engineered nano-porous materials (MOFs, nanoporous carbons) and bio-inspired catalysts for CO_2 sequestration and conversion	(Zhao et al., 2022)

algal biofuel systems can enable CO_2 fixation by microalgae. In addition, nano-enabled catalysts (such as nanozymes that are enzyme mimics) can reduce CO_2 to valuable chemicals or fuels. Chausali et al. (2023) observe that nanostructured materials (nanocomposites, MOFs, carbonaceous materials, etc.) have enormous possibilities in sequestration and reduction of greenhouse gases, biofuel production, and wastewater treatment, emphasizing their potential to fight climate change (Chausali et al., 2023).

- **Monitoring and Intelligent Sensing**: Extremely sensitive nanoscale sensors can identify pollutants, either chemical or biological, at extremely low concentrations in real time to provide timely interventions prior to the contamination spreading (Thakur & Kumar, 2022). For example, biosensors with nanoscale transducers can continuously monitor water quality or identify airborne pathogens, significantly enhancing environmental monitoring capacity. Such accurate surveillance is essential for the use of early warning systems and adaptive management of pollution (Pal et al., 2025; Ramakrishnan et al., 2024; Vikesland, 2018).
- **Sustainable Agriculture**: In the agrifood sector, nanobiotechnology is going to improve resource efficiency. Nano formulations of fertilizers and pesticides allow for targeted delivery of fertilizer or agrochemicals, employing mechanisms such as encapsulation or nanoparticle carriers to maximize crop uptake and minimize

environmental runoff (An et al., 2022). Engineered nanomaterials also have the ability to act as soil amendments or seed coating, promoting plant growth and microbial activity. These practices can improve yields while being in conjunction with reduced chemical inputs, thus improving food security without causing agricultural pollution (Beni & Jabbari, 2022).

- **Waste Valorization and Circular Economy**: Nanobiotechnology may make recycling and waste management possible by creating new routes for material reuse. For instance, waste plastic can be broken down into reusable monomers by nano catalysts, or agricultural waste can be processed into biochar. Biohybrid nanomaterials (such as bacteria that generate nanocellulose) can convert organic waste streams into useful biomaterials. Such processes are in line with a circular economy mentality, converting waste into resources while minimizing environmental impact (Wang et al., 2020), (Mohanty et al., 2024; Swain et al., 2025).

Together, these examples illustrate the promise of nanobiotechnology to address a range of environmental problems simultaneously. By taking advantage of the nanoscale phenomena and integrating biological systems, researchers are developing revolutionary solutions not yet available by conventional technologies alone. As outlined in a recent study, nanotechnology offers a wide range of applications in environment sectors, with its products and processes being among the most efficient tools for the achievement of sustainability goals. In brief, nanobiotechnology is extremely promising to enhance environmental sustainability by facilitating cleaner and more efficient management of pollution, reducing waste, and ensuring the sustainable utilization of resources, all while still delivering human well-being (Beni & Jabbari, 2022; De & Goswami, 2022; Show et al., 2023).

4 Green Synthesis of NPS

Green or "bio-" synthesis of NPs utilizes living organisms or their extracts to synthesize nanoscale particles from metal ions at mild conditions. Plant biomolecules (e.g., flavonoids, phenolics, proteins) and microbial metabolites (enzymes, exopolysaccharides) are used as reducing and capping agents in reality (Ovais et al., 2018). For instance, *Azadirachta indica* or *Moringa oleifera* plant extracts contain flavonoids and polyphenols that reduce Ag^+ or Zn^{2+} to metallic or metal oxide NPs (Patil et al., 2022; Perumalsamy et al., 2024). Likewise, bacteria and fungi are capable of extracellular (secreted reductases or proteins) or intracellular (enzymes precipitating metals in the cytosol) biosynthesis of NPs (Durán et al., 2011). This "biofactory" method has been proven successfully in many species. In green synthesis, there is no need for external chemical reductants—the organism's own biochemicals act as mediators of nucleation and growth (Alsaiari et al., 2023). For instance, in one report rosemary leaf extract performed triple duty as the reducing, stabilizing, and capping agent to synthesize ZnO NPs. These phytochemicals adsorb on nascent particles, regulating size and preventing aggregation (Uysal et al., 2024). In microbial synthesis, enzymes such as nitrate reductase (in bacteria) or phytochelatins (in fungi) reduce metal ions to NPs. Therefore, many plant species and microbes have been utilized to synthesize gold, silver, zinc oxide, and other NPs under eco-friendly conditions (Hussain et al., 2016).

4.1 Plant Extracts and Microorganisms Utilize

Plant extracts and microbial cultures provide a "green" "bottom-up" pathway to NPs. In plant-mediated synthesis, leaf, fruit, or root extracts—containing biomolecules—are combined with a metal precursor. Plant phytochemicals (flavonoids, phenolic acids, terpenoids) provide electrons to metal ions, reducing them to neutral atoms and nucleating NPs (Dikshit et al., 2021; Hussain et al., 2016; Krishan et al., 2025). For instance, sweet cherry fruit exudate reduces Ag^+ to Ag^0 to produce ~25 nm silver NPs with antifungal activity against fungal pathogens (Soltani Nejad et al., 2024). Microbial routes employ bacteria or fungi grown with a metal salt. Some microbes precipitate metals as NPs on cell walls or secreted polymers; others sequester reduced metal particles intracellularly. Microbes such as *Bacillus* spp. and *Pseudomonas* spp. have been encapsulated in biopolymers (alginate) with supporting nanostructures to slowly release and preserve viability of the cells, illustrating nanocarrier strategies may improve biocontrol agent delivery in soils (Saberi Riseh et al., 2021). In conclusion, plant and microbial systems are nanofactories: they supply non-toxic, renewable reductants and capping molecules that yield well-defined NPs without extreme reagents.

4.2 Benefits Compared with Classical Chemical Synthesis

Green synthesis has a few advantages over traditional chemical routes. Firstly, it is non-toxic and biocompatible by nature: using natural extracts or living cells, it can eliminate toxic reducing agents (e.g., hydrazine or sodium borohydride) and organic solvents that afflict chemical routes (Hussain et al., 2016). The obtained NPs are usually capped by biological molecules, stabilizing the NPs and minimizing cytotoxicity. Secondly, green routes are generally low-cost and less complicated. They demand only mild conditions (room temperature/pressure) and low-cost plant or microbial material, conserving energy input and raw material expense. Thirdly, green synthesis can provide controlled size and shapes: the diversity of biomolecules (proteins, polysaccharides, polyphenols) provides control over NP morphology and dispersion. Variations in plant extract concentration or pH, for instance, can modulate the diameter of Ag or ZnO NPs (Elango & Roopan, 2015; Gour & Jain, 2019; Hussain et al., 2016; Vijayaram et al., 2024).

Biology (green) courses have the following advantages:

- Non-toxic reagents: uses water-based microbial/plant extracts rather than toxic chemicals.
- Energy and cost savings: low pressure and no ambient high heat synthesis using low-cost biomass saves cost.
- Increased biocompatibility: the biologically synthesized NPs are coated with organic molecules, which can reduce toxicity to the environment and humans.
- Controlled morphology: natural capping agents allow formation of well-defined sized, stable NPs (Gour & Jain, 2019).

On the other hand, traditional chemical synthesis tends to utilize the strong chemical reductants and heat-driven processes that create undesired by-products and excessive energy consumption. As such, green protocols are compatible with the goal of sustainability because they create fewer by-products and use renewable biologic materials.

5 Case Studies of Green Synthesis

A recent study disclosed a novel green plant-mediated protocol for the synthesis of zinc oxide nanoparticles (ZnO NPs) from rosemary (*Rosmarinus officinalis*) leaf extract. In the process, an aqueous rosemary extract was mixed with a zinc salt, wherein the phytochemical compounds of the extract, like polyphenols and terpenoids, synergistically acted as reducing, stabilizing, and capping agents. Fourier-transform infrared spectroscopy confirmed the interaction of organic moieties of rosemary with the ZnO surface, thus establishing their role in stabilization of particles. The ZnO NPs thus formed were incorporated into a cross-linked polymer hydrogel. When applied to wheat plants, rosemary-mediated ZnO NPs (labeled as rosemary-ZnO-NPs) caused no harmful effects on seed germination but on root and shoot growth considerably at higher concentrations, thus establishing the crucial role of dosage. Interestingly, the rosemary-ZnO-NPs exhibited greater antibacterial activity toward *E. coli* than chemically synthesized ZnO NPs. This situation is a routine example that green plant-mediated synthesis of ZnO can provide functional nanomaterials with enhanced biological efficacy. Moreover, the use of rosemary extract avoids the utilization of toxic chemicals in the process and imparts the nanoparticles with a natural plant-derived capping (Uysal et al., 2024).

In another example of green synthesis, sweet cherry (*Prunus avium*) fruit exudate was employed to synthesize silver NPs. Here, the kernel juice of sweet cherries played the role of the reducing agent: its mixture of sugars, organic acids and phenolics reduced Ag^+ to Ag^0 in the absence of any chemical reagent from outside. Biosynthesized Ag NPs were of medium size ~ 25 nm, as evidenced using TEM. Sweet-cherry–derived AgNPs exhibited high antifungal activity: at concentrations as low as 5–40 μg/mL, they suppressed *Alternaria alternata* (a cherry fruit rot fungus) growth and harmed fungal hyphae. Of particular significance was the emphasis of this work on the eco-friendliness of the process: utilizing an agricultural waste product (cherry pulp/exudate) the process valorizes waste and does not employ toxic chemicals. This case study illustrates how fruit extracts—natural antioxidants in abundance—can produce bioactive metallic NPs under mild conditions (Soltani Nejad et al., 2024).

Another study involves the use of *Curcuma longa* flower extract in order to enable silver nanoparticle (AgNP) synthesis, which harbors immense antibacterial activity against pathogenic microorganisms such as *Staphylococcus aureus* and *Escherichia coli*. The synthesized nanoparticles were spherical with diameters varying from 5 to 10 nm (Rajak et al., 2023). Aqueous extract of *Satureja hortensis* L. has also been used in order to obtain silver-montmorillonite nanocomposites. The nanocomposites exhibit enhanced antibacterial activity due to synergistic interaction between silver nanoparticles and montmorillonite clay (Sedaghat & Khalaj, 2017). Aloe vera, neem, and onion extracts have been used, either alone or combined, in the synthesis of AgNPs. The AgNPs synthesized are extremely active against *Staphylococcus aureus*, thus establishing the versatility of plant extracts in the synthesis of AgNPs (Chand et al., 2019). *Coffee arabica* leaf extract has also been used in order to obtain stable AgNPs, which were later converted into low-cost biosensors for the determination of cysteine. The biosensors are extremely sensitive and selective in nature, thus establishing the integration of green synthesis approaches into analytical applications (Harsha Haridas et al., 2022). Therefore, the integration of green synthesis techniques allows for sustainable nanoparticle production (see Table 2).

6 Applications of Nanobiotechnology in Environmental Sustainability

6.1 Agriculture

6.1.1 Nanofertilizers and Their Role in Sustainable Farming

Nanofertilizers are NPs or nano-formulations of nutrients that are engineered to improve crop nutrient-use efficiency. Due to their small size and increased surface area, nanofertilizers can provide nutrients (N, P, K, micronutrients) in a controlled-release form, minimizing leaching and runoff. Several reviews indicate that nanofertilizers improve crop yield and soil fertility and decrease chemical fertilizer consumption. One of the meta-analyses estimated at up to ~30% improvement in nutrient-use efficiency with nanofertilizers compared to traditional fertilizers (Khundi et al., 2025). This translated to significantly higher yields (up to ~25% more maize) and enabled a decrease in nitrogen inputs by ~40% in field experiments. The improved performance is because nanoparticulate nutrients stick better to soil and plant surfaces and can be engineered to release slowly. In reality, nano-forms of Zn, Fe, or hydroxyapatite are being engineered to deliver trace elements and phosphorus with low losses. Nanofertilizers also have secondary benefits: soil microbial health and soil structure are improved in some studies because the load of pollution is decreased. In

Table 2 Selected case studies of green synthesis of nanoparticles and their applications

Biological source	Nanoparticle type	Synthesis method	Applications	References
Rosmarinus officinalis (Rosemary)	ZnO NPs	Leaf extract-mediated synthesis	Antimicrobial activity	(Uysal et al., 2024)
Prunus avium (Sweet cherry)	AgNPs	Fruit extract reduction	Antibacterial coating, packaging	(Soltani Nejad et al., 2024)
Curcuma longa (Turmeric flower)	AgNPs	Phytochemical reduction	Antibacterial, wound healing potential	(Rajak et al., 2023)
Satureja hortensis	Ag-Montmorillonite composite	Plant extract with clay support	Antimicrobial nanocomposite for surface treatment	(Sedaghat & Khalaj, 2017)
Pseudomonas sp. (PGPR)	AgNPs	Extracellular microbial synthesis	Biocontrol, biofilm disruption	(Jain et al., 2021)
Coffea arabica (Coffee leaf)	AgNPs	Green synthesis + biosensor dev	Biosensor for cysteine detection	(Harsha Haridas et al., 2022)
Trianthema portulacastrum	Ag, ZnO, TiO_2 NPs	Multimetal synthesis	Antioxidant, photocatalytic, antimicrobial	(Z. U. H. Khan et al., 2019)
Musa balbisiana, Azadirachta indica, Ocimum tenuiflorum	AgNPs	Microwave-assisted green synthesis	Antimicrobial agents for agriculture	(Banerjee et al., 2014)
Mahonia fortunei	AgNPs	Leaf extract-based synthesis	Management of *Pectobacterium*-induced soft rot	(Wei et al., 2022)
Aloe vera, Neem, Onion	AgNPs	Individual/combined extracts	Broad-spectrum antibacterial	(Chand et al., 2019)

the African context, these advantages are pertinent: increasing population pressures necessitate sustainable intensification, and nanofertilizers represent one pathway to improve yields without adding to agrochemical pollution (Khundi et al., 2025).

6.1.2 Nanoscale Delivery Systems for Biocontrol Agents

Nanotechnology also applies to biological pest control by facilitating delivery of microbial or botanical biopesticides. Useful microbes (e.g., *Bacillus*, *Pseudomonas*, entomopathogenic fungi) and plant extracts are prone to rapid degradation in the field or inhomogeneous dispersion. Nano-encapsulation can stabilize these agents and facilitate slow release. Studies have reported alginate-based nanofiber or nanoparticle carriers can contain bacteria and enable moisture-triggered release, with high biocontrol strain viability in soil. A dramatic report revealed that nano-encapsulation of Bacillus subtilis in combination with *Pseudomonas fluorescens* (with carbon nanotubes and silica NPs) promoted pistachio root growth dramatically in comparison with free bacteria. Other examples are alginate-gelatin microcapsules containing *Bacillus* strains enriched with TiO_2 NPs, which reduced seedling mortality caused by fungal attack (Ghannam et al., 2022; Moradipour et al., 2019; Saberi Riseh et al., 2021). In all cases, the nanoscale matrix (usually biodegradable polymers like alginate or chitosan) prevents the microbes from desiccation and UV inactivation. Similarly, DNA or protein-based biocontrol products (e.g., double-stranded RNA vaccines against insects) can be delivered by nanocarriers to facilitate uptake by plants or pests. These nanoscale delivery systems therefore maximize the efficacy and longevity of biological pesticides, which is consistent with sustainable integrated pest management. With nanotechnology, it is possible to optimize biopesticide formulations and displace more toxic chemical pesticides and reduce overall environmental impact (Saberi Riseh et al., 2021; Yang et al., 2022).

6.2 *Identification and Abatement of Pollution*

6.2.1 Environmental Monitoring with Nanoparticles

Nanomaterials have transformed pollutant detection by allowing highly sensitive, selective, and portable sensors. Nanoparticle-based biosensors can identify trace amounts of heavy metals, pesticides, or pathogens in water and air without sophisticated laboratory facilities. Nanosensors generally have a recognition element (e.g., an antibody or an enzyme) fused with a nanomaterial transducer (e.g., AuNPs, carbon nanotubes, quantum dots) to amplify the signal (Vikesland, 2018). Willner and Vkesland (2018) survey nanomaterial-enabled sensors detect at the nanomolar-to-picomolar range. That is, pollutants such as mercury, arsenic, or pesticide residues can be quantified at parts-per-billion or trillion level in situ. Since these sensors can be made in miniaturized formats (dipsticks, portable electrodes, optical fiber probes), they enable real-time field monitoring of environmental quality. In practice, NPs serve as colorimetric probes (e.g., AuNPs that color change on binding a toxin), electrochemical catalysts (e.g., CNT electrodes to accelerate current for metal ions), or fluorescence enhancers (quantum dots that glow upon binding a target) (Willner & Vikesland, 2018). The main advantage is the high surface-to-volume ratio of NPs, which allows numerous binding sites and rapid response times. Consequently, nano-enabled sensors convert laboratory-level sensitivity into point-of-use testing, enabling early warning of pollution and quick decision-making in environmental management (Pal et al., 2025; Vikesland, 2018).

6.2.2 Bioremediation Strategies that Include Microbial-Mediated NPs

Nanobiotechnology is also used to remediate chronic contamination by utilizing biologically synthesized nanomaterials. One approach is to use microorganisms to synthesize or enhance NPs to capture or degrade pollutants. Some bacteria and algae can biomineralize metals as nanoscale sulfides or oxides, in effect sequestering toxic elements (Mourdikoudis et al., 2021). Magnetotactic bacteria synthesize naturally intracellular magnetite (Fe_3O_4) NPs; the magnetosomes can adsorb heavy metals (e.g., cadmium) and enable magnetic recoverability of pollutant-containing biomass. More broadly, microbial systems can biosynthesize zero-valent metal NPs (primarily Fe^0) that are effective reductants. Green-synthesized iron NPs (commonly referred to as nZVI) have been utilized to detoxify chlorinated solvents and reduce Cr (VI) in soil and aqueous phases. In a recent review, microbial nanotechnology is reported to pave a path toward eco-friendly remediation of industrial effluent (Ray et al., 2025; Reddy Koteswara et al., 2025; Swain et al., 2025). Iron NPs are unique: their strong redox potential (due to reaction with water) and magnetic susceptibility make them effective and recoverable for pollutant cleanup (Akter et al., 2022; Li et al., 2021; Stefaniuk et al., 2016).

Another approach is to hybridize NPs with microbial biofilms. Nanoparticle-supported engineered biofilms are more effective at degrading organic contaminants. For example, photocatalytic bacteria-encapsulated TiO_2 NPs can oxidize dyes in wastewater under illumination (Limage et al., 2020; Safo et al., 2025). Metal–organic-framework (MOF) NPs are also being engineered to adsorb heavy metals, in certain cases combined with microbial enzymes. In all cases, synergy arises because nanoscale facilitates surface reactions and mass transfer, and the biological element provides specificity or regeneration. The outcome is a "green" nanobioremediation toolkit: pollutant removal systems that do not use harsh chemicals and leverage nature's processes. By combining microbial life with engineered NPs, these approaches seek to cleanup soils and waters sustainably. The combination of microbial life with designed NPs can effectively cleanup pollutants from soils and waters (Sha et al., 2022).

6.3 Energy Production

6.3.1 Nanotechnology in Renewable Energy Systems

Nanomaterials are central enablers for next-generation renewable energy technologies. In solar photovoltaics, for instance, nanostructured materials greatly enhance light harvesting and charge separation. A recent assessment of solar energy applications reported that the use of nanomaterials can enhance incident light absorption by as much as ninefold, resulting in ~ 10% efficiency gains for solar collectors compared to bulk designs. Nanoscale coatings, quantum dots or perovskite nanocrystals, and plasmonic metal NPs are also being applied to absorb more photons or enable multiple exciton generation in next-generation solar cells. Dye-sensitized and organic solar cells also utilize nanostructured TiO_2 and carbon nanomaterials to enable electron transport and suppress recombination (Safo et al., 2025).

In addition to the solar, nanotechnology also promotes bioenergy and hydrogen. Nano-catalysts (such as Ni, Pt, Fe oxide NPs on high-surface supports) can significantly speed up rates of catalytic reactions to produce biofuels and cleave water. For example, nanoengineered catalysts raise the yield of biohydrogen and biodiesel from biomass by reducing activation barriers. In fuel cells and batteries, nanomaterials (carbon nanotube electrodes, graphene current collectors, and nanoscale electrolytes) enhance conductivity and energy density, and thus make renewable energy storage more efficient (Safo et al., 2025; Sha et al., 2022; Zhao et al., 2022). In microbial fuel cells, nanostructured anodes (such as nanowires or gold NPs) allow large surface area for bacterial adhesion and electron transfer, and thus improve electrical output from waste biomass. In this way, on wind, solar, and bioenergy platforms, nanoengineering leads to enhanced efficiency, material usage and cost savings. These improvements are the secret to large-scale deployment of clean energy: by making components lighter and more active, nanotechnology brings us closer to low-carbon, low-cost energy systems.

6.3.2 Biofixation of CO_2 Improvement by Nanobiotechnology

Nanobiotechnology offers innovative strategies to enhance biological CO_2 sequestration through improved photosynthesis in plants and algae. Recent studies have demonstrated that nanofertilizers and foliar sprays containing nanoparticles (NPs) such as iron oxide and titanium dioxide can significantly boost chlorophyll production and photosynthetic efficiency in crops. For instance, the application of TiO_2 NPs has been shown to enhance photosynthetic rates and chlorophyll content in various plant species, leading to increased biomass production. Similarly, iron oxide NPs have been reported to improve seed germination, photosynthesis, and overall plant growth (Bhatia et al., 2025; Gohari et al., 2020).

In addition to enhancing plant photosynthesis, nanobiotechnology has been employed to improve CO_2 capture in microalgae. One approach involves the use of alginate nanofibers embedded with carbonic anhydrase (CA), an enzyme that facilitates the conversion of CO_2 to bicarbonate, thereby enhancing carbon fixation in microalgal cultures. Furthermore, nanoporous materials and dendrimer-based coatings have been developed to act as CO_2-absorbing substrates, providing a sustained release of CO_2 to support algal growth (Jun et al., 2020; Nyabadza et al., 2024).

These advancements suggest that nanobiotechnology can play a pivotal role in sustainable carbon management by augmenting the natural CO_2 sequestration capabilities of plants and microalgae. However, it is essential to conduct comprehensive studies to assess the long-term ecological impacts of these nanomaterials to ensure their safe and effective application in environmental remediation efforts.

7 Challenges and Considerations

Even though nanobiotechnology has great potential for attaining environmental sustainability, there are still a number of issues and problems that need to be resolved before it can be widely used.

7.1 *Nanotoxicity and Ecological Risks*

Unexpected repercussions could arise from the release of nanoparticles (NPs) into the environment. The long-term ecological implications of even biogenic NPs are not entirely understood, despite the fact that green synthesis techniques are less harmful than traditional chemical ways. For example, microbial communities, food chains, or soil health may be harmed by the buildup of metal oxide nanoparticles in terrestrial or aquatic environments (Das et al., 2024b; Das et al., 2025; Dey et al., 2025; Ray et al., 2025). Thorough ecotoxicological evaluations are necessary due to the possibility of bioaccumulation and trophic transmission.

7.2 *Lack of Regulatory Framework*

There are currently no widely recognized safety standards or regulations in the field of nanobiotechnology, particularly in low- and middle-income nations. To protect human health and the environment, robust frameworks that guarantee consistent testing, risk assessments, and labeling of nanoproducts are desperately needed.

7.3 Scalability of Green Synthesis

Green synthesis based on microbes and plants is environmentally friendly, however, it has problems with consistency and scalability. The practical relevance of biological synthesis techniques is limited since they are frequently slower, more susceptible to environmental factors, and yield different nanoparticle sizes and shapes based on the biomass composition.

7.4 Economic and Infrastructural Barriers

Adoption of nanobiotechnology solutions necessitates large investments in training, fabrication facilities, and research infrastructure. The cost and technological requirements for deployment continue to be prohibitive for many locations, particularly rural or undeveloped countries. Collaborations, financing plans, and public–private partnerships are necessary to close this gap.

7.5 Knowledge Gaps and Interdisciplinary Needs

Multidisciplinary knowledge in microbiology, materials science, agriculture, and toxicity is necessary for the efficient use of nanobiotechnology. The shift from lab-scale innovation to practical impact may be slowed down by a lack of cross-sectoral collaboration and a lack of awareness among practitioners.

8 Future Perspectives

The field of nanobiotechnology has the potential to significantly impact the development of a resilient and sustainable future. The field's development is highlighted by the following new directions. AI, machine learning, and advanced modeling can speed up toxicity assessment, synthesis optimization, and nanoparticle design. AI may make intelligent environmental monitoring and dynamic remediation techniques possible when combined with real-time nanosensor data. Synthetic biology and genetic engineering offer opportunities to develop microbes capable of biosynthesizing tailored nanoparticles or cooperative groups for carbon capture, pollutant degradation, or enhanced plant–microbe interactions in soil ecosystems. Future research will likely focus on using nanobiotechnology to enable closed-loop processes, including converting agro-industrial waste into useful nanomaterials, bioplastics, or biofertilizers, in order to minimize environmental effect and optimize resource efficiency. A future in nanobiotechnology that is socially conscious requires proactive stakeholder involvement, moral foresight, and inclusive regulatory frameworks. Gaining the trust of society will need open science, public education, and transparent governance.

9 Conclusion

A potent fusion of nanoscience and biotechnology, nanobiotechnology provides creative answers to some of the most important environmental problems of our day. It facilitates the shift to a resilient and circular green economy through applications in waste valorization, water purification, CO_2 sequestration, sustainable agriculture, and pollution remediation. The promise of environmentally friendly nanomaterials is demonstrated by the use of microorganisms and plants in green synthesis processes. However, major obstacles including nanotoxicity, regulatory ambiguity, and technical scalability must be addressed if nanobiotechnology is to reach its full potential. It also calls for a multidisciplinary strategy that incorporates ethical considerations, public involvement, policy, and scientific innovation. In the future, the environmental advantages of nanobiotechnology will be increased by combining it with cutting-edge technologies like artificial intelligence, synthetic biology, and sustainable manufacturing. In conclusion, by addressing both scientific and societal considerations, nanobiotechnology can emerge as a cornerstone of global efforts toward environmental sustainability and planetary health.

References

Ahmadi Dehrashid, A., Valizadeh, N., Gholizadeh, M. H., Ahmadi Dehrashid, H., & Nasrollahizadeh, B. (2022). Perspectives of climate change. In *Climate change* (pp. 369–388). Springer International Publishing. https://doi.org/10.1007/978-3-030-86290-9_21

Akter, M., Tajuddin Sikder, Md., & Atique Ullah, A. K. M. (2022). *Zero-valent iron and some other nanometal particles for environmental remediation* (pp. 539–568). https://doi.org/10.1007/698_2021_821

Allam, S. S. T., Gacem, A., Govind, G., Tummalacharla, P., Ramakrishnan, G., Yadav, K. K.,... & Das, A. P. (2025). Zero waste initiatives in industrial hydrocarbons and waste management. *Environmental Science and Engineering, Part, F218*, 105–126. https://doi.org/10.1007/978-3-031-81615-4_6

Alsaiari, N. S., Alzahrani, F. M., Amari, A., Osman, H., Harharah, H. N., Elboughdiri, N., & Tahoon, M. A. (2023). Plant and microbial approaches as green methods for the synthesis of nanomaterials: Synthesis, applications, and future perspectives. *Molecules, 28*(1), 463. https://doi.org/10.3390/molecules28010463

An, C., Sun, C., Li, N., Huang, B., Jiang, J., Shen, Y., Wang, C., Zhao, X., Cui, B., Wang, C., Li, X., Zhan, S., Gao, F., Zeng, Z., Cui, H., & Wang, Y. (2022). Nanomaterials and nanotechnology for the delivery of agrochemicals: Strategies towards sustainable agriculture. *Journal of Nanobiotechnology, 20*(1), 11. https://doi.org/10.1186/s12951-021-01214-7

Banerjee, P., Satapathy, M., Mukhopahayay, A., & Das, P. (2014). Leaf extract mediated green synthesis of silver nanoparticles from widely available Indian plants: Synthesis, characterization, antimicrobial property and toxicity analysis. *Bioresources and Bioprocessing, 1*(1), 3. https://doi.org/10.1186/s40643-014-0003-y

Beni, A. A., & Jabbari, H. (2022). Nanomaterials for environmental applications. https://doi.org/10.1016/j.rineng.2022.10046

Bhatia, A., Khatri, A., Yadav, M., Kumari, A., Mona, S., & Bhateria, R. (2025). Potential of iron oxide nanoparticles in enhancing growth and development of plants: A review. *Physiological and Molecular Plant Pathology, 139*, Article 102746. https://doi.org/10.1016/j.pmpp.2025.102746

Chand, K., Abro, M. I., Aftab, U., Shah, A. H., Lakhan, M. N., Cao, D., Mehdi, G., & Ali Mohamed, A. M. (2019). Green synthesis characterization and antimicrobial activity against *Staphylococcus aureus* of silver nanoparticles using extracts of neem, onion and tomato. *RSC Advances, 9*(30), 17002–17015. https://doi.org/10.1039/C9RA01407A

Chausali, N., Saxena, J., & Prasad, R. (2023). Nanotechnology as a sustainable approach for combating the environmental effects of climate change. *Journal of Agriculture and Food Research, 12*, Article 100541. https://doi.org/10.1016/j.jafr.2023.100541

Das, A. P., Mohanty, A. K., van Hullebusch, E. D., Figueiredo, G., Das, A. P., Mohanty, A. K., van Hullebusch, E. D., & Figueiredo, G. (2023). Bioremediation as an emerging technology for the removal of synthetic microplastic pollutants from marine ecosystem. *MarPB, 194*, Article 115297. https://doi.org/10.1016/J.MARPOLBUL.2023.115297

Das, A., Ray, S. K., Mohanty, M., Mohanty, J., Dey, S., & Das, A. P. (2024b). Ecotoxicity of microplastic wastes and their sustainable management: A review. *Environmental Chemistry and Ecotoxicology, 6*, 144–152. https://doi.org/10.1016/J.ENCECO.2024.05.003

Das, A., Dey, S., Gacem, A., Yadav, K. K., Nath, H., Ram, D. K., & Das, A. P. (2025). Revolutionizing solid hydrocarbon sourcing: Plastic's journey from waste to upcycled treasure. *Environmental Science and Engineering, Part, F218*, 331–351. https://doi.org/10.1007/978-3-031-81615-4_14

Das, A., Dey, S., Bala, K., Gnanasekaran, R., & Das, A. P. (2024a). Microbial bioremediation technology for sustainable treatment and management of synthetic microfiber waste. In *Sustainable microbial technology for synthetic and cellulosic microfiber bioremediation* (pp. 221–238). https://doi.org/10.1007/978-3-031-62600-5_12

De, B., & Goswami, T. K. (2022). Nanobiotechnology—A green solution. In *Biotechnology for zero waste* (pp. 379–396). Wiley. https://doi.org/10.1002/9783527832064.ch25

Dey, S., Tripathy, B., Kumar, M. S., & Das, A. P. (2023b). Ecotoxicological consequences of manganese mining pollutants and their biological remediation. *Environmental Chemistry and Ecotoxicology*. https://doi.org/10.1016/J.ENCECO.2023.01.001

Dey, S., Das, A., Mallick, K., Sahu, A., & Das, A. P. (2023a). *Environmental petroleum waste: Pollution, toxicity, sustainable remediation* (pp. 159–175). https://doi.org/10.1007/978-3-031-48220-5_7

Dey, S., Bissoyi, A., Koteswara Reddy, G., & Das, A. P. (2024a). Synthetic microfibers: An urban emerging pollutant demanding for sustainable environmental solutions. In *Sustainable microbial technology for synthetic and cellulosic microfiber bioremediation* (pp. 107–121). https://doi.org/10.1007/978-3-031-62600-5_7

Dey, S., Dash, A., & Das, A. P. (2024b). Bioremediation as a sustainable and circular solution for mining waste management and bioenergy generation. In *Sustainable management of mining waste and tailings: A circular economy approach*, 107–119. https://doi.org/10.1201/9781003442455-6

Dikshit, P., Kumar, J., Das, A., Sadhu, S., Sharma, S., Singh, S., Gupta, P., & Kim, B. (2021). Green synthesis of metallic nanoparticles: applications and limitations. *Catalysts, 11*(8), 902. https://doi.org/10.3390/catal11080902

Dudgeon, D., & Strayer, D. L. (2025). Bending the curve of global freshwater biodiversity loss: What are the prospects? *Biological Reviews, 100*(1), 205–226. https://doi.org/10.1111/brv.13137

Durán, N., Marcato, P. D., Durán, M., Yadav, A., Gade, A., & Rai, M. (2011). Mechanistic aspects in the biogenic synthesis of extracellular metal nanoparticles by peptides, bacteria, fungi, and plants. *Applied Microbiology and Biotechnology, 90*(5), 1609–1624. https://doi.org/10.1007/s00253-011-3249-8

El-Abeid, S. E., El-Tabakh, M. A. M., Shehata, A. Z. I., AbdelHamid, R. I., & Soliman, A. G. (2024). The docking and physiological characteristics as detectors of nanoparticle's role in plant responses to biotic stress. In *Nanoparticles in plant biotic stress management* (pp. 265–292). Springer Nature Singapore. https://doi.org/10.1007/978-981-97-0851-2_10

Elango, G., & Roopan, S. M. (2015). Green synthesis, spectroscopic investigation and photocatalytic activity of lead nanoparticles. *Spectrochimica Acta Part a: Molecular and Biomolecular Spectroscopy, 139*, 367–373. https://doi.org/10.1016/j.saa.2014.12.066

Elsaid, K., Olabi, A. G., Abdel-Wahab, A., Elkamel, A., Alami, A. H., Inayat, A., Chae, K.-J., & Abdelkareem, M. A. (2023). Membrane processes for environmental remediation of nanomaterials: Potentials and challenges. *Science of the Total Environment, 879*, Article 162569. https://doi.org/10.1016/j.scitotenv.2023.162569

Filonchyk, M., Peterson, M. P., Zhang, L., Hurynovich, V., & He, Y. (2024). Greenhouse gases emissions and global climate change: Examining the influence of CO_2, CH_4, and N_2O. *Science of the Total Environment, 935*, Article 173359. https://doi.org/10.1016/j.scitotenv.2024.173359

Ghamarpoor, R., Fallah, A., & Jamshidi, M. (2024). A review of synthesis methods, modifications, and mechanisms of ZnO/TiO_2 -Based photocatalysts for photodegradation of contaminants. *ACS Omega, 9*(24), 25457–25492. https://doi.org/10.1021/acsomega.3c08717

Ghannam, M. T., Selim, M. Y. E., Zekri, A. Y., & Esmail, N. (2022). Viscoelastic behavior of crude oil-gum emulsions in enhanced oil recovery. *Polymers, 14*(5), 1004. https://doi.org/10.3390/polym14051004

Ghosh, S., & Das, A. P. (2020). Microbial metagenomics: current advances in investigating microbial ecology and population dynamics. In S. K. Nayak & B. B. Mishra (Eds.), *Frontiers in soil and environmental microbiology* (1st ed., pp. 247–254). CRC Press.

Gohari, G., Mohammadi, A., Akbari, A., Panahirad, S., Dadpour, M. R., Fotopoulos, V., & Kimura, S. (2020). Titanium dioxide nanoparticles (TiO2 NPs) promote growth and ameliorate salinity stress effects on essential oil profile and biochemical attributes of Dracocephalum moldavica. *Scientific Reports, 10*(1), 912. https://doi.org/10.1038/s41598-020-57794-1

Gour, A., & Jain, N. K. (2019). Advances in green synthesis of nanoparticles. *Artificial Cells, Nanomedicine, and Biotechnology, 47*(1), 844–851. https://doi.org/10.1080/21691401.2019.1577878

Harsha Haridas, E. S., Bhattacharya, S., Ravi Varma, M. K., & Kumar Chandra, G. (2022). *Green synthesis of eco-friendly silver nanoparticles using Coffee arabica leaf extract and development of a cost-effective biosensor for cysteine.*

Hellweg, S., Benetto, E., Huijbregts, M. A. J., Verones, F., & Wood, R. (2023). Life-cycle assessment to guide solutions for the triple planetary crisis. *Nature Reviews Earth & Environment, 4*(7), 471–486. https://doi.org/10.1038/s43017-023-00449-2

Hussain, I., Singh, N. B., Singh, A., Singh, H., & Singh, S. C. (2016). Green synthesis of nanoparticles and its potential application. *Biotechnology Letters, 38*(4), 545–560. https://doi.org/10.1007/s10529-015-2026-7

Jain, M., Yadav, M., & Chaudhry, S. (2021). Copper oxide nanoparticles for the removal of divalent nickel ions from aqueous solution. *Toxin Reviews, 40*(4), 872–885. https://doi.org/10.1080/15569543.2020.1799407

Jun, S.-H., Yang, J., Jeon, H., Kim, H. S., Pack, S. P., Jin, E., & Kim, J. (2020). Stabilized and immobilized carbonic anhydrase on electrospun nanofibers for enzymatic CO_2 conversion and utilization in expedited microalgal growth. *Environmental Science & Technology, 54*(2), 1223–1231. https://doi.org/10.1021/acs.est.9b05284

Khan, F. S. A., Mubarak, N. M., Tan, Y. H., Khalid, M., Karri, R. R., Walvekar, R., Abdullah, E. C., Nizamuddin, S., & Mazari, S. A. (2021). A comprehensive review on magnetic carbon nanotubes and carbon nanotube-based buckypaper for removal of heavy metals and dyes. *Journal of Hazardous Materials, 413*, Article 125375. https://doi.org/10.1016/j.jhazmat.2021.125375

Khan, Z. U. H., Sadiq, H. M., Shah, N. S., Khan, A. U., Muhammad, N., Hassan, S. U., Tahir, K., safi, S. Z., Khan, F. U., Imran, M., Ahmad, N., Ullah, F., Ahmad, A., Sayed, M., Khalid, M. S., Qaisrani, S. A., Ali, M., & Zakir, A. (2019). Greener synthesis of zinc oxide nanoparticles using Trianthema portulacastrum extract and evaluation of its photocatalytic and biological applications. *Journal of Photochemistry and Photobiology B: Biology, 192*, 147–157. https://doi.org/10.1016/j.jphotobiol.2019.01.013

Khan, N. A., Singh, S., López-Maldonado, E. A., Méndez-Herrera, P. F., López-López, J. R., Baig, U.,... & Aljundi, I. H. (2023). Emerging membrane technology and hybrid treatment systems for the removal of micropollutants from wastewater. *Desalination, 565*, 116873. https://doi.org/10.1016/j.desal.2023.116873

Khundi, Q., Jiang, Y., Sun, Y., & Rui, Y. (2025). Nanofertilizers for sustainable African agriculture: A global review of agronomic efficiency and environmental sustainability. *Nanomaterials, 15*(5), 390. https://doi.org/10.3390/nano15050390

Krishan, B., Kumar, A., Azmi, W., & Dhiman, S. (2025). Biological activities of citrus fruit-derived copper oxide nanoparticles: Towards sustainable antimicrobial and antioxidant solutions. *Folia Microbiologica*. https://doi.org/10.1007/s12223-025-01266-4

Li, Q., Chen, Z., Wang, H., Yang, H., Wen, T., Wang, S., Hu, B., & Wang, X. (2021). Removal of organic compounds by nanoscale zero-valent iron and its composites. *Science of the Total Environment, 792*, Article 148546. https://doi.org/10.1016/j.scitotenv.2021.148546

Limage, R., Tako, E., Kolba, N., Guo, Z., García-Rodríguez, A., Marques, C. N. H., & Mahler, G. J. (2020). TiO_2 nanoparticles and commensal bacteria alter mucus layer thickness and composition in a gastrointestinal tract model. *Small (Weinheim an der Bergstrasse, Germany), 16*(21), 2000601. https://doi.org/10.1002/smll.202000601

Lu, Q.-B. (2022). Major contribution of halogenated greenhouse gases to global surface temperature change. *Atmosphere, 13*(9), 1419. https://doi.org/10.3390/atmos13091419

Matar, Z., Zainon Noor, Z., Al-Hindi, A., & Yuliarto, B. (2025). Recent advances in paper-based nano-biosensors for waterborne pathogen detection: Challenges and solutions. *Chemistry & Biodiversity*. https://doi.org/10.1002/cbdv.202403451

Mishra, S., & Das, A. P. (2023). Microbial remediation of synthetic microfiber contaminated wastewater. In M. P. Shah (Ed.), *Microbial technologies in industrial wastewater treatment* (pp. 337–350). Springer, Singapore. https://doi.org/10.1007/978-981-99-2435-6_17

Mishra, S., Das, A. P., & Seragadam, P. (2009). Microbial remediation of hexavalent chromium from chromite contaminated mines of Sukinda Valley, Orissa (India). *Journal of Environmental Research and Development*, 1122–1127.

Mishra, S., Dash, D., & Das, A. P. (2023). Aquatic microbial diversity on plastisphere: Colonization and potential role in microplastic biodegradation. *Geomicrobiology Journal*, 1–12. https://doi.org/10.1080/01490451.2023.2209750

Mohanty, M., Mohanty, J., Dey, S., Dutta, K., Shah, M. P., & Das, A. P. (2024). The face mask: A tale from protection to pollution and demanding sustainable solution. *Emerging Contaminants, 10*(2), Article 100298. https://doi.org/10.1016/J.EMCON.2023.100298

Moradipour, M., Saberi-Riseh, R., Mohammadinejad, R., & Hosseini, A. (2019). Nano-encapsulation of plant growth-promoting rhizobacteria and their metabolites using alginate-silica nanoparticles and carbon nanotube improves UCB1 pistachio micropropagation. *Journal of Microbiology and Biotechnology, 29*(7), 1096–1103. https://doi.org/10.4014/jmb.1903.03022

Mourdikoudis, S., Kostopoulou, A., & LaGrow, A. P. (2021). Magnetic nanoparticle composites: synergistic effects and applications. *Advanced Science, 8*(12). https://doi.org/10.1002/advs.202004951

Nag, M., Lahiri, D., Ghosh, S., Sarkar, T., Pati, S., Das, A. P., Ram, D. K., Bhattacharya, D., & Ray, R. R. (2023). Application of microorganisms in biotransformation and bioremediation of environmental contaminant: A review. *Geomicrobiology Journal*. https://doi.org/10.1080/01490451.2023.2261443

Nain, R., Patel, H., Chahar, M., Kumar, S., Rohilla, D., & Pal, M. (2025). Biosynthesized metallic nanoparticles for sustainable environmental remediation: Mechanisms, applications, and future perspectives. *Discover Chemistry, 2*(1), 124. https://doi.org/10.1007/s44371-025-00203-1

Nyabadza, A., Makhesana, M., Plouze, A., Kumar, A., Ramirez, I., Krishnamurthy, S., Vazquez, M., & Brabazon, D. (2024). Advanced nanomaterials and dendrimers in water treatment and the recycling of nanomaterials: A review. *Journal of Environmental Chemical Engineering, 12*(3), Article 112643. https://doi.org/10.1016/j.jece.2024.112643

Ovais, M., Khalil, A. T., Islam, N. U., Ahmad, I., Ayaz, M., Saravanan, M., Shinwari, Z. K., & Mukherjee, S. (2018). Role of plant phytochemicals and microbial enzymes in biosynthesis of metallic nanoparticles. *Applied Microbiology and Biotechnology, 102*(16), 6799–6814. https://doi.org/10.1007/s00253-018-9146-7

Pal, P., Pandey, B., Prakash, A., Sarangi, P. K., Rathore, S. S., Agarwal, V., Gupta, A. K., & Singh, A. K. (2025). *Exploration of biosensors in innovative applications and sustainable solutions for environmental monitoring and management* (pp. 561–591). https://doi.org/10.1007/978-981-97-7221-6_19

Patil, S. P., Chaudhari, R. Y., & Nemade, M. S. (2022). Azadirachta indica leaves mediated green synthesis of metal oxide nanoparticles: A review. *Talanta Open, 5*, Article 100083. https://doi.org/10.1016/j.talo.2022.100083

Paul, A., Dey, S., Ram, D. K., & Das, A. P. (2023a). *Hexavalent chromium pollution and its sustainable management through bioremediation.* https://doi.org/10.1080/01490451.2023.2218377

Paul, A., Dey, S., Ram, D. K., & Das, A. P. (2023b). Hexavalent chromium pollution and its sustainable management through bioremediation. *Geomicrobiology Journal*, 1–11. https://doi.org/10.1080/01490451.2023.2218377

Perumalsamy, H., Balusamy, S. R., Sukweenadhi, J., Nag, S., MubarakAli, D., El-Agamy Farh, M., Vijay, H., & Rahimi, S. (2024). A comprehensive review on Moringa oleifera nanoparticles: Importance of polyphenols in nanoparticle synthesis, nanoparticle efficacy and their applications. *Journal of Nanobiotechnology, 22*(1), 71. https://doi.org/10.1186/s12951-024-02332-8

Pradhan, G., Tripathy, B., Ram, D. K., Digal, A. K., & Das, A. P. (2023). Bauxite mining waste pollution and its sustainable management through bioremediation. *Geomicrobiology Journal*, 1–10. https://doi.org/10.1080/01490451.2023.2235353

Rajak, K. K., Pahilani, P., Patel, H., Kikani, B., Desai, R., & Kumar, H. (2023). *Green synthesis of silver nanoparticles using Curcuma longa flower extract and antibacterial activity.*

Ramakrishnan, S., Sathvara, P., Tripathi, S., & Jayaraman, A. (2024). *Water pollutants, sensor types, and their advantages and challenges* (pp. 78–101). https://doi.org/10.4018/979-8-3693-1930-7.ch005

Ramburrun, P., Khan, R. A., & Choonara, Y. E. (2022). Design, preparation, and functionalization of nanobiomaterials for enhanced efficacy in current and future biomedical applications. *Nanotechnology Reviews, 11*(1), 1802–1826. https://doi.org/10.1515/ntrev-2022-0106

Ray, S., Dey, S., Digal, A. K., Yadav, K. K., & Das, A. P. (2025). Sanitary waste and microplastic pollution in the Euro-Mediterranean region: Challenges and solutions. *Euro-Mediterranean Journal for Environmental Integration, 2025*, 1–23. https://doi.org/10.1007/S41207-025-00879-Y

Reddy Koteswara, G., Bhargavi, S., Yamuna, D., Anjani Devi, M., Jeevana Lakshmi Kumar, G., Manas, T., Adarsh, S., Bhavya Sai, V., Nath, H., Alreshidi, M. A., Togun, H., Ram, D. K., & Das, A. P. (2025). Recent advancements in the remediation of marine hydrocarbon and plastic pollutants. *Environmental Science and Engineering*, *Part*, *F218*, 127–142. https://doi.org/10.1007/978-3-031-81615-4_7

Saberi Riseh, R., Skorik, Y. A., Thakur, V. K., Moradi Pour, M., Tamanadar, E., & Noghabi, S. S. (2021). Encapsulation of plant biocontrol bacteria with alginate as a main polymer material. *International Journal of Molecular Sciences, 22*(20), 11165. https://doi.org/10.3390/ijms222011165

Safo, K., Rubangakene, N. O., Noby, H., & El-Shazly, A. H. (2025). Photocatalytic purification of dye-containing wastewater using a novel embedded hybrid TiO_2–slag catalyst heterojunction nanocomposite coupled with statistical models: A sustainable and techno-economic approach. *Water Science and Engineering, 18*(2), 151–164. https://doi.org/10.1016/j.wse.2025.02.003

Sahu, A., Dey, S., Mallick, K., Priyadarsini, S., Behera, I. D., & Das, A. P. (2024). A review on advanced technology for sustainable management of synthetic microplastic waste. In *Sustainable management of urban plastic waste through circular economic approaches* (pp. 262–279). https://doi.org/10.1201/9781003465447-15

Salehi, E., Khajavian, M., Sahebjamee, N., Mahmoudi, M., Drioli, E., & Matsuura, T. (2022). Advances in nanocomposite and nanostructured chitosan membrane adsorbents for environmental remediation: A review. *Desalination, 527*, Article 115565. https://doi.org/10.1016/j.desal.2022.115565

Sapna, K., Manzur Ali, P. P., & Mohamed Hatha, A. A. (2023). Nanomaterials in bioimaging and cell labeling. In *Applications of multifunctional nanomaterials* (pp. 499–523). Elsevier. https://doi.org/10.1016/B978-0-12-820557-0.00012-6

Saravanan, A., Kumar, P. S., Vo, D.-V.N., Jeevanantham, S., Karishma, S., & Yaashikaa, P. R. (2021). A review on catalytic-enzyme degradation of toxic environmental pollutants: Microbial enzymes. *Journal of Hazardous Materials, 419*, Article 126451. https://doi.org/10.1016/j.jhazmat.2021.126451

Saxena, V. (2025). Water quality, air pollution, and climate change: Investigating the environmental impacts of industrialization and urbanization. *Water, Air, & Soil Pollution, 236*(2), 73. https://doi.org/10.1007/s11270-024-07702-4

Saxena, N. (2023). Bio-nanotechnology in waste to energy conversion in a circular economy approach for better sustainability. In *Bionanotechnology towards green energy* (pp. 253–274). CRC Press. https://doi.org/10.1201/9781003316374-13

Sedaghat, S., & Khalaj, M. (2017). *Rapid green biosynthesis of Silver-montmorillonite nanocomposite using water extract of Satureja hortensis L and evaluation of the antibacterial properties.*

Selvinsimpson, S., & Chen, Y. (2023). Microbial-based synthesis of nanoparticles to remove different pollutants from wastewater. In *Environmental applications of microbial nanotechnology* (pp. 167–181). Elsevier. https://doi.org/10.1016/B978-0-323-91744-5.00001-1

Sha, M., Xu, W., Fang, Q., Wu, Y., Gu, W., Zhu, C., & Guo, S. (2022). Metal-organic-framework-involved nanobiocatalysis for biomedical applications. *Chem Catalysis, 2*(10), 2552–2589. https://doi.org/10.1016/j.checat.2022.09.005

Shahcheraghi, N., Golchin, H., Sadri, Z., Tabari, Y., Borhanifar, F., & Makani, S. (2022). Nano-biotechnology, an applicable approach for sustainable future. *3 Biotech, 12*(3), 65. https://doi.org/10.1007/s13205-021-03108-9

Show, P.-L., Chew, K. W., Ong, W.-J., Varjani, S., & Juan, J. C. (2023). New trends in nanobiotechnology. *Beilstein Journal of Nanotechnology, 14*, 377–379. https://doi.org/10.3762/bjnano.14.32

Shukla, B. K., Parashar, B., Patel, T., Gupta, Y., Verma, S., & Singh, S. (2025). Polymeric membranes in water treatment: Insights into contaminant removal mechanisms and advanced processes. *ASEC, 2024*, 69. https://doi.org/10.3390/engproc2025087069

Soltani Nejad, M., Najafabadi, N. S., Aghighi, S., Zargar, M., Bayat, M., & Pakina, E. (2024). Green synthesis of silver nanoparticles by sweet cherry and its application against cherry spot disease. *Heliyon, 10*(10), Article e31508. https://doi.org/10.1016/j.heliyon.2024.e31508

Stefaniuk, M., Oleszczuk, P., & Ok, Y. S. (2016). Review on nano zerovalent iron (nZVI): From synthesis to environmental applications. *Chemical Engineering Journal, 287*, 618–632. https://doi.org/10.1016/j.cej.2015.11.046

Sultana, A., Zare, M., Thomas, V., Kumar, T. S. S., & Ramakrishna, S. (2022). Nano-based drug delivery systems: Conventional drug delivery routes, recent developments and future prospects. *Medicine in Drug Discovery, 15*, Article 100134. https://doi.org/10.1016/j.medidd.2022.100134

Swain, S., Purnima Sudha, M., Dey, S., Badu, Y., Bhoi, S. R., Alreshidi, M. A., Qasim, M. T., Nath, H., Ram, D. K., & Das, A. P. (2025). Sources of micro and nano plastic hydrocarbon pollutants in marine bodies and their treatment. *Environmental Science and Engineering, Part, F218*, 45–67. https://doi.org/10.1007/978-3-031-81615-4_3

Thakur, A., & Kumar, A. (2022). Recent advances on rapid detection and remediation of environmental pollutants utilizing nanomaterials-based (bio)sensors. *Science of the Total Environment, 834*, Article 155219. https://doi.org/10.1016/j.scitotenv.2022.155219

Uysal, Y., Görkem Doğaroğlu, Z., Çaylali, Z., & Karakulak, D. S. (2024). Rosemary-mediated green synthesis of ZnO Nanoparticles and their Integration into hydrogel matrices: Evaluating effects on wheat growth and antibacterial properties. *Global Challenges, 8*(11). https://doi.org/10.1002/gch2.202400120

Valipour, M., Bateni, S. M., & Jun, C. (2021). Global surface temperature: A new insight. *Climate, 9*(5), 81. https://doi.org/10.3390/cli9050081

Vijayaram, S., Razafindralambo, H., Sun, Y.-Z., Vasantharaj, S., Ghafarifarsani, H., Hoseinifar, S. H., & Raeeszadeh, M. (2024). Applications of green synthesized metal nanoparticles—a review. *Biological Trace Element Research, 202*(1), 360–386. https://doi.org/10.1007/s12011-023-03645-9

Vikesland, P. J. (2018). Nanosensors for water quality monitoring. *Nature Nanotechnology, 13*(8), 651–660. https://doi.org/10.1038/s41565-018-0209-9

Wang, C., Lei, H., Qian, M., Huo, E., Zhao, Y., Zhang, Q., Mateo, W., Lin, X., Kong, X., Zou, R., & Ruan, R. (2020). Application of highly stable biochar catalysts for efficient pyrolysis of plastics: A readily accessible potential solution to a global waste crisis. *Sustainable Energy & Fuels, 4*(9), 4614–4624. https://doi.org/10.1039/D0SE00652A

Wei, Z., Xu, S., Jia, H., & Zhang, H. (2022). Green synthesis of silver nanoparticles from Mahonia fortunei extracts and characterization of its inhibitory effect on Chinese cabbage soft rot pathogen. *Frontiers in Microbiology, 13*. https://doi.org/10.3389/fmicb.2022.1030261

Willner, M. R., & Vikesland, P. J. (2018). Nanomaterial enabled sensors for environmental contaminants. *Journal of Nanobiotechnology, 16*(1), 95. https://doi.org/10.1186/s12951-018-0419-1

Yang, W., Wang, B., Lei, G., Chen, G., & Liu, D. (2022). Advances in nanocarriers to improve the stability of dsRNA in the environment. *Frontiers in Bioengineering and Biotechnology, 10*. https://doi.org/10.3389/fbioe.2022.974646

Zhao, W., Adeel, M., Zhang, P., Zhou, P., Huang, L., Zhao, Y., Ahmad, M. A., Shakoor, N., Lou, B., Jiang, Y., Lynch, I., & Rui, Y. (2022). A critical review on surface-modified nano-catalyst application for the photocatalytic degradation of volatile organic compounds. *Environmental Science: Nano, 9*(1), 61–80. https://doi.org/10.1039/D1EN00955A

Emerging Approach for Managing Organic Waste Through Advanced Biotechnological Methods

Maheswata Moharana, A. D. Shobhanam Shobhana, and Subrat Kumar Pattanayak

1 Introduction

The production of organic waste has significantly increased due to the world's population growth and rapid industrialization causing serious environmental and public health risks. Worldwide, over 38 billion tons of organic wastes are produced annually (Peng et al., 2023). Management of the waste include several methods to make the environment cleaner and waste free, such as processing, reusing/recycling, removal or degradation, and waste generation control (Mashifana, 2025). Organic wastes especially the food waste is generated from a variety of industrial, agricultural, and household sources due to changing lifestyles and the rapid urbanization of the world and its amount is continuously rising (Girotto et al., 2015; Noor et al., 2020; Sharma et al., 2019). Food waste produced during the processing, packaging, transportation, and storage presents serious disposal management issues and expenses, as well as environmental issues (Wani et al., 2024, Das et al., 2025; Ray et al., 2025). The food waste generates a significant amount of waste in the form of undesirable by-products, such as seeds, peels, bagasse, shells, trims, and stems (Torres-León et al., 2018). Food waste can be liquid, solid, or semi solid, and can include things like wastewater, fats, used oil, hazardous home materials, and many more (Lin et al. 2013). The extensive use of water for various purposes, such as cooking, cleaning, sanitation in household works and primary production, preparation and processing of different industrial operations results in the huge generation of liquid waste. Additionally, solid wastes are compacted with monosaccharides, cellulose, amylose, and

M. Moharana · A. D. Shobhanam Shobhana
Department of Chemistry, School of Basic Sciences and Humanities, Odisha University of Technology and Research, Bhubaneswar, India

S. K. Pattanayak (✉)
Department of Chemistry, National Institute of Technology, Raipur, India
e-mail: skpiitbbs@gmail.com

S. Ghosh et al. (eds.), *Advances in Organic Waste Conversion Through Industrial Biotechnology*, Sustainable Environmental Waste Management Strategies,
https://doi.org/10.1007/978-3-032-11377-1_13

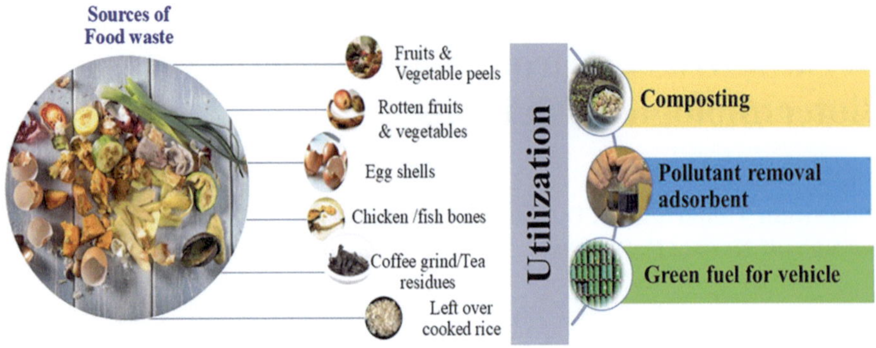

Fig. 1 Waste generated from different food products and their sustainable utilization

lignin, expressing nutrients in a polluted form (Anu et al., 2024). The impact of food waste on environment is eliminated by strategies that turn it into commodities with additional value and offer profitable product pathways. Food waste generated from different sources and their utilization to produce value-added products have depicted in Fig. 1.

Agricultural wastes are non-products of agricultural product generation and processing that may contain materials which are useful to humans but whose economic worth is less that the expenses of gathering, transporting, and processing them for useful purposes. Agricultural waste or agro-waste, includes crop waste such as corn stalks, sugarcane bagasse, drops and culls from fruits and vegetables, and prunings, animal waste including manure, dead animal bodies, food processing waste (Maji et al., 2020). It was estimated that although the amount of agro-waste production is few, it is widely believed that they account for a significant portion of the waste matter produced in the developed world. Agricultural crop wastes, livestock manure, and agro-industrial by-products have all naturally increased in quantity as a result of expanding agricultural production (Singh et al., 2021). Global agricultural waste is probably going to rise significantly if developing countries keep intensifying their farming practices (Yadav et al., 2024). Municipal solid waste is referred as garbage is generally coming from the products we use in our day-to-day life and then discard including packaging, glass clippings, clothing, furniture, bottles, food scarps, newspapers, appliances, paint, and batteries (Kadhila, 2019) (Fig. 2). Different sources such as our homes, hospitals, schools, and business sectors are generating the municipal waste. Organic components (biowaste) including lipids, cellulose, and carbohydrates, are the majority constituents of municipal solid waste (Haas et al., 2015; Vasileiadou, 2024). According to Central Pollution Control Board, the solid municipal waste generated in India on an average of 170,338 tons per day in the year, and it is expected to be rise significantly (Prajapati, 2022). Many nations implemented various strategies into practice and established government agencies specifically to deal with municipal solid waste (Guerrero et al., 2013). It was seen that the unscientific disposal of these wastes has a negative impact on human health and other aspects of the environment. Proper infrastructure, maintenance, and regular upgradations of processes are necessary for any municipal solid waste management operations (Halfawy, 2008). Because of the ongoing, uncontrolled expansion of metropolitan areas, this becomes more costly and complicated. The inability of managing municipal corporations to provide the required standard of public service in metropolitan areas in sometimes attributed to their low financial standing (Haque, 2001). There are four major functions of municipal solid waste management in India includes waste generation, collection, transportation and disposal. These wastes are typically disposed of in low-lying areas without any operational controls or precautions which now becomes the major environmental issues (Brand et al., 2018).

Conventional waste management techniques (Fig. 3), such landfilling and incineration, are becoming more widely acknowledged as unsustainable because of their adverse effects on the environment, which include inefficient resource utilization and pollution of air and water (Reddy Koteswara et al., 2025; Sahu & Das, 2024; Swain et al., 2025). New biotechnological methods, on the other hand, offer long-term waste valorization alternatives (Koutinas et al., 2014).

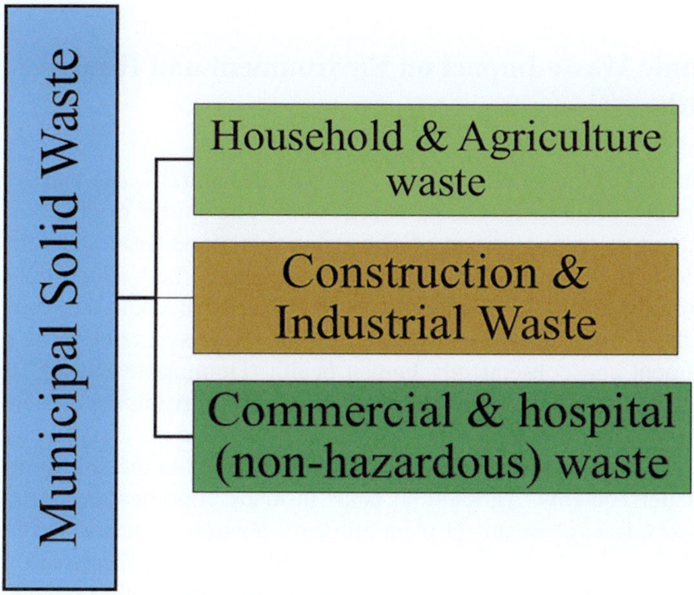

Fig. 2 Municipal solid waste generated through different sources

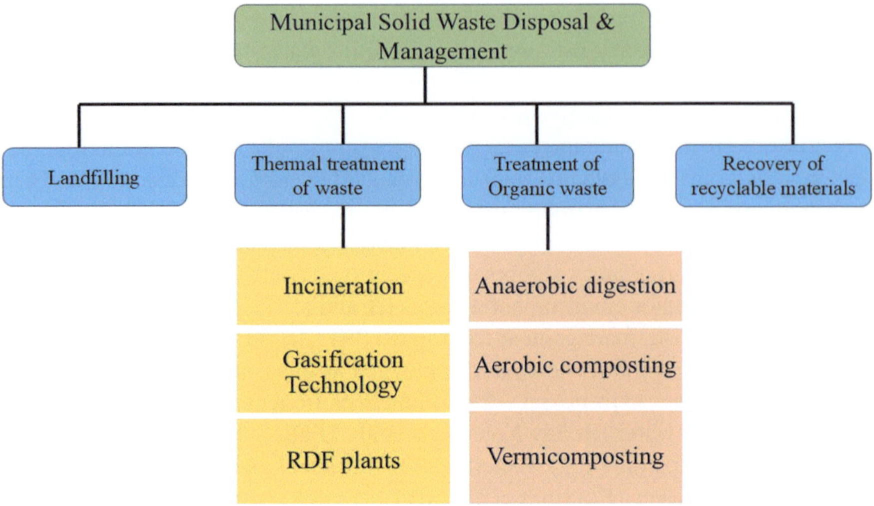

Fig. 3 Disposal and treatment strategies of municipal solid waste

2 Organic Waste Impact on Environment and Human Health

Organic waste or biodegradable waste is a type of natural waste that is generated by either plants or animals (Garcıa et al., 2005). The wastes originate from living beings and comes from a variety of sources, such as food waste, household waste, paper waste, human waste, slaughterhouse waste, sewage, herbal industries, markets, poultry (Ibrahim et al., 2016). Because of the unpleasant smell, coliform contamination, and potential to induce contamination, these wastes represent a major risk to environment and consequently human health (Domingo & Nadal, 2009). The emission of organic waste has increases because of the rising population, industrialization, and utilization of existing resources (Sharma et al., 2019) causing severe pollution of the environment. Organic wastes are added to the soil to increase the organic matter contents. Agricultural lands in many countries are fertilized using farm effluents, sewage treatment plant effluents, abattoir effluents, and food sector effluents (Mittal, 2004). The primary contaminant from the milk processing industry is whey, which causes environmental issues because of its high organic contents and lager volume (Buchanan et al., 2023). Lactose is the main constituent of whey (about 4–5% of its composition). It is the most polluting by-product of the dairy industry due to its high biological oxygen demand and chemical oxygen demand (Ahmad et al., 2019). By using these untreated and potentially contaminated wastes could be extremely harmful to farm animals and the environment. Furthermore, the production of fruit and vegetable crops in soil contaminated with slurry, untreated sewage sludge, or abattoir waste clearly increase the risk of developing Verocytotoxin generating E. coli, particularly when the products are consumed with inadequate processing (García & Heredia, 2017). In America, around 165 billion US dollar food wasted annually; when this garbage is dumped in landfills, it starts decompose and rot releasing methane (Aziz et al., 2020). Methane gas, which affects the earth's temperature and climate, is the primary factor of climate change and global warming (Kumar, 2018). There is a considerable waste of fresh water and a decline in the amount of groundwater as over 70% of the water utilized in agriculture is wasted. The application of organic manure improved the chemical properties of the soil and contributed significant amounts of phosphorus and potassium (Adekiya et al., 2020). On the other hand, adding a lot of manure raised the phosphorus availability in the soil initiating phosphorous leaching by incorporating manure into the soil (Van Es, 2004). Both the soil and plants are also impacted by the toxic elements, like heavy metals, which are produced when sewage sludge and municipal solid waste are added to the soil. According to the US Public Health Service, improper solid waste disposal has been associated to 22 human diseases (Hossain et al., 2011). Numerous reported studies reveals that waste combustion leads to heart attacks, emphysema, and asthma (Balmes, 2019). Municipal waste frequently contains human waste, which when mixed with untreated rotten garbage, attracts other animals and spreads illnesses like malaria and dengue fever (Krystosik et al., 2020). The landfills of organic waste provide a nesting environment to mosquitoes. In addition, workers of organic waste plants in several countries' experiences eye and skin irritation, nausea, vomiting, etc., along with organic dust toxic syndrome, allergic alveolitis, respiratory allergies from inhaled spores, dermal pulmonary and systematic infections, and contact allergies (Gowda et al., 2023; Sultana et al., 2017). Metals like chromium, manganese, and zinc, which were common in soil tainted by organic waste, posed a major hazard to human health (Ames et al., 2013; Oves et al., 2012). Landfills, a popular method of disposing of trash, contribute significantly to environmental contamination. Approximately 20% of the CH4 produced globally originates from these landfills, per one study (Bogner & Spokas, 1993). Other than methane, there are several gases produced due to the practice of landfills. For this reason, the environmental condition is concerning because the majority of landfills are responsible to rise pollution levels, periodic release of leachate and gas into the groundwater (Wang et al., 2024). The improper landfill management results in significant metal contamination of crops, soil, and water with open burning of wastes results in emissions of air pollutants like CO_2, SO_2, etc., into the atmosphere (Salami & Popoola, 2023). Environmental contamination caused by toxic metals is regarded as one of the most dangerous, especially for human health (Das et al., 2024a, 2024b; Dey et al., 2024a, 2024b; Mishra et al., 2018). The majority of research has shown that landfilling not only affect environment but also affect the human health as it has been associated with release of SoC into the atmosphere. The living beings especially the youngsters staying near landfilling areas were highly affected by the lung cancer which is caused by H_2S gas generated from the landfilling activity. It was noted that depending upon the contaminants, health effects can vary from acute intoxication to genotoxicity, mutagenicity, and endocrine-related toxicity (Manibusan & Touart, 2017; Mohanty et al., 2024; Priyadarshini et al., 2024). Certain pollutants dissolve and penetrate into the soil when they come into contact with water, contaminating the underwater level. Ammonium nitrogen, which can lead to eutrophication, which deteriorates the quality of the water (Akhtar et al., 2021). Additionally, some bacteria can produce toxic methyl mercury. In addition to these, sulfur oxides produce acid rain, volatile organic compounds produce dangerous ground-level ozone, particulate organic matter reduces photosynthetic rate and helps form photochemical smog, Nox produces photochemical smog and contributes to acid rain and phytotoxicity, and other gaseous emissions (Kumar & Kundu, 2020). Numerous studies have also revealed that landfills are more likely to experience health problems and have a higher chance of having infants with malformations. Measurements of emissions, soil gas concentrations, and volatile organic compounds provide the foundation for the risk factor assessment of landfilling on human health (Martí et al., 2014; Ray et al., 2024; Sahu et al., 2024).Toxic compounds may accumulate in the food chain through the plants and animals that consume it, posing serious health risks when untreated waste is dumped directly into rivers, lakes, and seas (Babaniyi et al., 2024). Exposure to hazardous waste may have an impact on human health, with children being particularly vulnerable to these contaminants (Sarigiannis, 2017). As chemical poisoning results from release of chemical waste into the environment, it has been proven that immediate exposure can cause illness through chemical exposure. Industrial and agricultural organic wastes have the potential to pose major health hazards. Additionally, disposing of hazardous, industrial, and municipal waste together can expose people to radioactive and chemical risks. Furthermore, unmanaged solid waste can obstruct storm water discharge, resulting in stagnant water bodies that act as disease breeding grounds (Chattopadhyay et al., 2020; Das et al., 2023, 2024a, 2024b). Furthermore, trash near water sources contaminates water bodies or groundwater supplies. Despite being essential to the aquatic environment, nutrients like phosphorous and nitrogen that is coming from fertilizers and manure can cause overfertilization and promote the eutrophication cycle (Khan & Mohammad, 2013), which is the natural aging process of lakes. This acceleration leads to an overabundance of aquatic vegetation, significant overall changes, and a general transformation of the biological community from a low productivity with a wide variety of species to high productivity with a large number of less desirable species (Wilson, 1997).

3 Emerging Techniques Foe Organic Waste Management

Waste management encompasses various methods to remove waste materials from the environment, such as processing, recusing/recycling, removal or degradation, and waste generation control. The primary goal of organic waste management strategies is to turn waste materials into value-added products while moving toward the ideas of the circular economy. These methods include biomass conversion, anaerobic digestion, bioremediation, and composting (Nilsen-Nygaard et al., 2021).

3.1 Anaerobic Digestion

In comparison with other methods, such as thermal, biological, and chemical, anaerobic digestion is a technology that shows potential efficacy for treating organic waste. There are four different biochemical reactions such as hydrolysis, acidogenesis, acetogenesis, and methanogenesis through which the anaerobic digestion can be processed (Uddin & Wright, 2023). Anaerobic digestion of organic waste from landfills releases CH_4 and CO_2, contaminating the environment. However, anaerobic digestion produces value-added products like biofuel and a number of commercial chemicals without oxygen by providing excellent microbial activity and overall process performance under controlled parameters, such as temperature, pH, nutrient content, and retention duration. The methane and hydrogen gas produced by this method are considered to be cleaner potential fuels than fossil fuels.

An additional advantage of this process that it does not depends on fossil fuels for energy consumption. Utilizing food waste can increase the buffering capacity which will provide a stable environment for AD at lower pH levels and speed up the hydrolysis of food waste (Ye et al., 2018). Although AD is a promising method to reduce pollution in the environment with simultaneously producing organic fertilizer and biogas, which are useful resources for farming. However, this approach has certain disadvantages, including extended retention periods and ammonia generation that reduces microbial activity.

3.2 Composting

In the field of waste management, composting is a ray of hope since it provides a long-term solution to the expanding problem of organic waste generation. Several countries especially in India, where environmental concerns are growing and waste management problems are major, composting shows as promising solution. Composting is an aerobic process in which microorganisms' breakdown complex degradable materials and transformed into organic and inorganic by-products. By-products are distinguished from native soil, coals, and peats by the presence of humic-like compounds. Composting is a safe and beneficial way to turn a variety of biodegradable wastes into products that can be used as biofertilizers and soil supplements. Nitrogen molecules undergo a variety of alterations during the composting process, including the biological processes of ammonification, nitrification, and denitrification as well as the physical–chemical processes of immobilization and loss by leaching or volatilization (Bernal et al., 2009). In the ammonification process, proteins and other nitrogen molecules are broken down by enzymes into amino acids, which are then converted into ammonia and ammonium ions. The total nitrogen and nitrate-nitrogen typically rise during composting process, whereas ammonia–nitrogen typically decreases. However, nitrogen losses from ammonia volatilization must be carefully considered, particularly when high ammonia concentrations and high pH levels are present in the composting system. Furthermore, nitrification may be inhibited by extended anaerobic conditions, which could delay compost maturation and increase the ammonium-to-nitrate ratio. Additionally, compared to conventional composting techniques, the employment of earthworms in vermicomposting can enhance nitrogen transformation processes through a particular enzyme system, resulting in higher quantities of absorbed nitrogen in the processed compost. Several forms of phosphorous such as organic P, water soluble P, inorganic P, and citric P are formed during composting process. It was reported in many studies that during composting, the available phosphorous forms may change considerably (Wei et al., 2015). For instance, availability of phosphorous decreases during thermophilic phase of composting and increases again at the compost maturation phase. For this reason, organic waste which is the feedstock for composting plays a crucial role in the changes of several forms of phosphorous. In the composting process, microbial activity is crucial for the transformation of phosphorous, whereas inoculation with particular microbial consortia such as *P. chrysosporium*, *T. viride*, and *P. aeruginosa* accelerated the composting process and increased the nutritional values of compost by increasing the enzymatic activity (Tian et al., 2023). Compared to the landfilling technique of waste disposal, which might be endangered to the ground water pollution, the composting process helps prevent pollution of the underground water. This is because composting reduces the number of chemical contaminants and bacteria. These are the harmful pathogenic microorganisms that are found in waste. Among the different methods of composting, vermicomposting is the composting of degradable organic materials with earthworms (Ali et al., 2015). In this process, earthworms can eat almost any type of organic materials and break it down. The excretory product generated from the earthworms called as castings are rich source of potassium, calcium, magnesium, and phosphorus which are essential source for increasing the fertility of the soil (Osman, 2012). Static composting is a traditional method where composting is carried out by aerobically using passive aeration (Fernandes & Sartaj, 1997). It is a time-consuming process having low operational and capital cost compared to other composting methods. In sheet composting, organic materials including leaves, grass clippings, weeds, garden waste, and vegetative food are thinly applied directly to the soil to create mulch. Instead of rotting in a heap or container, the organic materials are then tilled in using hoe, spade, or garden fork. This process is comparatively simpler and affordable. The Indian Indore composting method uses raw sources, including water, wood ash, dirt, plant waste, animal feces, and urine (Dubey et al., 2025). During the entire composting process, moistening compost is done for three times. This approach requires a lot work and time. Additionally, flies are a problem, along with wind and pest disturbances can cause nutrients to be lost (Stoffolano, 2022). Despite the numerous established advantages of composting that have already been discussed, this waste management technique presents a number of drawbacks, including its impact on climate change, release of CO_2 into the atmosphere, depletion of oxygen, and the generation of unbearable odors due to release of H_2S from anaerobic activity (Boon & Vincent, 2003). Due to these health effects, different organizations in various countries have implemented a number of rules governing to use these methods.

3.3 Bioremediation

Waste management approach "bioremediation" which uses plants, microbes, or their enzymes to breakdown or purify organic wastes in the contaminated areas. It is an environmentally friendly and frequently economical way to clean up the impacted areas by industrial operations, agricultural runoff, or other organic waste sources. Common organic wastes such as chlorinated hydrocarbon are relatively toxic and persistent in the environment (Ghosh & Das, 2020; Ghosh et al., 2016, 2020). Solvents of chlorinated hydrocarbon such as polychlorinated biphenyl, perchloroethylene, trichloroethylene, chloroform, and carbon tetrachloride, are widely known neurotoxins and carcinogens (Bale et al., 2011). In agriculture and forestry, a number of organochlorine pesticides, including dichlorodiphenyltrichloroethane, and hexachlorocyclohexane, and 2,4-dichlorophenoxyacetic acid have been widely used to control pests and fight disease (da Silva Júnior et al., 2024). These organic substances have the potential to cause cancer, birth defects, and reproductive damage, and other health risks. Bioremediation process changed these organic waste materials, and also other toxic molecules and heavy metals into less toxic and non-toxic materials (Dey et al., 2023; Ghosh et al., 2023; Paul et al., 2023). There are numerous benefits of using bioremediation techniques over alternative remediation approaches such as, the bioremediation process eliminates hazardous elements from polluted media instead of directly transferring them to the environment and bioremediation process can make treating hazardous waste sites much less expensive than any conventional methods. Bioremediation processes are associated with some of the limitations. In certain situations, the process has two main drawbacks like microorganisms cannot readily breakdown highly chlorinated contaminants, and microbial remediation of certain pollutants may result in even more hazardous and unstable chemicals than the initial pollutants (Kumari & Das, 2023; Mishra et al., 2023; Mishra & Das, 2022, 2023).

3.4 Biomass Conversion

Biomass conversion is one of the important strategies for managing organic waste which turns organic waste materials into useful resources including fuels electricity, and many more products. This process breaks down organic waste and transforms it into useful products using a variety of technologies, such as thermochemical and biochemical methods. There are several advantages of biomass conversion. It promotes economic growth by lowering dependency on fossil fuels by offering an alternative to conventional energy product techniques, reducing greenhouse gas emissions, and creating job opportunities. A number of biomass conversion processes, such as gasification, fermentation, and pyrolysis, have been developed to turn waste materials into useful products including chemicals, fertilizers, and biofuels. Biomass is a vast and mainly undiscovered resource with significant potential foe energy generation. It is mainly composed of forestry waste, urban organic matter, and agricultural wastes. Several studies on bioenergy are propelling the development of conversion technologies, increasing their effectiveness and economy (Chum et al., 2015). As most widely used renewable energy source in the world today, bioenergy accounts for more than 55% of all renewable energy and more than 6% of global energy consumption (Reid et al., 2020). Achieving the objectives of waste reduction, the circular economy, and a sustainable bioenergy system can be facilitated by the effective use of biomass resources and sustainable organic waste management techniques. However, the immediate and widespread use of waste and biomass residues as a source of bioenergy is hampered by high production costs and environmental concerns.

4 Advanced Biotechnological Methods

Different branches of science and engineering including biochemistry, microbiology, genetics, zoology, botany, physics, chemical engineering, food engineering, etc., provide the foundation of biotechnology. Biotechnology is the procedure which uses living cells or cellular components to generate products, enhance healthcare and agriculture, renew or modify things, or make living cells for particular purposes. In order to produce valuable products, high performance in chemical industries and economic growth require unrestricted energy sources, practical large-scale techniques, and inexpensive feedstocks. In terms of environmental issues, biotechnology finds use in the biological treatment of wastewater and the composting technique for the removal of solid waste. Air pollutants are also treated using biological techniques. Anaerobic treatment, oxidation ponds, activated sludge, trickling filters, and biofilters are biotechnology-based wastewater treatment techniques (Arora et al., 2023). Additionally, biotechnology applications in environmental engineering include biosorption, bio-trickling filters, and solid waste composting methods. Finding the right microbes to break down organic materials and finishing the treatment process under ideal circumstances are crucial for all these techniques.

4.1 Genetic Engineering

Among the several advanced biotechnological approaches, genetic engineering of microbes to transform biomass into a variety of bioproducts and biofuels. By using these methods, greenhouse emissions can be decreased, efforts to locate sustainable raw resources can be increased and valuable food, feed, and biofuel products can be produced. Genetic engineering is a strategy where a recombinant microorganism is used to overproduce an enzyme that can be used to purify organophosphate-containing wastes (Khaliq, 2023). The issues related to the release of genetically engineered organisms into the environment are resolved when waste materials are processed using enzymes instead of microbes. Since enzyme-based treatment methods are more environmentally friendly, they might be the first-time genetic engineering which has been used in situ waste management problems. Advanced biotechnology can improve the efficacy of microalgae and cyanobacterial in removing nutrients and accumulating biomass. These changes are based on mitochondrial, chloroplast, and nuclear genomes. The majority of the enzymes that generate secondary products are encoded by the nuclear genome. Targeting a particular gene alteration through transformation and selection techniques are essential phases in microalgae genetic engineering.

4.2 Bioplastic Products from Organic Waste Conversion

The use of bioplastics is gradually growing and proved to be environmentally friendly substitute for conventional plastic. Bioplastics produced from biomass, which includes naturally occurring polysaccharides like cellulose and starch found in plants, which also processed and combined with additives to accommodate a variety of applications (Abe et al., 2021). Examples of bioplastics made from bioderived monomers are, polyglycolic acid, made from condensation or polymerization of glycolic acid and is also made from biomass fermentation and bio-polyethylene is made by starting with ethylene through biomass fermentation. The microorganisms which polymerize and produce bioplastics are *Azotobacter beijernickii*, *Cupriavidus necator*, and strain of *Bacillus*, *Pseudomonas*, *Rhodococcus*, *Micrococcus*, and *Rhododococcus* (Longo et al., 2024). These microorganisms naturally produce bioplastic in response to the presence of excess carbon and nutrient-limited conditions. One of the most advanced biotechnologies for making bioplastic is microbial fermentation. However, a significant barrier to the commercialization of the process is the high cost of such products. Bioplastic is an appropriate alternative for the petrochemical plastics with several advantages such as non-hazardous, eco-friendly, economic, and biodegradability (Singh et al., 2022). Vegetable, fruit, and plant waste has been effectively processed to separate natural biopolymers such as cellulose, starch, pectin, and lignin. The production of bioplastic and materials for food packaging, household goods, etc., are made by combining various biopolymers. For example, citrus fruits are excellent source of pectin that are used in food processing industry for jelly making, food stabilizer, dietary fiber supplement, fat reduction activity, and as medicine in pharmaceutical industry. It has been established that banana peels work well as a substrate for the production of bioplastic. Banana peels have been successfully used in a no. of studies to produce biodegradable films. Using the alkaline leaching process, starch that has been recovered from leftover banana peels has been used to create bioplastics (Oyedeji et al., 2024). The potential of algae biomass as feedstock for the production of renewable energy. Bio-oil, biofuel, biofertilizer, pigments, animal feed additives, and antibiotics have all been reported to be produce through the processing of algae biomass with a particular emphasis on the production of biofuel from a variety of algae species. For example, bioplastics synthesized from *Spirulina platensis* is a high protein context alga (Zeller et al., 2013).

4.3 Bioreactors for Managing Organic Waste

Among the bioreactor systems of organic waste management, bioreactor landfill is a modern, constructed version of a conventional anaerobic/aerobic landfill in which fresh or wastewater is injected to moisten the waste and the leachate is recirculated to maximize waste degradation process (Salimov et al. 2021). Lechate recirculation keeps the landfilled waste at an ideal moisture content and make it easier for microorganisms and nutrients to cycle into the waste mass. The bioreactor processes to be executed, the cycling of microbes and nutrients is meant to improve microbial processes for transformation and stabilization of readily and moderately degradable organic wastes. Bioreactor reusing the generated leachate to be recirculated back into the reactor helps speed up the biodegradation process. The amount and quality of leachate generated can be enhanced with this recirculation leachate, and methane gas production can production can be maximized for subsequent use as an alternate renewable energy source (Zhang et al., 2023). When the waste deposited contains a significant amount of biodegradable organic wastes, the bioreactor approach may be used. Anaerobic, aerobic, semi-aerobic, and hybrid landfill designs are all possible for bioreactor landfills. Lechate recirculation, landfill gas collection, and air injection system operations, layouts, and arrangements are the primary differences among these bioreactor landfill designs (Zhang et al., 2023).The majority of bioreactor landfills are run in anaerobic environments.

4.4 Enzymatic Hydrolysis

Enzymatic hydrolysis uses enzymes to break down complex organic compounds into simpler, useable components, making it an efficient and sustainable way to manage organic waste (Jordan et al., 2007). In addition to encouraging a circular economy and lowering dependency on landfills and incineration, this technique can recover valuable products from waste streams, such as proteins, fatty acids, and carbohydrates. The enzymatic hydrolysis is considered efficient if the conditions for producing and using enzymes are optimized. This method focuses on different biotechnological routes for utilization of organic waste compounds by emphasizing the necessity of enzymes, incorporating accessory enzymes to enhance yields during the enzymatic hydrolysis of biomass (Nargotra et al., 2024). During processing of food waste, the enzyme hydrolysis followed through pretreatment procedures as food waste has a diverse composition as well as complicated chains. Understanding the chemical structure of food waste materials is essential for using enzymes that are specific to each substrate. Certain hydrolytic enzymes such a lipases, amylases, and proteases, are used in the enzymatic process to catalyze the breakdown of lipids, proteins, and carbohydrates into simpler and more usable components (Bilal & Iqbal, 2020). Protease hydrolyzes lipids into glycerol and free fatty acids, lipases hydrolyze proteins into peptides and amino acids, and amylases catalyze the breakdown of complex carbohydrates into simple sugars (García-Cano et al., 2020). Sugars are particularly interesting because fermentation processes can use them as a platform to produce biproducts. Bioethanol, a sustainable energy source that has drawn a lot of attention due to its role in lowering greenhouse gas emissions (Von Blottnitz & Curran, 2007). Using microbial processes, bioethanol is produced by fermenting food waste that is high in carbohydrates such as fruit feels and starchy leftovers. The resulting bioethanol can help fight climate change worldwide by serving as a clean fuel substitute for energy production (Panahi et al., 2022). By using organic waste as a fuel, biohydrogen production from food waste provides a sustainable method for producing energy. Biodegradable materials are transformed into hydrogen gas using different techniques of enzymatic hydrolysis such as fermentation, photofermentation, or hybrid approaches, which lessen waste disposal problems and advance the development of renewable energy (Zaaba & Jaafar, 2020).

5 Conclusions and Future Prospects

With increasing urbanization, industry, and population expansion, managing organic waste has become a crucial concern. Because of their detrimental effects on the environment, including as methane emissions, groundwater contamination, and public health hazards, conventional techniques like landfilling and open dumping are becoming more widely acknowledged as unsustainable. For the treatment of organic waste, the creation and use of cutting-edge biotechnological techniques present viable, sustainable, and environmentally beneficial options. Technologies including enzymatic degradation, composting, anaerobic digestion, and microbial conversion have greatly improved waste decomposition efficiency and the recovery of important by-products like organic acids, biogas, and biofertilizers. Utilizing naturally existing or genetically modified microbes and enzymes, biotechnological treatments can convert waste into valuable resources. For example, anaerobic digestion contributes to circular bioeconomy models by producing biogas, a renewable energy source, in addition to reducing waste volume. Similarly, by creating nutrient-rich compost, aerobic composting and vermicomposting enhance soil health. Recent innovations such as microbial fuel cells, algal bioconversion, further demonstrate the versatility and potential of biotechnology in waste valorization. The construction of microbial consortia with improved metabolic capacities, which speed up waste breakdown and increase product yields, has also been made possible by the merging of molecular biology, genomics, and synthetic biology. With the help of interdisciplinary research and policy integration, biotechnological instruments will continue to be developed and improved upon in the future of organic waste management. The development of metabolic engineering and synthetic biology is anticipated to produce specially engineered microorganisms that can transform complex waste substrates into valuable bioproducts like medicines, biofuels, bioplastics, and enzymes. Waste treatment systems will be more flexible and effective if strong microbial consortia that can operate in a variety of environmental settings are developed. The combination of digital innovation, biotechnology, and environmental practices provides a revolutionary approach to organic waste management. These new biotechnological technologies have enormous potential to move society toward a circular, zero-waste economy that puts equal emphasis on economic growth and environmental preservation, provided that stakeholders work together and make smart investments.

References

Abe, M. M., Martins, J. R., Sanvezzo, P. B., Macedo, J. V., Branciforti, M. C., Halley, P., et al. (2021). Advantages and disadvantages of bioplastics production from starch and lignocellulosic components. *Polymers, 13*(15), 2484.

Adekiya, A. O., Ejue, W. S., Olayanju, A., Dunsin, O., Aboyeji, C. M., Aremu, C., et al. (2020). Different organic manure sources and NPK fertilizer on soil chemical properties, growth, yield and quality of okra. *Scientific Reports, 10*(1), 16083.

Ahmad, T., Aadil, R. M., Ahmed, H., ur Rahman, U., Soares, B. C., Souza, S. L., et al. (2019). Treatment and utilization of dairy industrial waste: A review. *Trends in Food Science and Technology, 88*, 361–372.

Akhtar, A. B. T., Naseem, S., Yasar, A., & Naseem, Z. (2021). Persistent organic pollutants (POPs): sources, types, impacts, and their remediation. In *Environmental pollution and remediation* (pp. 213–246). Springer.

Ali, U., Sajid, N., Khalid, A., Riaz, L., Rabbani, M. M., Syed, J. H., & Malik, R. N. (2015). A review on vermicomposting of organic wastes. *Environmental Progress and Sustainable Energy, 34*(4), 1050–1062.

Ames, B. N., Magaw, R., & Gold, L. S. (2013). Ranking possible carcinogenic hazards. In *Readings in risk* (pp. 76–100). RFF Press.

Anu, A., Kuhad, R. C., Rapoport, A., Kumar, V., Singh, D., et al. (2024). Biological pretreatment of lignocellulosic biomass: An environment-benign and sustainable approach for conversion of solid waste into value-added products. *Critical Reviews in Environmental Science and Technology, 54*(10), 771–796.

Arora, K., Kumar, V., Nayak, S. J., Surya, A., & Sugumar, S. (2023). Recent Advances in Biological Remediation of Volatile Organic Compounds (VOCs) and heavy metals. *Bio-Nano Filtration in Industrial Effluent Treatment*, 49–68.

Aziz, H. A., Rosli, N. A., & Hung, Y. T. (2020). Landfill methane emissions. In *Handbook of environment and waste management: Acid rain and greenhouse gas pollution control* (pp. 397–454).

Babaniyi, B. R., Olamide, I. G., Fagbamigbe, D. E., Adebomi, J. I., & Areo, I. F. (2024). Environmental pollution and the entrance of toxic elements into the food chain. In *Phytoremediation in food safety* (pp. 109–124). CRC Press.

Bale, A. S., Barone, S., Jr., Scott, C. S., & Cooper, G. S. (2011). A review of potential neurotoxic mechanisms among three chlorinated organic solvents. *Toxicology and Applied Pharmacology, 255*(1), 113–126.

Balmes, J. R. (2019). Household air pollution from domestic combustion of solid fuels and health. *Journal of Allergy and Clinical Immunology, 143*(6), 1979–1987.

Bernal, M. P., Alburquerque, J. A., & Moral, R. (2009). Composting of animal manures and chemical criteria for compost maturity assessment. A review. *Bioresource Technology, 100*(22), 5444–5453.

Bilal, M., & Iqbal, H. M. (2020). State-of-the-art strategies and applied perspectives of enzyme biocatalysis in food sector—current status and future trends. *Critical Reviews in Food Science and Nutrition, 60*(12), 2052–2066.

Bogner, J., & Spokas, K. (1993). Landfill CH4: Rates, fates, and role in global carbon cycle. *Chemosphere, 26*(1–4), 369–386.

Boon, A. G., & Vincent, A. J. (2003). Odour generation and control. *Handbook of Water and Wastewater Microbiology*, 545–557.

Brand, J. H., Spencer, K. L., O'shea, F. T., & Lindsay, J. E. (2018). Potential pollution risks of historic landfills on low-lying coasts and estuaries. *Wiley Interdisciplinary Reviews: Water, 5*(1), Article e1264.

Buchanan, D., Martindale, W., Romeih, E., & Hebishy, E. (2023). Recent advances in whey processing and valorisation: Technological and environmental perspectives. *International Journal of Dairy Technology, 76*(2), 291–312.

Chattopadhyay, P., Chattopadhyay, P., & Palit, D. (2020). Effect of environmental pollution on health and its prevention: an overview. *Environmental and Sustainable Development Through Forestry and Other Resources*, 229–266.

Chum, H. L., Nigro, F., McCormick, R., Beckham, G., Seabra, J., Saddler, J., et al. (2015). Conversion technologies for biofuels and their use. *Bioenergy and Sustainability: Bridging the Gaps, 72*, 374–467.

da Silva Júnior, A. H., De Oliveira, C. R. S., Leal, T. W., Mapossa, A. B., Fiates, J., De Souza, A. A. U., et al. (2024). Organochlorine pesticides remediation techniques: Technological perspective and opportunities. *Journal of Hazardous Materials Letters, 5*, 100098.

Das, A. P., Mohanty, A. K., van Hullebusch, E. D., Figueiredo, G., Das, A. P., Mohanty, A. K., van Hullebusch, E. D., & Figueiredo, G. (2023). Bioremediation as an emerging technology for the removal of synthetic microplastic pollutants from marine ecosystem. *MarPB, 194*, Article 115297. https://doi.org/10.1016/J.MARPOLBUL.2023.115297

Das, A. P., Nayak, B., & Behera, I. D. (2024a). Applications of microorganisms in advanced geomicrobiological and biotechnological exploration. *Geomicrobiology Journal, 41*(4), 309–311. https://doi.org/10.1080/01490451.2024.2327059/ASSET//CMS/ASSET/DE2A1F67-08C1-475A-ABC7-2B3F4C9E8BCF/01490451.2024.2327059.FP.PNG

Das, A., Dey, S., & Das, A. P. (2024b). Ecotoxicological impacts of synthetic microfiber pollutants and development of sustainable mitigation strategies. *Environmental Chemistry and Ecotoxicology*. https://doi.org/10.1016/J.ENCECO.2024.12.003

Das, A., Dey, S., Gacem, A., Yadav, K. K., Nath, H., Ram, D. K., & Das, A. P. (2025). Revolutionizing solid hydrocarbon sourcing: Plastic's journey from waste to upcycled treasure. *Environmental Science and Engineering, Part, F218*, 331–351. https://doi.org/10.1007/978-3-031-81615-4_14

Dey, S., Tripathy, B., Kumar, M. S., & Das, A. P. (2023). Ecotoxicological consequences of manganese mining pollutants and their biological remediation. *Environmental Chemistry and Ecotoxicology*. https://doi.org/10.1016/J.ENCECO.2023.01.001

Dey, S., Bissoyi, A., Koteswara Reddy, G., & Das, A. P. (2024). Synthetic microfibers: An urban emerging pollutant demanding for sustainable environmental solutions. *Sustainable Microbial Technology for Synthetic and Cellulosic Microfiber Bioremediation*, 107–121. https://doi.org/10.1007/978-3-031-62600-5_7

Dey, S., Dash, A., & Das, A. P. (2024). Bioremediation as a sustainable and circular solution for mining waste management and bioenergy generation. *Sustainable Management of Mining Waste and Tailings: A Circular Economy Approach*, 107–119. https://doi.org/10.1201/978 1003442455-6/BIOREMEDIATION-SUSTAINABLE-CIRCULAR-SOLUTION-MINING-WASTE-MANAGEMENT-BIOENERGY-GENERATION-SUDESHNA-DEY-AKANSHYA-DASH-ALOK-PRASAD-DAS

Domingo, J. L., & Nadal, M. (2009). Domestic waste composting facilities: A review of human health risks. *Environment International, 35*(2), 382–389.

Dubey, M., Kumar, S., & Ram, M. (2025). Methods of composting in organic farming. In *Organic production of vegetable crops* (pp. 289–326). Apple Academic Press.

Fernandes, L., & Sartaj, M. (1997). Comparative study of static pile composting using natural, forced and passive aeration methods. *Compost Science and Utilization, 5*(4), 65–77.

García, S., & Heredia, N. (2017). Microbiological safety of fruit and vegetables in the field, during harvest, and packaging: A global issue. In *Global food security and wellness* (pp. 27–48). Springer.

Garcıa, A. J., Esteban, M. B., Marquez, M. C., & Ramos, P. (2005). *Biodegradable municipal solid waste: Characterization and potent*

García-Cano, I., Rocha-Mendoza, D., Kosmerl, E., Zhang, L., & Jiménez-Flores, R. (2020). Technically relevant enzymes and proteins produced by LAB suitable for industrial and biological activity. *Applied Microbiology and Biotechnology, 104*(4), 1401–1422.

Ghosh, S., Mohanty, S., Nayak, S., Sukla, L. B., & Das, A. P. (2016). Molecular identification of indigenous manganese solubilising bacterial biodiversity from manganese mining deposits. *Journal of Basic Microbiology, 56*(3), 254–262. https://doi.org/10.1002/JOBM.201500477

Ghosh, S., & Das, A. P. (2020). Microbial metagenomics: current advances in investigating microbial ecology and population dynamics. In S. K. Nayak & B. B. Mishra (Eds.), *Frontiers in soil and environmental microbiology* (1st ed., pp. 247–254). CRC Press.

Ghosh, S., Gandhi, M., van Hullebusch, E. D., & Das, A. P. (2020). Proteomic insights into Lysinibacillus sp.-mediated biosolubilization of manganese. *Environmental Science and Pollution Research, 28*(30), 40249–40263. https://doi.org/10.1007/S11356-020-10863-4

Ghosh, S., Tripathy, B., Dey, S., & Das, A. P. (2023). Comparative investigation of fungal and bacterial manganese biomining mechanisms. *Manganese Mining Microorganisms*, 99–115. https://doi.org/10.1016/b978-0-12-822147-1.00016-9

Girotto, F., Alibardi, L., & Cossu, R. (2015). Food waste generation and industrial uses: A review. *Waste Management, 45*, 32–41.

Gowda, B., Gurusiddappa, L. H., & Kalikeri, S. (2023). Study on occupational health hazards of municipal solid waste workers-a review. *World Journal of Environmental Biosciences, 12*(1–2023), 24–31.

Guerrero, L. A., Maas, G., & Hogland, W. (2013). Solid waste management challenges for cities in developing countries. *Waste Management, 33*(1), 220–232.

Haas, W., Krausmann, F., Wiedenhofer, D., & Heinz, M. (2015). How circular is the global economy? An assessment of material flows, waste production, and recycling in the European Union and the world in 2005. *Journal of Industrial Ecology, 19*(5), 765–777.

Halfawy, M. R. (2008). Integration of municipal infrastructure asset management processes: Challenges and solutions. *Journal of Computing in Civil Engineering, 22*(3), 216–229.

Haque, M. S. (2001). The diminishing publicness of public service under the current mode of governance. *Public Administration Review, 61*(1), 65–82.

Hossain, M. S., Santhanam, A., Norulaini, N. N., & Omar, A. M. (2011). Clinical solid waste management practices and its impact on human health and environment–a review. *Waste Management, 31*(4), 754–766.

Ibrahim, M. H., Quaik, S., & Ismail, S. A. (2016). Introduction to organic wastes and its management. In *Prospects of organic waste management and the significance of earthworms* (pp. 1–21). Springer.

Jordan, S. N., & Mullen, G. J. (2007). Enzymatic hydrolysis of organic waste materials in a solid–liquid system. *Waste Management, 27*(12), 1820–1828.

Kadhila, T. (2019). Implementation of a municipal solid waste management system in Swakopmund, Namibia (Doctoral dissertation, Stellenbosch: Stellenbosch University).

Khaliq, N. (2023). Microbial enzymes as a robust process to mitigate pollutants of environmental concern. In *Microbial biomolecules* (pp. 241–267). Academic.

Khan, M. N., & Mohammad, F. (2013). Eutrophication: Challenges and solutions. *Eutrophication: Causes, consequences and control* (Vol. 2, pp. 1–15). Springer, Netherlands.

Koutinas, A. A., Vlysidis, A., Pleissner, D., Kopsahelis, N., Garcia, I. L., Kookos, I. K., et al. (2014). Valorization of industrial waste and by-product streams via fermentation for the production of chemicals and biopolymers. *Chemical Society Reviews, 43*(8), 2587–2627.

Krystosik, A., Njoroge, G., Odhiambo, L., Forsyth, J. E., Mutuku, F., & LaBeaud, A. D. (2020). Solid wastes provide breeding sites, burrows, and food for biological disease vectors, and urban zoonotic reservoirs: A call to action for solutions-based research. *Frontiers in Public Health, 7*, 405.

Kumar, R., & Kundu, S. (2020). Microbial bioremediation and biodegradation of hydrocarbons, heavy metals, and radioactive wastes in solids and wastewaters. In *Microbial bioremediation and biodegradation* (pp. 95–112). Springer.

Kumar, A. (2018). Global warming, climate change and greenhouse gas mitigation. In *Biofuels: Greenhouse gas mitigation and global warming: Next generation biofuels and role of biotechnology* (pp. 1–16). Springer.

Kumari, S., & Das, S. (2023). Bacterial enzymatic degradation of recalcitrant organic pollutants: Catabolic pathways and genetic regulations. *Environmental Science and Pollution Research, 30*(33), 79676–79705.

Lin, C. S. K., Pfaltzgraff, L. A., Herrero-Davila, L., Mubofu, E. B., Abderrahim, S., Clark, J. H., et al. (2013). Food waste as a valuable resource for the production of chemicals, materials and fuels. Current situation and global perspective. *Energy and Environmental Science, 6*(2), 426–464.

Longo, A., Fanelli, F., Villano, M., Montemurro, M., & Rizzello, C. G. (2024). Bioplastic production from agri-food waste through the use of haloferax mediterranei: A comprehensive initial overview. *Microorganisms, 12*(6), 1038.

Maji, S., Dwivedi, D. H., Singh, N., Kishor, S., & Gond, M. (2020). Agricultural waste: Its impact on environment and management approaches. In *Emerging eco-friendly green technologies for wastewater treatment* (pp. 329–351). Springer.

Manibusan, M. K., & Touart, L. W. (2017). A comprehensive review of regulatory test methods for endocrine adverse health effects. *Critical Reviews in Toxicology, 47*(6), 440–488.

Martí, V., Jubany, I., Pérez, C., Rubio, X., De Pablo, J., & Giménez, J. (2014). Human health risk assessment of a landfill based on volatile organic compounds emission, immission and soil gas concentration measurements. *Applied Geochemistry, 49*, 218–224.

Mashifana, T. (2025). 2 Waste management principles and their. municipal solid, agricultural, and mining waste in Sub-Saharan Africa: Sustainable and circular valorization opportunities, 16.

Mishra, S., & Das, A. P. (2022). Treatment of the wastewater polluted with synthetic microfiber released from washing machine. In B. B. Das, H. Hettiarachchi, P. K. Sahu, & S. Nanda (Eds.), *Lecture notes in civil engineering* (Vol. 207, pp. 109–117). Springer Science and Business Media Deutschland GmbH. https://doi.org/10.1007/978-981-16-7509-6_9/COVER

Mishra, S., & Das, A. P. (2023). Microbial remediation of synthetic microfiber contaminated wastewater. In M. P. Shah (Ed.), *Microbial technologies in industrial wastewater treatment* (pp. 337–350). Springer. https://doi.org/10.1007/978-981-99-2435-6_17

Mishra, S., Bharagava, R. N., More, N., Yadav, A., Zainith, S., Mani, S., & Chowdhary, P. (2018). eavy metal contamination: an alarming threat to environment and human health. In *Environmental biotechnology: For sustainable future* (pp. 103–125). Springer Singapore.

Mishra, S., Dash, D., & Das, A. P. (2023). Aquatic microbial diversity on plastisphere: colonization and potential role in microplastic biodegradation. *Geomicrobiology Journal*, 1–12. https://doi.org/10.1080/01490451.2023.2209750

Mittal, G. S. (2004). Characterization of the effluent wastewater from abattoirs for land application. *Food Reviews International, 20*(3), 229–256.

Mohanty, J., Dey, S., Bala, K., & Das, A. P. (2024). Recent advances in microplastic pollution for its sustainable management. *Sustainable Management of Urban Plastic Waste Through Circular Economic Approaches*, 27–48. https://doi.org/10.1201/9781003465447-3

Nargotra, P., Ortizo, R. G. G., Wang, J. X., Tsai, M. L., Dong, C. D., Sun, P. P., et al. (2024). Enzymes in the bioconversion of food waste into valuable bioproducts: a circular economy perspective. *Systems Microbiology and Biomanufacturing, 4*(3), 850–868.

Nilsen-Nygaard, J., Fernández, E. N., Radusin, T., Rotabakk, B. T., Sarfraz, J., Sharmin, N., et al. (2021). Current status of biobased and biodegradable food packaging materials: Impact on food quality and effect of innovative processing technologies. *Comprehensive Reviews in Food Science and Food Safety, 20*(2), 1333–1380.

Noor, T., Javid, A., Hussain, A., Bukhari, S. M., Ali, W., Akmal, M., & Hussain, S. M. (2020). Types, sources and management of urban wastes. In *Urban ecology* (pp. 239–263). Elsevier.

Osman, K. T. (2012). Plant nutrients and soil fertility management. *Soils: Principles, properties and management* (pp. 129–159). Springer, Netherlands.

Oves, M., Khan, M. S., Zaidi, A., & Ahmad, E. (2012). Soil contamination, nutritive value, and human health risk assessment of heavy metals: an overview. Toxicity of heavy metals to legumes and bioremediation, 1–27.

Oyedeji, S., Patel, N., Krishnamurthy, R., & Fatoba, P. O. (2024). *Agricultural wastes to value-added products: Economic and environmental perspectives for waste conversion.*

Panahi, H. K. S., Dehhaghi, M., Guillemin, G. J., Gupta, V. K., Lam, S. S., Aghbashlo, M., & Tabatabaei, M. (2022). Bioethanol production from food wastes rich in carbohydrates. *Current Opinion in Food Science, 43*, 71–81.

Paul, A., Dey, S., Ram, D. K., & Das, A. P. (2023). Hexavalent chromium pollution and its sustainable management through bioremediation. *Geomicrobiology Journal*, 1–11. https://doi.org/10.1080/01490451.2023.2218377

Peng, X., Jiang, Y., Chen, Z., Osman, A. I., Farghali, M., Rooney, D. W., & Yap, P. S. (2023). Recycling municipal, agricultural and industrial waste into energy, fertilizers, food and construction materials, and economic feasibility: A review. *Environmental Chemistry Letters, 21*(2), 765–801.

Prajapati, U. B. (2022). Socio-economic perspective of river health: A case study of river Ami, Uttar Pradesh, India. In *Ecological significance of river ecosystems* (pp. 167–186). Elsevier.

Priyadarshini, S., Jagatee, S., & Das, A. P. (2024). Synthetic fabrics and microfiber pollution–an assessment of their global impact. *Environmental Science and Engineering, Part, F2333*, 137–157. https://doi.org/10.1007/978-3-031-51792-1_8

Ray, S., Dey, S., Digal, A. K., Yadav, K. K., & Das, A. P. (2025). Sanitary waste and microplastic pollution in the Euro-Mediterranean region: Challenges and solutions. *Euro-Mediterranean Journal for Environmental Integration, 2025*, 1–23. https://doi.org/10.1007/S41207-025-008 79-Y

Ray, S. K., Dey, S., Das, A. P., & John, A. (2024). Microplastic waste management through greener technologies: A review. *Sustainable Management of Urban Plastic Waste Through Circular Economic Approaches*, 237–261. https://doi.org/10.1201/9781003465447-14

Reddy Koteswara, G., Bhargavi, S., Yamuna, D., Anjani Devi, M., Jeevana Lakshmi Kumar, G., Manas, T., Adarsh, S., Bhavya Sai, V., Nath, H., Alreshidi, M. A., Togun, H., Ram, D. K., & Das, A. P. (2025). Recent advancements in the remediation of marine hydrocarbon and plastic pollutants. *Environmental Science and Engineering*, *Part F218*, 127–142. https://doi.org/10.1007/978-3-031-81615-4_7

Reid, W. V., Ali, M. K., & Field, C. B. (2020). The future of bioenergy. *Global Change Biology, 26*(1), 274–286.

Sahu, S., & Das, A. P. (2024). Microplastics degradation mechanism by aquatic plastiomes: A review. *Environmental Science and Engineering, Part, F2827*, 213–226. https://doi.org/10.1007/978-3-031-50840-0_12

Sahu, A., Dey, S., Mallick, K., Priyadarsini, S., Behera, I. D., & Das, A. P. (2024). A review on advanced technology for sustainable management of synthetic microplastic waste. *Sustainable Management of Urban Plastic Waste Through Circular Economic Approaches*, 262–279. https://doi.org/10.1201/9781003465447-15

Salami, L., & Popoola, L. T. (2023). A comprehensive review of atmospheric air pollutants ssessment around landfill sites. *Air, Soil and Water Research, 16*, 11786221221145380.

Salimov, O. U., Imomov, S. J., Shodiyev, E. B., Azimov, Z. X., Ergashov, Z. J., Juraev, T. K., & Sabirov, K. N. (2021). Physical-mechanical properties of organic waste reduced to bioreactor.

In *IOP conference series: Earth and environmental science* (Vol. 868, No. 1, p. 012088). IOP Publishing.

Sarigiannis, D. A. (2017). Assessing the impact of hazardous waste on children's health: The exposome paradigm. *Environmental Research, 158*, 531–541.

Sharma, B., Vaish, B., Monika, Singh, U. K., Singh, P., & Singh, R. P. (2019). Recycling of organic wastes in agriculture: an environmental perspective. *International Journal of Environmental Research, 13*(2), 409–429.

Singh, R., Das, R., Sangwan, S., Rohatgi, B., Khanam, R., Peera, S. P. G., et al. (2021). Utilisation of agro-industrial waste for sustainable green production: A review. *Environmental Sustainability, 4*(4), 619–636.

Singh, S. S., Thakur, A., Sandilya, S., & Kumar, A. (2022). Recent advances in bioplastics: Synthesis and emerging perspective. *The Iranian Journal of Chemistry and Chemical Engineering Research Article, 41*(8).

Stoffolano, Jr, J. G. (2022). Synanthropic flies—a review including how they obtain nutrients, along with pathogens, store them in the crop and mechanisms of transmission. *Insects, 13*(9), 776.

Sultana, M. S., Rana, S., Yamazaki, S., Aono, T., & Yoshida, S. (2017). Health risk assessment for carcinogenic and non-carcinogenic heavy metal exposures from vegetables and fruits of Bangladesh. *Cogent Environmental Science, 3*(1), 1291107.

Swain, S., Purnima Sudha, M., Dey, S., Badu, Y., Bhoi, S. R., Alreshidi, M. A., Qasim, M. T., Nath, H., Ram, D. K., & Das, A. P. (2025). Sources of micro and nano plastic hydrocarbon pollutants in marine bodies and their treatment. *Environmental Science and Engineering, Part, F218*, 45–67. https://doi.org/10.1007/978-3-031-81615-4_3

Tian, X., Gao, R., Li, Y., Liu, Y., Zhang, X., Pan, J., et al. (2023). Enhancing nitrogen conversion and microbial dynamics in swine manure composting process through inoculation with a microbial consortium. *Journal of Cleaner Production, 423*, 138819.

Torres-León, C., Ramírez-Guzman, N., Londoño-Hernandez, L., Martinez-Medina, G. A., Díaz-Herrera, R., Navarro-Macias, V., et al. (2018). Food waste and byproducts: An opportunity to minimize malnutrition and hunger in developing countries. *Frontiers in Sustainable Food Systems, 2*, 52.

Uddin, M. M., & Wright, M. M. (2023). Anaerobic digestion fundamentals, challenges, and technological advances. *Physical Sciences Reviews, 8*(9), 2819–2837.

Van Es, H. M., Schindelbeck, R. R., & Jokela, W. E. (2004). Effect of manure application timing, crop, and soil type on phosphorus leaching. *Journal of Environmental Quality, 33*(3), 1070–1080.

Vasileiadou, A. (2024). From organic wastes to bioenergy, biofuels, and value-added products for urban sustainability and circular economy: A review. *Urban Science, 8*(3), 121.

Von Blottnitz, H., & Curran, M. A. (2007). A review of assessments conducted on bio-ethanol as a transportation fuel from a net energy, greenhouse gas, and environmental life cycle perspective. *Journal of Cleaner Production, 15*(7), 607–619.

Wang, Y., Fang, M., Lou, Z., He, H., Guo, Y., Pi, X., et al. (2024). Methane emissions from landfills differentially underestimated worldwide. *Nature Sustainability, 7*(4), 496–507.

Wani, N. R., Rather, R. A., Farooq, A., Padder, S. A., Baba, T. R., Sharma, S., et al. (2024). New insights in food security and environmental sustainability through waste food management. *Environmental Science and Pollution Research, 31*(12), 17835–17857.

Wei, Y., Zhao, Y., Xi, B., Wei, Z., Li, X., & Cao, Z. (2015). Changes in phosphorus fractions during organic wastes composting from different sources. *Bioresource Technology, 189*, 349–356.

Wilson, D. S. (1997). Biological communities as functionally organized units. *Ecology, 78*(7), 2018–2024.

Yadav, C., Yadav, P., Joshi, A., Meena, M., Harish, & Arora, J. (2024). Agricultural waste and its impact on the environment. In *Transforming agriculture residues for sustainable development: From waste to wealth* (pp. 3–19). Springer Nature Switzerland.

Ye, M., Liu, J., Ma, C., Li, Y. Y., Zou, L., Qian, G., & Xu, Z. P. (2018). Improving the stability and efficiency of anaerobic digestion of food waste using additives: A critical review. *Journal of Cleaner Production, 192*, 316–326.

Zaaba, N. F., & Jaafar, M. (2020). A review on degradation mechanisms of polylactic acid: Hydrolytic, photodegradative, microbial, and enzymatic degradation. *Polymer Engineering and Science, 60*(9), 2061–2075.

Zeller, M. A., Hunt, R., Jones, A., & Sharma, S. (2013). Bioplastics and their thermoplastic blends from Spirulina and Chlorella microalgae. *Journal of Applied Polymer Science, 130*(5), 3263–3275.

Zhang, G., Liu, K., Lv, L., Gao, W., Li, W., Ren, Z., et al. (2023). Enhanced landfill process based on leachate recirculation and micro-aeration: A comprehensive technical, environmental, and economic assessment. *Science of The Total Environment, 857*, 159535.

Overview of Microbial Waste Management of Organic Waste Through Industrial Biotechnology

Nisha, Ishfaq Ahmad Baba, Hadiya Amin Kantroo, Zahoor Ahmad, Himanshu Jangid, Gaurav Kumar, Alok Prasad Das, and Anuprita Ray

1 Introduction

The high urbanization, industrialization, and global population increase have contributed immensely to the generation of huge quantities of organic waste. Organic wastes like food waste, crop residues, municipal solid waste (MSW), animal manure, and agro-industrial by-products make up more than 50% of world solid waste (Hajam et al., 2023). According to the United Nations Environment Program Food Waste Index Report 2021, nearly 931 million tons of food waste was generated globally in the year 2019 alone (Environment, 2024). These quantities of organic waste, if not processed, create severe environmental, economic, and health issues. Anaerobic breakdown of organic wastes in landfills leads to the generation of methane (CH_4), 25 times more powerful than carbon dioxide (CO_2), and these contribute immensely to global warming and climate change. In addition, leachate generated from wastes

---

The original version of the chapter has been revised. A correction to this chapter can be found at https://doi.org/10.1007/978-3-032-11377-1_21

Nisha · H. Jangid · A. Ray (✉)
School of Bioengineering and Biosciences, Lovely Professional University, Jalandhar, Punjab, India
e-mail: anuprita.ray88@gmail.com

I. A. Baba · H. A. Kantroo · Z. Ahmad
Clinical Microbiology PK/PD Division, CSIR - Indian Institute of Integrative Medicine, Srinagar 190005, India

Academy of Scientific and Innovative Research (AcSIR), Ghaziabad, India

G. Kumar
Amity Institute of Microbial Technology, Amity University Rajasthan, Jaipur, India

A. P. Das
Department of Life Sciences, Rama Devi Women's University, Bhubaneswar, Odisha, India

S. Ghosh et al. (eds.), *Advances in Organic Waste Conversion Through Industrial Biotechnology*, Sustainable Environmental Waste Management Strategies,
https://doi.org/10.1007/978-3-032-11377-1_14

during decomposition can pollute soil and groundwater, while unhygienic accumulation acts as breeding sites for vectors and pathogens, aggravating public health risks (Behera & Das, 2025; Das et al., 2025; Kumar et al., 2024; Priyadarshini et al., 2024a, 2024b; Sahu et al., 2024a).

Traditional methods of organic waste disposal such as incineration and landfilling are increasingly viewed as unsustainable. These are energy-intensive, land-hungry, and capable of secondary pollution. There is thus global pressure for low-cost, environmentally friendly, and sustainable practices of organic waste management (Abubakar et al., 2022). In response to these concerns, waste valorization—transferring waste to value-added products—has become a concept, particularly in the planning of a circular bioeconomy. The basis of this philosophy is the promise of industrial biotechnology, which employs microbial systems and enzyme-based processes for the bioconversion of organic waste to value-added bioproducts such as biofertilizers, bioplastics, biofuels, enzymes, organic acids, and biogas (Pal et al., 2024).

Microorganisms constitute the center of degradation and transformation of organic matter processes. Microorganisms such as bacteria, fungi, actinomycetes, and archaea degrade complex polymers like cellulose, hemicellulose, proteins, and lipids. Different microbial processes such as aerobic composting, anaerobic digestion, solid-state fermentation, and submerged fermentation are widely applied for the bioconversion of organic residues to bioenergy and biobased products (Das & Ghosh, 2023c, 2023e; Duan et al., 2021; Nag et al., 2023; Satapathy et al., 2024). For instance, anaerobic digestion of food and agricultural residues by methanogenic consortia generates biogas, a gas mixture of methane and carbon dioxide, as a renewable energy source (Chen et al., 2021). Similarly, aerobic composting, which is fungus and thermophilic bacteria-mediated, degrades organic matter into nutrient-rich compost, enhancing soil fertility and health (Duan et al., 2022).

Biotechnology in the industrial sector leverages these microbial processes on an industrial scale through the optimization of operational conditions and the application of advanced technologies such as metabolic engineering, synthetic biology, and bioreactor design. Genetically engineered microbial strains have also been developed to enhance substrate specificity, stress resistance, and yield of target bioproducts (Ezeako et al., 2025). For example, certain engineered Escherichia coli and Saccharomyces cerevisiae strains have been designed to produce high levels of bioethanol and organic acids from lignocellulosic waste (Panahi et al., 2022). Application of enzyme technology, e.g., cellulases and amylases of microbial origin, also facilitates the hydrolysis of complex polymers into fermentable sugars, which can be converted to industrial chemicals and biofuels (Mokrani et al., 2024). The chapter provides an exhaustive discussion of the trend, microbial mechanisms, and biotechnological applications of the microbial management of organic waste. It includes the major microbial systems, industrial processes, and value-added products from waste, case studies, technology innovations, and challenges.

2 Microbial Decomposition of Organic Waste

Microbial decomposition is the foundation for organic waste management, based on microorganisms' natural capacity for degrading complex organic material into simpler, stable end products. It is a biological process essential for biogeochemical carbon and nutrient cycling, and key to environmentally friendly waste transformation processes. Organic wastes—e.g., crop husks, food residues, sewage sludge, and animal manures—are rich in carbohydrates, proteins, fats, lignin, and cellulose. These macromolecules act as carbon and energy substrates for a vast range of microorganisms, consisting of bacteria, fungi, actinomycetes, and archaea. Hydrolysis, acidogenesis, acetogenesis, and methanogenesis are the usual stages of breaking down organic matter with the help of microbe shown in Table 1.

The first step in digestion is hydrolysis, turning things like carbohydrates, proteins, and lipids into sugars, amino acids, and fatty acids that solution can dissolve. The changes in the cell happen because of hydrolytic enzymes present outside the cells. These enzymes work on lipids, carbohydrates and proteins; examples include esterases which split ester bonds, glycosidases for carbohydrate bonds and peptidases for cutting links in proteins (Das & Ghosh, 2023d;; Das et al., 2024a, 2024b, 2024c, 2024d; Mohanty et al., 2024a, 2024b; Zhang et al., 2007). The particular bacteria known as hydrolytic bacteria make these crucial enzymes to help start decomposition. Several bacteria commonly recognized as hydrolytic are *Clostridium, Proteus vulgaris, Peptococcus, Bacteroides, Bacillus, Vibrio, Acetivibrio cellulolyticus, Staphylococcus,* and *Micrococcus* (Amani et al., 2010). Some examples are *Cellulomonas* species, which have cellulases to break down polysaccharides into single sugars called monosaccharides. For instance, Bacillus can produce proteases that change proteins into amino acids, and *Mycobacterium* can produce enzymes called lipases that break down lipids into fatty acids. A lot of the time, experts regard hydrolysis as a key step because breaking down large organic molecules by bacteria through hydrolysis is relatively slow compared to the other processes (Westerholm et al., 2019).

After hydrolysis, fermentative microorganisms use the simple soluble compounds created in phase one and make new substances like volatile fatty acids (like propionate and butyrate), lactate, alcohols, and carbon dioxide gas (CO_2) (Das et al., 2024a; Mohanty et al., 2024a, 2024b; Sahu et al., 2024b). Different microbial taxa help drive specific types of fermentations. Saccharomyces organisms carry out the process of alcohol fermentation, whereas *Butyribacterium* and *Clostridium* mainly make butyrate. *Lactobacillus* and *Streptococcus* facilitate the synthesis of lactic acid, while *Clostridium* is also involved in the formation of propionate. Acetate, which plays a major part in microbial degradation, is produced by bacterial species such as *Acetobacterium, Clostridium, Eubacterium, and Sporomusa* (Amani et al., 2010; Harirchi et al., 2022; Li et al., 2011).

Acetogenesis is the third phase of microbial degradation and it relies on acetogenic bacteria to transform volatile fatty acids (VFAs) and alcohols into acetate, hydrogen (H_2), and carbon dioxide (CO_2). Acetogen is the name for several species of bacteria.

Table 1 Degradation of microorganisms involved in organic waste management

Degradation stage	Microorganisms	Role/function	Substrates/Products	References
Hydrolysis	*Bacillus paralicheniformis, Bacillus velezensis, Bacillus subtilis, Bacillus amyloliquefaciens, Penicillium spp.*	Secrete hydrolases (cellulases, proteases, lipases) to break down macromolecules	Carbohydrates → sugars, proteins → amino acids, lipids → fatty acids	Jodhani et al., (2024), Mironov et al. (2024a)
Acidogenesis	*Clostridium spp., Enterobacter spp., Escherichia coli, Lactobacillus spp., Streptococcus spp.*	Convert monomers to volatile fatty acids (VFAs), alcohols, CO_2, and H_2 via fermentation	Sugars, amino acids → VFAs, ethanol, lactate, H_2, CO_2	Jodhani et al., (2024), Rózsenberszki et al., (2025)
Acetogenesis	*Syntrophomonas wolfei, Syntrophobacter spp., Pelotomaculum spp., Clostridium aceticum*	Convert VFAs and alcohols to acetate, CO_2, and H_2	Butyrate, propionate → Acetate, H_2, CO_2	Amin et al., (2021b)
Methanogenesis	*Methanosaeta concilii, Methanothrix soehngenii, Methanobacterium spp., Methanobrevibacter spp., Methanosarcina spp.*	Produce methane from acetate and/or $CO_2 + H_2$	Acetate or $CO_2 + H_2 \rightarrow CH_4 + CO_2$	da Motta et al. (2022), Pei et al., (2024)
Composting (Aerobic)	*Bacillus spp., Penicillium spp., Actinomycetes, Pseudomonas spp.*	Degrade organic matter aerobically; involved in humus formation	Organic matter → CO_2, water, stabilized humus	Mironov et al. (2024)

As another example, *Syntrophobacter wolinii* transforms butyrate into acetate while *Smithella propionica* uses propionate to do the same. More examples of acetogenic bacteria are *Syntrophobacter fumaroxidans, Syntrophomonas wolfei, Pelotomaculum thermopropionicum,* and *Pelotomaculum schinkii,* all of which help to break down VFAs into formate, H_2 and CO_2. *Clostridium aceticum* is also known for being able to directly generate acetate by using hydrogen and carbon dioxide (Harirchi et al., 2022; Wainaina et al., 2019).

Methanogenesis is the final stage of the organic matter decomposition process in which methane is produced mainly from acetate and hydrogen gas. Methanogenic archaea play a key role in this step since they are using the terminal products of acetogenesis. Hydrogen accumulation could repress the activity of acetate-forming bacteria, and thus the role of methanogens becomes decisive for process equilibrium. Methanogens are classified according to their metabolic types: (i) acetoclastic methanogens, e.g., *Methanothrix soehngenii* and *Methanosaeta concilii,* that oxidize acetate to methane and carbon dioxide; (ii) hydrogenotrophic methanogens, i.e., *Methanobacterium bryantii, Methanobacterium thermoautotrophicum,* and *Methanobrevibacter arboriphilus,* that reduce carbon dioxide with hydrogen to form methane; and (iii) formate-using methanogens, i.e., *Methanobacterium formicicum, Methanobrevibacter smithii,* and *Methanococcus voltae,* which produce methane from formate, hydrogen, and carbon dioxide(Agrawal et al., 2021; Amin et al., 2021a).

3 Biotechnological Approaches in Waste Management

The global surge in waste generation, driven by population growth, urbanization, and industrialization, poses significant environmental and public health challenges. In 2020, around 2.24 billion tons of municipal solid waste were generated globally, and this figure is projected to rise to 3.88 billion tons by 2050 (Solid Waste Management, n.d.). Traditional waste management practices, such as landfilling and incineration, contribute to greenhouse gas emissions, soil contamination, and air pollution (Mishra et al., 2024b; Ray et al., 2024; Tripathy et al., 2024). Biotechnological approaches offer sustainable alternatives by harnessing biological systems such as microorganisms, enzymes, and bioprocesses to degrade, transform, or valorize waste (Ezeorba et al., 2024). These methods align with the principles of a circular economy, converting waste into valuable resources like biofuels, bioproducts, and fertilizers. Three key biotechnological strategies for waste management include enzyme-based degradation, fermentation and composting technologies, and anaerobic digestion with biogas production (Dey & Das, 2024a, 2024b, 2024c; Priyadarshini et al., 2024a). Each approach is examined for its mechanisms, applications, advancements, challenges, and future potential, supported by recent research.

3.1 Enzyme-Based Degradation

Enzyme-based degradation employs specific enzymes to break down complex waste materials into simpler, biodegradable components (Kaushal et al., 2021). Enzymes such as cellulases, lipases, proteases, laccases, and cutinases target diverse substrates, including plastics, lignocellulosic biomass, and organic pollutants. This approach is particularly promising for managing recalcitrant wastes that resist conventional treatments.

Mechanisms and Applications

- **Plastic Degradation**: The discovery of *Ideonella sakaiensis*, a bacterium that produces PETase and MHETase enzymes, marked a breakthrough in plastic waste management. PETase hydrolyzes polyethylene terephthalate (PET) into mono(2-hydroxyethyl) terephthalic acid, which MHETase further degrades into recyclable monomers (Yoshida et al., 2016). Engineered PETase variants with enhanced thermal stability and catalytic efficiency can degrade PET films in days, offering a scalable solution for plastic recycling (Das et al., 2024a, 2024b, 2024c, 2024d; Dey et al., 2023; Sahu & Das, 2024; Tournier et al., 2020).
- **Lignocellulosic Waste**: Agricultural residues, such as corn stover and wheat straw, are rich in cellulose and hemicellulose. Cellulases (endoglucanases, exoglucanases, and β-glucosidases) and hemicellulases degrade these polymers into fermentable sugars, which serve as feedstocks for bioethanol or bioproduct synthesis (Saini et al., 2015). Fungal enzymes from *Trichoderma reesei* are widely used due to their high activity (Kansoh et al., 1999).
- **Pollutant Remediation**: Laccases from white-rot fungi, such as *Pleurotus ostreatus*, oxidize phenolic compounds, dyes, and polycyclic aromatic hydrocarbons in industrial wastewater. These enzymes are eco-friendly alternatives to chemical treatments, reducing secondary pollution (Mayolo-Deloisa et al., 2020). For example, laccase-mediated degradation of textile dyes achieves over 90% decolorization in optimized conditions.

Recent advances in enzyme engineering, including directed evolution and computational protein design, have improved enzyme stability and substrate specificity. A thermostable cutinase from *Thermobifida fusca* degrades PET at 70 °C, enhancing process efficiency (Sabari et al., 2025). Immobilization techniques, such as enzyme encapsulation in metal–organic frameworks or silica nanoparticles, increase enzyme reusability and resistance to environmental stressors (Zdarta et al., 2018). However, challenges persist, including high production costs, limited enzyme stability under industrial conditions (e.g., extreme pH or temperature), and difficulties in scaling up enzymatic processes. Additionally, the heterogeneity of waste streams requires enzyme cocktails tailored to multiple substrates, increasing complexity. Future research should focus on developing cost-effective enzyme production systems using recombinant microorganisms, such as *Escherichia coli* or *Pichia pastoris*. Integrating enzymatic degradation with microbial bioprocesses, such as fermentation, could enable sequential waste valorization. For instance, enzymatic hydrolysis of lignocellulosic waste followed by microbial fermentation could produce biofuels and biochemicals. Exploring novel enzymes from extremophiles, which thrive in harsh environments, may also yield robust biocatalysts for industrial applications.

3.2 *Fermentation and Composting Technologies*

Fermentation and composting are microbial-driven processes that transform organic waste into valuable products, such as biofuels, organic acids, and soil amendments. These technologies are widely applied to food waste, agricultural residues, and municipal solid waste, addressing both waste reduction and resource recovery. Fermentation involves anaerobic or aerobic microbial metabolism to convert organic substrates into high-value products.

- **Bioethanol Production**: *Saccharomyces cerevisiae* ferments sugars from food waste or lignocellulosic biomass into ethanol, a renewable fuel. Pretreatment methods, such as acid or enzymatic hydrolysis, enhance substrate accessibility, achieving ethanol yields of up to 0.45 g/g substrate (Zabed et al., 2017). Food waste, rich in carbohydrates, is an ideal feedstock due to its abundance and low cost (Saravanan et al., 2021).
- **Lactic Acid Production**: Lactic acid bacteria, such as *Lactobacillus rhamnosus*, ferment food waste into lactic acid, a precursor for polylactic acid (PLA), a biodegradable plastic. Optimized fermentation at low pH reduces contamination risks and improves yields (Ojo & de Smidt, 2023).
- **Solid-State Fermentation (SSF)**: SSF employs fungi like *Aspergillus niger* to degrade lignocellulosic waste under low-moisture conditions, producing enzymes (e.g., cellulases), organic acids, or biofertilizers. SSF is energy-efficient and suitable for solid waste streams (Bamidele et al., 2025).

However, fermentation faces challenges like substrate heterogeneity, inhibitor accumulation (e.g., furfural from lignocellulosic pretreatment), and low product yields. Co-fermentation of mixed waste streams, such as food waste and sewage sludge, balances nutrient profiles and improves yields.

Composting: Composting is an aerobic process where microorganisms, including bacteria (*Bacillus subtilis*), actinomycetes, and fungi (*Trichoderma harzianum*), decompose organic matter into humus-rich compost (Greff et al., 2022).

- **Process Optimization**: Controlled parameters, such as aeration (0.4–0.6 L/kg/min), moisture (50–60%), and carbon-to-nitrogen ratio (25–30:1), enhance microbial activity and reduce composting time to 3–4 weeks. In-vessel composting systems improve process control and minimize odors (Noor et al., 2024).
- **Vermicomposting**: Earthworms, such as *Eisenia fetida*, accelerate decomposition by fragmenting organic matter and stimulating microbial diversity. Vermicompost

exhibits higher nutrient content and microbial activity than traditional compost, improving soil fertility (Pathma & Sakthivel, 2012)
- **Biochar Integration**: Adding biochar to composting enhances carbon sequestration, reduces ammonia emissions, and improves compost maturity, making it a promising strategy for sustainable waste management (Sebahire et al., 2024).

Nevertheless, composting is limited by odor emissions, pathogen survival, and long processing times. Innovations include microbial inoculants (e.g., *Bacillus* spp.) to accelerate decomposition and biofilters to mitigate odors. Integrating fermentation with composting, where fermentation residues are composted, creates a closed-loop system for waste valorization.

Genetic engineering of microbial consortia could enhance fermentation efficiency by optimizing metabolic pathways. For example, CRISPR-edited *S. cerevisiae* strains with improved xylose fermentation could increase bioethanol yields from lignocellulosic waste. In composting, smart monitoring systems using IoT and machine learning can optimize process parameters in real time. Hybrid systems combining fermentation and composting could maximize resource recovery, producing biofuels and compost from the same waste stream (Das & Ghosh, 2023b; Ghosh & Das, 2020).

Anaerobic Digestion and Biogas Production

Anaerobic digestion (AD) is a multi-stage process where microorganisms degrade organic waste in the absence of oxygen, producing biogas (50–70% methane, 30–50% carbon dioxide) and nutrient-rich digestate. AD is widely used for managing sewage sludge, food waste, livestock manure, and industrial organic waste (Okopi et al., 2024).

Process and Microbial Dynamics

AD involves four stages:

1. **Hydrolysis**: Hydrolytic bacteria (*Clostridium* spp., *Bacteroides* spp.) break down complex polymers into sugars, amino acids, and fatty acids (Menzel et al., 2020).
2. **Acidogenesis**: Acidogenic bacteria convert these intermediates into volatile fatty acids, alcohols, and hydrogen (Marquez et al., 2014).
3. **Acetogenesis**: Acetogenic bacteria produce acetate, hydrogen, and carbon dioxide (Gould, 2015).
4. **Methanogenesis**: Methanogenic archaea (*Methanobacterium* spp., *Methanosarcina* spp.) convert acetate and hydrogen into methane (Fu et al., 2021).

Biogas is a renewable energy source for electricity, heat, or vehicle fuel, while digestate serves as a fertilizer, supporting sustainable agriculture (Kabeyi & Olanrewaju, 2022).

Technological Advances

- **Co-digestion**: Combining substrates, such as food waste and manure, balances carbon-to-nitrogen ratios, enhancing biogas yields by 20–50% compared to mono-digestion (Rajput et al., 2021). For example, co-digestion of food waste and dairy manure achieves methane yields of 300–400 mL/g volatile solids (El-Mashad & Zhang, 2010).
- **Pretreatment Methods**: Thermal (100–180 °C), chemical (alkali/acid), or enzymatic pretreatments improve substrate degradability, increasing methane output by 15–30% (Saelor et al., 2024). Enzymatic pretreatment with cellulases is particularly effective for lignocellulosic waste (Ilić et al., 2023).
- **Bioelectrochemical Systems**: Coupling AD with microbial electrosynthesis or microbial fuel cells stimulates methanogenic activity via electron transfer, boosting methane production by up to 25% (Wang & Zhou, 2024).
- **Biogas Upgrading**: Technologies like pressure swing adsorption and membrane separation remove carbon dioxide, producing biomethane (>95% methane) for grid injection or vehicle fuel (Werkneh, 2022).

However, AD faces challenges like low biogas yields, process instability due to ammonia accumulation or volatile fatty acid imbalances, and high capital costs for large-scale systems. Real-time monitoring with pH, volatile fatty acid, and methane sensors, coupled with machine learning, optimizes process stability. Two-stage AD, separating hydrolysis/acidogenesis from methanogenesis, improves efficiency and reduces inhibition. Decentralized AD systems, designed for small-scale applications in rural or urban settings, minimize environmental impact and promote sustainable waste management (Roopnarain et al., 2021). Integrating AD with other biotechnologies, such as microalgae cultivation for carbon dioxide capture or pyrolysis for biochar production, could create closed-loop systems. For example, microalgae grown on AD digestate can produce biofuels, while biochar enhances digestate quality as a soil amendment. Scaling up decentralized AD systems, supported by policy incentives and modular reactor designs, could address waste management in resource-constrained regions. Additionally, synthetic biology approaches, such as engineering methanogens for higher methane yields, hold promise for next-generation AD systems.

4 Microbial Conversion of Organic Waste into Value-Added Products

Microbial conversion, facilitated by bacteria, fungi, and other microorganisms, provides an eco-friendly alternative by transforming organic waste into valuable products such as biofertilizers, compost, biofuels, biopolymers, and bioplastics (Gaur et al., 2024).

4.1 Biofertilizers and Compost

Biofertilizers and compost are produced through microbial decomposition of organic waste, enhancing soil fertility and structure. Composting involves aerobic degradation by microorganisms such as *Bacillus subtilis*, *Pseudomonas fluorescens*, and *Aspergillus niger*, which break down complex organic matter into humus-like substances. Biofertilizers, enriched with nitrogen-fixing (*Rhizobium*, *Azotobacter*), phosphate-solubilizing (*Bacillus megaterium*), or potassium-mobilizing (*Frateuria aurantia*) bacteria, improve nutrient availability for plants (Rath et al., 2022). Composting occurs in controlled environments with optimal temperature (50–60 °C), moisture (50–60%), and aeration. Organic waste, such as food scraps and agricultural residues, is mixed with bulking agents (e.g., straw) to facilitate microbial activity. The resulting compost is used in agriculture to improve soil health, reduce chemical fertilizer dependency, and enhance crop yields. Biofertilizers are applied directly to soil or seeds, promoting sustainable farming practices. For instance, *Rhizobium*-based biofertilizers can increase legume yields by 10–20% (Manea et al., 2024).

4.1.1 Biofuels

Biogas: Biogas, primarily methane (60–70%) and carbon dioxide, is produced through anaerobic digestion of organic waste by methanogenic archaea (*Methanobacterium*, *Methanosarcina*) and hydrolytic bacteria (*Clostridium*, *Bacteroides*). The process involves four stages: hydrolysis, acidogenesis, acetogenesis, and methanogenesis. Substrates include food waste, manure, and crop residues. Biogas is used for electricity generation, heating, or as a vehicle fuel, with global production reaching 59 billion cubic meters in 2022 (Goswami et al., 2016).

Bioethanol: Bioethanol is generated via fermentation of lignocellulosic waste (e.g., corn stover, sugarcane bagasse) by *Saccharomyces cerevisiae* or *Zymomonas mobilis*. Pretreatment (e.g., acid hydrolysis) breaks down cellulose and hemicellulose into fermentable sugars. Bioethanol is blended with gasoline, reducing fossil fuel dependency. Current challenges include high pretreatment costs and competition with food crops.

Biodiesel: Biodiesel is produced from lipid-rich organic waste (e.g., waste cooking oil) via transesterification, often catalyzed by lipases from *Candida antarctica* or *Rhizopus oryzae*. Microalgae, such as *Chlorella vulgaris*, cultivated on organic waste, also serve as biodiesel feedstocks due to their high lipid content (20–50% dry weight). Biodiesel is a renewable diesel substitute, but scalability and feedstock availability remain limitations (Bayrakci Ozdingis & Kocar, 2018).

4.2 Biopolymers and Bioplastics

Biopolymers, such as polyhydroxyalkanoates (PHA) and polylactic acid (PLA), and bioplastics are synthesized by microorganisms using organic waste as a carbon source (Behera et al., 2022a). *Cupriavidus necator* and *Pseudomonas putida* produce PHA by fermenting sugars or fatty acids from waste streams (Bellini et al., 2022). PLA is derived from lactic acid produced by *Lactobacillus* species during carbohydrate fermentation (Nicolescu et al., 2023). These biopolymers are biodegradable, offering alternatives to petroleum-based plastics (Dey et al., 2024a, 2024b, 2024c). PHA production involves microbial fermentation under nutrient-limited conditions, followed by polymer extraction. Organic waste, such as food waste or glycerol from biodiesel production, serves as a substrate. PLA production requires lactic acid fermentation, purification, and polymerization. Bioplastics are used in packaging, medical implants, and agricultural films. In 2023, global bioplastic production reached 2.2 million tons, with PHA and PLA constituting 30% of the market (Bala et al., 2024; Chaudry et al., 2025; John & Das, 2024; Mishra et al., 2024a).

Bioplastics reduce plastic pollution and reliance on fossil fuels. However, high production costs and limited mechanical properties compared to conventional plastics hinder widespread adoption. Advances in synthetic biology, such as CRISPR-engineered microbes, could enhance yield and reduce costs.

Microbial conversion of organic waste into biofertilizers, compost, biofuels, and bioplastics offers a sustainable pathway to address waste management and resource scarcity. Integrating these processes into a biorefinery framework could maximize resource recovery and economic viability (Dey et al., 2024a, 2024b, 2024c; Paul et al., 2023; Pradhan et al., 2023). For instance, anaerobic digestion residues can be used as biofertilizers, while fermentation byproducts can feed bioplastic production. However, challenges such as process optimization, scalability, and regulatory frameworks must be addressed. Emerging technologies, including microbial consortia engineering and bioreactor design, promise to enhance efficiency. By fostering interdisciplinary research and policy support, microbial conversion can drive a circular economy, transforming organic waste into valuable resources.

5 Industrial Applications and Case Studies

Microbes are natures scavengers, with metabolic diversity. The properties such as secretion of degradation enzymes and other proteins can be leveraged to develop sustainable industrial practices (Kondakindi et al., 2024). In the sections below, we briefly explain microbial waste management models and case studies impacting biotechnology-based waste management practices.

5.1 Successful Microbial Waste Management Models

Microbial waste management models utilize the capability of micro-organisms such as actinomycetes, bacteria, fungi to decompose biodegradable wastes into products of daily use like biofuels and fertilizers (Kumar Mahish et al., 2024).

Case 1—Composting and Vermicomposting: Bacteria such as *Bacillus subtilis* and fungi such as *Aspergillus niger*, have ability to produce nutrient-rich compost. *Pseudomonas* break down cellulose, ensuring efficient decomposition, while as Rot fungi—*Phanerochaete chrysosporium* degrades lignin. Earthworms are successful contributors of vermicompost (Palaniveloo et al., 2020). A successful story of biotech-based waste management model comes from the city of San Francisco's composting program. Under this program 80% of organic waste is diverted from landfills. San Francisco, aiming for zero waste, exceeded its 75% diversion goal early, recovering over 80% of waste and halving disposal. In 2003, it adopted the Environment Code, grounded in the Precautionary Principle, and enforced a 2006 ordinance mandating recovery of construction and demolition debris, significantly advancing sustainable waste management practices (US EPA 2013). This not only adds to vermicompost but also reduces greenhouse gas emissions and supports local farming. Reduces landfill volume, mitigates methane emissions, and produces compost that enhances soil fertility.

Case 2: Anaerobic Digestion (AD): Anaerobic digestion (AD) transforms organic solid waste into biofuels and fertilizers through microbes such as *Methanobacterium*, reducing greenhouse gas emissions and promoting sustainability. It converts organic waste to biogas and nutrient-rich digestate (Prasanna Kumar et al., 2024). Lemvig Biogas, Denmark's largest thermophilic plant, processes 200,000 tons of organic waste annually, produces renewable energy, generating 33 million kWh electricity and 55 million kWh heat annually, supporting 22,000 people and 3,000 households while reducing pollutants (Simic, 2023). This reduces the burden on the use of fossil fuels, and in the meantime helps reduce biomass and enhance agricultural activity. Some of the key microbes that play role in this are methanogens and hydrolytic bacteria such as *Clostridium.* They break down the organic matter like polymers into simple monomers by enzymatic degradation (Wang et al., 2023).

Case 3: Microalgae-Based Wastewater Treatment: Agricultural wastewater, rich in nitrogen and phosphorus, presents opportunities for nutrient removal and microalgae cultivation. Integrating microalgae growth with wastewater treatment offers low-cost production, nutrient recycling, and value-added bioproducts. Microalgae efficiently uptake nutrients, yielding high-protein, lipid, and carbohydrate biomass. Despite challenges in dewatering and extraction, wastewater-grown microalgae show economic promise. This approach supports sustainable water treatment and bioproduct generation, aligning with environmental goals and reducing treatment costs. Microalgae such as *Chlorella vulgaris* play such roles while producing biomass for biofuels or fertilizers (Alavianghavanini et al., 2024).

Algae Systems LLC in U.S uses offshore photobioreactors (PBRs) to treat municipal wastewater, achieving 75% nitrogen, 93% phosphorus, and 92% BOD removal. It reduces sludge accumulation and leads to biofuel production. The key roles are played by *Cyanobacteria* and green algae have ability to assimilate pollutants, while as symbiotic bacteria enhance degradation (Novoveská et al., 2016).

5.2 Commercial Case Studies of Biotech-Based Waste Management

Microbial biotechnology offers translation of research into commercial applications with scalable and economic viability in waste management. Various commendable efforts have been made in translating the research into commercial tool.

Case Study I: Organica Biotech (Cleanmaxx Sewage): Organica Biotech has revolutionized sanitation in India with its innovative bio-toilet solutions. Their Bioclean Bio-toilet & Bio-digester, an award-winning technology, effectively degrades waste using microbial digestion under diverse conditions. An Indian company, developed Cleanmaxx Sewage, which involves usage of microbial consortium for sewage treatment plants. The product contains adaptable bacteria that degrade high organic loads in wastewater used in municipal and industrial sewage treatment plants across India, Cleanmaxx enhances the activated sludge process, improving effluent quality and reducing sludge volume. The technology treats millions of liters of sewage daily, enabling water reuse and protecting public health. It is cost-effective compared to chemical treatments and aligns with India's Swachh Bharat Mission. Microbial consortia outperform single-strain solutions due to synergistic degradation pathways, making them ideal for complex waste streams (Team, 2020).

Case Study II: Mint Innovation (E-Waste Recycling): Electronic waste poses serious health and environmental risks, with 52 million tons generated annually. Traditional recycling methods harm ecosystems, while bioleaching offers a sustainable solution (Das et al., 2023). Using bacteria and fungi, companies like Mint Innovation and N2S recover precious metals without toxic chemicals. Projects like REEgains promote recycling and inclusivity, driving a $4.5 trillion circular economy by 2030. Bioleaching ensures safer recycling, reduces waste, and creates jobs, emphasizing collaboration with electronic manufacturers for widespread adoption. Mint Innovation, a New Zealand-based startup, uses microbial bioleaching to recover precious metals (e.g., gold, copper) from electronic waste (e-waste). The microorganisms of bacterial origin like *Acidithiobacillus ferrooxidans* are used to leach metals from e-waste for conversion to reusable materials. Gold recovered from circuit boards in pilot facilities of Auckland. Mint Innovation's reduces the environmental footprint of e-waste (Caballar, 2020). It is an ecologically sustainable, and offers low-cost. This process based on microbial bioleaching offers an alternative to smelting (Das & Ghosh, 2018; Ghosh et al., 2016).

Case Study III: Novozymes (Enzyme-Based Waste Valorization): Overview: Food and agricultural wastes are toxic to environment. In an interesting case study, Novozymes, a Danish biotech company, offers solutions for producing useful materials like biofuels and biochemicals from them. To ensure that this purpose is met with efficiency, the use of microbial enzymes, e.g., cellulases, lipases are instrumental. In collaboration with another recycling company—Beta Renewables, they supply enzymes for a second-generation bioethanol production plant in Italy. The waste generates a tremendous amount of lignocellulosic content which amounts to 270,000 tons of biomass annually. This huge amount of biomass in turn leads to production of 75 million liters of bioethanol yearly. reducing greenhouse gas emissions by 90% and overcoming the burden of using fossil fuels. This process ensures that microbial solutions remain in use against the overburdened crop residues for creation of circular economy and making the microbial solutions effective market competitors.

6 Challenges and Future Prospects

Microbial biotechnology offers an effective, low cost, and ecologically sustainable solutions towards organic waste management. It not only provides natural end to waste management, but also provides an economy based on viable outcomes. The use of microbial biotechnology in industrial applications faces technical issues based on scalability and regulatory challenges based on bio accumulative practices. Advanced innovative practices have led to industrial interventions in biotechnology, which are paving the way for bioeconomy driven innovations. Keeping in with existing technological and regulatory limitations, many challenges are driving the future development of microbial biotechnological waste management practices.

6.1 Limitations of Microbial Waste Management

The adoption of waste management based on microbial biotechnology is hindered by bio accumulation and scalability issues. Following issues must be considered before use.

6.1.1 Slow Degradation Rates

The enzyme mediated degradation of complex organic waste is a slow and time-consuming process. The breakdown of lignocellulose, plastics requires weeks or months. One such example is the use of white rot fungi for lignin breakdown. It takes a huge amount of time which can range between several days, while keeping under consideration the use of optimal temperature, pH, and pressure (Sánchez,

2009). Such a process will massively impact the outcome and cause a whooping increase in operational costs. It also impacts industrial scalability and diminishes high-throughput settings. Petroleum waste management using biological methods is often met with slow breakdown of polycyclic aromatic hydrocarbons (PAHs) (Abdel-Shafy & Mansour, 2016). This is because PAHs are difficult to breakdown and by taking microbial route, it becomes even more difficult.

6.1.2 Environmental Sensitivity

It is an established fact that microbes work at optimal conditions of temperature, pH, oxygen, and nutrient availability. Any change in such conditions impacts the efficiency of microbes (Das & Ghosh, 2023a; Das et al., 2024a, 2024b, 2024c, 2024d). The same concept is applicable to the microbes used in industrial practices. Change of optimal to suboptimal conditions, e.g., aerobic to anaerobic or vice-versa causes reduction of enzymatic activity and in turn on degradation efficiency of microbes and its implementation on large scale (Robinson, 2015). The drop-down effect of temperature on enzymatic activity will lead to use of costly heating arrangements in winters, hence increasing operational costs (Das et al., 2024b;; Priyadarshini et al., 2024b). This can be understood from the use of microalgae-based wastewater treatment. Microalgae work effectively in light and moderate temperature, hence its use in organic waste treatment is less effective in low-light or cold environments (Singh & Singh, 2015).

1. **Regulatory and Societal Barriers**: Genetically modified microorganisms (GMMs) is often seen with skepticism due its potential ecological risks and ethical reasons. This leads to its strict regulations (Bawa & Anilakumar, 2013). A major redline in ethical use of GMMs is unintended gene transfer from modified organisms to the natural strains (Philips et al., 2022). This can cause a whirlwind effect on the natural microorganisms by changing their genetic makeup and giving rise to resistant strains. These challenges cost adoption of such technology and thus delay its commercial aspect. The regulatory guidelines such as Cartagena Protocol on Biosafety create hurdles in the deployment of Genetically Engineered Microorganisms (GEMs) (Gaugitsch, 2016). Allonnia is a company that adopts and deploys the GEMs for reduction of pollution is often met with documentation and biosafety regulations impacting its deliverables (Han et al., 2022).
2. **Scalability and Economic Viability**: One of the biggest operational hurdles is scaling of microbial processes from lab to industrial levels. The microbial stability, optimization, and management of bioreactors with heterogeneous waste streams is a challenge involving high capital. These challenges deter investment in the places where infrastructure for industrial development is ongoing (Fradinho et al., 2021). E-waste bioremediation involving microbes is still an underdevelopment and will incur huge costs till its feasible conditions outside lab are met (Khan et al., 2024).

6.2 *Innovations and Emerging Trends in Industrial Biotechnology*

Biotechnological waste management is advancing with ever increasing innovations. Some of the innovative practices that are evolving for development of economy based on biological practices.

1. **Synthetic Biology and CRISPR-Cas**: One would never have thought of editing the gene but thanks to CRISPR-Cas gene editing, these editing can be done precisely by this innovative technique. Through this technique the modifications to microbial genomes can be done, for increase in the yield of degradative enzymes or production of highly efficient enzymes (Rafeeq et al., 2023). Toluene is an environmental toxin produced by industrial and petroleum waste. Such pollutants need to be taken out. CRISPR-Cas has the capability to produce Engineered *Bacterium* which can degrade such xenobiotics and produce an alternative to the petroleum products (Ishikawa et al., 2021). An example in this case are polyhydroxyalkanoates (PHAs), which can be alternative to petroleum-based plastics, thus offering bioplastics as an alternative to chemically derived plastics (Behera et al., 2022b). These microbes can be engineered for adaptation to sub-optimal conditions, scalability, and specificity. Allonnia–US-based biotech company in collaboration with Ginkgo Bioworks spin off is involved in the development of synthetic microbes which can mitigate environmental pollution by detecting and degrading pollutants like FAS, heavy metals, and hydrocarbons (July 2021, n.d.).
2. **Microbial Consortia and Co-culture Systems**: Microalgae-bacteria consortia offer sustainable wastewater treatment, with omics techniques revealing mechanisms, interactions, and enhancing organic pollutant removal efficiency (Vo et al., 2024). Microbial consortia (e.g., microalgae-bacteria systems) utilize the capability of their enzymes to degrade pollutants into simple molecules. The simple molecules can be nitrogen, phosphorus, and other nutrients of microbial use. These molecules can be used by microalgae/bacteria for their own metabolic pathways. This leads to a productive cycle of enzyme mediated degradation by bacteria and production of valuable nutrients for assimilation by microalgae and biomass formation (Gururani et al., 2022). Use of dual strain systems enhances the degradation efficiency as well as generation of value biodegradable biomass (Rathour et al., 2024). One such successful use of dual strain consortium is use of microalgae-bacterial consortium in South Korea. In a synergistic interaction between these two strains, it is useful in the treatment of textile wastewater. This not only generates a valuable biomass but also is time restrained and within 48 h of treatment, the dyes and organic pollutants are broken down for further processing (Gururani et al., 2022).
3. **Nanotechnology Integration**: One of the emerging trends in waste management is use of microbial nanoparticles (e.g., Fe_3O_4–SiO_2). These materials are tailored using microbes and hence enhance microbial activity. The enhanced microbial activity is due improved enzyme stability, as the same can be utilized

for enzyme stability. The microbial nanomaterials are stable and allow effective pollutant adsorption (Sadegh et al., 2017). Such nanomaterials can be used for immobilization of ecologically suitable enzymes and hence sustain the long-term degradation and turnover capacity of enzymes. The economic costs are lowered as they enable continuous processing. They also ensure that microbes sustain a toxic environment. Immobilized lignin peroxidase nanoparticles are useful in the treatment of textile wastewater, achieving over 90% of organic pollutant removal in 24 h (Bhatt et al., 2022).

4. **Droplet Microfluidics**: In an advanced technique era, droplet microfluidics is based on high-throughput screening of microbial consortium for the efficient production of enzymes used for degradation and byproduct generation in a limited time and space. It makes possible to isolate and test thousands of microbes in a single attempt (Das & Ghosh, 2022;; Das & Ghosh, 2023a; Das et al., 2024a, 2024b, 2024c, 2024d; Hosokawa et al., 2015). It not only accelerates the development of novel strain consortium to be used in bioremediation but also reduces timelines and improving the cost-effectiveness of the process. *Pseudomonas* strains were identified using this approach to degrade plastic waste, advancing bioplastic recycling (Arvind et al., 2024).
5. **Bio-Based Plastic Production**: Microbes like *Ralstonia eutropha* 5119 have ability to produce biodegradable plastics like PHAs by metabolizing the organic waste i.e.-by using low-cost lignocellulosic biomass hydrolysates (LBHs) like barley, pine, and Miscanthus. Despite fermentation inhibitors (e.g., furfural, vanillin), the strain co-metabolizes them for PHA production. This highlights its potential for cost-effective, sustainable bioplastic synthesis from agricultural residues with minimal environmental impact, adding to their dual property of waste management and pollution control. Biodegradable bioplastics lead to reduction of gaseous emissions and plastic pollution (Bhatia et al., 2019). Being pure biodegradable materials, PHAs ensure the preservation of fossils. Bacteria like *Cupriavidus necator* and *Bacillus megaterium* can accumulate PHAs using agro-wastes like rice hulls and sugarcane bagasse, which can reduce agricultural waste and replace the single-use plastics (de Oliveira Schmidt et al., 2022).
6. **Smart Probiotics and Containment Strategies**: Engineered CRISPR-based kill switches in *E. coli* Nissle 1917 ensure microbial biocontainment, preventing proliferation while degrading pollutants effectively. Thus engineered "smart probiotics" prevent unintended proliferation of GEMs as these microbes are designed to for specific purpose, i.e., to degrade specific pollutants and then deactivate (Rottinghaus et al., 2022). Addressing the regulatory issues concerning their ecological threat perception, it becomes necessary to enable the safe deployment of GEMs. Synthetic biology transforms biological systems, impacting many scientific fields. Initially focused on cost-effective microbial factories to reduce oil dependency, it now directly aids environmental protection. Genetically engineered microbes and plants, combined with computational tools, provide remediation systems to address pollutants. Synthetic biology offers cost-effective solutions for pollution monitoring and targeted remediation, addressing major environmental risks and health threats. These measures ensure no ecological

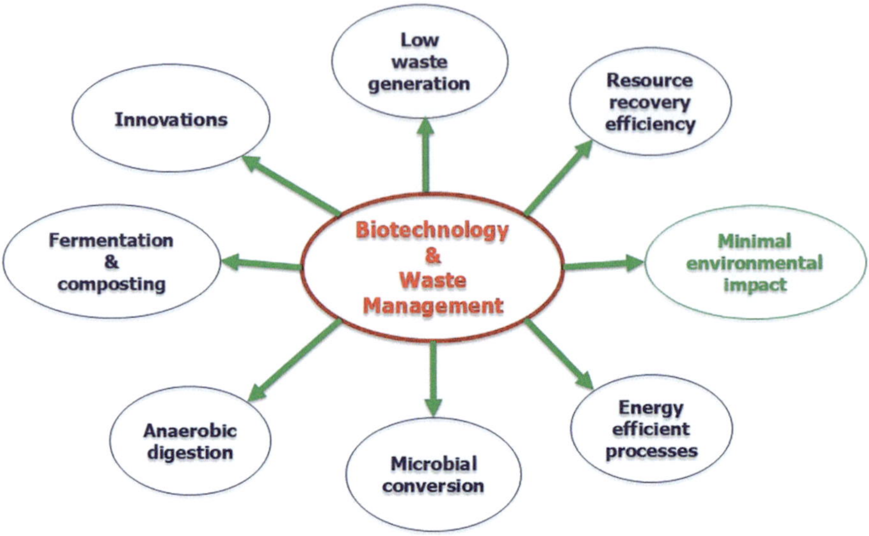

Fig. 1 This image represents the scope of biotechnology-based waste management

disruption (Aminian-Dehkordi et al., 2023). The take away information from this book chapter can be summarized in Fig. 1.

7 Conclusion

Industrial biotechnology-based microbial waste management is an efficient, sustainable, and scalable process for organic waste treatment and valorization. The synergistic use of microbial systems—actinomycetes, archaea, fungi, and bacteria—enables the transformation of complex organic substrates into valuable end-products such as biofertilizers, biofuels, compost, and biochemicals. These biological transformations not only reduce the environmental footprint of conventional disposal processes but also enable the enhancement of circular bioeconomies. Industrial biotechnology continues to enhance the efficiency and applicability of microbial treatment of waste with improvements in new technologies in microbial strain improvement, enzyme engineering, and bioprocess optimization. Future research needs to address the design of robust microbial consortia, adaptation of microbial function to waste streams, and scale-up of biotechnological interventions to industrial scales. In general, the use of microbial solutions for organic waste management is a pillar to success in sustainable environmental and economic development.

References

Abdel-Shafy, H. I., & Mansour, M. S. M. (2016). A review on polycyclic aromatic hydrocarbons: Source, environmental impact, effect on human health and remediation. *Egyptian Journal of Petroleum, 25*(1), 107–123. https://doi.org/10.1016/j.ejpe.2015.03.011

Abubakar, I. R., Maniruzzaman, K. M., Dano, U. L., AlShihri, F. S., AlShammari, M. S., Ahmed, S. M. S., Al-Gehlani, W. A. G., & Alrawaf, T. I. (2022). Environmental sustainability impacts of solid waste management practices in the global South. *International Journal of Environmental Research and Public Health, 19*(19), 12717. https://doi.org/10.3390/ijerph191912717

Agrawal, R., Verma, A., Verma, S., & Varma, A. (2021). Industrial methanogenesis: Biomethane production from organic wastes for energy supplementation. In R. Prasad, V. Kumar, J. Singh, & C. P. Upadhyaya (Eds.), *Recent developments in microbial technologies* (pp. 99–115). Springer Nature. https://doi.org/10.1007/978-981-15-4439-2_4

Alavianghavanini, A., Shayesteh, H., Bahri, P. A., Vadiveloo, A., & Moheimani, N. R. (2024). Microalgae cultivation for treating agricultural effluent and producing value-added products. *Science of the Total Environment, 912*, Article 169369. https://doi.org/10.1016/j.scitotenv.2023.169369

Amani, T., Nosrati, M., & Sreekrishnan, T. R. (2010). Anaerobic digestion from the viewpoint of microbiological, chemical, and operational aspects—a review. *Environmental Reviews, 18*(NA), 255–278. https://doi.org/10.1139/A10-011

Amin, F. R., Khalid, H., El-Mashad, H. M., Chen, C., Liu, G., & Zhang, R. (2021a). Functions of bacteria and archaea participating in the bioconversion of organic waste for methane production. *Science of the Total Environment, 763*, Article 143007. https://doi.org/10.1016/j.scitotenv.2020.143007

Amin, F. R., Khalid, H., El-Mashad, H. M., Chen, C., Liu, G., & Zhang, R. (2021b). Functions of bacteria and archaea participating in the bioconversion of organic waste for methane production. *The Science of the Total Environment, 763*, Article 143007. https://doi.org/10.1016/j.scitotenv.2020.143007

Aminian-Dehkordi, J., Rahimi, S., Golzar-Ahmadi, M., Singh, A., Lopez, J., Ledesma-Amaro, R., & Mijakovic, I. (2023). Synthetic biology tools for environmental protection. *Biotechnology Advances, 68*, Article 108239. https://doi.org/10.1016/j.biotechadv.2023.108239

Arvind, M., D.r., J., Prasad, E., Ahmed, S., Srinath, S., Arvind, M., D.r, J., Prasad, E., Ahmed, S., & Srinath, S. (2024). Biodegradation of plastic by Pseudomonas aeruginosa. *International Journal of Science and Research Archive, 11*(2), Article 2. https://doi.org/10.30574/ijsra.2024.11.2.0507

Bala, K., Anshul, Verma, S., Dogra, A., Das, A. P., Kumar, R., & Nandini. (2024). Sustainable management for urban plastic waste generation. *Sustainable Management of Urban Plastic Waste Through Circular Economic Approaches*, 211–236. https://doi.org/10.1201/9781003465447-13

Bamidele, M. O., Bamikale, M. B., Cárdenas-Hernández, E., Bamidele, M. A., Castillo-Olvera, G., Sandoval-Cortes, J., & Aguilar, C. N. (2025). Bioengineering in solid-state fermentation for next sustainable food bioprocessing. *Next Sustainability, 6*, Article 100105. https://doi.org/10.1016/j.nxsust.2025.100105

Bawa, A. S., & Anilakumar, K. R. (2013). Genetically modified foods: Safety, risks and public concerns—a review. *Journal of Food Science and Technology, 50*(6), 1035–1046. https://doi.org/10.1007/s13197-012-0899-1

Bayrakci Ozdingis, A. G., & Kocar, G. (2018). Current and future aspects of bioethanol production and utilization in Turkey. *Renewable and Sustainable Energy Reviews, 81*, 2196–2203. https://doi.org/10.1016/j.rser.2017.06.031

Behera, I. Dipamitra., & Das, A. Prasad. (2025). *Environmental hydrocarbon pollution and zero waste approach towards a sustainable waste management*. 436.

Behera, S., Priyadarshanee, M., Vandana, & Das, S. (2022a). Polyhydroxyalkanoates, the bioplastics of microbial origin: Properties, biochemical synthesis, and their applications. *Chemosphere, 294*, 133723. https://doi.org/10.1016/j.chemosphere.2022.133723

Behera, S., Priyadarshanee, M., Vandana, & Das, S. (2022b). Polyhydroxyalkanoates, the bioplastics of microbial origin: Properties, biochemical synthesis, and their applications. *Chemosphere, 294*, 133723. https://doi.org/10.1016/j.chemosphere.2022.133723

Bellini, S., Tommasi, T., & Fino, D. (2022). Poly(3-hydroxybutyrate) biosynthesis by *Cupriavidus necator*: A review on waste substrates utilization for a circular economy approach. *Bioresource Technology Reports, 17*, Article 100985. https://doi.org/10.1016/j.biteb.2022.100985

Bhatia, S. K., Gurav, R., Choi, T.-R., Jung, H.-R., Yang, S.-Y., Moon, Y.-M., Song, H.-S., Jeon, J.-M., Choi, K.-Y., & Yang, Y.-H. (2019). Bioconversion of plant biomass hydrolysate into bioplastic (polyhydroxyalkanoates) using *Ralstonia eutropha* 5119. *Bioresource Technology, 271*, 306–315. https://doi.org/10.1016/j.biortech.2018.09.122

Caballar, R. D. (2020). A New Zealand startup is using microbes to suck solid gold out of E-waste. *OneZero*. https://onezero.medium.com/a-new-zealand-startup-is-using-microbes-to-suck-solid-gold-out-of-e-waste-48814d425eef

Chaudry, S., Hurtado-McCormick, V., Cheng, K. Y., Willis, A., Speight, R., & Kaksonen, A. H. (2025). Microalgae to bioplastics–routes and challenges. *Cleaner Engineering and Technology, 25*, Article 100922. https://doi.org/10.1016/j.clet.2025.100922

Chen, Y.-Y., Soma, Y., Ishikawa, M., Takahashi, M., Izumi, Y., Bamba, T., & Hori, K. (2021). Metabolic alteration of *Methylococcus capsulatus* str. Bath during a microbial gas-phase reaction. *Bioresource Technology, 330*, 125002. https://doi.org/10.1016/j.biortech.2021.125002

da Motta, I. G. B., Santana, L. A. R., Pereira, H. P., de Paula, V. R., Martins, M. F., da Costa Carneiro, J., & Otenio, M. H. (2022). Population dynamics of methanogenic archea in co-digestion systems operating different industrial residues for biogas production. *Sustainability, 14*(18), Article 18. https://doi.org/10.3390/su141811536

Das, A. P., & Ghosh, S. (2018). Bioleaching of Manganese from mining waste materials. *Materials Today: Proceedings, 5*(1), 2381–2390. https://doi.org/10.1016/J.MATPR.2017.11.459

Das, A., & Ghosh, S. (2022). Role of microorganisms in extenuation of mining and industrial wastes. *Geomicrobiology Journal, 39*, 173–175. https://doi.org/10.1080/01490451.2022.2038953

Das, A. P., Mallavarapu, M., & Ghosh, S. (2023). Ecotoxicity of mining pollutants on the environment and their remediation. *Environmental Chemistry and Ecotoxicology, 5*, 165–167. https://doi.org/10.1016/J.ENCECO.2023.08.002

Das, A., Ray, S. K., Mohanty, M., Mohanty, J., Dey, S., & Das, A. P. (2024d). Ecotoxicity of microplastic wastes and their sustainable management: A review. *Environmental Chemistry and Ecotoxicology, 6*, 144–152. https://doi.org/10.1016/J.ENCECO.2024.05.003

Das, A., Dey, S., Gacem, A., Yadav, K. K., Nath, H., Ram, D. K., & Das, A. P. (2025). Revolutionizing solid hydrocarbon sourcing: Plastic's journey from waste to upcycled treasure. *Environmental Science and Engineering, Part, F218*, 331–351. https://doi.org/10.1007/978-3-031-81615-4_14

Das, A. P., & Ghosh, S. (2023a). Evolution of biomining technology. *Manganese Mining Microorganisms*, 5–19. https://doi.org/10.1016/B978-0-12-822147-1.00012-1

Das, A. P., & Ghosh, S. (2023b). Metagenomic insights into unculturable microbial diversity. *Manganese Mining Microorganisms*, 63–76. https://doi.org/10.1016/B978-0-12-822147-1.00002-9

Das, A. P., & Ghosh, S. (2023c). Microorganisms in manganese biomining processes. *Manganese Mining Microorganisms,* 55–61. https://doi.org/10.1016/B978-0-12-822147-1.00013-3

Das, A. P., & Ghosh, S. (2023d). Mixed culture bioleaching: an insight into manganese biomining process and efficacy. *Manganese Mining Microorganisms*, 129–141. https://doi.org/10.1016/B978-0-12-822147-1.00001-7

Das, A. P., & Ghosh, S. (2023e). Role of metabolites and metabolomics for understanding of manganese bioleaching. *Manganese Mining Microorganisms*, 117–127. https://doi.org/10.1016/B978-0-12-822147-1.00011-X

Das, A., Dey, S., Bala, K., Gnanasekaran, R., & Das, A. P. (2024a). Microbial bioremediation technology for sustainable treatment and management of synthetic microfiber waste. *Sustainable Microbial Technology for Synthetic and Cellulosic Microfiber Bioremediation*, 221–238. https://doi.org/10.1007/978-3-031-62600-5_12

Das, A. P., Nayak, B., & Behera, I. D. (2024). Applications of microorganisms in advanced geomicrobiological and biotechnological exploration. *Geomicrobiology Journal, 41*(4), 309–311. https://doi.org/10.1080/01490451.2024.2327059/ASSET//CMS/ASSET/DE2A1F67-08C1-475A-ABC7-2B3F4C9E8BCF/01490451.2024.2327059.FP.PNG

Das, A. Prasad., Behera, I. Dipamitra., & Das, N. Prasad. (2024). *Renewable energy generation and value addition from environmental microfiber pollution through advanced greener solution.* 278.

de Oliveira Schmidt, V. K., Santos, E. F. dos, de Oliveira, D., Ayub, M. A. Z., Cesca, K., Cortivo, P. R. D., de Andrade, C. J., & Hickert, L. R. (2022). Production of polyhydroxyalkanoates by bacillus megaterium: Prospecting on rice hull and residual glycerol potential. *Biomass, 2*(4), Article 4. https://doi.org/10.3390/biomass2040026

Dey, S., Das, A., Mallick, K., Sahu, A., & Das, A. P. (2023). Environmental petroleum waste: Pollution, toxicity, sustainable remediation (pp. 159–175). https://doi.org/10.1007/978-3-031-48220-5_7

Dey, S., Bissoyi, A., Koteswara Reddy, G., & Das, A. P. (2024). Synthetic microfibers: An urban emerging pollutant demanding for sustainable environmental solutions. *Sustainable Microbial Technology for Synthetic and Cellulosic Microfiber Bioremediation*, 107–121. https://doi.org/10.1007/978-3-031-62600-5_7

Dey, S., Dash, A., & Das, A. P. (2024). Bioremediation as a sustainable and circular solution for mining waste management and bioenergy generation. *Sustainable Management of Mining Waste and Tailings: A Circular Economy Approach*, 107–119. https://doi.org/10.1201/9781003442455-6/BIOREMEDIATION-SUSTAINABLE-CIRCULAR-SOLUTION-MINING-WASTE-MANAGEMENT-BIOENERGY-GENERATION-SUDESHNA-DEY-AKANSHYA-DASH-ALOK-PRASAD-DAS

Dey, S., Dash, N., Ray, S. K., Priyadarsini, S., Chhotaray, C., & Das, A. P. (2024). Occurrence and quantification of tire wear plastics pollutants from highway city dust of Bhubaneswar City of Odisha. *Sustainable Microbial Technology for Synthetic and Cellulosic Microfiber Bioremediation*, 77–89. https://doi.org/10.1007/978-3-031-62600-5_5

Duan, Y., Awasthi, M. K., Wu, H., Yang, J., Li, Z., Ni, X., Zhang, J., Zhang, Z., & Li, H. (2022). Biochar regulates bacterial-fungal diversity and associated enzymatic activity during sheep manure composting. *Bioresource Technology, 346*, Article 126647. https://doi.org/10.1016/j.biortech.2021.126647

Duan, H., Ji, M., Xie, Y., Shi, J., Liu, L., Zhang, B., & Sun, J. (2021). Exploring the microbial dynamics of organic matter degradation and humification during co-composting of cow manure and bedding material waste. *Sustainability, 13*(23), Article 23. https://doi.org/10.3390/su132313035

El-Mashad, H. M., & Zhang, R. (2010). Biogas production from co-digestion of dairy manure and food waste. *Bioresource Technology, 101*(11), 4021–4028. https://doi.org/10.1016/j.biortech.2010.01.027

Environment, U. N. (2024). *Global Waste Management Outlook 2024 | UNEP - UN Environment Programme*. https://www.unep.org/resources/global-waste-management-outlook-2024

Ezeako, E. C., Nwiloh, B. I., Odo, M. C., & Ozougwu, V. E. (2025). Harnessing synthetic biology for sustainable industrial innovation: Advances, challenges, and future direction. *Biochemical Engineering Journal, 221*, Article 109777. https://doi.org/10.1016/j.bej.2025.109777

Ezeorba, T. P. C., Okeke, E. S., Mayel, M. H., Nwuche, C. O., & Ezike, T. C. (2024). Recent advances in biotechnological valorization of agro-food wastes (AFW): Optimizing integrated approaches for sustainable biorefinery and circular bioeconomy. *Bioresource Technology Reports, 26*, Article 101823. https://doi.org/10.1016/j.biteb.2024.101823

Fradinho, J., Allegue, L. D., Ventura, M., Melero, J. A., Reis, M. A. M., & Puyol, D. (2021). Up-scale challenges on biopolymer production from waste streams by Purple Phototrophic Bacteria mixed cultures: A critical review. *Bioresource Technology, 327*, Article 124820. https://doi.org/10.1016/j.biortech.2021.124820

Fu, S., Angelidaki, I., & Zhang, Y. (2021). *In situ* biogas upgrading by CO_2-to-CH_4 bioconversion. *Trends in Biotechnology, 39*(4), 336–347. https://doi.org/10.1016/j.tibtech.2020.08.006

Gaugitsch, H. (2016). Under the cartagena protocol on biosafety—where is the roadmap for risk assessment taking us? *Frontiers in Bioengineering and Biotechnology, 3*, 212. https://doi.org/10.3389/fbioe.2015.00212

Gaur, S., Kaur, M., Kalra, R., Rene, E. R., & Goel, M. (2024). Application of microbial resources in biorefineries: Current trend and future prospects. *Heliyon, 10*(8), Article e28615. https://doi.org/10.1016/j.heliyon.2024.e28615

Ghosh, S., Mohanty, S., Akcil, A., Sukla, L. B., & Das, A. P. (2016). A greener approach for resource recycling: Manganese bioleaching. *Chemosphere, 154*, 628–639. https://doi.org/10.1016/J.CHEMOSPHERE.2016.04.028

Ghosh, S., & Das, A. P. (2020). Microbial metagenomics: current advances in investigating microbial ecology and population dynamics. In S. K. Nayak & B. B. Mishra (Eds.), *Frontiers in soil and environmental microbiology* (1st ed., pp. 247–254). CRC Press.

Goswami, R., Chattopadhyay, P., Shome, A., Banerjee, S. N., Chakraborty, A. K., Mathew, A. K., & Chaudhury, S. (2016). An overview of physico-chemical mechanisms of biogas production by microbial communities: A step towards sustainable waste management. *3 Biotech, 6*(1), 72. https://doi.org/10.1007/s13205-016-0395-9

Gould, M. C. (2015). Chapter 18—bioenergy and anaerobic digestion. In A. Dahiya (Ed.), *Bioenergy* (pp. 297–317). Academic. https://doi.org/10.1016/B978-0-12-407909-0.00018-3

Greff, B., Szigeti, J., Nagy, Á., Lakatos, E., & Varga, L. (2022). Influence of microbial inoculants on co-composting of lignocellulosic crop residues with farm animal manure: A review. *Journal of Environmental Management, 302*, Article 114088. https://doi.org/10.1016/j.jenvman.2021.114088

Gururani, P., Bhatnagar, P., Kumar, V., Vlaskin, M. S., & Grigorenko, A. V. (2022). Algal consortiums: A novel and integrated approach for wastewater treatment. *Water, 14*(22), Article 22. https://doi.org/10.3390/w14223784

Hajam, Y. A., Kumar, R., & Kumar, A. (2023). Environmental waste management strategies and vermi transformation for sustainable development. *Environmental Challenges, 13*, Article 100747. https://doi.org/10.1016/j.envc.2023.100747

Han, P., Teo, W. Z., & Yew, W. S. (2022). Biologically engineered microbes for bioremediation of electronic waste: Wayposts, challenges and future directions. *Engineering Biology, 6*(1), 23–34. https://doi.org/10.1049/enb2.12020

Harirchi, S., Wainaina, S., Sar, T., Nojoumi, S. A., Parchami, M., Parchami, M., Varjani, S., Khanal, S. K., Wong, J., Awasthi, M. K., & Taherzadeh, M. J. (2022). Microbiological insights into anaerobic digestion for biogas, hydrogen or volatile fatty acids (VFAs): A review. *Bioengineered, 13*(3), 6521–6557. https://doi.org/10.1080/21655979.2022.2035986

Hosokawa, M., Hoshino, Y., Nishikawa, Y., Hirose, T., Yoon, D. H., Mori, T., Sekiguchi, T., Shoji, S., & Takeyama, H. (2015). Droplet-based microfluidics for high-throughput screening of a metagenomic library for isolation of microbial enzymes. *Biosensors and Bioelectronics, 67*, 379–385. https://doi.org/10.1016/j.bios.2014.08.059

How is biotechnology used in sewage treatment. (2025). https://organicabiotech.com/how-is-biotechnology-used-in-sewage-treatment/

Ilić, N., Milić, M., Beluhan, S., & Dimitrijević-Branković, S. (2023). Cellulases: From lignocellulosic biomass to improved production. *Energies, 16*(8), Article 8. https://doi.org/10.3390/en16083598

Ishikawa, M., Kojima, T., & Hori, K. (2021). Development of a biocontained toluene-degrading bacterium for environmental protection. *Microbiology Spectrum, 9*(1), https://doi.org/10.1128/spectrum.00259-21

Jodhani, S., Sebastian, J., Lee, J., Venkiteshwaran, K., Lee, H.-S., Singh, V., Ormeci, B., & Hussain, A. (2024). Acidogenic fermentation of food waste for the production of short-chain fatty acids: The impact of inoculum type and inoculum heat pretreatment. *Fermentation, 10*(3), Article 3. https://doi.org/10.3390/fermentation10030162

John, A., & Das, A. P. (2024). Government strategies for achieving circular economy for plastic waste management in Southeast Asia. *Sustainable Management of Urban Plastic Waste Through Circular Economic Approaches*, 72–82. https://doi.org/10.1201/9781003465447-6

Kabeyi, M. J. B., & Olanrewaju, O. A. (2022). Biogas production and applications in the sustainable energy transition. *Journal of Energy, 2022*(1), 8750221. https://doi.org/10.1155/2022/8750221

Kansoh, A. L., Essam, S. A., & Zeinat, A. N. (1999). Biodegradation and utilization of bagasse with *Trichoderma reesie. Polymer Degradation and Stability, 63*(2), 273–278. https://doi.org/10.1016/S0141-3910(98)00105-0

Kaushal, J., Khatri, M., & Arya, S. K. (2021). Recent insight into enzymatic degradation of plastics prevalent in the environment: A mini—review. *Cleaner Engineering and Technology, 2*, Article 100083. https://doi.org/10.1016/j.clet.2021.100083

Khan, S., Alam, F. S., & Borah, D. (2024). Microbial cell factories in the remediation of e-wastes: An insight. *Biotechnology for the Environment, 1*(1), 4. https://doi.org/10.1186/s44314-024-00003-4

Kondakindi, V. R., Pabbati, R., Erukulla, P., Maddela, N. R., & Prasad, R. (2024). Bioremediation of heavy metals-contaminated sites by microbial extracellular polymeric substances–a critical view. *Environmental Chemistry and Ecotoxicology, 6*, 408–421. https://doi.org/10.1016/j.enceco.2024.05.002

Kumar, D. J. P., Mishra, R. K., Chinnam, S., Binnal, P., & Dwivedi, N. (2024). A comprehensive study on anaerobic digestion of organic solid waste: A review on configurations, operating parameters, techno-economic analysis and current trends. *Biotechnology Notes, 5*, 33–49. https://doi.org/10.1016/j.biotno.2024.02.001

Kumar Mahish, P., Kumar Verma, D., Ghritlahare, A., Arora, C., & Otero, P. (2024). Microbial bioconversion of food waste to bio-fertilizers. *Sustainable Food Technology, 2*(3), 689–708. https://doi.org/10.1039/D3FB00041A

Li, Y., Park, S. Y., & Zhu, J. (2011). Solid-state anaerobic digestion for methane production from organic waste. *Renewable and Sustainable Energy Reviews, 15*(1), 821–826. https://doi.org/10.1016/j.rser.2010.07.042

Manea, E. E., Bumbac, C., Dinu, L. R., Bumbac, M., & Nicolescu, C. M. (2024). Composting as a sustainable solution for organic solid waste management: current practices and potential improvements. *Sustainability, 16*(15), Article 15. https://doi.org/10.3390/su16156329

Marquez, G. P. B., Santiañez, W. J. E., Trono, G. C., Montaño, M. N. E., Araki, H., Takeuchi, H., & Hasegawa, T. (2014). Seaweed biomass of the Philippines: Sustainable feedstock for biogas production. *Renewable and Sustainable Energy Reviews, 38*, 1056–1068. https://doi.org/10.1016/j.rser.2014.07.056

Mayolo-Deloisa, K., González-González, M., & Rito-Palomares, M. (2020). Laccases in food industry: Bioprocessing, potential industrial and biotechnological applications. *Frontiers in Bioengineering and Biotechnology, 8*. https://doi.org/10.3389/fbioe.2020.00222

Menzel, T., Neubauer, P., & Junne, S. (2020). Role of microbial hydrolysis in anaerobic digestion. *Energies, 13*(21), Article 21. https://doi.org/10.3390/en13215555

Mironov, V., Zhukov, V., Efremova, K., & Brinton, W. F. (2024). Enhancing aerobic composting of food waste by adding hydrolytically active microorganisms. *Frontiers in Microbiology, 15*, 1487165. https://doi.org/10.3389/fmicb.2024.1487165

Mishra, S., Dash, D., & Das, A. P. (2024a). Aquatic microbial diversity on plastisphere: Colonization and potential role in microplastic biodegradation. *Geomicrobiology Journal, 41*(4), 312–323. https://doi.org/10.1080/01490451.2023.2209750

Mishra, S., Dash, D., & Das, A. P. (2024b). Isolation and molecular characterization of polyester-degrading bacterial strains from municipal wastewater and their application in microplastic degradation in contaminated wastewater. *Geomicrobiology Journal*. https://doi.org/10.1080/01490451.2024.2376101

Mohanty, J., Dey, S., Bala, K., & Das, A. P. (2024). Recent advances in microplastic pollution for its sustainable management. *Sustainable Management of Urban Plastic Waste Through Circular Economic Approaches*, 27–48. https://doi.org/10.1201/9781003465447-3

Mohanty, M., Tripathy, B., Koteswara Reddy, G., & Das, A. P. (2024). Greener technologies for establishing restraint over microplastic pollution. *Sustainable Microbial Technology for Synthetic and Cellulosic Microfiber Bioremediation*, 239–258. https://doi.org/10.1007/978-3-031-62600-5_13

Mokrani, S., Nabti, E.-H., Mokrani, S., & Nabti, E.-H. (2024). Recent status in production, biotechnological applications, commercial aspects, and future prospects of microbial enzymes: A comprehensive review. *International Journal of Agricultural Science and Food Technology, 10*(1), 006–020.

Nag, M., Lahiri, D., Ghosh, S., Sarkar, T., Pati, S., Das, A. P., Ram, D. K., Bhattacharya, D., & Ray, R. R. (2023). Application of microorganisms in biotransformation and bioremediation of environmental contaminant: A review. *Geomicrobiology Journal*, 1–18. https://doi.org/10.1080/01490451.2023.2261443

Nicolescu, C. M., Bumbac, M., Buruleanu, C. L., Popescu, E. C., Stanescu, S. G., Georgescu, A. A., & Toma, S. M. (2023). Biopolymers produced by lactic acid bacteria: Characterization and food application. *Polymers, 15*(6), Article 6. https://doi.org/10.3390/polym15061539

Noor, R. S., Shah, A. N., Tahir, M. B., Umair, M., Nawaz, M., Ali, A., Ercisli, S., Abdelsalam, N. R., Ali, H. M., Yang, S. H., Ullah, S., & Assiri, M. A. (2024). Recent trends and advances in additive-mediated composting technology for agricultural waste resources: A comprehensive review. *ACS Omega, 9*(8), 8632–8653. https://doi.org/10.1021/acsomega.3c06516

Novoveská, L., Zapata, A. K. M., Zabolotney, J. B., Atwood, M. C., & Sundstrom, E. R. (2016). Optimizing microalgae cultivation and wastewater treatment in large-scale offshore photobioreactors. *Algal Research, 18*, 86–94. https://doi.org/10.1016/j.algal.2016.05.033

Ojo, A. O., & de Smidt, O. (2023). Lactic acid: A comprehensive review of production to purification. *Processes, 11*(3), Article 3. https://doi.org/10.3390/pr11030688

Okopi, S., Li, Y., & Xu, F. (2024). Biomass digestion. In M. A. Abraham (Ed.), *Encyclopedia of sustainable technologies* (2nd ed., pp. 236–251). Elsevier. https://doi.org/10.1016/B978-0-323-90386-8.00051-6

Pal, P., Singh, A. K., Srivastava, R. K., Rathore, S. S., Sahoo, U. K., Subudhi, S., Sarangi, P. K., & Prus, P. (2024). Circular bioeconomy in action: Transforming food wastes into renewable food resources. *Foods, 13*(18), Article 18. https://doi.org/10.3390/foods13183007

Palaniveloo, K., Amran, M. A., Norhashim, N. A., Mohamad-Fauzi, N., Peng-Hui, F., Hui-Wen, L., Kai-Lin, Y., Jiale, L., Chian-Yee, M. G., Jing-Yi, L., Gunasekaran, B., & Razak, S. A. (2020). Food waste composting and microbial community structure profiling. *Processes, 8*(6), Article 6. https://doi.org/10.3390/pr8060723

Panahi, H. K. S., Dehhaghi, M., Dehhaghi, S., Guillemin, G. J., Lam, S. S., Aghbashlo, M., & Tabatabaei, M. (2022). Engineered bacteria for valorizing lignocellulosic biomass into bioethanol. *Bioresource Technology, 344*, Article 126212. https://doi.org/10.1016/j.biortech.2021.126212

Pathma, J., & Sakthivel, N. (2012). Microbial diversity of vermicompost bacteria that exhibit useful agricultural traits and waste management potential. *Springerplus, 1*(1), 26. https://doi.org/10.1186/2193-1801-1-26

Paul, A., Dey, S., Ram, D. K., & Das, A. P. (2023). Hexavalent chromium pollution and its sustainable management through bioremediation. *Geomicrobiology Journal*, 1–11. https://doi.org/10.1080/01490451.2023.2218377

Pei, S., Fan, X., Qiu, C., Liu, N., Li, F., Li, J., Qi, L., & Wang, S. (2024). Effect of biochar addition on the anaerobic digestion of food waste: Microbial community structure and methanogenic pathways | Water Science & Technology | IWA Publishing. *Water Science and Technology, 90*(3). https://doi.org/10.2166/wst.2024.199

Philips, J. G., Martin-Avila, E., & Robold, A. V. (2022). Horizontal gene transfer from genetically modified plants—Regulatory considerations. *Frontiers in Bioengineering and Biotechnology, 10*, Article 971402. https://doi.org/10.3389/fbioe.2022.971402

Pradhan, G., Tripathy, B., Ram, D. K., Digal, A. K., & Das, A. P. (2023). Bauxite mining waste pollution and its sustainable management through bioremediation. *Geomicrobiology Journal*, 1–10. https://doi.org/10.1080/01490451.2023.2235353

Prasanna Kumar, D. J., Mishra, R. K., Chinnam, S., Binnal, P., & Dwivedi, N. (2024). A comprehensive study on anaerobic digestion of organic solid waste: A review on configurations, operating parameters, techno-economic analysis and current trends. *Biotechnology Notes, 5*, 33–49. https://doi.org/10.1016/j.biotno.2024.02.001

Priyadarshini, B., Santhosh Kumar, M., & Das, A. P. (2024a). Microbial degradation of synthetic plastics released from COVID-19 generated single use disposable face masks. *Environmental Science and Engineering, Part, F2827*, 1–16. https://doi.org/10.1007/978-3-031-50840-0_1

Priyadarshini, S., Jagatee, S., & Das, A. P. (2024b). Synthetic fabrics and microfiber pollution–an assessment of their global impact. *Environmental Science and Engineering, Part, F2333*, 137–157. https://doi.org/10.1007/978-3-031-51792-1_8

Rajput, A. A., Zeshan, & Hassan, M. (2021). Enhancing biogas production through co-digestion and thermal pretreatment of wheat straw and sunflower meal. *Renewable Energy, 168*, 1–10. https://doi.org/10.1016/j.renene.2020.11.149

Rath, P. P., Das, K., & Pattanaik, S. (2022). Microbial activity during composting and plant growth impact: A review. *Journal of Pure & Applied Microbiology, 16*(1). https://microbiologyjournal.org/microbial-activity-during-composting-and-plant-growth-impact-a-review/

Rathour, R. K., Sharma, D., Ullah, S., Mahmoud, E.-H.M., Sharma, N., Kumar, P., Bhatt, A. K., Ahmad, I., & Bhatia, R. K. (2024). Bacterial–microalgal consortia for bioremediation of textile industry wastewater and resource recovery for circular economy. *Biotechnology for the Environment, 1*(1), 6. https://doi.org/10.1186/s44314-024-00005-2

Ray, S. K., Dey, S., Das, A. P., & John, A. (2024). Microplastic waste management through greener technologies : A review. *Sustainable Management of Urban Plastic Waste Through Circular Economic Approaches*, 237–261. https://doi.org/10.1201/9781003465447-14

Robinson, P. K. (2015). Enzymes: Principles and biotechnological applications. *Essays in Biochemistry, 59*, 1–41. https://doi.org/10.1042/bse0590001

Roopnarain, A., Rama, H., Ndaba, B., Bello-Akinosho, M., Bamuza-Pemu, E., & Adeleke, R. (2021). Unravelling the anaerobic digestion 'black box': Biotechnological approaches for process optimization. *Renewable and Sustainable Energy Reviews, 152*, Article 111717. https://doi.org/10.1016/j.rser.2021.111717

Rottinghaus, A. G., Ferreiro, A., Fishbein, S. R. S., Dantas, G., & Moon, T. S. (2022). Genetically stable CRISPR-based kill switches for engineered microbes. *Nature Communications, 13*(1), 672. https://doi.org/10.1038/s41467-022-28163-5

Rózsenberszki, T., Kalauz-Simon, V., Fejes, R., Bakonyi, P., Koók, L., Kurdi, R., Nemestóthy, N., & Bélafi-Bakó, K. (2025). Carboxylic acid production from kitchen waste and sewage sludge digestate inoculation via acidogenic fermentation at high organic load. *Waste and Biomass Valorization*. https://doi.org/10.1007/s12649-025-02887-z

Sabari, V. L. D., Rajmohan, G., S b, R., S, S., Nagasubramanian, K., G, S. K., & Venkatachalam, P. (2025). Improving the binding affinity of plastic degrading cutinase with polyethylene terephthalate (PET) and polyurethane (PU); an in-silico study. *Heliyon, 11*(2), e41640. https://doi.org/10.1016/j.heliyon.2025.e41640

Sadegh, H., Ali, G. A. M., Gupta, V. K., Makhlouf, A. S. H., Shahryari-ghoshekandi, R., Nadagouda, M. N., Sillanpää, M., & Megiel, E. (2017). The role of nanomaterials as effective adsorbents and their applications in wastewater treatment. *Journal of Nanostructure in Chemistry, 7*(1), 1–14. https://doi.org/10.1007/s40097-017-0219-4

Saelor, S., Kongjan, P., Prasertsan, P., Mamimin, C., & O-Thong, S. (2024). Enhancing thermophilic methane production from oil palm empty fruit bunches through various pretreatment methods: A comparative study. *Heliyon, 10*(20), e39668. https://doi.org/10.1016/j.heliyon.2024.e39668

Sahu, S., & Das, A. P. (2024). *Microplastics degradation mechanism by aquatic plastiomes: A review*. 213–226. https://doi.org/10.1007/978-3-031-50840-0_12
Sahu, A., Dey, S., Mallick, K., Priyadarsini, S., Behera, I. D., & Das, A. P. (2024a). A review on advanced technology for sustainable management of synthetic microplastic waste. *Sustainable Management of Urban Plastic Waste Through Circular Economic Approaches*, 262–279. https://doi.org/10.1201/9781003465447-15
Sahu, A., Dey, S., Mallick, K., Priyadarsini, S., Behera, I. D., & Das, A. P. (2024b). A review on advanced technology for sustainable management of synthetic microplastic waste. *Sustainable Management of Urban Plastic Waste Through Circular Economic Approaches*, 262–279. https://doi.org/10.1201/9781003465447-15
Saini, J. K., Saini, R., & Tewari, L. (2015). Lignocellulosic agriculture wastes as biomass feedstocks for second-generation bioethanol production: Concepts and recent developments. *3 Biotech, 5*(4), 337–353. https://doi.org/10.1007/s13205-014-0246-5
Sánchez, C. (2009). Lignocellulosic residues: Biodegradation and bioconversion by fungi. *Biotechnology Advances, 27*(2), 185–194. https://doi.org/10.1016/j.biotechadv.2008.11.001
Saravanan, A., Senthil Kumar, P., Khoo, K. S., Show, P.-L., Femina Carolin, C., Fetcia Jackulin, C., Jeevanantham, S., Karishma, S., Show, K.-Y., Lee, D.-J., & Chang, J.-S. (2021). Biohydrogen from organic wastes as a clean and environment-friendly energy source: Production pathways, feedstock types, and future prospects. *Bioresource Technology, 342*, Article 126021. https://doi.org/10.1016/j.biortech.2021.126021
Satapathy, S., Kar, D., Priyadarsini, S., Das, A. P., & Kuanar, A. (2024). Application of microorganisms in emerging waste management strategies and remediation of synthetic and cellulosic microfibers. *Sustainable Microbial Technology for Synthetic and Cellulosic Microfiber Bioremediation*, 183–206. https://doi.org/10.1007/978-3-031-62600-5_10
Sebahire, F., Faridullah, F., Irshad, M., Bacha, A. U. R., Hafeez, F., & Nduwamungu, J. (2024). Effect of biochar on composting of cow manure and kitchen waste. *Land, 13*(10), Article 10. https://doi.org/10.3390/land13101545
Simic, M. (2023). Lemvig Biogas. *TrackMyElectricity*. https://www.trackmyelectricity.com/powerplants/lemvig-biogas/
Singh, S. P., & Singh, P. (2015). Effect of temperature and light on the growth of algae species: A review. *Renewable and Sustainable Energy Reviews, 50*, 431–444. https://doi.org/10.1016/j.rser.2015.05.024
Team, E. (2020). CleanMaxx is designed to solve majority of wastewater problems. *B2B Purchase*. https://b2bpurchase.com/cleanmaxx-is-designed-to-solve-majority-of-wastewater-problems-priyanka-khaire-organica-biotech-pvt-ltd/
Tournier, V., Topham, C. M., Gilles, A., David, B., Folgoas, C., Moya-Leclair, E., Kamionka, E., Desrousseaux, M.-L., Texier, H., Gavalda, S., Cot, M., Guémard, E., Dalibey, M., Nomme, J., Cioci, G., Barbe, S., Chateau, M., André, I., Duquesne, S., & Marty, A. (2020). An engineered PET depolymerase to break down and recycle plastic bottles. *Nature, 580*(7802), 216–219. https://doi.org/10.1038/s41586-020-2149-4
Tripathy, B., Sahoo, P. P., Sundaray, H., & Das, A. P. (2024). In-vitro biodegradation of discarded marine microplastics across the eastern coast of the Bay of Bengal, India using Exiguobacterium sp. *Environmental Chemistry and Ecotoxicology, 6*, 236–247. https://doi.org/10.1016/J.ENCECO.2024.07.003
Vo, T. P., Danaee, S., Chaiwong, C., Pham, B. T., Poddar, N., Kim, M., Kuzhiumparambil, U., Songsomboon, C., Pernice, M., Ngo, H. H., Ralph, P. J., & Vo, P. H. N. (2024). Microalgae-bacteria consortia for organic pollutants remediation from wastewater: A critical review. *Journal of Environmental Chemical Engineering, 12*(6), Article 114213. https://doi.org/10.1016/j.jece.2024.114213
Wainaina, S., Lukitawesa, Kumar Awasthi, M., & Taherzadeh, M. J. (2019). Bioengineering of anaerobic digestion for volatile fatty acids, hydrogen or methane production: A critical review. *Bioengineered, 10*(1), 437–458. https://doi.org/10.1080/21655979.2019.1673937

Wang, H., & Zhou, Q. (2024). Potential application of bioelectrochemical systems in cold environments. *Science of the Total Environment, 927*, Article 172385. https://doi.org/10.1016/j.scitotenv.2024.172385

Wang, Z., Hu, Y., Wang, S., Wu, G., & Zhan, X. (2023). A critical review on dry anaerobic digestion of organic waste: Characteristics, operational conditions, and improvement strategies. *Renewable and Sustainable Energy Reviews, 176*, Article 113208. https://doi.org/10.1016/j.rser.2023.113208

Werkneh, A. A. (2022). Biogas impurities: Environmental and health implications, removal technologies and future perspectives. *Heliyon, 8*(10), Article e10929. https://doi.org/10.1016/j.heliyon.2022.e10929

Westerholm, M., Schnürer, A., Westerholm, M., & Schnürer, A. (2019). Microbial responses to different operating practices for biogas production systems. In *Anaerobic digestion*. IntechOpen. https://doi.org/10.5772/intechopen.82815

Zabed, H., Sahu, J. N., Suely, A., Boyce, A. N., & Faruq, G. (2017). Bioethanol production from renewable sources: Current perspectives and technological progress. *Renewable and Sustainable Energy Reviews, 71*, 475–501. https://doi.org/10.1016/j.rser.2016.12.076

Zdarta, J., Meyer, A. S., Jesionowski, T., & Pinelo, M. (2018). A general overview of support materials for enzyme immobilization: Characteristics, properties, practical utility. *Catalysts, 8*(2), Article 2. https://doi.org/10.3390/catal8020092

Zhang, B., He, P., Lü, F., Shao, L., & Wang, P. (2007). Extracellular enzyme activities during regulated hydrolysis of high-solid organic wastes. *Water Research, 41*(19), 4468–4478. https://doi.org/10.1016/j.watres.2007.06.061

Zero-Waste Manufacturing: A Green Chemistry Approach

Prasann Kumar

1 Introduction

Zero-waste manufacturing is an emerging concept in which waste generation in every stage of a product's lifecycle is reduced or eliminated. The strict definition of zero-waste manufacturing is that all materials introduced into a system are finally reused, recycled, or biodegraded by the industrial system, with no residual waste released externally to the environment. It derives from the circular economy model, which is based on continuously cycling resources within a closed loop, reducing the environmental impact and improving resource productivity (Höfer et al., 2024; Das et al., 2025; G et al., 2025; Waseem et al., 2024). Zero-waste production is based on the idea that waste is not a necessary production outcome, but a design error. Conventional-linear manufacturing extract, make, use, and dispose of is, by nature, unsustainable because it depends on limited resources and produces pollution. On the contrary, zero-waste systems are conceived based on a life cycle view, focusing on eco-design, clean production, and end-of-life reclamation. Zero-waste industries also emphasize the design of barrier-free products so that they can be easily disassembled, their materials captured, and parts reused. Technological advancements offer a huge leap forward in the pursuit of zero-waste manufacturing (Koteswara et al., 2025; Ray et al., 2025). Sophisticated IT developments and data analytics, machine learning, and advanced automation are being used to monitor material flows, forecast waste hotspots, and optimize process parameters on the fly (Abdellatief & Tawalbeh, 2025; Zhou et al., 2025). In addition, connecting with Industry 4.0 solutions like the digital twins and smart sensors makes tracking resource efficiency and waste generation in real-time easy, which helps manufacturers take corrective

P. Kumar (✉)
Department of Agronomy, School of Agriculture, Lovely Professional University, Phagwara, Punjab, India
e-mail: prasann0659@gmail.com

S. Ghosh et al. (eds.), *Advances in Organic Waste Conversion Through Industrial Biotechnology*, Sustainable Environmental Waste Management Strategies,
https://doi.org/10.1007/978-3-032-11377-1_15

measures then and there. In the systems thinking approach, zero waste manufacturing isn't just about reducing physical waste; it's about reducing energy and water wastage and emissions. In such systems, energy recovery from unavoidable by-products that cannot be avoided, recycled, or reused (e.g. heat integration or biogas production from organic waste) is a key feature. Additionally, encouraging industrial symbiosis where one industry's waste becomes another's resource means thinking in zero waste terms on a macro scale. The shift towards zero-waste production does come with challenges. We need a change in manufacturing culture, significant investments in capital for cleaner processes, and policies to push this forward. However, the broader economic, environmental, and social benefits, such as lower regulatory risk, the best leverage of the corporate image, and good community relations, make it a good fit for the direction of modern industry. In synergy with zero-waste pathways, green chemistry represents the science underlying sustainable manufacturing by engendering the ability to design chemical products and processes that minimize or eliminate the origin and generation of hazardous substances. Based on the twelve guiding principles of Anastas and Warner, green chemistry focuses on waste prevention, atom economy, safer solvents, energy efficiency, and renewable feedstocks. In sustainable manufacturing, green chemistry can create less hazardous materials and processes (Sahu & Das, 2024; Swain et al., 2025). Eco-friendly strategies, such as replacing volatile organic solvents with water or ionic liquids, can minimize emissions, maximize occupational safety, and reduce operating costs (Khaled et al., 2025; Perera & Ranjith, 2025). At the concept level, catalysis, an idea central to green chemistry, makes reactions more selective, reduces the number of raw materials used, and prevents unwanted by-products. Furthermore, green chemistry favours using renewable resources by promoting bio-based feedstocks and enzymatic or microbial catalysis, in which only mild conditions are needed and less contamination of the environment is produced. This aligns with international sustainability directives like the EU's REACH, the US EPA's Safer Choice Programme, and the United Nations Sustainable Development Goals, specifically SDG 12. However, green chemistry can bring innovation in product design—biodegradable products, non-toxic pharmaceuticals, benign dyes, and green products. Still, it can also do what consumers need it to do. Notwithstanding the advantages, green chemistry is hampered by scale-up, cost efficacy, and efficiency-related challenges, which need economic analysis based on lifecycle, policies, incentives, and research and development to be addressed. To encourage the implementation of industrial sustainability, several waste minimization concepts provide practical guidelines for decreasing the amount and toxicity of waste. These approaches are usually organized in a pollution prevention hierarchy: first, source reduction, then reuse, recycling, energy recovery, and finally, safe disposal. Source reduction is the most favoured method, which means modifying the product or process to use less material so that hazardous material is no longer a part. Prime examples include lean manufacturing, process redesign, and substituting less hazardous inputs (Sai Bharadwaj et al., 2025; Sheldon, 2024). Process modification and optimization, such as introducing the closed-loop water system, increased heat integration, and improved process conditions, often result in significant waste reduction. Processes such as membrane filtration and supercritical fluid extraction reduce solvent use and make it possible to recover useful by-products. Material substitution is a second important strategy of replacing toxic or non-renewable raw materials with safer and more sustainable alternatives (e.g. lead stabilizers with calcium-zinc, chrome-based with electrodeposited coatings). Reuse and recycle measures like solvent recovery, metal scrap recycling, plastic regrind recovery, and disposal costs are reduced by saving raw material input costs. Close-the-loop recycling guarantees that waste is returned to the production chain and follows the rules of circular manufacturing. Energy and resource recovery technologies, including anaerobic digestion, pyrolysis, and incineration with energy recovery, can recover waste in an energy format and are alternatives to sending waste to landfills or using fossils. Likewise, nutrient recovery from industrial wastewater would benefit agriculture and food processing industries. Routine waste audits and use of environmental management systems, including ISO 14001, provide systematic processes for evaluating inefficiencies, establishing reduction goals, and documenting them (Sakthimurugan et al., 2025; Shen et al., 2024). Employee involvement via education and awareness programmes is also important because human behaviour plays an important role in the outcome of waste. This requires appropriate handling of the materials, regular maintenance, and waste separation. Additionally, incorporating Design for Environment (DfE) and Life Cycle Assessment (LCA) considerations into product design helps manufacturers measure environmental factors throughout the product lifecycle, enabling better-informed sustainable design decisions. When used together, such waste minimization methods decrease environmental load while providing economic benefits, including resource, raw material, and garbage cost savings. Combining zero-waste manufacturing, green chemistry, and a whole basket of waste minimization approaches is important in developing sustainable industrial ecosystems. These are mutually reinforcing and supportive concepts: zero-waste manufacturing supplies the systems perspective, green chemistry provides the enabling science, and waste reduction provides the practical roadmap (Abu Jadayil et al., 2025; Thamaraiselvan & Wickramasinghe, 2025). Combining these connected methods is key to environmentally sound, market-ready manufacturing in a world of tightening environmental restrictions and informed consumers. Connecting innovation with environmental stewardship allows businesses to achieve sustainability goals and underwrite long-term competitive advantage, resilience, and shared value generation (Fig. 1).

2 Principles of Green Chemistry in Zero-Waste Manufacturing

Implementing green chemistry in zero-waste manufacturing is a paradigm shift in industrial ecology. Through the prevention of waste, atom economy, use of renewable feedstocks, designing of safer chemicals, and energy efficiency, industries can

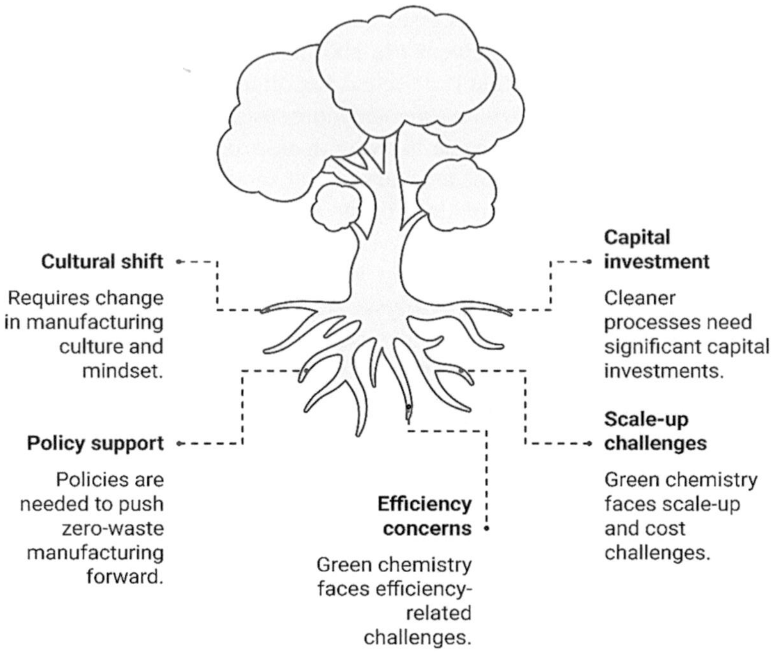

Fig. 1 Hindered adoption of zero-waste manufacturing

significantly minimize their environmental impact while at the same time being economically competitive. This paradigm is based on preventing waste, rather than cleaning it up after its creation, and represents a push for waste elimination at the source (Ahmed et al., 2024; Dash et al., 2025; Davies & Hirst, 2025). This strategy helps reduce environmental pollution and decrease waste treatment and management costs. To the extent that processes can be changed to prevent the production of hazardous substances, the manufacturers can have cleaner production lines and safer workplaces (Das et al., 2024a, 2024b; Dey et al., 2024a, 2024b). Atom economy, another principle of green chemistry, aims to achieve maximum utilization of all materials used (including the reactants and catalysts) in the final product process. There's a high atom economy, meaning a relatively small amount of resources are wasted in preparing the desired product, and little, if any, by-product is formed. This efficiency equates to less waste and less environmental impact, both objectives of zero-waste manufacturing. Transition to renewable feedstocks is crucial as well. Bio-based materials like those made from plant-based compounds diminish reliance on finite fossil resources and aid in more responsible material sourcing. These sustainably grown and often renewable materials cause biodegradable and non-toxic products, reducing the environmental burden. Creating safe chemicals and products is essential to reducing health and environmental risks (Dey et al., 2024a, 2024b; Mohanty et al., 2024a, 2024b). Manufacturers can avoid accidents and ease health and safety regulation bureaucracy by choosing inherently lower-risk materials (Ariwaodo & Olaniyan, 2024; Fuente-Ballesteros et al., 2025; Zamani et al., 2024). This concept promotes innovation to conceive alternatives that are functionally equivalent but AES-sparing. Energy savings and process optimization are also critical. By developing methods that consume less energy, particularly processes carried out at ambient temperatures and pressures, greenhouse gas emissions and the cost of the processes are reduced. Improvement of catalysis technologies and process advances promotes efficient catalytic reactions, consistent with green chemistry and zero waste production aims. When these principles are applied interactively, industries can make the systemic shift that allows for sustainability. The interplay of green chemistry and zero-waste manufacturing mitigates environmental issues and provides economic savings and social dividends. With a growing worldwide concern about natural resources exhaustion and pollution problems, it is necessary to implement these integrated approaches to ensure long-term sustainable development in the manufacturing industry.

Such a holistic model will guarantee manufacturing sustainability and induce systems thinking into the product's design, supply chain management, and life cycle analysis. Integrating the principles of green chemistry into the stages of zero-waste manufacturing, from raw material choice to end-of-life product recovery, is a manufacturing framework that conforms to nature's circular logic: what is an output becomes an input, and nothing is thrown away. The cornerstone of green chemistry is waste prevention, which encourages designers and engineers to look at each material's mechanism and life cycle to minimize the environmental burden of products and processes before they enter the environment (Priyadarshini et al., 2024; Ray et al., 2024; Sahu et al., 2024). Instead of treating the symptoms of pollution after they already exist (as violations of the endangered species act address species loss instead of preventing it), green chemistry focuses on generating less waste (source reduction) through design innovations, replacing highly toxic or bio-persistent substances with less harmful and biodegradable substances, and eliminating unnecessary chemical synthesis as a step in a process. Atom economy also provides a rational measure of the efficiency of a chemical transformation. A high atom-economy process is one in which the bulk of the atoms of the starting materials end up in the product, thereby minimizing waste. It is essential in chemical synthesis, materials science, and pharmaceutical production, as even minor improvements in atom efficiency can cause significant savings of material consumption and waste hazard (Ferro-Falla et al., 2025; Madzík et al., 2024; Murphy, 2025). Manufacturers can achieve significantly increased yield and a lowered environmental footprint of production by developing synthetic routes with fewer reaction steps, more selective catalysts, or fewer auxiliary substances (solvents, reagents).

Related to atom economy is the use of renewable feedstocks. Up to now, industrial production, the old-fashioned kind, used materials from petrochemical sources, which are either non-renewable or subject to geopolitics, all with massive carbon emissions. Green chemistry calls for the switch to renewable resources such as lignocellulosic agro-wastes, algae, and non-food oil seeds. These renewable feedstocks can support decentralized production systems, reduce transport fuels, and stimulate local economies. In addition, bio-based feedstocks are frequently biodegradable and less ecotoxic, which aligns with end-of-life options such as composting or microbial degradation in soil. With the focus of future generations of biorefineries to fractionate biomass to high-value chemicals, fuels, and materials using green solvents, enzymatic catalysts, and energy-efficient processes, it is the ideal representation of the zero-waste and green chemistry philosophies. This change is critical to the innovation of safer chemicals and products. The green chemistry concept known as 'design for degradation' (DfD) suggests that when these chemicals are finally used to make products, it is essential for them to decompose into harmless compounds, rather than persist and/or accumulate in the environment and living organisms (Becker et al., 2022; Guidara et al., 2024; Punj et al., 2023; Ramadan & Shehata, 2025). Scientists use computational toxicology, structure–activity relationship models, and predictive analytics to design molecules that work well and play nicely. For example, substituting conventional plasticizers, such as phthalates, for bio-based ones, or developing surfactants based on amino acid derivatives, can ensure product performance without posing a health and environmental risk over time. This is how product innovation becomes an instrument for sustainability, allowing industries to avoid risks and not just control them.

The energy conservation principle is also fundamental, especially in energy-intensive industries like metallurgy, cement, petrochemical, and electronics. In zero waste manufacturing, minimizing the energy input can reduce the greenhouse gas emissions and the total material throughput needed to supply the energy generation. With the latest trends in catalysis, reaction engineering, and non-solvent methodology, green chemistry also helps to become energy-efficient. Photochemical89, 90, and microwave-assisted91 reactions, for example, are frequently observed to occur under milder conditions and with more control over selectivity than classical thermal processes (Davies & Hirst, 2025; Perera & Ranjith, 2025; Sakthimurugan et al., 2025; Sheldon, 2024). In addition, continuous-flow reactors offer superior control over reaction conditions over batch processes, and result in increased product yields and less energy consumption. When coupled with renewable energy sources, like solar thermal or wind, renewable electricity, and green chemical processes, can result in almost net-zero emissions, perfectly fitting carbon–neutral and zero waste objectives. One of the key operational philosophies, process optimization, increases the efficacy of green chemistry and assists in achieving the goals of zero-waste manufacturing. Lean manufacturing principles, such as those defined in Six Sigma and Kaizen and those addressed in total productive maintenance, can be extended to embrace green chemistry metrics to pinpoint quality, cost, and environmental gaps. LCA and cradle-to-grave analysis are an inclusive assessment of ecological burdens throughout the supply chain, including extraction, processing, use, and disposal. These tools help manufacturers make the most sustainable choices at each decision point, encouraging a culture of continuous improvement where thinking on a project level extends beyond the organizational mindset (Davies & Hirst, 2025; Fuente-Ballesteros et al., 2025; Shen et al., 2024; Tawalbeh, 2025; Thamaraiselvan & Wickramasinghe, 2025).

In addition, the confluence of digitalization and green chemistry has pushed the boundaries of waste-free manufacturing. Industry 4.0 solutions such as digital twins, process analytics based on AI, and material tracking via blockchain can help streamline supply chains, reduce overproduction, and comply with green design standards. For example, with the help of predictive models, it is possible to detect waste streams in good time before they occur. Machine learning algorithms can potentially correct reaction conditions on the fly, optimizing reaction selectivity and minimizing the formation of side products. These solutions deliver process-relevant data that drives progressive implementation of green chemistry within the most complicated manufacturing enterprises. In pursuit of these principles, interdisciplinarity is essential (Dash et al., 2025; Perera & Ranjith, 2025; Shen et al., 2024; Thamaraiselvan & Wickramasinghe, 2025; Zamani et al., 2024). Chemists, chemical engineers, industrial ecologists, supply chain professionals, and sustainability advocates must collaborate to embed green chemistry in all manufacturing areas. This multidisciplinary concept helps translate molecular-level decisions to macroscopic considerations such as logistics, regulatory compliance, and market forces. For example, in converting to a green cleaning product containing safer solvents, new solvents may require different construction materials for equipment, staff training, and disposal, involving coordination among several departments'. Intersectoral partnerships between academia, industry, and government can also be critical in scaling green chemistry solutions, offering funding, technical know-how, and technological and policy incentives that drive a shift from traditional to sustainable manufacturing processes. Green chemistry and zero waste are gaining increasing recognition in public policy and regulations. Changing social, political, and legislative influences on these systems are driving the implementation of extended producer responsibility regulations (EPR), green procurement policies, and requirements for waste reduction that encourage design for the environment. International contracts such as the Paris Climate Accord and the UN Sustainable Development Goals (notably SDG 12 and SDG 13) also require companies to integrate sustainability into their activities. Achieving this often requires attention to the practices recommended by green chemistry: safe materials, waste reduction, energy efficiency, and sustainable feedstocks. Furthermore, regulations constrain demand for greener alternatives, providing an early advantage to those who pioneer them and stimulating innovation across the industry (Dash et al., 2025; Fuente-Ballesteros et al., 2025; Madzík et al., 2024; Sakthimurugan et al., 2025; Sheldon, 2024).

Consumer drive is also a significant force for change. Today's consumers are increasingly influenced by environmental activism and social media and opt for environmentally friendly, ethically sourced, and transparently made products. Marketing words like 'green chemistry certified', 'biodegradable', or 'zero waste' not only sell products but also show a company is willing to put its money where its mouth is when it comes to sustainability. Companies using green chemistry, a zero-waste approach, can stand out in crowded markets, appeal to environmentally conscious consumers, and create long-term brand loyalty. This market pressure also strengthens the business case for scaling up the deployment of sustainable chemistry. Ultimately, the 12 principles of green chemistry are a guide and an ethical compass for zero-waste manufacturing. They offer a way for industries to reconceptualize not just the how of production but the why, such that the loftiest concerns of ecological stewardship,

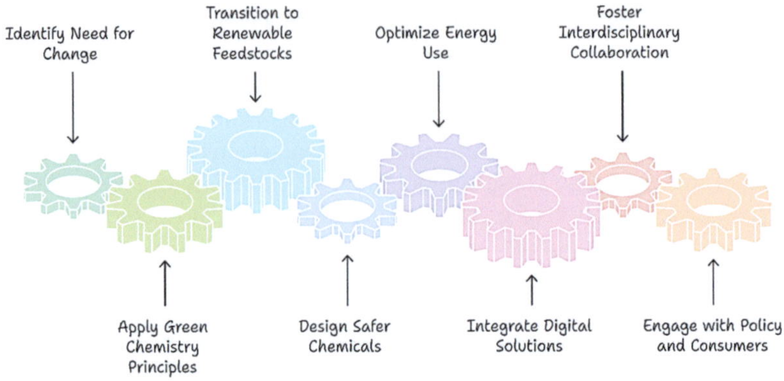

Fig. 2 Implementing green chemistry in manufacturing

public health, and intergenerational equity can be embedded in practices and prices (Dash et al., 2025; Sheldon, 2024; Shen et al., 2024; Thamaraiselvan & Wickramasinghe, 2025; Zhou et al., 2025). They are the principles of manufacturing becoming cleaner, regenerative, and securing a future where human well-being and planetary boundaries intersect. Suppose industries combine these values into research, design, and operations. In that case, they can become the agents of positive change, driving action and a shift in thinking from a vision of a linear economy to one that is circular, regenerative, and sustainable. In this noble pursuit, green chemistry is as much a science as the alchemy of a better world (Fig. 2).

3 Zero-Waste Strategies in Industrial Manufacturing

Industrial manufacturing is increasingly influenced by zero waste strategies by introducing process intensification, green synthesis, biodegradable and recyclable materials, alternative solvents, solvent-free processes, and advanced catalysis. Together, these strategies seek to prevent environmental harm and reduce resource utilization, while promoting sustainable industrial processes (Das et al., 2023, 2024a, 2024b; Ghosh et al., 2016, 2020).

In this context, process intensification and green synthesis are key in converting conventional processes into sustainable ones. Process intensification involves transforming chemical processes into more effective and efficient unit operations to use less energy, be safer, and produce less waste. These objectives have been accomplished using the microreactor technology, ultrasound, and microwave-assisted reactions. In addition, green synthesis also addresses chemical procedures that minimize or eliminate the consumption and production of hazardous substances. Both techniques significantly impact zero-waste production by maximizing conversion and minimizing waste (Abdellatief & Tawalbeh, 2025; Abu Jadayil et al., 2025; Ahmed et al., 2024; Davies & Hirst, 2025; Fuente-Ballesteros et al., 2025; Shen et al., 2024).

Using compostable and recyclable materials is another building block of zero-waste programmes. They can also dramatically decrease the waste they send to landfills by choosing natural, biodegradable, or recyclable materials that can be reprocessed into something new. For example, using biodegradable materials, including bagasse-based packaging, in the food industry has value in reducing plastic waste. Equally, designing for simple take-apart products simplifies recycling, as materials can be easily extracted for reuse.

Industrial chemistry, solvents, and solvent-free processes. Alternative solvents are also vital to reduce the environmental burden originating from manufacturing because traditional solvents are usually volatile and toxic; they risk harming health and the environment (Ghosh & Das, 2020; Mishra et al., 2023). Proponents of green chemistry recommend using safer alternatives, including supercritical and ionic liquids, with similar solvent characteristics but without the associated risks. In addition, the solvent-free procedures do not wash away any solvents, so no waste or exposure to hazardous substances is involved. Such procedures also add to safety and economics by streamlining the purification procedures and minimizing solvent recovery requirements.

The role of catalysis is to increase the efficiency and selectivity of chemical reactions, which is necessary for zero-waste manufacturing (Abu Ariwaodo & Olaniyan, 2024; Davies & Hirst, 2025; Jadayil et al., 2025; Khaled et al., 2025; Shen et al., 2024). Responses with the help of catalysts can be carried out under much milder conditions, require less activation energy, and generate fewer by-products. The catalyst design evolution, including the nanocatalyst, enables higher surface productivity during the reactions, higher selectivity, and more sustainable reactions on average. Moreover, recyclable, biodegradable catalysts can meet the requirements of green chemistry and will not cause environmental contamination due to the catalyst.

Any integration of such strategies within industrial manufacturing has to be considered from a broader perspective, including the whole product and process life cycle. Life Cycle Assessment (LCA) tools can reveal where gains can best be made and embed sustainability considerations throughout production. Moreover, there is a need for partnership between stakeholders such as researchers, industry players, and policy actors to foster innovation and best practices in the industry.

In summary, industrial manufacturing zero-waste approaches are multivariate and encompass process design, materials selection, solvent economy, and catalysis. By implementing these methods, sectors can achieve a sustainable operation that reduces environmental impact and improves competitiveness and efficiency in the global market (Mishra & Das, Mohanty et al., 2024a, 2024b; Pradhan et al., 2023).

In addition, applying zero waste to industrial production requires a change in generational philosophy and the introduction of circular economy principles. At the heart of the effort is the reengineering of production systems in which every output of one process is an input to another, forming a closed-loop system of making things. For instance, by-products or side streams like heat, gases, or solid residues can be repurposed as raw materials for other added-value products or energy sources in the chemical industry. This innovative interplay decreases the input of raw materials and the output of waste, and reduces the production cost. A case in point is the utilization of glycerol, a waste product of biodiesel manufacture, in synthesizing epichlorohydrin or as a raw material for producing biopolymers, highlighting how industrial symbiosis can contribute towards a zero-waste economy.

In addition, using digital technologies, such as AI, machine learning, and Internet of Things (IoT) in the industrial process has opened new dimensions to reach the zero-waste goal. These tools provide real-time monitoring, predictive maintenance, and process optimization, ultimately minimizing wasted resources and improving manufacturing efficiency (Ahmed et al., 2024; Fuente-Ballesteros et al., 2025; Khaled et al., 2025; Waseem et al., 2024). For example, AI-based models can forecast the best reaction conditions for chemical synthesis and, in turn, reduce experimental waste and energy use. Likewise, sensors based on IoT can track material flow and energy usage so that a warning can be sounded for the inefficient production line or interference in the operation before waste is accumulated significantly. Innovative factories like these require these feedback loops to ensure that the waste remains and that continuous improvement is achieved.

The transition to biodegradable and recyclable materials necessitates the development of new material science frameworks that achieve environmental compatibility while still offering performance. Currently, attention is being drawn to the development of polysaccharides, polylactic acid (PLA), polyhydroxyalkanoates (PHA), and other such natural polymers that can degrade under managed conditions, to attenuate the long-term consequences, in the environment, of the presence of non-degradable synthetic polymers. These materials are used in consumer products, packaging products, and the like, even in durable products such as automobile parts or building materials, with the needed degree of biodegradability and mechanical performance (Ahmed et al., 2024; Höfer et al., 2024; Zamani et al., 2024). Moreover, innovation in recycling technologies such as chemical recycling (in which polymers are broken down into monomers for repolymerization) provides a closed-loop solution to the end-of-life of plastics disposal issue.

Alternative solvents and solvent-free applications also enhance the sustainability and safety of the manufacturing environment. Supercritical fluids such as carbon dioxide ($scCO_2$) have been used in various applications, including but not limited to extraction and polymerization, due to their ability to tune solvation, non-toxicity, and easy removal after reaction. Deep eutectic solvents (DES), mainly based on biodegradable and low-toxic substances, are gaining importance as a new generation of environmentally benign and sustainable solvent systems with broad spectral solvation potential (Abu Ariwaodo & Olaniyan, 2024; Jadayil et al., 2025; Sheldon, 2024). However, over the past two decades, mechanochemistry using mechanical force rather than solvents to drive chemical reactions has emerged as a powerful new approach to pharmaceutical synthesis and materials science, greatly minimizing the toxic use of solvents and energy-hungry reaction conditions.

The momentum of catalytic technology achievements drives the precision and sustainability of industrial chemical-producing transformations. Enzymatic catalysis functions at moderate temperatures and pH levels. It has significant industrial applications in the textile, food, new pharmaceutical, and biopolymer industries as an environmentally friendly alternative for conventional chemical processes. Catalysts based on a transition metal have been developed, particularly on transition metals such as Fe, Ni, Cu, being modified to be highly active and selective, and favouring the formation of by-products with greater selectivity. Moreover, catalysts supported on recoverable materials like silica, alumina, or biochar from agricultural waste facilitate catalyst recovery and reuse in line with the circular economy concept (Abdellatief & Ahmed et al., 2024; Fuente-Ballesteros et al., 2025; Shen et al., 2024; Tawalbeh, 2025). Photocatalysis and electrocatalysis are catching attention as they drive reactions with renewable energy input, such as sunlight or electricity, contributing to a lower carbon footprint of industrial activities.

An essential aspect of zero-waste manufacturing is efficient supply chain logistics and material flows. Chores of the Day Instead of the traditional AP classes, my highschooler has been taking 'Crappy Little Jobs I' (designed to clean out bathroom windows for once) and 'Shit You Don't Want to Do' to learn all the latest manufacturing principles. Lean manufacturing. Lean is a production value that generates less solid waste. This includes techniques such as just-in-time production (minimizing inventory waste), reducing waste in the recycling process for the end of a product's life, and designing readily disassembled and recycled products (Ariwaodo & Olaniyan, 2024; Khaled et al., 2025; Madzík et al., 2024; Thamaraiselvan & Wickramasinghe, 2025). Furthermore, digital twin technologies, which replicate reality, enable producers to simulate different process layouts and material flows to determine an ideal zero-waste way forward without needing physical actions. These helps avoid expensive mistakes and waste of material in mass production.

The people and the policy aspects are instrumental in realizing manufacturing that delivers zero waste. To foster workforce readiness and create a sustainability-based culture in organizations capable of having employees at all levels internalize and deliver on the zero-waste process. Another strategic factor is the relevant policy architecture because such an institutional policy system can promote the effective adoption of green manufacturing technologies. International governments provide tax credits, research grants, and subsidies to encourage cleaner production methods and levy increasingly stringent environmental regulations to drive out polluting technology. International standards like ISO 14001 (environmental management systems) and ISO 50001 (energy management) continue to help industries methodically minimize waste and enhance environmental performance.

Zero-waste manufacturing in industry is a challenge that requires holistic and multidisciplinary perspectives to respond to technological innovation, material science, process optimization, digital enablement, and effective policy. Implementation of process intensification and green synthesis simplifies manufacturing routes and minimizes resource consumption (Abdellatief & Tawalbeh, 2025; Ariwaodo & Olaniyan, 2024; Dash et al., 2025; Sai Bharadwaj et al., 2025; Thamaraiselvan & Wickramasinghe, 2025; Zamani et al., 2024). By employing biodegradable or recyclable materials, products and by-products do not remain in the environment after their useful life has expired. Alternative solvents and solvent-less methods can replace chemical reagents, resulting in safer and cleaner production in these applications. Catalysis provides efficient and selective reactivity, preventing the formation of undesired by-products and achieving high yields. Together with a life cycle perspective, digital technologies, and policy context, these strategies make a strong foundation for turning industrial manufacturing into a zero-waste and sustainable activity. Firm-level activities provide incentives that help protect the environment and foster innovation, economic dynamism, and sustained industrial competitiveness in the long term, in the presence of global sustainability challenges (Fig. 3).

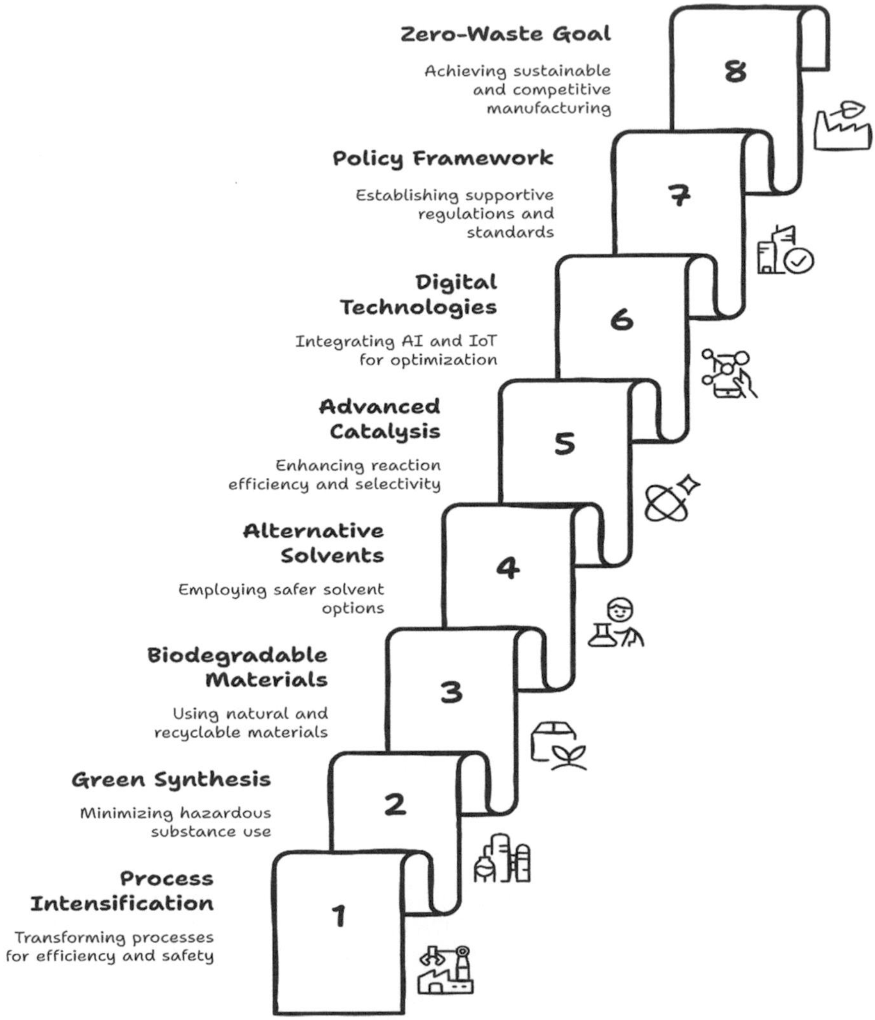

Fig. 3 Achieving zero-waste manufacturing

4 Case Studies in Zero-Waste Manufacturing

Green chemical synthesis, sustainable materials preparation, and circular economy concepts are combining to rewrite industrial manufacturing, and the paradigm for the future is being driven by environmental concern and resource conservation. Environmentally benign chemical synthesis, particularly in biorefineries and the pharmaceutical industry, focuses on processes with lower toxic, energy, and waste inputs (Dash et al., 2025; Perera & Ranjith, 2025; Shen et al., 2024; Thamaraiselvan & Wickramasinghe, 2025; Zamani et al., 2024). Biorefineries convert biomass to fuels, chemicals, and materials, and are an offering that is renewable and not based on fossil materials. In drug synthesis, green chemistry principles make the synthesis process more sustainable. Responses such as solvent replacement, catalytic reactions, and biocatalysis are used to minimize environmental implications and optimize productivity.

Sustainable materials processing centres on developing biodegradable plastics and composites to replace non-biodegradable materials. Biodegradable polymer composites (BPCs) are increasingly considered a viable solution to plastic pollution, when mechanical performances and biodegradation rates appeal to the packaging and other industries. However, more recent research has called into question the safety of some bioplastics, including starch-based products that can have the same health risks as regular plastic. This highlights the importance of characterizing biodegradable polymers to support their safety and environmental compatibility (Davies & Hirst, 2025; Höfer et al., 2024; Sakthimurugan et al., 2025; Sheldon, 2024; Thamaraiselvan & Wickramasinghe, 2025; Zamani et al., 2024).

Industrial applications of Circular Economy models are designed to promote recycling, reuse, and sustainable design, supporting the closing of the product life cycle loop. Circular strategies require rethinking business models to optimize resources and minimize waste. For example, they are testing circular business models that deliver competitive advantage by lessening the use of raw materials and creating new revenue sources. Transition to circular economy also requires embracing maintenance, repair, remanufacturing, and recycling practices, which will support decarbonization goals and add value to businesses.

Green chemical synthesis, sustainable materials, and circular economy nexus: a value chain approach—technology innovation, policy support and stakeholders' collaboration. The development of green chemistry technologies, particularly continuous reactions and synthetic biology, makes industrial processes more sustainable. And, as we transition to these models of the circular economy, we need to be able to measure and assess it (quantitatively or not), so businesses can make informed (data-based) decisions. Adopting these integrated strategies would enable industries to take steps towards more sustainable operations, which involve less environmental pollution and greater efficiency and competitiveness in the global arena. Part 6: Conforming integrated green synthesis, sustainable materials, and circular economy: towards large-scale implementation of the green synthesis approach: enabler through industrial transformation and technology standardization. Industrial ecosystems must move away from the traditional linear pattern of extract-product-use-throwaway to adopt the cyclic pattern of production outputs used in the next step in the system. In this line, the green chemical synthesis is vital in establishing such a closed-loop cycle, especially when combined with process intensification and catalysis with high selectivity (Ariwaodo & Olaniyan, 2024; Perera & Ranjith, 2025; Shen et al., 2024; Thamaraiselvan & Wickramasinghe, 2025; Zamani et al., 2024). For instance, such processes may convert lignocellulosic biomass, algal biomass, and artefacts of agriculture into chemistries such as ethanol, lactic acid, and succinic acid as precursors to polymers, solvents, and pharmaceuticals. This does away with the need to use petroleum-based compounds, thus lowering the carbon footprint and toxic emissions from the entire value chain.

One of the most critical flow chemistry applications is seen in the pharmaceutical industry, where batch processing has historically led to large amounts of waste; continuous flow synthesis and biocatalytic cascades are greener alternatives. Engineered enzymes and microbial fermentation can improve yield, selectivity, and scalability, and reduce energy consumption and solvent volumes. These technologies are in line with the concepts of green chemistry (e.g. atom economy, minimal use of auxiliary substances, inherently safer design) and facilitate economic manufacturing in agreement with the more stringent (future) requirements concerning the environment. Real-time analytics and process control also facilitate green pharmaceutical synthesis through in-process modification to optimize reaction conditions to minimize by-products and waste.

Concurrently, the catalysis and implementation of bio-degradable plastics and composites have been accelerating, especially in the context of the present plastics pollution crisis. Bio-based polymers, including polyhydroxyalkanoates (PHAs), polylactic acid (PLA), and bio-PET (polyethylene terephthalate), have been developed with similar or better properties than traditional plastics (Ahmed et al., 2024; Punj et al., 2023; Shen et al., 2024; Zamani et al., 2024). They decompose in the environment under controlled conditions and start to disappear in ecosystems, thereby decreasing the chances of the formation of microplastics. In addition, by adding natural fibres (e.g. hemp, jute, flax) and agricultural residues in composite matrices, the mechanical performances are improved, the environmental impact is reduced, and the waste streams are valorized from a circular economy perspective. However, any change to more sustainable materials must also consider the lifecycle: where they are sourced, how they are processed, where they end up at the end of their useful lives, and potential toxicity. Innovative environmental analyses, that is, approaches going beyond Life Cycle Assessment (LCA), guarantee that these substitutes offer real contributions to sustainability (Ahmed et al., 2024; Dash et al., 2025; Khaled et al., 2025; Sakthimurugan et al., 2025; Thamaraiselvan & Wickramasinghe, 2025; Zamani et al., 2024; Zhou et al., 2025).

Circular economy frameworks offer ways to incorporate these material innovations in wider industrial operations. Companies can use business models such as product-service systems, leasing, reverse logistics, and EPR to govern material flows and value material resources with a high recovery rate. For example, electronic and industrial equipment firms are moving towards modular designs that make it easy to disassemble, upgrade, reuse, and replace components. Such practices help to alleviate the need for raw materials, prolong the life of products, and open up new markets for remanufactured products. Construction The construction industry uses materials passports and building information modelling (BIM) to follow the journey of materials and to design for reuse, to help reduce demolition waste, and deliver circular building design.

Digital technologies such as blockchains, IoT, and AI are also making possible and transparent circular economy operations. Blockchain technology can create trusted and immutable records of product journeys, increasing trust and traceability in the secondary materials markets. IoT-connected sensors can track product use, forecast maintenance, and alert that it has reached the end-of-life, triggering pre-emptive recovery and recycling. Advanced analytics fuelled by AI can accelerate the optimization of supply chains, predict demand for secondary materials, and enhance waste sorting and material recovery processes. Together with green chemical processes and sustainable material advancements, these technologies help form the basis for a digital, zero-waste industrial future.

Policy drivers and regulatory incentives are fundamental to the process. They need strong government policies to ensure: support for the R&D of green chemistry, subsidies for the uptake of sustainable materials, and the enforcement of circular economy requirements. Legislation about extended producer responsibility (EPR) can encourage manufacturers to design for longevity and recyclability, and green public procurement can also stimulate market demand for sustainable products. International collaboration is essential to harmonize standards, enable circular trade beyond borders, and help developing countries shift production towards green industrial models.

Just as crucial is teaching and mentoring the next generation of green chemists, engineers, and practitioners of the circular economy. University and technical curricula need to change to include interdisciplinary training for systems thinking, environmental science, materials chemistry, and industrial engineering. Workforce development programmes, certifications, and dissemination channels can also hasten the adoption of best practices within and between industries. Industry-academia relationships, public–private partnerships, and global innovation networks are key for translating scientific advances into commercially feasible and socially inclusive solutions.

This green and circular BPM lifecycle will only be scalable and impactful if performance metrics and reporting frameworks are internationally agreed on and broadly accepted. Environmental, social, and governance (ESG) reporting, in alignment with initiatives like the Global Reporting Initiative (GRI) and Task Force on Climate-related Financial Disclosures (TCFD), can assist companies to set their baseline progress and to communicate with stakeholders transparently. Circularity metrics, including the material circularity index (MCI), product-level reuse rate, and recycling efficiency, can offer practical starting points for evaluating how well a company practices circular principles (Ariwaodo & Olaniyan, 2024; Perera & Ranjith, 2025; Thamaraiselvan & Wickramasinghe, 2025; Waseem et al., 2024; Zamani et al., 2024).

Combined with carbon accounting, water use, and energy intensity measurements, they provide a 360° perspective on industrial sustainability.

On the horizon, integrating green chemistry synthesis with sustainable materials development and circular economy design will define the future of industrial manufacturing. The industries that shift to this model will shrink their ecological footprint and open new economic doors; they will be more resilient to shocks in supply chains and enjoy closer interaction with consumers who care about the environment. This positive impact of cleaner production can be even more effective when interconnected to industrial symbiosis networks, where waste streams or by-products from a particular company are used as raw materials from another, developing clusters of companies that can transfer their waste to other types of products. For example, one possible adjacent algae biorefinery can use carbon dioxide from a cement plant. Concurrently, the waste heat of an industrial furnace can aid a local district heating system or hydroponic greenhouses (Abu Dash et al., 2025; Jadayil et al., 2025; Khaled et al., 2025; Shen et al., 2024; Zamani et al., 2024; Zhou et al., 2025).

A genuinely zero-waste industrial future need not be a pipe dream, but a reality. If all stakeholders, governments, scientists, manufacturers, and consumers, work together, we can level up and make systemic change a reality. Green chemical synthesis supplies the methods and practices needed for cleaner and more efficient manufacturing, sustainable materials constitute a framework for developing solutions to minimize environmental burden, and circular economy models ensure resources are used for as high value as possible for as long as possible. These elements compose a sustainable and regenerative industrial model framework and integrate economic prosperity with environmental responsibility and social health (Fig. 4).

5 Challenges and Opportunities

The need for zero-waste production is also becoming widely regarded as paramount for sustainable industry. However, such a transition is hampered by a complex set of technological and economic barriers, policy provisions, and the necessity for integrating new and innovative sustainable trends (Abdellatief & Tawalbeh, 2025; Höfer et al., 2024; Perera & Ranjith, 2025; Sai Bharadwaj et al., 2025). One of the significant technical challenges towards zero-waste manufacturing remains incorporating advanced technologies to facilitate and promote the use of more advanced concepts towards resource efficiency and waste reduction. Most current manufacturing systems are developed on linear models, which are not prepared to be circular, and their retrofitting is technically complex and economically infeasible. Standardized technology and operational procedures for waste minimization are also unavailable, making it challenging to transfer zero-waste practices to other industries.

Financially, the startup costs for zero-waste manufacturing can be significant. Small and medium-sized businesses (SMEs) can have difficulty finding spare cash to invest in new technology, training staff, and redesigning processes. Furthermore, the ROI in zero-waste programmes tends to be over a longer time horizon, which can

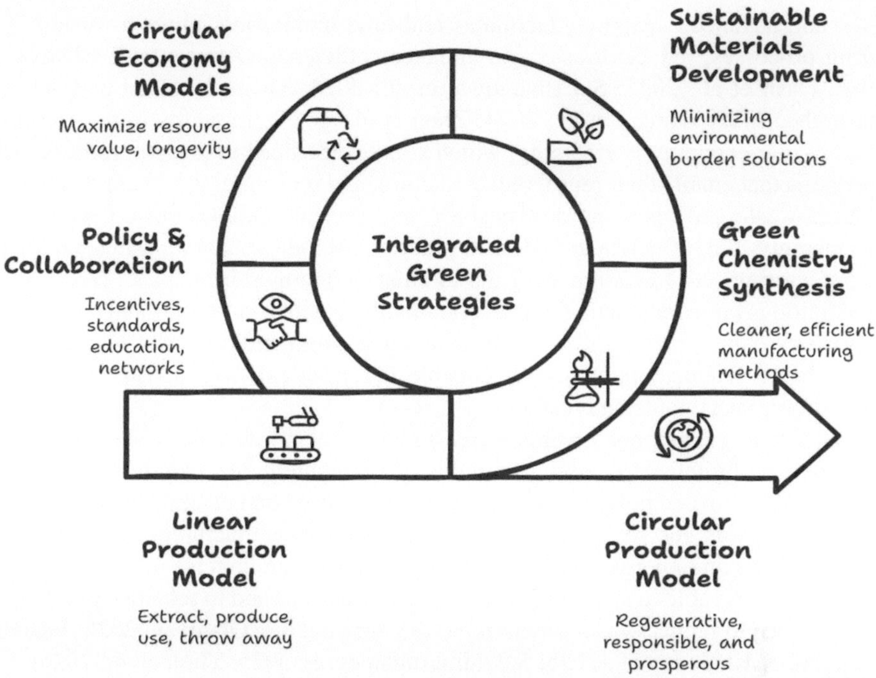

Fig. 4 Achieving sustainable industrial manufacturing

dissuade short-term, financially motivated businesses. (2011) demonstrates the role of policy and regulatory levers in encouraging zero-waste manufacturing through setting targets and standards, using incentives, and driving the uptake of sustainable approaches. Governments may introduce laws setting waste volume reduction goals, requiring the party responsible to use recycled materials, and punishing those who fail to comply (Abdellatief & Dash et al., 2025; Khaled et al., 2025; Shen et al., 2024; Tawalbeh, 2025; Thamaraiselvan & Wickramasinghe, 2025; Zamani et al., 2024). Extending producer responsibility (EPR) policies, for example, puts the onus on manufacturers throughout the entire lifecycle of their products, and in doing so, offers an incentive for products to be designed for easier recycling and reuse.

In addition, financial incentives and tax breaks can defray some of the costs associated with switching over to zero-waste manufacturing. These policies can help more companies generate a return on investment for the sustainable tech and practices we need by lowering the economic hurdles. In addition, public–private partnership alliances may encourage sharing information and establishing sector-level standards as a consultative means to sustainability. The new application, One Click LCA Industrial, improves the construction sector, which is undergoing considerable changes as sustainable industrial practices gain greater visibility. The adoption of Industry 4.0 technologies, including Internet of Things (IoT), artificial intelligence

(AI), and advanced analytics, facilitates real-time monitoring, improves manufacturing processes, reduces waste, and enhances efficiency (Ariwaodo & Olaniyan, 2024; Dash et al., 2025; Sai Bharadwaj et al., 2025; Thamaraiselvan & Wickramasinghe, 2025; Zamani et al., 2024; Zhou et al., 2025). In addition, the concept of a circular economy is emerging, which stresses the need to design products and processes that enable their reuse, remanufacture, and recycling. It's an approach that reduces waste and opens up new business and revenue opportunities. Companies increasingly look at new materials and processes (biodegradable plastics, additive manufacturing, etc.) to align with the circular economy's principles. Zero-waste production is far from practically or economically feasible, but policy and regulation play a key part in making the shift to zero-waste production happen. Overcoming these obstacles and adopting new sustainable directions can help industries progress and become sustainable and resilient.

Landscaping is tipping towards increasingly circular, low-impact operations. Of these trends, digitalization alongside Industry 4.0 technologies like the IoT, AI and machine learning are poised to transform waste tracking and resource management. by smart sensors and real-time analytics that enable manufacturers in monitoring material flows and energy consumptions by offering more precise information and by making them able to take immediate corrective actions and to schedule predictive maintenance to reduce waste generation (Abu Ariwaodo & Olaniyan, 2024; Jadayil et al., 2025; Khaled et al., 2025; Sakthimurugan et al., 2025; Zhou et al., 2025). In addition, blockchain technology is being widely investigated to add transparency and traceability across the supply chain and the end of life, to make responsible material sourcing, manufacturing, post-consumption, and recycling more accountable and compliant with the circular economy.

Three-dimensional printing, also known as additive manufacturing, is yet another radical trend that plays into the potential to achieve zero waste. Since additive manufacturing allows materials only to be deposited where needed, it results in much less offcut and scrap material than subtractive processes. This also enables on-demand part customization, significantly reducing the amount of inventory that must be kept on hand and associated risks for overproduction and obsolescence. In addition, progress in materials science facilitates the development of new generations of sustainable materials such as biodegradable polymers, biobased composites, and materials made from agricultural waste, which are now increasingly used in sectors like packaging, automotive, and electronics.

Green chemistry represents a cornerstone of sustainable manufacturing, particularly in the pharmaceutical industry, where the environmental impact of synthesis can be substantial. Developments in catalytic systems, solvent-free reactions, and biocatalysis also minimize the use of hazardous waste and enhance atom economy by reducing raw materials. In sustainable materials production, bio-refineries are a key model that transforms biomass into fuels, chemicals, and materials to imitate the natural cycle and reduce emissions (Dash et al., 2025; Höfer et al., 2024; Madzík et al., 2024; Sai Bharadwaj et al., 2025; Thamaraiselvan & Wickramasinghe, 2025).

All these systems are designed to take the maximum benefit out of biomass feedstocks, combining energy production, reactants synthesis, and waste valorization as a closed loop.

Simultaneously, green supply chain management is popularizing approaches such as supplier development, eco-design, product life extension, and reverse logistics. Manufacturers are increasingly working with suppliers to lower this environmental impact and address greener transportation and the ability to recycle or compost the product. Changing consumer preferences towards the demand for sustainable products has led to this change, further enabling industries to align brand strategies with environmental stewardship. Life Cycle Assessment (LCA) is commonly used to analyse the environmental impacts of products and processes quantitatively, so decision-making is oriented towards more sustainable paths.

Consumer awareness, activism, and public pressure force manufacturers to switch to zero-waste production. Social media and digital channels are spreading news of good or bad environmental practices so quickly now that consumers can make informed decisions immediately. In return, companies are accelerating their efforts to certify their sustainability and use of sea labels and show transparency in reporting to prove adherence to zero waste principles.

Although these nascent movements offer future hope, it will take ongoing innovation, education, and systemic change to pave the way towards manufacturing with zero waste on a larger scale. Educational institutions and research institutes are the key drivers towards the knowledge of sustainability and the development of new materials, processes, and testing methods (Ahmed et al., 2024; Dash et al., 2025; Sakthimurugan et al., 2025; Sheldon, 2024; Zamani et al., 2024; Zhou et al., 2025). Interdisciplinary efforts between chemists, engineers, economists, and policymakers are required to explore the cross-cutting issues of zero-waste manufacturing. In addition, workforce training and reskilling are critical to ensure employees have the knowledge and skills to manage sustainable manufacturing systems successfully.

Intersecting the circular economy with digital infrastructure, policy support offers excellent potential for re-imagining industrial manufacturing. Therefore, countries and regions that place sustainability at the centre of their industrial strategies will likely fare best in the new green economy. With increased environmental consciousness and a changing regulatory environment, zero-waste manufacturing will become a moral necessity and a strategic imperative for long-term business survival and ecological sustainability (Fig. 5).

6 Future Perspectives and Conclusions

Green chemistry now represents a significant and rapidly expanding area of sustainable manufacturing, providing a science-based approach to achieve zero-waste production processes. Green chemistry is evolving beyond conventional environmental thinking into an entirely new process design methodology that transcends environmental compliance and maximizes resource and energy efficiency, so that

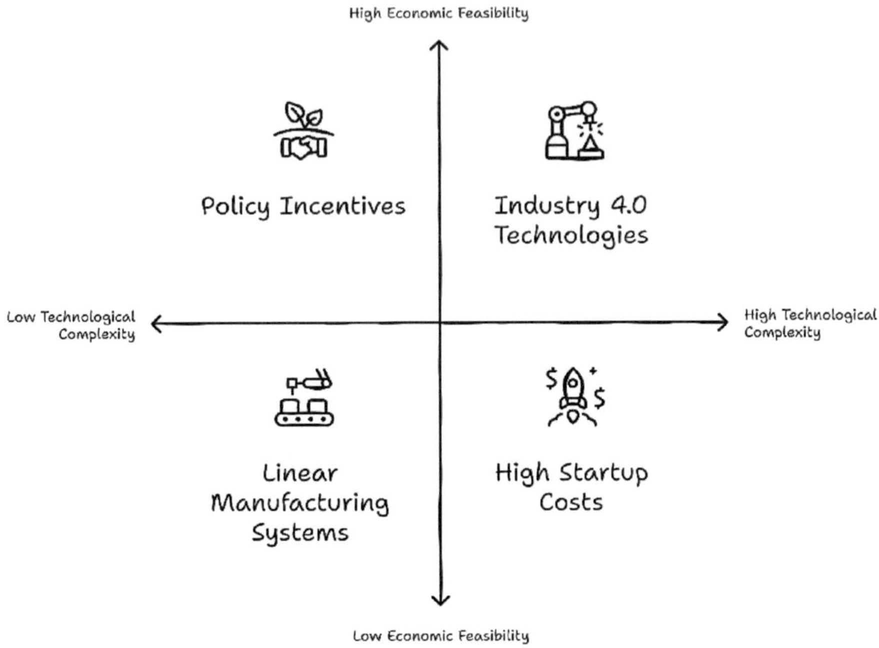

Fig. 5 Challenges and opportunities in zero waste manufacturing

materials and processes' overall economic and ecological impacts are dramatically reduced, if not eliminated. Catalysis, bio-based feedstocks, and solvent replacements lead to the invention of chemical reactions that are highly efficient and selective, and that significantly decrease or even prevent the production of toxic waste at source. For example, when biocatalysts or enzyme-catalyzed reactions are used, it is possible to carry out reactions under milder conditions with less energy requirements and more efficient atom economy. At the same time, solvent-free and aqueous-phase reaction methods are being increasingly developed to eliminate conventional VOC liquids, which cause pollution and waste. These improved processes are more environmentally sound and offer economic advantages regarding reduced raw material use and waste disposal costs. Green chemistry has become an attractive concept for the future industry.

Deeper integration of digital tools, including AI and IoT, is also a key enabler of scaling innovative processes to achieve zero waste. They enable real-time monitoring and management of material flows, energy consumption, and waste emissions in complex industrial environments at unprecedented process performance and optimization. Predictive modelling deployed by AI-based analytics can predict where waste events will happen and where and when physical process inefficiencies will emerge, so manufacturers can proactively change their parameter set materials properties. Real-time IoT sensor networks establish feedback loops that make it possible to increase the traceability and transparency of supply chains with informed decision-making on the use of resources and waste management. Moreover, digital twins of the real world, i.e. digital replicas of physical processes, are used to create simulations and optimize production systems to minimize trial-and-error experimentation and the resulting wastefulness. This coupling of digital tools creates a flexible, nimble manufacturing system that incorporates waste prevention in everyday activity and brings the industry closer to zero waste.

At the heart of these trends is a narrative that gets itself back: It's about closing the loop and transforming towards a circular and regenerative economy. Contrary to the 'take-make-dispose' linear model, which has operated the world's production for centuries, the circular economy is a ballgame that exclusively aims to keep the value of products, materials, and resources as long as possible and to reuse, remanufacture, and recycle everything. This paradigm shift at end-of-life demands products and processes to be developed with recovery in mind, and materials to retain a higher level of quality when they are recovered and recycled back into production cycles. Recent advances in material science have increased the range of recyclable and biodegradable materials available to the industry, making products more versatile and circular. Biodegradable polymers from renewable resources, such as these, also have the additional benefit of reducing dependence on fossil resources and having environmentally benign disposal pathways. While there has also been progress in chemical recycling technologies, which have lifted the earlier limitations on material purity and economic feasibility, they enable polymers, plastics, and even composites to be reverted to monomers from which they can be reused.

However, beyond technology shifts, closing the loop requires systemic shifts in industrial organization, supply chain operations, and consumer engagement. Extended producer responsibility (EPR) schemes are being developed in many countries to support the principle that manufacturers should bear part of, if not all, the post-consumer responsibility for their goods. Circular business models such as product-as-service (manufacturers retain ownership of the product and provide it as a service) drive resource-efficient design and product end-of-life recovery. Simultaneously, consumer understanding and engagement are vital to the success of circular systems, and education and infrastructure to make product return and material separation straightforward are required.

Integrating green chemistry, digitalization, and circular economy is a holistic route towards sustainable industrial ecosystems with zero waste and regenerative resource utilization. However, it will take collective action to achieve that vision on several fronts. Continued R&D is needed to extend the frontiers of green chemistry further, and the focus should be on the scalability, cost reduction, and integration of new technologies into the existing industrial landscape. Digital technologies should be tailored in their interoperability and accessibility for different manufacturing sectors, while data security and ethical aspects must be managed. Policy frameworks need to be adapted to include explicit incentives and norms that promote circularity and the reduction of waste, within which different stakeholders such as governments, industry, academia, and civil society can collaborate.

Education and workforce training are also critical to this evolution. Training programmes must train engineers, chemists, and managers with a cross-disciplinary understanding of chemistry, data science, environmental science, and systems thinking. Establishing a culture of sustainability in organizations is necessary to integrate the zero-waste concept into day-to-day activities and long-term strategic planning.

Manufacturing of the future will meet at the crossroads of state-of-the-art green chemistry, advanced digital technologies, and sustainable circular economy models. These elements complement one another in the design and operation of production systems with a low impact on the environment, efficient resource use, and, where possible, regeneration of natural capital. Although barriers to technology uptake, economic viability, and policy harmonization persist, the drive towards zero-waste manufacturing is unstoppable and picking up steam. Sectors that adopt these innovations and cultivate collaborative ecosystems will be able to handle the increasing market requirements for sustainable products and processes. Finally, the cross-fertilization of these approaches signals the arrival of a new manufacturing age, moving beyond the traditional state of affairs to yield resilient, equitable, and environmentally virtuous industrial systems for the benefit of the generations.

Conflicts None.

References

Abdellatief, T. M. M., & Tawalbeh, M. (2025). *3.17—Nanomaterials and nanocomposites from green polymers* (P. A. G. B. T.-C. G. M. (First E. Olabi (ed.); pp. 316–329). Elsevier. https://doi.org/10.1016/B978-0-443-15738-7.00113-6

Abu Jadayil, W., Shwaheen, G., Ramadan, M., & Alkhedher, M. (2025). Beyond energy: Review of innovations in green technologies for resilience and environmental sustainability. *Energy Reports, 13*, 5410–5426. https://doi.org/10.1016/j.egyr.2025.04.067

Ahmed, M., Abdullah, Eiman, E., Al-Ahmary, K. M., Aftab, F., Sohail, A., Raza, H., & Ali, I. (2024). Advances in green liquid chromatography for pharmaceutical analysis: A comprehensive review on analytical greenness to sustainable chemistry approaches. *Microchemical Journal, 205*, 111400. https://doi.org/10.1016/j.microc.2024.111400

Ariwaodo, C. A., & Olaniyan, O. F. (2024). Fleshy fruit waste and the green chemistry of its conversion to valuable products for humans and animals. *Food Chemistry Advances, 4*, 100634. https://doi.org/10.1016/j.focha.2024.100634

Becker, J., Manske, C., & Randl, S. (2022). Green chemistry and sustainability metrics in the pharmaceutical manufacturing sector. *Current Opinion in Green and Sustainable Chemistry, 33*, 100562. https://doi.org/10.1016/j.cogsc.2021.100562

Das, A. P., Mohanty, A. K., van Hullebusch, E. D., Figueiredo, G., Das, A. P., Mohanty, A. K., van Hullebusch, E. D., & Figueiredo, G. (2023). Bioremediation as an emerging technology for the removal of synthetic microplastic pollutants from marine ecosystem. *MarPB, 194*, Article 115297. https://doi.org/10.1016/J.MARPOLBUL.2023.115297

Das, A., Dey, S., & Das, A. P. (2024a). Ecotoxicological impacts of synthetic microfiber pollutants and development of sustainable mitigation strategies. *Environmental Chemistry and Ecotoxicology*. https://doi.org/10.1016/J.ENCECO.2024.12.003

Das, A. P., Nayak, B., & Behera, I. D. (2024b). Applications of microorganisms in advanced geomicrobiological and biotechnological exploration. *Geomicrobiology Journal, 41*(4), 309–311. https://doi.org/10.1080/01490451.2024.2327059/ASSET//CMS/ASSET/DE2A1F67-08C1-475A-ABC7-2B3F4C9E8BCF/01490451.2024.2327059.FP.PNG

Das, A., Dey, S., Gacem, A., Yadav, K. K., Nath, H., Ram, D. K., & Das, A. P. (2025). Revolutionizing solid hydrocarbon sourcing: Plastic's journey from waste to upcycled treasure. *Environmental Science and Engineering, Part, F218*, 331–351. https://doi.org/10.1007/978-3-031-81615-4_14

Dash, M., Dubey, A. K., Choudhary, T., Liu, Y., Nanda, H. S., & Pati, S. (2025). Critical metal extraction from spent battery cathodes and anticipated developments using next generation green solvents for achieving a net-zero future. *Chemical Engineering Journal, 507*, 160324. https://doi.org/10.1016/j.cej.2025.160324

Davies, J. C., & Hirst, J. D. (2025). *Software tools for green and sustainable chemistry* (B. B. T.-E. of G. C. (First E. Török (ed.); pp. 414–425). Elsevier. https://doi.org/10.1016/B978-0-443-15742-4.00049-1

Dey, S., Bissoyi, A., Koteswara Reddy, G., & Das, A. P. (2024). Synthetic microfibers: An urban emerging pollutant demanding for sustainable environmental solutions. *Sustainable Microbial Technology for Synthetic and Cellulosic Microfiber Bioremediation*, 107–121. https://doi.org/10.1007/978-3-031-62600-5_7

Dey, S., Dash, A., & Das, A. P. (2024). Bioremediation as a sustainable and circular solution for mining waste management and bioenergy generation. *Sustainable Management of Mining Waste and Tailings: A Circular Economy Approach*, 107–119. https://doi.org/10.1201/978 1003442455-6/BIOREMEDIATION-SUSTAINABLE-CIRCULAR-SOLUTION-MINING-WASTE-MANAGEMENT-BIOENERGY-GENERATION-SUDESHNA-DEY-AKANSHYA-DASH-ALOK-PRASAD-DAS

Ferro-Falla, J., Rowles, L. S., Arriaga, F. D., & Plazas-Tuttle, J. (2025). Examining green and chemical methods for zero-valent iron nanoparticle synthesis in heavy metal adsorption. *Water Cycle*. https://doi.org/10.1016/j.watcyc.2025.05.003

Fuente-Ballesteros, A., Ares, A. M., & Bernal, J. (2025). Paving the way towards green contaminant analysis: Strategies and considerations for sustainable analytical chemistry. *Green Analytical Chemistry, 12*, 100221. https://doi.org/10.1016/j.greeac.2025.100221

Ghosh, S., Mohanty, S., Nayak, S., Sukla, L. B., & Das, A. P. (2016). Molecular identification of indigenous manganese solubilising bacterial biodiversity from manganese mining deposits. *Journal of Basic Microbiology, 56*(3), 254–262. https://doi.org/10.1002/JOBM.201500477

Ghosh, S., & Das, A. P. (2020). Microbial metagenomics: current advances in investigating microbial ecology and population dynamics. In S. K. Nayak & B. B. Mishra (Eds.), *Frontiers in soil and environmental microbiology* (1st ed., pp. 247–254). CRC Press.

Ghosh, S., Gandhi, M., van Hullebusch, E. D., & Das, A. P. (2020). Proteomic insights into Lysinibacillus sp.-mediated biosolubilization of manganese. *Environmental Science and Pollution Research, 28*(30), 40249–40263. https://doi.org/10.1007/S11356-020-10863-4

Guidara, M., Yaich, H., Petrut, R., Richel, A., Blecker, C., Attia, H., & Garna, H. (2024). Manufacturing bioprocess of novel cream formulations based on polysaccharides "ulvans" from green seaweed Ulva lactuca: Effect of extraction procedures and concentrations on the optical and rheological properties of ulvan creams. *Sustainable Chemistry and Pharmacy, 42*, 101852. https://doi.org/10.1016/j.scp.2024.101852

Höfer, R., & Selig, M. B. T.-R. M. in M. S. and M. E. (2024). *Green chemistry and green polymer chemistry in the circular economy*. Elsevier. https://doi.org/10.1016/B978-0-323-95486-0.00046-6

Khaled, K. A. M., Olabi, A. G., Azzam, C. R., Aboel-Ainin, M. A., & Shehata, N. (2025). *2.18—Tools and techniques for green chemicals and advanced materials selection* (P. A. G. B. T.-C. G. M. (First E. Olabi (ed.); pp. 264–277). Elsevier. https://doi.org/10.1016/B978-0-443-15738-7.00109-4

Madzík, P., Falát, L., Yadav, N., Lizarelli, F. L., & Čarnogurský, K. (2024). Exploring uncharted territories of sustainable manufacturing: A cutting-edge AI approach to uncover hidden research avenues in green innovations. *Journal of Innovation and Knowledge*, *9*(3), 100498. https://doi.org/10.1016/j.jik.2024.100498

Mishra, S., Dash, D., & Das, A. P. (2023). Aquatic microbial diversity on plastisphere: colonization and potential role in microplastic biodegradation. *Geomicrobiology Journal*, 1–12. https://doi.org/10.1080/01490451.2023.2209750

Mohanty, M., Mohanty, J., Dey, S., Dutta, K., Shah, M. P., & Das, A. P. (2024b). The face mask: A tale from protection to pollution and demanding sustainable solution. *Emerging Contaminants, 10*(2), Article 100298. https://doi.org/10.1016/J.EMCON.2023.100298

Mohanty, J., Dey, S., Bala, K., & Das, A. P. (2024). Recent advances in microplastic pollution for its sustainable management. *Sustainable Management of Urban Plastic Waste Through Circular Economic Approaches*, 27–48. https://doi.org/10.1201/9781003465447-3

Murphy, M. A. (2025). *Historical background and development of green and sustainable chemistry* (B. B. T.-E. of G. C. (First E. Török (ed.); pp. 1–16). Elsevier. https://doi.org/10.1016/B978-0-443-15742-4.00093-4

Perera, M. A. G. P., & Ranjith, P. G. (2025). Greener horizons: Revolutionizing construction materials with waste-based innovations—An experimental study. *Journal of Building Engineering, 103*, 112211. https://doi.org/10.1016/j.jobe.2025.112211

Pradhan, G., Tripathy, B., Ram, D. K., Digal, A. K., & Das, A. P. (2023). Bauxite mining waste pollution and its sustainable management through bioremediation. *Geomicrobiology Journal*, 1–10. https://doi.org/10.1080/01490451.2023.2235353

Priyadarshini, S., Jagatee, S., & Das, A. P. (2024). Synthetic fabrics and microfiber pollution–an assessment of their global impact. *Environmental Science and Engineering, Part, F2333*, 137–157. https://doi.org/10.1007/978-3-031-51792-1_8

Punj, N., Ahmi, A., Tanwar, A., & Abdul Rahim, S. (2023). Mapping the field of green manufacturing: A bibliometric review of the literature and research frontiers. *Journal of Cleaner Production, 423*, 138729. https://doi.org/10.1016/j.jclepro.2023.138729

Ramadan, R., & Shehata, N. (2025). *2.08—Green electronic materials* (P. A. G. B. T.-C. G. M. (First E. Olabi (ed.); pp. 107–127). Elsevier. https://doi.org/10.1016/B978-0-443-15738-7.00064-7

Ray, S., Dey, S., Digal, A. K., Yadav, K. K., & Das, A. P. (2025). Sanitary waste and microplastic pollution in the Euro-Mediterranean region: Challenges and solutions. *Euro-Mediterranean Journal for Environmental Integration, 2025*, 1–23. https://doi.org/10.1007/S41207-025-00879-Y

Ray, S. K., Dey, S., Das, A. P., & John, A. (2024). Microplastic waste management through greener technologies : A review. *Sustainable Management of Urban Plastic Waste Through Circular Economic Approaches*, 237–261. https://doi.org/10.1201/9781003465447-14

Reddy Koteswara, G., Bhargavi, S., Yamuna, D., Anjani Devi, M., Jeevana Lakshmi Kumar, G., Manas, T., Adarsh, S., Bhavya Sai, V., Nath, H., Alreshidi, M. A., Togun, H., Ram, D. K., & Das, A. P. (2025). Recent advancements in the remediation of marine hydrocarbon and plastic pollutants. *Environmental Science and Engineering*, *Part F218*, 127–142. https://doi.org/10.1007/978-3-031-81615-4_7

Sahu, S., & Das, A. P. (2024). Microplastics degradation mechanism by aquatic plastiomes: A review. *Environmental Science and Engineering, Part, F2827*, 213–226. https://doi.org/10.1007/978-3-031-50840-0_12

Sahu, A., Dey, S., Mallick, K., Priyadarsini, S., Behera, I. D., & Das, A. P. (2024). A review on advanced technology for sustainable management of synthetic microplastic waste. *Sustainable Management of Urban Plastic Waste Through Circular Economic Approaches*, 262–279. https://doi.org/10.1201/9781003465447-15

Sai Bharadwaj, A. V. S. L., Nayak, R. R., Koteswararao, J., Sampath, C., Gaddala, B., Pawar, B. G., & Gupta, N. K. (2025). Environmental life-cycle assessment and green principles in process intensification: A review of novel catalysts from solid waste. *Chemical Engineering and Processing—Process Intensification, 211*, 110208. https://doi.org/10.1016/j.cep.2025.110208

Sakthimurugan, V., Lakshmikanth, G., Balaji, N., Roopashree, R., Kumar, D., & Devarajan, Y. (2025). Green hydrogen revolution: Advancing electrolysis, market integration, and sustainable energy transitions towards a net-zero future. *Results in Engineering*, *26*, 104849. https://doi.org/10.1016/j.rineng.2025.104849

Sheldon, R. A. (2024). Green chemistry and biocatalysis: Engineering a sustainable future. *Catalysis Today*, *431*, 114571. https://doi.org/10.1016/j.cattod.2024.114571

Shen, Y., Du, W., Sui, X., Liu, X., Tang, B. Z., & Yang, J. (2024). Green manufacturing process design for infusible acrylic resin composites: A data-guided life cycle management model incorporating material-process-property-energy-emission relationships. *Composites Part A: Applied Science and Manufacturing*, *181*, 108146. https://doi.org/10.1016/j.compositesa.2024.108146

Swain, S., Purnima Sudha, M., Dey, S., Badu, Y., Bhoi, S. R., Alreshidi, M. A., Qasim, M. T., Nath, H., Ram, D. K., & Das, A. P. (2025). Sources of micro and nano plastic hydrocarbon pollutants in marine bodies and their treatment. *Environmental Science and Engineering, Part, F218*, 45–67. https://doi.org/10.1007/978-3-031-81615-4_3

Thamaraiselvan, C., & Wickramasinghe, R. (2025). Recent progress on sustainable membrane manufacturing with green solvents and biopolymers. *Current Opinion in Chemical Engineering*, *47*, 101092. https://doi.org/10.1016/j.coche.2025.101092

Waseem, M., Rizwan Javed, M., Ali, K., Saleem, M., Faisal Manzoor, M., Farhan, M., Mugabi, R., Sharma, A., & Ahmad Nayik, G. (2024). Microwave-sonication synergistic extraction of dairy waste proteins: A review of green approach for dairy waste proteins valorization. *Ultrasonics Sonochemistry*, *111*, 107111. https://doi.org/10.1016/j.ultsonch.2024.107111

Zamani, P., Zabihi, O., Ahmadi, M., Zamani, M. R., Zohuriaan-Mehr, M. J., Kannangara, T., Joseph, P., & Naebe, M. (2024). Assessing sustainability and green chemistry in synthesis of a Vanillin-based vitrimer at scale: Enabling sustainable manufacturing of recyclable carbon fiber composites. *Composites Part A: Applied Science and Manufacturing*, *179*, 108016. https://doi.org/10.1016/j.compositesa.2024.108016

Zhou, J., Li, B., Du, Y., Li, Y., Dong, J., Chang, C., Wen, J., & Zheng, W. (2025). Review of the research progress of magnesium oxysulfate cement and its recent application in green manufacturing. *Journal of Cleaner Production*, *490*, 144751. https://doi.org/10.1016/j.jclepro.2025.144751

The Role of Nanobiotechnology in Environmental Sustainability

Saniya Kadankandy Ramesh, Shahzad Malak, and Michael Warren Gonzales Ceballos

1 Introduction

1.1 Context and Motivation

The current era faces significant environmental and resource challenges stemming from a growing global population and increased industrial activity. This has resulted in an unprecedented surge in organic waste generation, posing substantial environmental and societal problems. Organic waste, including agricultural residues, food by-products, and municipal waste, constitutes a large portion of global solid waste, around 46% (Bian et al., 2024). Improper handling of this massive waste stream, through burning, unregulated dumping, or landfilling, causes severe environmental damage (Bala et al., 2023, Das et al., 2025; G et al., 2025; Ray et al., 2025). These

S. K. Ramesh · S. Malak · M. W. G. Ceballos (✉)
Faculté Des Sciences et Technologies, La Rochelle Université, La Rochelle, France
e-mail: michaelwarren.c@mail.ucv.es

Facultad de Veterinaria y Ciencias Experimentales, Universidad Católica de Valencia San Vicente Mártir, València, Valencia, Spain

S. K. Ramesh
e-mail: saniya.kadankandy_ramesh@etudiant.univ-lr.fr; sakara@mail.ucv.es

S. Malak
e-mail: shahzad.malak@mail.ucv.es

S. K. Ramesh
Central Marine Fisheries Research Institute, Kochi, Kerala, India

S. Malak
Thomas More University of Applied Sciences, Geel, Belgium

M. W. G. Ceballos
Centre of Marine Sciences (CCMAR/CIMAR LA), University of Algarve, Faro, Portugal

S. Ghosh et al. (eds.), *Advances in Organic Waste Conversion Through Industrial Biotechnology*, Sustainable Environmental Waste Management Strategies,
https://doi.org/10.1007/978-3-032-11377-1_16

practices significantly contribute to greenhouse gas emissions, exacerbating climate change, and lead to widespread pollution of soil, air, and water (Bala et al., 2023). These environmental impacts directly threaten human health and undermine the long-term sustainability of ecosystems.

Conventional waste disposal is insufficient, necessitating a shift towards sustainable resource recovery. Agro-residues, rich in nutrients, should be viewed as valuable raw materials rather than waste (Bala et al., 2023). Simultaneously, increasing global demand for vital resources like clean water, energy, healthcare, and agricultural supplies strains Earth's limited reserves (Aslam et al., 2025). The agri-food system is crucial in these interconnected challenges, where resource scarcity, climate change, and ecosystem degradation are intertwined with food production, distribution, and management policies (Flores-Contreras et al., 2024).

In response to these multifaceted crises, the circular economy model has risen to prominence, emphasizing resource efficiency, renewable energy integration, and waste reduction through the revaluation of waste (Flores-Contreras et al., 2024). Within this context, nanobiotechnology stands out as a crucial interdisciplinary field. It provides cutting-edge tools at the nexus of sustainability and industrial biotechnology, enabling the efficient and economical transformation of complex, low-value organic waste into valuable products (Bala et al., 2023). This capability to transform waste into new resources directly addresses resource scarcity and facilitates the closure of material loops, moving beyond simple waste management to a deeper, more transformative role in the philosophy and practice of resource utilization. The practical implementation of circular economy principles is significantly advanced by nanobiotechnology.

1.2 Scope and Objectives

This report provides a comprehensive examination of nanobiotechnology's critical role in fostering environmental sustainability, specifically through its applications in organic waste Valori zation and environmental remediation. Nanobiotechnology, involving the manipulation of materials at the 1–100 nm scale (Aslam et al., 2025), is a transformative technology with applications spanning biotechnology and biodiversity conservation (Udoh et al., 2025). Recognized for its role in modern agriculture and sustainable development (Bala et al., 2023), it offers a dual benefit: environmental pollutant detoxification (e.g., wastewater treatment and dye degradation Aslam et al. 2025) and the valorization of a wide array of organic wastes into valuable products (Bala et al., 2023).

Nanobiotechnology offers both direct and indirect support for the United Nations Sustainable Development Goals (SDGs). This review focuses particularly on its contributions to SDG 12: Responsible Consumption and Production, which centers on sustainable resource management and waste reduction. Nanobiotechnology advances SDG 12 through improved e-waste management, the promotion of sustainable practices, and the encouragement of collaborative research and development

(Raman et al., 2024). Moreover, the field plays a vital role in achieving SDG 13: Climate action by enabling the transformation of organic waste into biofuels and by enhancing agricultural sustainability through the use of nano-fertilizers and nano-pesticides. These applications help lower greenhouse gas emissions and lessen the impacts of climate change (Udoh et al., 2025). Additionally, the use of nanobiotechnology in environmental monitoring and pollution control further bolsters climate action initiatives (Udoh et al., 2025).

The influence of nanobiotechnology extends beyond direct contributions to SDG 12 and 13, demonstrating a broader, interconnected impact across the SDGs. Developments in agricultural nanobiotechnology, such as nano-fertilizers and nano-pesticides, enhance nutrient use and pest control, increasing crop production and decreasing environmental harm (Kashezheva et al., 2024), thereby indirectly furthering SDG 2 (zero hunger) and SDG 6 (clean water and sanitation) by minimizing agricultural pollution. Furthermore, nanobiotechnology's role in producing biofuels and clean energy supports SDG 7 (affordable and clean energy), while its application in sustainable waste management contributes to SDG 11 (sustainable cities and communities), and its use in environmental cleanup aids SDG 15 (life on land) (Kashezheva et al., 2024). This multifaceted influence establishes nanobiotechnology as a versatile tool capable of simultaneously advancing multiple sustainable development targets.

Moreover, the rapid growth of the nanotechnology market underscores a powerful economic impetus driving its adoption for sustainability. Projections indicate that the global market for nanotechnology products could reach USD 1 trillion to USD 3 trillion by 2024, creating over 2 million new jobs worldwide (Khatoon & Velidandi, 2025). This implies that the sustainable solutions offered by nanobiotechnology are increasingly becoming economically viable and attractive, which is expected to accelerate their integration into industrial processes and further contribute to sustainability efforts through market forces, complementing policy mandates.

2 Nanobiotechnology for Organic Waste Valorization

2.1 Nanocatalysts in Waste Conversion

2.1.1 Types of Nanocatalysts: Metallic Nanoparticles and Enzyme-Mimetic Nanomaterials (Nanozymes)

Nanocatalysts, characterized by their dimensions within the 1–100 nm range, exhibit an exceptionally elevated surface area-to-volume ratio, a property that differentiates them from macroscopic materials (Kapoor et al., 2024). This unique characteristic engenders substantial enhancements in catalytic activity, selectivity, and efficacy. As a result, nanocatalysts are capable of augmenting the rates of chemical reactions, frequently necessitating reduced temperatures and energy input (Kapoor et al.,

2024). These properties make nanocatalysts ideal for various applications in green chemistry and environmental protection (Kashezheva et al., 2024). Nanocatalysts enhance catalytic efficiency through optimized active site exposure, faster mass transfer, and potential quantum effects due to their nanoscale design. This allows for energy savings by enabling operation under milder conditions, which is vital for sustainability.

Some types of nanocatalysts are employed in waste conversion are metal-based catalysts and nanozymes.

Metal-Based Catalysts

Metallic nanocatalysts, including noble metals like platinum (Pt) and palladium (Pd), and transition metals such as nickel (Ni), cobalt (Co), and iron (Fe), play a significant role in organic waste valorization. Noble metals exhibit high efficiency in hydrogenation and dehydrogenation reactions, essential for biofuel synthesis (Kapoor et al., 2024). Transition metals offer cost-effective alternatives for large-scale applications, widely used in pyrolysis, gasification, and hydrogenation for converting biomass into biofuels and platform chemicals (Kapoor et al., 2024). To enhance stability and dispersion, these metallic nanoparticles are often immobilized on support materials like biochar or activated carbon, forming supported catalysts that prevent aggregation and deactivation during operation (Kapoor et al., 2024).

Enzyme-Mimetic Nanomaterials (Nanozymes)

Nanozymes are a class of nanomaterials that possess intrinsic enzyme-like catalytic activities, bridging the fields of nanocatalysis, enzymology, and biology (Mansur & Mansur, 2025). A key development in this area is biogenic metallic nanoparticles, synthesized using biological systems like plants, bacteria, and fungi (Ngcongco et al., 2023). These biogenic nanozymes offer significant benefits over natural enzymes, including reduced production costs, simplified processes, enhanced stability, and greater durability. This makes them well suited for harsh environmental conditions where natural enzymes might denature (Ngcongco et al., 2023). For example, ferrihydrite nanoparticles from Comamonas testosteroni and magnetosomes from Magnetospirillum gryphiswaldense exhibit peroxidase-like activity used in detection and diagnostics (Ngcongco et al., 2023). Plant extracts, containing functional groups such as phenolic acids and polyphenols, can also synthesize and stabilize metallic nanoparticles, improving their catalytic effectiveness (Ngcongco et al., 2023). The capacity to synthesize these catalysts from biological sources further integrates them into a sustainable, bio-based economy, reducing dependence on traditional chemical synthesis methods and their environmental impact.

2.1.2 Applications in Waste Conversion

Nanocatalysts are revolutionizing the conversion of organic waste into valuable products, particularly in the hydrolysis of lignocellulosic biomass and the transesterification of lipids for biodiesel production.

Due to its intricate structure, lignocellulosic biomass, an abundant and renewable resource, is naturally difficult to break down (Luque et al., 2018). Nanobiotechnology provides efficient, economical, and eco-friendly methods to address this issue (Sankaran et al., 2021). Nanomaterials, owing to their small dimensions, can effectively enter the cell walls of lignocellulosic biomass and engage with its constituents (cellulose and hemicellulose) to yield simple and complex sugars (Sankaran et al., 2021). This step is essential for the effective breakdown of biomass into fundamental chemical building blocks. The nano-shear hybrid alkaline (NSHA) pretreatment method significantly enhances the enzymatic breakdown of biomass by efficiently removing lignin and exposing cellulose and hemicellulose (Sankaran et al., 2021). This process utilizes a nanomixing reactor to combine high-speed shear and chemical reagents with mild heat, achieving rapid results. By directly addressing the structural recalcitrance of biomass, this advancement is crucial for the economic viability of lignocellulosic biofuels and biochemicals, opening up a substantial, previously untapped resource.

Nanocatalysts significantly accelerate the transesterification reaction, a primary method for biodiesel production, by enhancing catalytic activity, increasing surface area, and providing superior thermal stability (Pandit et al., 2023). These catalysts improve efficiency and economics through reduced catalyst weight, lower reaction temperatures, and optimized oil-to-alcohol ratios (Pandit et al., 2023). Their strong saponification resistance ensures high biodiesel yields (Pandit et al., 2023). Metal oxide nanocatalysts are highly preferred for maximizing biodiesel production (Pandit et al., 2023). For instance, CaO nanoparticles have achieved 97.61% biodiesel yield from soybean oil and can be reused for multiple cycles (Pandit et al., 2023). Carbon nanocatalysts derived from carbon nanotubes (CNTs), graphene, and reduced graphene oxides offer high surface area and biocompatibility for applications in the transesterification of triglycerides and fatty acids (Pandit et al., 2023). Other nanocatalysts, including magnetic nanocatalysts, zeolite nanocatalysts, and nanobiocatalysts, are also known to increase the biodiesel yield (Pandit et al., 2023). The adaptability and specificity of nanobiotechnology allow for the creation of customized catalytic systems suitable for diverse organic waste, including challenging streams that conventional methods struggle with due to impurities or complex compositions. This versatility is essential for a robust and comprehensive circular economy capable of processing a wide range of waste materials effectively.

2.2 *Nanostructured Bioreactors*

2.2.1 Immobilized Enzymes or Nanofibers for Continuous-Flow Systems

A significant advancement in industrial biotechnology is the creation of nanostructured bioreactors, especially those employing enzymes immobilized on nanomaterials, for continuous-flow systems. Enzyme immobilization, which involves attaching enzymes to a support material, is essential for enhancing their catalytic performance, stability, and potential for reuse—all of which are vital for widespread industrial use (Sankaran et al., 2021).

Nanomaterials serve as superior support structures for enzymes due to their high surface area-to-volume ratios, unique physicochemical properties, and ability to provide a biocompatible microenvironment (Sun et al., 2023). Common nanomaterials utilized for enzyme immobilization include but not limited to:

- **Graphene and Its Derivatives**

Graphene (GR), graphene oxide (GO), and reduced graphene oxide (RGO) offer excellent electrical conductivity and high specific surface area, making them effective transducer materials and ideal substrates for enzyme protection (Xu et al., 2020).

- **Carbon Nanotubes (CNTs)**

Both single-walled (SWCNTs) and multi-walled (MWCNTs) possess high biocompatibility and electrical conductivity, allowing for high enzyme loading and enhanced electron transfer (Sun et al., 2023).

- **Metal–Organic Frameworks (MOFs)**

These porous materials can accommodate specific enzymes within their precisely designed cavity structures, enabling controlled distribution and enhanced substrate channeling (Xu et al., 2020).

- **Silica (SiO_2) and Polymers**

These materials, along with DNA nanostructures, are also widely used as versatile matrices for enzyme immobilization (Sankaran et al., 2021).

Nanomaterials enhance enzyme stability by providing protection against inactivation factors, leading to prolonged biocatalytic activity (Sankaran et al., 2021; Sun et al., 2023). Immobilizing multiple enzymes on nanomaterials facilitates substrate channeling, increasing reaction rates by minimizing diffusion limitations (Xu et al., 2020). For biosensor applications, nanomaterials improve electron transfer, increasing sensitivity (Sun et al., 2023). Enzyme immobilization on solid supports, especially magnetic nanoparticles, enables easy separation and reuse, reducing costs and promoting sustainability (Sankaran et al., 2021). This reusability enhances the industrial appeal and aligns with circular economy principles.

2.2.2 Nano-Enhanced Anaerobic Digestion for Biogas Production

Anaerobic digestion (AD), a recognized biotechnological method, transforms organic waste into renewable bioenergy, mainly biogas. However, the effectiveness of AD can be hampered by factors like the slow breakdown of input materials and the existence of inhibiting compounds (Jeyakumar & Vincent, 2022). Nanomaterials have emerged as effective additives in AD, substantially enhancing sludge breakdown, shortening the initial inactive period, and boosting methane production (Jeyakumar & Vincent, 2022). These materials possess unique physical characteristics, such as a large surface area, ideal particle size, and catalytic activity. These features allow them to penetrate microbial cell membranes, thereby promoting microbial activity and increasing methane formation (Jeyakumar & Vincent, 2022).

Various types of nanomaterials contribute to nano-enhanced anaerobic digestion (see Table 1):

Table 1 Impact of nanomaterials in nano-enhanced anaerobic digestion

Nanomaterial type	Specific example	Concentration/ Dosage	Feedstock	Observed enhancement	Key mechanism
Metallic NPs	NZVI	10 mg/g TSS	Waste activated sludge (WAS)	120% increase in methane production (Jeyakumar & Vincent, 2022)	Activates enzymes, lowers ORP, eliminates H2S
Metal oxide NPs	Fe_3O_4	1 g/L	WAS	21.66% increase in biogas production (Jeyakumar & Vincent, 2022)	Large surface area, high tensile strength, easy recovery
Carbon-based NM	Activated carbon	Not specified	Not specified	Mitigates ammonia inhibition, enhances CH_4 generation (Jeyakumar & Vincent, 2022)	Adsorbs gaseous pollutants, provides an immobilization matrix
Nanocomposites	nZVI-BC	Not specified	Not specified	Increased methane yields, improved process stability (Jeyakumar & Vincent, 2022)	Enhances the synthesis/ degradation of organic acids
Trace elements	Pseudo-ginseng residues (PGR)	10% (TS basis)	Wheat straw	28% increase in biogas, 37% increase in methane (Xi et al., 2021)	Acts as a trace element, a co-factor for enzymes

- **Metallic Nanoparticles**

Metals such as nickel (Ni), copper (Cu), silver (Ag), cobalt (Co), and gold (Au) enhance anaerobic digestion by serving as vital trace elements and co-factors for essential enzymes that support microbial growth and metabolism throughout the AD process (Jeyakumar & Vincent, 2022). For instance, Ni is critical for methane production because of its function in methyl co-enzyme M reductase. Nano zero valent iron (NZVI) is especially effective in lowering the oxidative-reductive potential, thus establishing a more suitable anaerobic environment (Jeyakumar & Vincent, 2022). It also eliminates H_2S and activates key enzymes involved in methanogenesis. Research indicates that adding 10 mg/g of total suspended solids (TSS) of NZVI can boost methane production by 120% compared to a control group (Khan et al., 2022). These nanomaterials function as active bio-stimulants, precisely controlling and optimizing the intricate microbial populations engaged in anaerobic digestion. This leads to improved efficiency, stability, and resilience in the conversion of waste to energy.

- **Metal Oxide Nanomaterials**

Nanoparticles of TiO_2, ZnO, MgO, Fe_2O_3, and CuO have unique physical and chemical characteristics that improve biogas production (Jeyakumar & Vincent, 2022). Iron oxide nanoparticles (Fe_3O_4, Fe_2O_3), for example, have large surface area and high tensile strength, aiding in increasing AD efficiency and are easily recovered (Jeyakumar & Vincent, 2022). The addition of 1 g/L of Fe_3O_4 enhanced biogas production by 21.66% (Khan et al., 2022).

- **Carbon-Based Nanomaterials**

Highly conductive materials such as graphene, activated carbon, biochar, and carbon nanotubes (CNTs) create optimal conditions for microbial growth and proliferation (Jeyakumar & Vincent, 2022). These materials function as effective immobilization matrices for microorganisms, leading to increased microbial activity and promoting direct interspecies electron transfer (DIET), which consequently enhances methane production. The porous nature of activated carbon allows it to adsorb gaseous contaminants and reduce ammonia inhibition, thereby further improving CH_4 generation (Jeyakumar & Vincent, 2022).

- **Nanocomposite Materials**

Encapsulating or binding multiple chemicals in nanomaterials addresses nanoparticle aggregation's cytotoxic effects and enhances the connection between microbes and nanoparticles, thus boosting reactivity and overall performance. For example, biochar-supported nanoscale zerovalent iron (nZVI-BC) enhances process stability and results in higher methane yield (Jeyakumar & Vincent, 2022).

Nanomaterials address a critical challenge in anaerobic digestion, which is the presence of inhibitory substances. Activated carbon adsorbs gaseous pollutants, preventing ammonia-related inhibition, while NZVI removes H_2S (Jeyakumar & Vincent, 2022). Similarly, biochar reduces ammonia inhibition. These applications show that nanomaterials not only boost biogas production rate and yield but also enhance the stability and resilience of anaerobic digestion by counteracting common inhibitors.

Anaerobic co-digestion of wheat straw with herbal-extraction process residues (HPR) showed notable improvements with nano-additive treatment. Specifically, the addition of 10% pseudo-ginseng residues (PGR) resulted in a 28% increase in biogas productivity (337 ml/g TS) and a 37% increase in methane production (178 ml/g TS) compared to the control (Xi et al., 2021). This case study underscores the promising role of nano-additives in enhancing biogas output from various organic waste materials.

2.3 High-Value Products

2.3.1 Bioplastics: Nano-Reinforced PHB/PHA Composites

Polyhydroxyalkanoates (PHAs), such as polyhydroxybutyrate (PHB) and its copolymer poly(3-hydroxybutyrate-co-3-hydroxyvalerate) (PHBV), are naturally occurring, biodegradable, and biocompatible polyesters produced by diverse microorganisms utilizing organic waste (Arriaga et al., 2025). These biopolymers hold significant potential as sustainable alternatives to conventional petroleum-based plastics, especially in packaging applications, owing to their comparable characteristics and natural biodegradability (Vicente et al., 2023). The process of valorizing agri-food waste into PHAs generally entails a biotransformation where dark fermentation yields volatile fatty acids (VFAs), which are then converted into PHAs by microorganisms in environments with limited nutrients but abundant carbon sources (Arriaga et al., 2025).

Several obstacles impede the widespread industrial use of PHAs, despite their environmental benefits. These include high production costs (2–5 times that of synthetic plastics), competition with food for raw materials, and some undesirable physical characteristics like PHB's brittleness (Vicente et al., 2023). To address these limitations, especially regarding mechanical properties, integrating nanofillers into PHB/PHA composites has become a vital approach (Emaimo et al., 2022). This is crucial for resolving the typical "performance-cost" dilemma of bioplastics. Nanobiotechnology enhances the value of bioplastics by improving key attributes, which could justify higher manufacturing expenses or allow their application in more rigorous industries. This, in turn, can accelerate market uptake and enhance their contribution to a circular economy (Koteswara et al., 2025; Sahu & Das, 2024; Swain et al., 2025).

Nano-reinforcement significantly enhances the mechanical, thermal, and electrical properties of PHA composites:

- **Cellulose Nanocrystals (CNCs)**

Surface-modified cellulose nanocrystals (CNCs) significantly enhance the mechanical properties of polyhydroxyalkanoate (PHA) composites (Jo et al., 2021). Unmodified, hydrophilic CNCs tend to clump together due to their incompatibility with the hydrophobic PHA matrix (Jo et al., 2021). To overcome this, surface modification techniques like double silanization using tetraethyl orthosilicate (TEOS) and methyltrimethoxysilane (MTMS) are employed, making the CNCs hydrophobic and improving their dispersion in PHA (Jo et al., 2021). Incorporating 5% freeze-dried, double-silanized CNCs into PHA composites has resulted in a remarkable increase in ductility, with elongation at break improving by up to 301%, accompanied by only a slight decrease in Young's modulus (Emaimo et al., 2022). Furthermore, these modified CNCs can serve as nucleating agents, accelerating crystallization and reducing spherulite size, thereby enhancing the composite's elasticity and plasticity (Jo et al., 2021).

- **Graphene Nanoplatelets (GNPs) and Carbon Nanofibers (CNFs)**

Both GNPs and CNFs significantly improve the thermal stability of nanocomposites. Higher loadings of GNPs provide better thermal stability by creating a larger barrier to the diffusion of degradation products (Emaimo et al., 2022).

- **Magnetic Fe_3O_4–rGO Fillers**

Composite fillers have been incorporated into electrospun PHB scaffolds to enhance their properties. This addition results in larger fiber diameters and a reduction in crystallinity, leading to improved ductility and mechanical strength of the scaffolds (Pryadko et al., 2022). Furthermore, these fillers introduce magnetic characteristics and a piezoresponse, which can be beneficial for simulating biological stimuli in tissue engineering.

- **Natural Rubber (HMW-NR)**

Incorporating high-molecular weight natural rubber into PHBV can enhance its thermal balance and melting power, improving processability (Emaimo et al., 2022).

The above-mentioned types of nanofillers, their properties, and mechanisms are summarized in Table 2.

By tailoring the properties of PHAs through nano-reinforcement, these bioplastics can become more competitive with conventional plastics, expanding their range of applications and accelerating their adoption within a circular bioeconomy model.

Table 2 Nanofillers for enhanced bioplastic (PHB/PHA) properties

Nanofiller type	Specific modification/ Composition	Enhanced property	Observed improvement	Key mechanism/Role
Cellulose Nanocrystals (CNCs)	Double salinization (TEOS and MTMS)	Ductility	Up to 301% increase in elongation at break	Nucleating agent, improved dispersibility, and hydrophobicity
Graphene Nanoplatelets (GNPs)	Not specified	Thermal sstability	Significant improvement, higher loadings provide better stability	Barrier to diffusion of degradation products, uniform dispersion
Carbon Nanofibers (CNFs)	Not specified	Thermal stability	Significant improvement	Not specified, but similar to GNPs in effect
Magnetic Fe_3O_4–rGO fillers	Composite filler	Ductility, mechanical strength, piezoresponse	Increased fiber diameters, decreased crystallinity, enhanced ductility	Preserves magnetic properties, provides piezoresponse
Natural rubber (HMW-NR)	Gel extracted from NR	Processability, Thermal Balance	Superior thermal balance and melting power	Improves processability of PHBV blends

2.3.2 Biofuels: Nanocatalysts in Syngas Fermentation and Algal Biofuel Extraction

Nanocatalysts in Syngas Fermentation

Synthesis gas (syngas), primarily composed of carbon monoxide (CO), hydrogen (H_2), and carbon dioxide (CO_2), can be generated from industrial waste gases or via biomass gasification (Neto et al., 2024). Syngas fermentation (SNF) is a promising biological approach that employs specialized microorganisms to convert these waste gas streams into value-added products such as biofuels (ethanol, hydrogen, methane) and industrial chemicals (Neto et al., 2024). However, a significant bottleneck to the industrial-scale implementation of SNF is the inherently low gas-to-liquid mass transfer rate, which restricts the bioavailability of carbon and electron donors essential for microbial catalysis.

Nanoparticles (NPs) have emerged as a transformative solution to this challenge, enabling enhanced gas–liquid mass transfer through multiple synergistic mechanisms. (a) Nanostructured materials, including carbon nanotubes, provide a substantially increased surface area for gas absorption, thereby improving the interfacial contact and process efficiency (Dhakal & Acharya, 2021). (b) The "shuttling" or "grazing" effect of NPs facilitates the directed transport of gaseous substrates into the fermentation medium, promoting more efficient substrate uptake by microbial consortia (Dhakal & Acharya, 2021). (c) NPs influence local hydrodynamics at the gas–liquid interface, effectively lowering the resistance to mass transfer by altering fluid flow behavior (Dhakal & Acharya, 2021).

In particular, magnetic nanoparticles (MNPs) offer unique advantages in the context of organic waste valorization. (d) Due to their high surface-area-to-volume ratio, MNPs serve as microcarriers that support the attachment and growth of microbial cells, thereby increasing cell density and enhancing catalytic throughput. (e) Their magnetic responsiveness allows external magnetic fields to be applied, which improves micro-scale mixing and gas dispersion throughout the bioreactor environment (Dhakal & Acharya, 2021).

Moreover, (f) NPs can act as robust enzyme carriers, stabilizing and activating key enzymatic systems involved in syngas metabolism—such as hydrogenases and carbon monoxide dehydrogenases—thus sustaining catalytic activity under operational conditions (Dhakal & Acharya, 2021). For example, silver (Ag) NPs have demonstrated enhanced hydrogen yields, while manganese dioxide (MnO_2), nickel oxide (NiO), iron oxide (Fe_3O_4), and zinc oxide (ZnO) NPs have been shown to significantly boost bioethanol production from agricultural waste substrates (Sajeev et al., 2023). Collectively, these nano-enabled enhancements address critical limitations of traditional SNF systems, supporting the advancement of scalable, efficient, and economically viable biofuel production platforms.

Nanocatalysts in Algal Biofuel Extraction

Microalgae have emerged as a promising non-food biomass source for sustainable biofuel production due to their high lipid content and rapid growth rate (Arya et al., 2021). Recent advances have highlighted the role of nanocatalysts in enhancing both the extraction and transesterification of algal lipids into biodiesel, offering notable improvements in catalytic efficiency, reusability, and environmental compatibility (Arya et al., 2021; Verma et al., 2023). These nanocatalysts provide enhanced stability and activity, while their recyclable nature contributes to cost-effectiveness and scalability.

Several nanocatalyst systems have demonstrated particular efficacy in algal biofuel applications. (a) Photochemically synthesized nanocatalysts such as strontium titanate ($SrTiO_3$), ferric oxide (Fe_2O_3), and potassium hydroxide (KOH)-supported magnetic alumina–ferric oxide composites have shown promise in facilitating efficient transesterification reactions (Arya et al., 2021; Verma et al., 2023). Beyond their role in downstream processing, (b) bio-nanoparticles contribute to upstream processes by promoting algal biomass productivity. Specifically, metallic nanoparticles leveraging localized surface plasmon resonance (LSPR) can enhance light penetration and utilization in dense algal cultures, resulting in increased chlorophyll synthesis and cell proliferation (Kokkinos et al., 2021). (c) Nanoparticles can also alleviate nutrient limitations and improve CO_2 bioavailability, facilitating enhanced photosynthetic carbon fixation (Pramanik et al., 2023).

An important strategy for improving biodiesel yield involves the immobilization of lipase enzymes onto nanostructured supports. (d) Enzyme immobilization techniques utilize nanoparticles to entrap or anchor lipases, thereby improving enzyme stability, reusability, and facilitating efficient separation from the reaction medium (Kokkinos et al., 2021). The high surface area of nanomaterials allows greater enzyme loading and accelerates mass transfer kinetics, mitigating limitations observed with free lipase systems. Free enzymes typically suffer from low operational stability, difficulties in recovery, and deactivation in the presence of alcohols used in transesterification (Kokkinos et al., 2021).

Immobilized lipases exhibit superior performance characteristics, including (e) enhanced tolerance to pH fluctuations, increased thermal and functional stability, improved substrate specificity, and prolonged catalytic lifetime. Various nanomaterials have been successfully employed for lipase immobilization, including magnetic nanoparticles such as Fe_3O_4, silica-based nanoparticles, metal–organic frameworks (MOFs), carbon nanotubes, and hybrid nanoflower composites (Kokkinos et al., 2021; Pramanik et al., 2023; Verma et al., 2023). These systems have achieved high biodiesel conversion efficiencies and demonstrated strong potential for repeated use without significant loss of activity (Kokkinos et al., 2021).

Collectively, nanocatalyst-based strategies—spanning both upstream algal cultivation and downstream lipid conversion—represent a powerful and integrated approach for advancing the economic and environmental sustainability of algal biofuel technologies.

3 Environmental Bioremediation Via Nanobiotechnology

3.1 Nano Adsorbents for Pollutant Removal

Advanced products such as detergents, sanitary items, cosmetics, shampoos, and medicines are essential for a modern lifestyle. Due to the high persistence of several antibiotics and personal care products in drinking water, effluent treatment plants and seawater are indicators of spreading pollutants that directly damage the aquatic and terrestrial ecosystems. In view of these factors establishing novel, economically viable technology for the remediation and substantial reduction of contaminants is vital for a healthy and clean environment (Das et al., 2024a, 2024b; Dey et al., 2024a, 2024b). Strategic use of nanotechnological markers as prospective nanomaterials that can help by addressing current methods for detecting and preventing hazardous contaminants in water is one potential mitigation approach (Aguilar-Pérez et al., 2021; Ali et al., 2020). Nanomaterials are used as adsorbents due to their specificity, high surface area to volume ratio, selectivity, durability, and reusability features (Rasheed, 2022). Nanoadsorbents exhibit distinct physical and chemical attributes compared to traditional adsorbents, increasing their applications. Nanoadsorbents, including carbon tubes, metal, magnetic nanoparticles, graphene oxide (GO), polymer, metal–organic frameworks (MOFs), and nanocomposites, are found to be effective for pollutant remediation in the environment (El-sayed, 2020; Gómez-Pastora et al., 2014).

3.1.1 Graphene Oxide, Magnetic NPs for Heavy Metal/Organic Contaminant Capture

Yao and his coworkers studied the combined adsorption effect of heavy metals and antibiotics by GO (graphene oxide). Through the use of computational and experimental techniques, interactions between the pollutants and GO were revealed. It has been seen that heavy metals tend to increase the adsorption of antibiotics dramatically, while antibiotics showed a minor promotion for heavy metal adsorption. Contaminant coordination caused the complex formation, which adsorbs on GO with heavy metals acting as a bridge between the antibiotic and the GO. After six adsorption and desorption cycles, it has high adsorption activity, suggesting that GO is a feasible candidate for the adsorption of mixed pollutants (Yao et al., 2020).

He et al. fabricated sulfonated GO-based membranes (SGO@UiO-66-TFN) that are hydrophilic and repel contaminants, thereby enhancing the removal of heavy metals. The in-situ growth method was used to integrate the nanomaterial UiO-66 into SGO nanosheets. Further, the stability of this nanomaterial was improved by soaking in polyamide, which results in crosslinking between the SGO@UiO-66 and m-phenylenediamine. These GO-based membranes also showed the high removal of heavy metals copper (Cu^{2+}) and Lead (Pb^{2+}), which was greater than 99.4% in a 2-h period and increased to 97.5% in a 10 h period (He et al., 2020).

Yang and his team worked on synthesizing a hybrid nanocomposite of MOFs/GO, UiO-67/GO by the solvothermal method for capturing organophosphorus-based pesticides. To evaluate the GO-based adsorbent's adsorption capacity by equilibrium and kinetic adsorption isotherms. At pH 4, the adsorption capacity was found to be 482.69 mg/g for glyphosate. The extraordinary feature of this nano adsorbent was that it showed stability and high adsorption performance over a wide pH range (2–8). Moreover, GO's high surface area and abundance of Zr–OH groups also increased the adsorption of targeted organophosphorus pollutants (Yang et al., 2017).

Kasula et. al crafted GO-mobilized Ag–MOFs and Cu–MOFs by the ultrasonication process and evaluated the adsorption behaviour of Ag–MOFs, GO–Ag–MOFs, Cu–MOFs, and GO–Cu–MOFs for organic dye methylene blue (MB) under various conditions. The effect of temperature (24–45 °C), pH (3–10), and concentration of adsorbate was studied. Results revealed that the adsorption ability of GO-mobilized MOFs was 10 ten folds greater than pristine MOFs in all conditions, with adsorption percentages of 98% by GO–Ag–MOFs, 90% by Ag–MOFs, 75% by GO–Cu–MOFs, and 65% by Cu–MOFs, respectively. Similarly, the as-prepared materials also showed adsorption retention up to 80% over 3 cycles for MB, which showed the scalability and stability of these GO-mobilized MOFs (Kasula et al., 2022).

Massoud and his team prepared polyacrylamide acrylic acid p(AM-AA)-graphene oxide using a template polymerization technique with the utilization of gamma radiation. The resulting material was examined for the removal of heavy metal ions Sr^{2+}

and Co^{2+} from water. Experiments revealed that the presence of GO with p(AM-AA) caused the adsorption capacity to increase from 210 and 192 to 296 mg/g and 260 mg/g for Co^{2+} and Sr^{2+}, respectively (Massoud et al., 2020).

Wang et al. designed magnetic graphene oxide-based nanocomposite (Fe_3O_4@C-GO-MOF) which served as a magnetic solid phase extraction adsorbent for the specific adsorption and effective detection of traces of Lead ions. The nanocomposite Fe_3O_4@C-GO-MOF had a high surface area and porosity, which contributed to the high adsorption capacity of 343 mg/g 4.8 for Pb (II). Besides, this nanocomposite could be separated from the solution due to its magnetic nature and showed high regenerability and stability over five cycles. Hence, it has been seen as a potent candidate for the selective Lead adsorption (Wang et al., 2022).

Jin and his team synthesized a novel composite magnetic-CNT (carbon nanotubes)-Chitosan for the removal of heavy metals and organic pollutants. Being magnetically responsive, the prepared composite can be recovered easily by an external magnetic field. Different techniques like FTIR (Fourier transform infrared spectroscopy), TGA (thermogravimetric analysis), XRD (X-Ray diffraction), and zeta potential have been used to understand the interaction and adsorption mechanism internally. For organic contaminant removal, the adsorption for Rhodamine B (Rh B) was found to be greater than methyl orange due to π-π interaction between the aromatic ring of Rh B and CNT. In case of adsorption of heavy metals, the adsorption trend was found to be $Fe^{3+} > Cu^{2+} > Pb^{2+} > Zn^{2+}$. Moreover, in terms of recyclability, it possessed great potential and displayed as a potent material for the adsorption of heavy metals and organic wastes (Jin et al., 2021).

Let et al. constructed microporous polymer incorporated magnetic Fe_3O_4 nanoparticle (120-TMA@Fe) composite by post polymerization, followed by chemical modifications. This magnetic polymer adsorption activity was evaluated for organic dyes, antibiotics, and plasticizers. Experiments revealed the adsorption capacity values of 120-TMA@Fe composite for methyl orange and plastic bisphenol, and ibuprofen were found to be 357, 555, and 285 mg/g, respectively. Furthermore, this composite exhibited stable and uniform performance under harsh conditions, with the remarkable regeneration performance over 10 cycles (Let et al., 2021).

Chu and his team developed MOF-based magnetic microspheres with a large surface area of 1002 m^2/g, using ZIF-8 and magnetic Fe_3O_4 nanoparticles, which demonstrated an exceptionally high adsorption for multiple adsorbates, including dyes and heavy metals as malachite green (MG), tetracycline (TC), and Cu (II). The quantitative values for adsorption capacities for MG, Cu (II), and TC were found to be 1823, 2305, and 909 mg/g, respectively. Additionally, following five cycles of adsorption and desorption with nearly constant saturation magnetization, the adsorption activity continues to be satisfactory (Chu et al., 2023).

Mahmoodi and his coworkers designed a magnetic bio-based (ZIF-67@Fe_2O_3@eggshell) nanocomposite for the adsorption of environmental contaminants. A novel adsorbent having a high surface area (1263.9 m^2/g) was created by stabilizing the zeolitic imidazole framework crystals on the magnetic eggshell membranes' surface. Experimental studies showed the maximum adsorption capacity of 250.18 mg/g for black red 18 and 344.82 mg/g for Cu^{2+}(II), respectively, hence proving the excellent removal of dye molecules and heavy metal ions simultaneously (Mahmoodi et al., 2019).

3.1.2 Regenerability and Scalability Challenges

Magnetic nanoadsorbents exhibit outstanding adsorption capacity and reusability under controlled laboratory settings; however, these results are rarely replicated at commercial scale, due to mechanical instability, chemical degradation, and process inefficiencies over time (You et al., 2021). Magnetic nanoadsorbents often suffer from reduced adsorption performance upon reuse. For instance, ibuprofen removal efficiency declined by 42% after the second ethanol regeneration cycle, suggesting incomplete desorption of the pollutantethanol (Peralta et al., 2021). Similarly, naphthalene removal dropped to 54% by the tenth cycle (Aliannejadi et al., 2019). Chemical methods such as acid or solvent washing are the most widely used techniques for regenerating spent adsorbents due to their simplicity, though alternatives like thermal, microbial, supercritical fluid, and microwave-assisted regeneration have been explored. However, regeneration success depends heavily on environmental and material-specific variables such as pH, temperature, and the pollutant's molecular structure (Campos et al., 2019; Gagliano et al., 2020; Meng et al., 2018; Momina et al., 2018).

Despite functional improvements, the high cost of synthesis and regeneration for magnetic Nano adsorbents especially those with advanced surface modifications limits their feasibility in real-world wastewater treatment (Augusto et al., 2019; Neha et al., 2021).

3.2 Precision Degradation of Pollutants

Emerging organic pollutants (EOPs), originating from pharmaceuticals, herbicides, pesticides, and various industrial products or byproducts, are remarkably resistant to natural biological breakdown. As a result, they persist in the environment for long periods, often ending up in our water sources and the air we breathe (Bhattu & Singh, 2023). The removal of pollutants from wastewater largely depends on their physical state, with solid and oily wastes typically eliminated through coagulation–flocculation processes (Ma et al., 2024a, 2024b), sedimentation, (Devendrapandi et al., 2024). electro-coagulation, (Ma et al., 2024a, 2024b). and chemical oxidation, Booton et al. (2024). methods. Shi et al. (2020). They investigated the removal of organic pollutants from wastewater using a red mud/graphitic carbon nitride (g–C_3N_4) composite (RM-CN) as a sorbent. Key properties such as high surface area, substantial porosity, and abundant functional groups are essential for enhancing the effectiveness of sorbent active sites (Opoku et al., 2017; Son et al., 2021). g–C_3N_4, functioning as a semiconductor, and red mud (RM), which consists of various metal oxides such as Fe_2O_3 and TiO_2, act together as photocatalysts. Thanks to their inherent adsorptive properties, this composite is effectively applied for treating wastewater contaminated with antibiotics and dyes. Photocatalytic degradation, utilizing heterogeneous, readily available, and economically viable catalysts under light exposure, has become a widely adopted approach for eliminating organic pollutants from both air and aquatic environments (Jabbar et al., 2024; Mohanty et al., 2024; Priyadarshini et al., 2024; Ray et al., 2024; Sahu et al., 2024).

3.2.1 Photocatalytic NPs (TiO_2, ZnO) for Pesticide/Pharmaceutical Breakdown

ZnO and TiO_2 are broadly used as photocatalysts for the breakdown of organic and inorganic contaminants. Beyond this, they also serve roles in water-splitting electrodes (Louis et al., 2024), hydrogen generation (Omar et al., 2024), carbon dioxide reduction (Montalvo et al., 2023), solar energy harvesting (Sharifi Rad et al., 2023), drug delivery systems (Esquivel-Castro et al., 2023), sensors (Yao et al., 2023), and energy storage technologies (El Gohary et al., 2023; Okai Amu-Darko et al., 2023), As n-type semiconductor photocatalysts, both ZnO and TiO_2 demonstrate the capability to degrade a diverse range of organic pollutants in environments ranging from atmospheric air to wastewater. TiO_2 can crystallize in three distinct forms: anatase, rutile, and brookite. Each structure features a central Ti^{4+} ion surrounded by six oxygen atoms arranged octahedrally. The band gap energies and exciton binding energies are relatively high for these materials ZnO has values of 3.3 eV and 60 meV, while TiO_2 exhibits 3.1 eV and 53 meV, respectively (Yıldırım, 2022). The photons required to excite electrons from the valence band (VB) to the conduction band (CB) fall within the ultraviolet range, which defines the activation window for these materials (Chen et al., 2017; Ghamarpoor et al., 2023a, 2023b; Karthikeyan et al., 2020; Saldaña-Ramírez et al., 2020). Other semiconductors such as CeO, CdS, and ZnS have also demonstrated promising photocatalytic activity. These materials offer advantages over many traditional conductive or insulating compounds, primarily due to their suitable band gap energies, which help reduce the recombination rate of photoinduced charge carriers (Ghamarpoor et al., 2023a, 2023b; Moradi et al., 2023). ZnO and TiO_2, in particular, possess several key features as semiconductor oxides, including well-aligned electronic structures, extended excited-state lifetimes, and strong light absorption that can be extended from the UV to the visible spectrum with minor chemical modifications (Chen et al., 2008; Han et al., 2012). Given the restricted utilization of UV light, modifying these semiconductors to become responsive in the visible range is crucial. Various structural modification techniques have been explored to achieve this shift. These include doping with transition metals (Ghamarpoor et al., 2024; Pérez-González et al., 2015; Wang et al., 2016), introducing metal ions (Kumar & Rao, 2017), or nonmetal ions (Patil et al., 2019), and forming composites with other semiconductors that have narrower band gaps (Rehman et al., 2009), Surface modification strategies such as decorating photocatalyst surfaces with noble metals (Singhal & Kumar, 2017) and incorporating elements like nitrogen, iodine, sulfur, and carbon into the TiO_2 and ZnO lattices allowing their p orbitals to interact with the O 2pz orbital have also been employed (Ding et al., 2022; Farhadian et al., 2019). Further enhancement can be achieved by doping with transition metals such as V, Cr, Fe, Mg, Ag, Co, Zn, and Mo (Inturi et al., 2014; Sukhadeve et al., 2023) applying synthetic techniques like surface coatings and sensitization, or introducing specific impurities (Kim et al., 2005). These strategies help either narrow the intrinsic band gap or increase surface area, thereby extending light absorption and improving photocatalytic performance under visible light (Xiao et al., 2017). Various synthesis techniques have been employed to develop TiO_2, ZnO, and their modified derivatives, which exhibit remarkable properties such as high porosity, low density, and large specific surface area. These include hydrothermal (Wang et al., 2023), solvothermal (Kovács et al., 2022), sol–gel (Ahmad et al., 2022; Shawky & Alshaikh, 2022), chemical deposition (Saeed et al., 2022), and sol–hydrothermal (Li et al., 2022), methods—all known for producing materials with excellent electronic and optical characteristics as well as enhanced structural permeability. Hu et al. (2016), synthesized hollow black titanium spheres with mesoporous architectures using a multi-step process that involved a solvothermal synthesis under template-free conditions, followed by an encapsulation step in reflux with an amine compound, and subsequent high-temperature hydrogenation. The resulting material displayed high crystallinity, mesoporous walls, and surface-disordered Ti^{3+} frameworks.

To address the rapid recombination of photoinduced electron–hole pairs in titanium-based photocatalysts (Kozlova et al., 2021), incorporated platinum (Pt), which acts as an electron trap and enhances photocatalytic efficiency by prolonging the lifetime of charge carriers. This improvement is attributed to the formation of Schottky barriers at the metal–semiconductor interface. Maheswari et al. engineered composites by dispersing CuO and ZnO within TiO_2 matrices to form similar Schottky junctions (Maheswari et al., 2022). These composites demonstrated enhanced antioxidant and photocatalytic performance, particularly in the photodegradation of methylene blue (MB). Dong et al. developed a TiO_2/graphene-modified (GM) composite aimed at the degradation of pharmaceutical residues in medical wastewater (Dong et al., 2022). A key benefit of this composite was its ease of separation from solution and superior reusability compared to unmodified TiO_2. he degradation of organic pollutants typically proceeds through a series of photooxidation reactions on the surface of photocatalysts, initiated by the excitation of electrons from the valence band (VB) to the conduction band (CB) upon exposure to light, leaving behind positive holes in the VB (Xiao et al., 2016), Once photoexcited, these electrons may follow two primary pathways: (1) recombination with holes in the VB. or (2) interaction with oxygen molecules adsorbed on the catalyst surface to initiate photooxidation reactions that contribute to pollutant breakdown. Electrons and holes, as mobile charge carriers, migrate to the catalyst surface where they, respectively, drive reduction and oxidation reactions. Specifically, electrons reduce surface-bound O_2 to generate superoxide anions ($O_2\cdot-$), while holes oxidize water molecules adsorbed on the surface to form hydroxyl radicals ($OH\cdot$). These reactive oxygen species (ROS), formed indirectly through the photoreaction process, further degrade the organic contaminants into harmless end-products such as water (H_2O) and carbon dioxide (CO_2).

In a related study, Qiang et al. fabricated a $RuTe_2$/black TiO_2 composite and elucidated a degradation mechanism wherein photoexcited electrons transition from the VB to the CB of black TiO_2, and subsequently transfer to the CB of $RuTe_2$ (Qiang et al., 2022). The more negative potential of $RuTe_2$ facilitates this electron transfer, thereby enhancing photocatalytic efficiency.

3.2.2 Synergy with Microbial Consortia (Nanoparticle-Microbe Hybrid Systems)

Nanobioremediation combines microbial systems with nanomaterials to form a highly effective platform for addressing persistent pollution (Usman et al., 2020). The unique properties of nanoparticles such as high surface area, enhanced catalytic activity, and adsorption capacity contribute to increased contaminant accessibility and faster reaction kinetics (Narain, 2020). n parallel, microorganisms especially bacteria and fungi provide the biological machinery capable of transforming or completely mineralizing complex organic pollutants through natural enzymatic processes (Liu et al., 2014). This integrated approach takes advantage of the complementary capabilities of both components, allowing for enhanced performance and greater flexibility in environmental remediation applications (Partovinia & Rasekh, 2018).

Microbial Consortia: Functional Advantages

Employing diverse microbial consortia rather than single strains significantly enhances bioremediation efficiency by providing broader metabolic and genetic capabilities that can target multiple contaminants or their break-down products. These communities allow for syntrophic cooperation, where some microbes degrade the primary pollutant while others metabolize intermediate compounds, leading to more complete detoxification (Massot et al., 2022). For instance, microbial consortia of actinobacteria have demonstrated the ability to simultaneously degrade contaminants such as lindane, phenanthrene, and textile dyes in complex environments (Antezana et al., 2023; Elumalai et al., 2021). Moreover, microorganisms within consortia often form biofilms structured communities encased in extracellular polysaccharides which enhance their ability to survive and remain active in toxic or fluctuating environmental conditions. Biofilms confer protective advantages and support microbial viability under stressors such as pH shifts, nutrient scarcity, and the presence of toxic pollutants (Fazeli-Nasab et al., 2022; Yan & Bassler, 2019).

Types of Nanoparticle–Microbe Interactions

Nanoparticles can interact with microorganisms in two primary ways: through cell immobilization or cell decoration, each offering distinct benefits for remediation applications (Giese et al., 2020; Khatoon & Rai, 2018; Ranmadugala et al., 2018). Immobilization involves anchoring microbial cells onto larger solid matrices such as inorganic supports or polymeric materials to enhance biomass retention, minimize cell washout, and improve operational stability in bioreactors. This method is particularly useful in large scale or continuous treatment systems, as it facilitates microbial reuse and resistance to physical and chemical stresses (Bouabidi et al., 2019; Klein & Wagner, 1983; Suzana et al., 2013). However, immobilization can also limit nutrient and oxygen diffusion due to mass transfer barriers, which can affect microbial activity (El-Mansi, 2007; Lagarde & Jaffrezic-Renault, 2011; Mehrotra et al., 2021). On the other hand, cell decoration refers to the attachment or synthesis of nanoparticles directly on microbial cell surfaces or biofilms, typically at the nanoscale (1–100 nm), increasing reactivity and mass transfer efficiency (Bouabidi et al., 2019; Giese et al., 2020; Salam & Das, 2015). This strategy improves catalysis and pollutant interaction while maintaining microbial mobility and activity, often leading to improved degradation outcomes (Bouabidi et al., 2019; Giese et al., 2020). Additionally, the method of application, whether simultaneous (nanoparticles and microbes added together) or sequential (one after the other) significantly influences treatment performance, as sequential application often reduces nanotoxicity and enhances pollutant accessibility (Horváthová et al., 2019a).

Examples of Applications: Combined Systems in Action

A hybrid nanobioremediation system using *Pseudomonas stutzeri* in combination with nano-zero-valent iron (nZVI) was successfully applied for the degradation of tetrabromodiphenyl ether (t-BDE), a common polybrominated diphenyl ether (PBDE) pollutant. Initially, the nZVI dominated the process through strong adsorption and reductive debromination, temporarily inhibiting the bacterium's biosorption activity. However, *P. stutzeri* adapted over time, regaining dominance in the later stages of treatment and facilitating biodegradation through enzymatic mechanisms (Huang et al., 2016). In a similar case, *Pseudomonas putida* was paired with Pd/nZVI to enhance t-BDE degradation while minimizing toxic by-product formation; the bacteria's acidic metabolites helped reduce nanoparticle passivation, sustaining catalytic activity (Jia et al., 2019). When applied in natural environments, such as mangrove sediments contaminated with deca-BDE, nZVI improved degradation efficiency in the presence of indigenous microbial communities, even though microbial growth was partially inhibited at the beginning. Over the course of treatment, the native microbes recovered, illustrating their capacity to adapt and re-establish synergistic degradation with the nanoparticles (Pan et al., 2021). In another example, combining *Microbacterium* with sulfur-doped nZVI (S-nZVI) enabled the simultaneous removal of deca-BDE and heavy metals such as Cr^{6+} and Ni^{2+}, reaching degradation efficiencies as high as 95%. In that case, the nanoparticles played a dual role: reducing metal toxicity through adsorption and contributing to the chemical degradation of the organic pollutant via redox reactions (Wei et al., 2022a, 2022b).

Polychlorinated Biphenyls (PCBs)

Polychlorinated biphenyls (PCBs), known for their environmental persistence and toxicity, have been effectively treated using combined nano-bio systems. In one study, a sequential treatment involving Pd/nZVI followed by the PCB-degrading bacterium *Burkholderia xenovorans* LB400 achieved a dechlorination rate of 89% for the PCB mixture Aroclor 1248 within six days, with subsequent microbial degradation removing 90% of the resulting biphenyls within 24 h (Le et al., 2015). This demonstrates the effectiveness of combining catalytic nanoparticle action with enzymatic biodegradation in sequential fashion.

A separate comparative study showed that the order in which microbes and nanoparticles are applied significantly influences overall remediation efficiency—treatments where nanoparticles were introduced first (nanobioremediation) achieved up to 99% pollutant removal, while the reversed order (bionanoremediation) yielded a maximum of 77%. This improvement is attributed to the ability of nanoparticles to modify the contaminant structure, increasing its bioavailability before microbial action. These studies highlight that both the treatment sequence and the microbial configuration (single strains vs. consortia) are critical parameters for maximizing PCB removal efficiency (Horváthová et al., 2019a, 2019b). In real-world scenarios, PCB-contaminated sediments often acting as pollutant reservoirs due to the hydrophobic nature of PCBs have been effectively treated using Pd/nZVI and bacterial cultures in ex situ systems (Jönsson et al., 2003). These treatments not only reduced the PCB load but also preserved microbial diversity, with some native populations being slightly inhibited while others were stimulated. Moreover, bimetallic nanoparticles accelerated the degradation of less-chlorinated congeners, while more heavily chlorinated forms like PCB 101 remained partially recalcitrant, indicating that multi-stage or adapted microbial strategies might be needed for full remediation (He et al., 2009).

Dioxins

Dioxins, such as 2,3,7,8-tetrachlorodibenzo-p-dioxin (TCDD), are among the most toxic and persistent organic pollutants, known to cause severe health and ecological effects, including carcinogenicity and immune system disruption (Domingo et al., 2020; Zanaroli et al., 2012). Although conventional biodegradation methods have attempted to reduce their toxicity, these approaches often fall short of achieving complete mineralization due to the compound's high stability and lipophilicity (Andrada Suarez et al., 2025; Takada et al., 1996). A breakthrough in nanobioremediation was achieved through a two-step hybrid system where Pd/nZVI was first applied under anaerobic conditions to dechlorinate TCDD, followed by aerobic biodegradation using *Sphingomonas wittichii* RW1, which resulted in full degradation of the pollutant. The first phase reduced TCDD to monochlorinated derivatives in just 10 h, demonstrating the effectiveness of Pd-enhanced nZVI in rapidly breaking halogen bonds. In the second step, the bacterium mineralized the intermediate products, such as monochlorinated dioxins, producing catechol as a benign end product under aerobic conditions within 18 h. This study highlights the synergistic potential of nanomaterials and microbes, where the chemical reactivity of the nanoparticle accelerates contaminant transformation, and microbial metabolism ensures its safe and complete breakdown (Bokare et al., 2012). However, the authors caution that nanomaterials like Pd/nZVI must be carefully assessed for environmental toxicity and compatibility, as some may interfere with detoxification pathways or bioaccumulate in unintended ecosystems (Vannuccini et al., 2015; Wei et al., 2022a, 2022b).

Environmental and Practical Considerations

Despite the strong potential of nanobioremediation systems, the toxicity of certain nanoparticles, particularly metal-based ones like nZVI, remains a critical limitation, as they can adversely affect microbial viability and system stability (Pan et al., 2021). For example, studies have shown that excessive concentrations of nZVI can inhibit microbial growth in both engineered systems and natural environments, underscoring the importance of careful dose optimization (Farré et al., 2011; Jia et al., 2019). In some cases, these nanoparticles become passivated through oxidation or form surface hydroxide layers, reducing their catalytic performance and diminishing the expected synergy with microbial partners (Peng et al., 2013). To mitigate such drawbacks, researchers have explored the use of natural or modified nanomaterials, such as nanoclays and polymer-coated nanoparticles, which offer similar catalytic or adsorptive benefits with reduced ecological risk. For instance, coating nZVI with carboxymethyl cellulose (CMC) has been shown to lower its antimicrobial effects while maintaining its reactivity, thereby improving compatibility with microbial cultures in hybrid systems (Liang et al., 2021; Singh et al., 2013). These strategies not only improve the biocompatibility of the system but also expand the applicability of nanobioremediation in ex situ settings such as wastewater treatment plants and bioreactors, where environmental containment and recovery of nanomaterials are more feasible. Overall, the success of these technologies hinges on system-specific design, including nanoparticle selection, treatment sequence, and microbial community structure—all of which must be optimized to minimize risks and maximize remediation outcomes (Andrada Suarez et al., 2025; Jain et al., 2021; Liang et al., 2021; Singh et al., 2013).

3.3 Nano-Biosensors for Real-Time Monitoring

The rapid industrialization and urban growth have intensified environmental pollution, calling for technologies that enable sensitive, real-time monitoring across water, air, and soil ecosystems (Kalyani et al., 2021). Traditional methods like GC, HPLC, and ICP-MS, while accurate, are resource-intensive, require laboratory infrastructure, and are not suitable for real-time, on-site detection (Thakur & Kumar, 2022). To overcome these limitations, biosensors incorporating nanomaterials have emerged as promising tools due to their high sensitivity, portability, and cost-effectiveness for decentralized environmental applications (Nelson, 2023). These biosensors combine biological elements such as enzymes, antibodies, aptamers, or nucleic acids with nanostructures like gold nanoparticles, graphene, and carbon nanotubes to identify trace pollutants such as methane, heavy metals, and VOCs.

For example, gold nanoparticle-based sensors have detected mercury ions down to 0.005 ppb, while graphene modified biosensors have shown capabilities for lead ion detection at just 0.01 ppb (Zhang et al., 2021). Integration with Internet of Things (IoT) platforms enhances the functionality of these biosensors, allowing remote data access, real-time analysis, and predictive environmental responses (Khanam et al., 2020). This technological fusion enables efficient detection and management of pollutants in waste streams, such as methane and VOCs, which are critical for smart waste systems (Kumar et al., 2022). Additionally, these innovations support multiple UN sustainable development goals (SDGs) by aiding efforts in clean water access, pollution control, and sustainable urban management (Xiong et al., 2022). Thanks to their compact size, rapid response, and ability to detect multiple analytes simultaneously, nanomaterial-enabled biosensors are increasingly vital in the design of intelligent, real-time environmental monitoring solutions (Jain et al., 2022; Nelson, 2023).

3.3.1 Detection of Methane, VOCs, and Toxic Ions in Waste Streams

The detection of harmful gases such as methane is a critical component of waste stream monitoring due to its potent greenhouse effects and risk of explosion in confined environments (Sharma et al., 2025). Metal oxide semiconductor (MOS)-based gas sensors have shown great potential for detecting methane owing to their high sensitivity and operational stability under ambient conditions (Fu et al., 2023). These sensors work by detecting changes in electrical resistance triggered by methane's interaction with the sensing material, usually nanostructured oxides like SnO_2 or ZnO (Ejeian et al., 2018). These VOC sensors often employ materials such as graphene, carbon nanotubes, or gold nanoparticles, which enhance surface interactions and signal transduction mechanisms. Additionally, the detection of toxic metal ions like mercury (Hg^{2+}), lead (Pb^{2+}), arsenic (As^{3+}), and cadmium (Cd^{2+}) in aqueous waste streams is enabled by electrochemical biosensors and optical nanobiosensors that employ specific biological recognition elements such as DNA aptamers and enzymes (Li et al., 2018). These biosensors provide high selectivity and can detect ions at very low concentrations, making them ideal for continuous environmental monitoring of industrial and municipal wastewater (Odobašić et al., 2019). Collectively, these nanobiosensing technologies allow for real-time, on-site detection of multiple hazardous analytes in complex waste matrices, significantly advancing the capabilities of modern environmental remediation strategies.

3.3.2 IoT Integration for Smart Waste Management

The evolution of smart wearable sensors has opened new frontiers in real-time environmental monitoring, particularly in waste management systems that rely on the Internet of Things (IoT). These wearable gas sensors are increasingly being integrated into larger sensor networks to provide instantaneous feedback on air quality, toxic gas exposure, and environmental conditions an essential advancement in occupational health and smart infrastructure planning (Das et al., 2023, 2024a, 2024b; Patel et al., 2022). Their connectivity capabilities, especially with smartphones via Bluetooth and wireless networks, allow for remote access to data and cloud-based analytics, enhancing responsiveness and facilitating data-driven decisions in waste management. Additionally, advancements in microelectronic components and power efficiency have brought the vision of real time, on-node environmental processing into reach, providing a backbone for smart, connected waste handling systems. Such integration allows emergency responders, environmental inspectors, and facility managers to act on live pollutant data collected from sensors embedded in waste processing environments, garments, or infrastructure. These developments are contributing significantly to the advancement of IoT ecosystems by supporting the deployment of decentralized, smart environmental sensing systems (Dai et al., 2020; Hooshmand et al., 2023; Kang et al., 2022).

The integration of biosensors with Internet of Things (IoT) platforms has transformed environmental monitoring by enabling real-time data acquisition and remote-sensing capabilities. This convergence empowers smart waste management systems to continuously track pollutant levels, ensuring swift detection and timely intervention to mitigate environmental risks. Nanomaterials such as graphene, gold nanoparticles, and carbon nanotubes significantly enhance the sensitivity, selectivity, and responsiveness of biosensors, making them ideal for IoT-based deployments (Khanam et al., 2020; Kumar et al., 2022). These sensors can detect minute concentrations of hazardous pollutants, including heavy metals and volatile organic compounds, which are common in waste streams (Dey et al., 2023; Ghosh et al., 2023; Paul et al., 2023). The seamless transmission of data from these nanosensors to centralized monitoring systems enhances decision-making in waste management and supports the automation of remediation responses. Moreover, IoT-linked biosensors reduce the dependence on labor-intensive sampling and laboratory analysis by enabling in situ monitoring, which is both cost effective and efficient (Kumar et al., 2022; Wang et al., 2024a, 2024b). This synergy between nanotechnology and digital connectivity is pivotal for achieving sustainable development goals related to clean water, climate action, and responsible consumption (Huang et al., 2023).

4 Sustainability and Policy Frameworks

4.1 Alignment with Global Sustainability Agendas

4.1.1 The European Green Deal and Circular Bioeconomy: A Synergistic Approach

Safe and sustainable nanomaterials are vital for achieving crucial European policy goals, including the European Green Deal, the Circular Economy Action Plan, and the Chemicals Strategy for Sustainability. This alignment emphasizes the "Safe- and Sustainable-by-Design" (SSBD) concept in nanotechnology, actively promoted by organizations like the Joint Research Centre (JRC) and DG Research and Innovation.

This principle requires that materials and products are safe for both human health and the environment. Furthermore, they must be inherently compatible with a climate-neutral, circular, and resource-efficient economy, maintaining optimal properties and functionality throughout their entire life cycle.

Integrating safety, circularity, and functionality from the outset of nanomaterial design, the SSBD concept offers a comprehensive, systems-based strategy. This holistic approach is crucial for fostering a sustainable, equitable, and inclusive economy within the European Union. This aligns well with the circular bioeconomy principles, which prioritize the environmentally sound use of lignocellulosic biomass and waste streams to create valuable industrial products. These processes often utilize advanced biotechnology and microbial enzymes to convert waste into resources, thereby lessening dependence on fossil fuels (Ghosh & Das, 2020; Ghosh et al., 2016, 2020; Pranaw et al., 2022).

Overarching policy frameworks exert a significant top-down influence by directly catalyzing technological research and innovation. For instance, the European Green Deal's ambitious environmental and economic targets establish a regulatory and market landscape. This actively incentivizes and guides nanotechnology development toward specific sustainability goals, notably circularity and inherent safety (Pranaw et al., 2022). This dynamic highlights the power of strong, visionary policy in shaping technological paths. It fosters an environment where sustainable nanomaterials are not just an option but a primary developmental focus, thus accelerating the shift towards a more sustainable future.

4.1.2 Contributions to the United Nations Sustainable Development Goals (SDGs)

Microbial nanotechnology, a field that involves the synthesis of metal, metal oxide, and other nanoparticles using microorganisms such as bacteria, fungi, viruses, and algae, offers powerful and novel pathways to address several United Nations Sustainable Development Goals (UN SDGs) (Singh et al., 2025). This biological synthesis of nanoparticles is especially beneficial due to its environmentally friendly, more sustainable, and faster nature compared to conventional physical and chemical techniques, which typically require aggressive chemicals and significant energy consumption (Singh et al., 2025).

The contributions of nanobiotechnology align with several specific SDGs, demonstrating its broad impact:

- *SDG 3 (Good Health and Well-being)*

Nanobiotechnology supports healthy lives and well-being by diminishing water-borne illnesses like diarrhea, dysentery, cholera, skin problems, and lung infections. Progress in nanomedicine, such as nanovaccines, precise drug delivery, and improved diagnostics, further contributes to this (Sanni et al., 2024).

- *SDG 6 (Clean Water and Sanitation)*

Biogenic nanomaterials offer highly efficient wastewater treatment by removing various contaminants such as heavy metals, dyes, and microplastics. These nanomaterials can also adjust the pH of purified water to safe drinking levels (Sanni et al., 2024).

- *SDG 12 (Responsible Consumption and Production)*

Employing reactive green nanostructures to detoxify industrial wastewater by converting hazardous materials aligns with the promotion of responsible consumption and production. Furthermore, the conversion of agricultural waste biomass into useful adsorbents supports circular economy principles (Sanni et al., 2024).

- *SDG 14 (Life Below Water)*

Biogenic nanoparticles offer a way to protect aquatic life by efficiently capturing pollutants such as microplastics, which accumulate in fish. This reduces harm to aquatic ecosystems and safeguards human consumers dependent on these resources (Sanni et al., 2024).

Nanobiotechnology's diverse applications extend beyond specific objectives to encompass food security, sustainable energy, and bioremediation (Singh et al., 2025). This wide-ranging utility positions it as a key "cross-cutting enabler" for multiple UN Sustainable Development Goals (SDGs). Its advancements offer synergistic benefits, meaning progress in this field can simultaneously address various aspects of sustainable development. This comprehensive impact underscores the need for strategic investment and the integration of nanobiotechnology into broader international sustainable development efforts. By offering multi-faceted solutions, it holds the potential to accelerate the achievement of the 2030 Agenda.

4.2 Life-Cycle Assessment (LCA) of Nanobiotechnologies: Navigating Efficiency and Ecotoxicity Trade-Offs

4.2.1 Methodologies and Inherent Challenges in Nanomaterial LCA

Life-Cycle Assessment (LCA) is a vital, systematic approach for the quantitative determination and evaluation of a product's or process's environmental sustainability. This method relies on thorough mass and energy balances (Bippus et al., 2024). The European Platform on LCA (EPLCA) champions LCA as a key integrated environmental assessment tool. EPLCA aims to support the European Green Deal's goals and wider EU policymaking by highlighting LCA's role in guiding sustainable development (Bippus et al., 2024).

Significant obstacles hinder comprehensive Life-Cycle Assessment (LCA) studies for nanomaterials. A primary issue is the widespread lack of data, especially regarding the usage and disposal of these materials. Furthermore, standardized metrics for evaluating the environmental consequences of nanoparticle releases are notably absent (Arvidsson et al., 2015). This scarcity of data is exacerbated by the fact that the production of many nanomaterials consumes more energy than that of their conventional counterparts, further complicating their environmental impact (Arvidsson et al., 2015). Additional challenges stem from the nascent stage of many nanotechnology products, difficulties in scaling up production, and the crucial need to define appropriate functional units for meaningful comparative analyses (Arvidsson et al., 2015). The inherent uncertainties associated with the developing nature of these technologies necessitate thorough sensitivity analyses to guarantee the reliability of LCA findings (Arvidsson et al., 2015). A fundamental systemic problem is the severe lack of complete and standardized data across the entire lifecycle of nanomaterials, particularly concerning their environmental release, fate, and long-term effects. These data gap severely restricts the ability to perform dependable LCAs, which, in turn, significantly obstructs effective risk assessment, regulatory creation, and well-informed decision making. Without these essential data, policymakers face difficulties in creating clear, evidence-based regulations for the safe and sustainable application of nanomaterials, potentially impeding responsible innovation and public confidence (Karn et al., 2009). This situation underscores the urgent requirement for focused, interdisciplinary research to produce these missing data and to develop standardized measurement and characterization methods that can keep up with the rapid progress in nanotechnology.

A fundamental problem is the significant lack of standardized data across the entire nanomaterial lifecycle, particularly concerning their environmental release, fate, and long-term effects. This deficiency severely limits the ability to conduct thorough and reliable life cycle assessments (LCAs), thereby hindering effective risk evaluation, regulatory development, and informed decision-making. Without this crucial data, policymakers face difficulties in establishing clear, evidence-based guidelines for the safe and sustainable application of nanomaterials, potentially hindering responsible innovation and undermining public trust. Resolving this issue demands immediate, collaborative, and interdisciplinary research to produce the necessary data and develop standardized measurement and characterization techniques that can keep up with the swift advancements in nanotechnology.

Balancing Performance Gains with Potential Environmental Toxicity

Nanomaterials possess unique physicochemical properties, such as a large surface area, strong catalytic activity, specific light absorption, and magnetic characteristics, leading to significant efficiency improvements in diverse applications (Roy et al., 2021). These properties facilitate their use in advanced electronics, energy storage, targeted drug delivery, and efficient industrial catalysis (Wang et al., 2024a, 2024b). However, the growing use of nanomaterials also raises serious ecotoxicity concerns. Release during manufacturing, use, and disposal can harm microorganisms, plants, and animals. Nanomaterial ecotoxicity depends on factors like surface charge, size, shape, and agglomeration, which change based on the environment. Smaller nanoparticles, with their high reactivity and ability to penetrate cell membranes, can be more toxic due to enhanced interaction with biological systems (F. Wang et al., 2024a, 2024b).

Nanoparticle toxicity is complex, involving direct cell membrane damage, increased intracellular Reactive oxygen species (ROS) leading to oxidative stress, cellular inflammation, and altered gene expression (Wang et al., 2024a, 2024b). Organic nanoparticles are typically less toxic and often biodegradable. However, inorganic and carbon-based nanoparticles can cause significant cytotoxic effects, requiring thorough evaluation and control (Eker et al., 2024).

These observations highlight a fundamental design paradox in nanotechnology: the very attributes that confer superior functionality and efficiency (e.g., high surface area, enhanced reactivity, small size) are frequently the same ones that contribute to their potential environmental toxicity (Kumari et al., 2023). Therefore, research in this field requires a careful balancing of priorities. Instead of solely focusing on maximizing efficiency, a strategic approach that emphasizes "Safe- and Sustainable-by-Design" principles is crucial for proactively reducing potential risks. The primary objective transitions from optimizing a single factor to achieving the best outcome across multiple dimensions—performance, safety, and sustainability—ultimately developing solutions that are both effective and environmentally friendly.

4.2.2 Case Study: Cradle-To-Gate Analysis of Nanocatalyst Production (e.g., Iron Oxide Nanoparticles)

A cradle-to-gate life-cycle assessment (LCA) comprehensively evaluated the environmental sustainability of iron oxide nanoparticles (IONPs) produced through two different synthesis approaches: a green method using Cymbopogon citratus extract and the conventional coprecipitation method. The study compared the environmental impacts of these methods based on the functional unit of 1 g of IONPs produced, ensuring a standardized evaluation (Patiño-Ruiz et al., 2021).

Life-cycle assessment (LCA) results clearly showed that the green synthesis method consistently had significantly lower overall environmental impacts across multiple categories when compared to the coprecipitation method (Patiño-Ruiz et al., 2021). Table 3 shows a comparison between the environmental impacts of the synthesis of iron nanoparticles via the green method and the coprecipitation method. Marine aquatic ecotoxicity (MAE) was the most significant impact category for both synthesis methods, highlighting it as a key concern. However, the coprecipitation method displayed a considerably higher MAE value (1.2×10^{-8} kg 1,4-DB equiv) in contrast to green synthesis (7.6×10^{-10} kg 1,4-DB equiv), clearly indicating a substantial environmental benefit of using the green synthesis approach (Patiño-Ruiz et al., 2021).

Both chemical and green synthesis methods for nanocatalysts have ethanol and electricity consumption during purification as major environmental hotspots. Although green synthesis lowers overall ecological impact by using less harmful starting materials, even "green" compounds like *Cymbopogon citratus* extract and sodium carbonate have environmental costs from their production (Patiño-Ruiz et al.,

Table 3 Comparative environmental impacts of iron oxide nanoparticle synthesis methods (Green vs. Coprecipitation)

Aspect	Green synthesis	Coprecipitation method
Overall environmental impact	Significantly lower	Higher
Highest impact category	Marine Aquatic Ecotoxicity (MAE)	Marine Aquatic Ecotoxicity (MAE)
MAE Value	7.6×10^{-10} kg 1,4-DB equiv.	1.2×10^{-8} kg 1,4-DB equiv.
Key Environmental Hotspots	Ethanol usage, Electricity usage	Ethanol usage, Electricity usage
Notes	Utilizes Cymbopogon citratus extract	Utilizes iron (II) salt precursor, NaOH

2021). To further decrease the environmental impact of nanocatalyst production, it is essential to optimize processes like ethanol recovery and adopt more sustainable production methods for all raw materials.

While "green synthesis" signifies progress over traditional methods, its environmental impact isn't necessarily zero. A complete life-cycle assessment (LCA) is crucial for identifying and mitigating potential "hidden burdens" throughout the supply chain, even with seemingly eco-friendly materials. True sustainability in nanobiotechnology necessitates a comprehensive approach that optimizes every stage, from material sourcing to disposal. Therefore, regulatory agencies and consumers should move beyond simple "green" labels, demanding transparency and full lifecycle accountability to ensure genuine environmental advantages.

4.3 Regulatory and Ethical Considerations

4.3.1 OECD/FDA Guidelines for Nano-Waste Disposal

The widespread application of nanomaterials across industrial, commercial, and medical fields, evident in everyday items like sunscreens and textiles, is leading to their increased presence in global waste (OECD, 2016). This has become a significant concern for international organizations. An OECD report, "Nanomaterials in Waste Streams: Current Knowledge on Risks and Impacts," points out that while existing advanced waste treatment can partially manage nanomaterials, substantial uncertainties remain about their final disposal and their long-term effects on environmental and human health. Consequently, there is a pressing need for more research in this relatively uninvestigated area (OECD, 2016).

Proper nanowaste disposal is challenging because engineered nanomaterials (ENMs) have unique properties, and effective treatment methods are not well understood for many types of this waste (Zahra et al., 2022). A significant gap exists in standardized information and optimized procedures for nanowaste identification, monitoring, and management. While common waste disposal methods like landfilling and incineration can create considerable environmental hazards without careful oversight, recycling offers a more sustainable solution (Zahra et al., 2022). Nanowaste management faces key obstacles, including difficulties in quantifying, detecting, and treating it due to nanomaterials' high reactivity, their potential to disperse in the environment via air or water, and their capacity to enter the food chain (Zahra et al., 2022). A crucial lack of knowledge pertains to the eventual fate and long-term environmental impact of nanowaste, especially once it mixes with other waste. This represents a major uncertainty in the life-cycle management of nanomaterials. Despite the growing production and use of nanomaterials, this knowledge deficit presents a considerable long-term environmental threat. Uncontrolled or poorly managed releases could lead to unexpected ecological and human health issues, potentially compromising the sustainability benefits of nanobiotechnology. This necessitates an immediate shift in research and funding focus toward thorough end-of-life studies and the creation of strong, specialized waste management systems for nanomaterials.

The Food and Drug Administration (FDA) oversees nanomedicine products by categorizing them within current regulatory structures for drugs, biologics, and devices (Regulating Nanomedicine at the Food and Drug Administration, 2019). These are frequently classified as "combination products" depending on their main function. While the FDA's Nanotechnology Task Force determined in 2007 that entirely new regulations were unnecessary, substantial issues remain. These involve concerns about whether current frameworks are suitable for products with unique nanoscale mechanisms, the possibility of new risks (such as nanoparticle toxicity and long-term impacts), and whether current labeling requirements sufficiently inform consumers about the presence and effects of nanotechnology (Regulating Nanomedicine at the Food & Drug Administration, 2019).

The worldwide regulatory environment for nanotechnology, particularly in agriculture, differs considerably across nations, each having its own set of rules (Kumari et al., 2023). For instance, in the United States, the Environmental Protection Agency (EPA) regulates pesticides, while the U.S. Department of Agriculture (USDA) oversees agricultural biotechnology. Within the European Union, the European Chemicals Agency (ECHA) and the European Food Safety Authority (EFSA) manage nanotechnology in food and pesticides (Kumari et al., 2023). Although international guidelines like ISO/TS 80,004–1 exist to ensure safety and quality (Kumari et al., 2023), the absence of consistent global standards creates regulatory ambiguity, impeding the approval and broad use of new nanomaterials (Amutha et al., 2024). This overall situation results in a disjointed and varied global regulatory system, akin to a "regulatory patchwork." This inconsistency poses major obstacles to the ethical global implementation, trade, and uniform risk management of nanotechnologies. It may encourage "regulatory arbitrage," potentially shifting development and production to areas with looser regulations, which could jeopardize global safety. Ultimately, this fragmentation prevents the creation of universally recognized safety standards and best practices, thereby delaying responsible innovation and fair global access to the advantages of nanotechnology.

4.3.2 Addressing Global Equity and Social Justice in Technology Access and Impact

Nanobiotechnology's ethical considerations heavily involve societal divisions and global equity. Significant concerns exist that its progress and application may worsen current societal inequalities, potentially increasing the divide between privileged and underprivileged populations both domestically and internationally (Bruce, 2006).

Despite substantial government investment in nanoscience and nanotechnology research in many developing countries, the social and ethical implications of these technologies have received less attention in terms of study, discussion, and public debate compared to developed nations. For example, research involving practitioners in India indicated that while a large majority (95%) acknowledged ethical concerns in nanotechnoscience, a smaller proportion (60%) could provide concrete illustrations. These worries encompassed potential negative impacts on the environment and human health, as well as the risk of exacerbating socioeconomic disparities (Patra et al., 2010).

Nanotechnology's considerable costs may result in its dominance by large corporations, fostering economic inequality (Patra et al., 2010). While advancements like extended lifespans and targeted drug delivery offer benefits, they could exacerbate global disparities by primarily serving wealthy nations leading this research (Patra et al., 2010). A key ethical dilemma concerns whether human enhancement technologies genuinely improve humanity and whether these advancements will inherently favor the wealthy, raising significant equity and justice issues (Bruce, 2006). The overarching purpose and regulation of nanobiotechnology research must prioritize broader societal needs, including vulnerable populations in developing nations, rather than solely focusing on measurable functions or commercial profits (Bruce, 2006).

5 Challenges and Future Directions

The integration of nanobiotechnology into organic waste valorization and environmental remediation represents a paradigm shift toward sustainable industrial processes. However, its widespread adoption is hindered by multifaceted challenges that span technical, economic, and regulatory domains (Bhandari et al., 2023). Addressing these limitations is imperative to unlock the full potential of nanobiotechnology while ensuring alignment with global sustainability agendas such as the United Nations Sustainable Development Goals (SDGs) and the European Green Deal (Pathak et al., 2024).

5.1 Technical Barriers: Stability, Scalability, and Green Synthesis

A primary challenge lies in the industrial-scale stability and recoverability of nanomaterials. While lab-scale studies demonstrate remarkable efficiency in nanocatalysts (metal–organic frameworks (MOFs) for enzymatic hydrolysis or magnetic nanoparticles for heavy metal adsorption), their performance often degrades under real-world conditions due to agglomeration, chemical fouling, or mechanical shear in continuous-flow systems (El-Ramady et al., 2023). For instance, TiO_2 nanoparticles used in photocatalytic wastewater treatment exhibit reduced activity after repeated cycles due to surface passivation by organic residues (Mahlambi et al., 2015; Najafidoust et al., 2020; Solano et al., 2019). Innovations in surface functionalization (PEGylation, bio-inspired coatings or antimicrobial peptides like nisin) and immobilization techniques (embedding in porous silica or polymeric matrices) could mitigate these issues, but scalability remains unproven (Ceballos et al., 2025; Khan et al., 2020).

Equally critical is the energy and resource intensity of nanomaterial synthesis. Traditional methods like chemical vapor deposition or sol–gel processes often require high temperatures, toxic solvents, and non-renewable precursors, undermining the net sustainability benefits (Esposito, 2019). Emerging green synthesis approaches—such as plant extract-mediated nanoparticle fabrication or microbial biosynthesis—offer lower-energy alternatives, but their yields and reproducibility are inconsistent (Bhardwaj et al., 2021). Recent advances in mechanochemical synthesis (ball-milling) and flow chemistry could bridge this gap, yet life-cycle assessments (LCAs) are urgently needed to compare their environmental footprints with conventional methods (Hauschild et al., 2018).

5.2 Economic Feasibility: Cost–Benefit Trade-Offs and Market Adoption

The economic viability of nanobiotechnology hinges on its cost-competitiveness with established waste treatment technologies, such as anaerobic digestion or activated sludge processes. While nano-enabled systems can achieve higher conversion efficiencies, their high capital expenditures (CAPEX)—driven by nanomaterial synthesis and reactor retrofitting—often outweigh operational savings (GadelHak et al., 2023).

To accelerate market penetration, public–private partnerships (PPPs) and circular business models are essential. For example, leasing models for nanocatalysts (where manufacturers retain ownership and ensure recovery) could reduce upfront costs for end users. Policy incentives, such as tax credits for nano-enabled carbon capture or subsidies for bio-nanocomposite production, could further stimulate adoption (Li et al., 2019).

5.3 *Emerging Innovations: AI, Biohybrids, and Precision Remediation*

The convergence of artificial intelligence (AI) and nanobiotechnology is poised to revolutionize waste-specific nanomaterial design. Machine learning algorithms can predict optimal nanoparticle properties—such as size, charge, and hydrophobicity—for targeted applications, including lignocellulosic waste breakdown and microplastic detection (Abdelaziz et al., 2022). For instance, neural networks trained on comprehensive datasets of nanoparticle-organic interactions have successfully identified gold-doped cerium oxide (Au–CeO_2) as a highly effective catalyst for lignin depolymerization, reducing experimental screening time (Yu et al., 2025). This approach not only accelerates material discovery but also reduces the environmental and economic costs associated with traditional trial-and-error methods.

Biohybrid nanosystems represent another frontier in precision remediation, merging the specificity of synthetic biology with the versatility of nanotechnology. Virus-like particles (VLPs) can be engineered to encapsulate enzymes tailored for the degradation of specific pollutants, enhancing both stability and delivery efficiency (Zeltins, 2013). Meanwhile, living functionalized biofilms integrated with quantum dots have shown promise for real-time monitoring of chemical oxygen demand (COD), enabling dynamic assessment of wastewater treatment processes (Huang et al., 2020). These hybrid systems offer unparalleled specificity and adaptability but still face significant hurdles related to genetic stability, biocompatibility, and the scalability of bioculturing techniques.

6 Further Studies

6.1 *Industrial-Scale Nanobioremediation*

Nanobioremediation, an area utilizing nanomaterials for environmental cleanup, is gaining attention for industrial uses, especially in treating challenging wastewater like that from the textile sector. Textile wastewater is a major pollutant because of dangerous dyes and high levels of organic and inorganic salts (Adak et al., 2023). Nanomaterials provide effective ways to eliminate these pollutants. A particularly promising method involves filters made from nanocellulose. Nanocellulose is well suited for dye adsorption due to its biodegradability, mechanical durability, and large surface area (Tavakolian et al., 2020).

- **Nanocellulose/Titanium Dioxide (NC/TiO_2) Nanocomposites**

These composites have shown enhanced adsorption and photocatalytic properties, making them favorable for treating textile industry wastewater (Tavakolian et al., 2020). Studies have demonstrated high removal efficiencies for various pollutants:

- Turbidity: Up to 97.48% (Industry A) and 100% (Industry B) removal efficiency (Tavakolian et al., 2020).
- Chemical Oxygen Demand (COD): Up to 87.18% (Industry A) and 92.03% (Industry B) removal efficiency (Tavakolian et al., 2020).
- Color: Up to 76.60% (Industry A) and 95.82% (Industry B) photocatalytic degradation efficiency (Tavakolian et al., 2020).

- **Electrosterically Stabilized Nanocrystalline Cellulose (ENCC)**

A high negative charge density is characteristic of hairy nanocellulose derived from wood pulp, granting it substantial adsorption capabilities for positively charged dyes such as methylene blue. Studies have indicated that this enzyme-modified nanocellulose (ENCC) can adsorb up to 1400 mg of dye per gram of adsorbent through an ion-exchange process (Tavakolian et al., 2020). To improve its usability for larger applications, composite hydrogel beads consisting of sodium alginate and ENCC (ALG-ENCC beads) have been created. These beads retain a high methylene blue removal capacity of 1250 mg/g and can be utilized in fixed-bed columns, suggesting potential for industrial implementation (Tavakolian et al., 2020).

- **Nanofiltration (NF) Membranes**

Nanofiltration stands out as an energy-efficient and reliable technology for treating textile wastewater. Its capability to effectively remove dyes, COD, and conductivity makes it highly significant. Pilot studies have explored the use of nanofiltration systems for treating fabric dyeing wastewater and recovering salts (Khan et al., 2025). Notably, NF membranes have demonstrated high COD retention (80–100%) and color retention rates (>90%) across different dye solutions (Adak et al., 2023; Tavakolian et al., 2020).

6.2 Urban Waste-To-Energy Plants

Urban waste-to-energy facilities frequently struggle with landfill leachate treatment. This complex wastewater contains high levels of organic matter (COD), ammonia, nitrogen, salinity, and toxins (Khan et al., 2025). The presence of emerging contaminants like nanomaterials and microplastics adds to the treatment complexity (Yildirim et al., 2023). Nano-functionalized membranes are proving to be a crucial technology for addressing this challenging waste stream.

Membrane technology, including reverse osmosis (RO), nanofiltration (NF), ultrafiltration (UF), and membrane bioreactors (MBRs), is a versatile and efficient approach for removing contaminants from landfill leachate (Xiang et al., 2025).

- **Nanomaterial-Based Assemblies**

To improve leachate treatment, assemblies integrating nanomaterials with microbes, chemical catalysts, or enzymes are utilized (Vaidh et al., 2022). Nanomaterials such as nanoparticles, nanomembranes, and nanofilters are well-suited for this purpose because of their large surface area, rapid reaction rates, and controllable pore dimensions (Vaidh et al., 2022).

- **Photocatalytic Self-Cleaning Polysulfone (PSF) Membranes with WO3 Nanoparticles**

Membranes incorporating WO3 nanoparticles have been created to enhance the effectiveness of leachate treatment. Research indicates that adding 2 wt% WO_3 improves the membrane's pore size, porosity, and hydrophilicity (Vaidh et al., 2022). Under UV irradiation, studies have shown that COD removal increased from 54.91% to 77.45% with this addition, indicating a self-cleaning capability (Vaidh et al., 2022).

- **Biologically Activated Carbon Membranes (Hybrid Ultrafiltration)**

This method combines adsorption, biodegradation, and membrane filtration for landfill leachate treatment. It has demonstrated high efficiency, achieving 95–98% removal of Total Organic Carbon (TOC) from leachate (Vaidh et al., 2022).

- **Membrane Bioreactors (MBRs)**

Membrane bioreactors (MBRs), frequently combined with biological methods, demonstrate exceptional efficiency in removing ammonia and nitrogen from landfill leachate, achieving over 90% removal rates (Mojiri et al., 2021). Although MBR technology surpasses traditional wastewater treatment, issues such as membrane fouling, which raises operational expenses, require attention through enhanced design and appropriate membrane selection.

Incorporating nano-functionalized membranes and sophisticated membrane technologies is essential for sustainable development and circular economy practices in urban waste management. This integration ensures that treated leachate has a limited impact on the environment and public health (Yildirim et al., 2023).

7 Conclusion

Nanobiotechnology stands as an interdisciplinary field, positioned at the confluence of sustainability science and industrial biotechnology to provide innovative solutions for intricate environmental problems. This review emphasizes its core strength: a dual capability for organic waste valorization and environmental pollutant remediation. This dual functionality underscores nanobiotechnology's comprehensive approach, tackling both resource recovery and pollution control, issues frequently handled as separate endeavors. The significant estimated production of 350,000 tons of nanomaterials in 2016 (Grillo et al., 2018) highlights the critical need for responsible development, including thorough assessment of their environmental effects and interactions within intricate ecosystems. To realize this potential fully and responsibly, a strategic roadmap encompassing targeted research, robust economic models, sound economic strategies, enhanced collaborations, and adaptable policies.

Strategic priorities for future nanobiotechnology research include: enhancing nanomaterial stability, creating effective recovery and reuse systems, and developing sustainable synthesis methods using green chemistry. Utilizing bio-waste for nanoparticle synthesis, for example, presents an environmentally friendly and cost-effective alternative (Flores-Contreras et al., 2024). Simultaneously, thorough cost–benefit analyses compared to traditional methods are essential to establish the economic viability of nanobiotechnology. The projected USD 1–3 trillion global market for nanotechnology by 2024 (Khatoon & Velidandi, 2025), fueled by industry investment and academic partnerships, highlights the economic necessity for scaling these technologies. To facilitate the transition from lab to industry, it is crucial to foster more public–private collaborations to secure necessary funding, infrastructure, and expertise. Ultimately, incorporating AI for nanomaterial design and investigating biohybrid nanosystems will expedite the creation of more efficient and targeted solutions for specific wastes, addressing current limitations and expanding the potential of the field. Biohybrid systems, combining biological components with nanomaterials, hold promise for targeted pollutant removal and ecosystem restoration (Malara & Frontera, 2022).

Sustainable and equitable technological progress is crucial. Policymakers should collaborate with researchers and industry to establish flexible regulations and ethical principles for nano-waste management and technology implementation. This necessitates thorough Lifecycle Assessments (LCAs) to evaluate environmental effects comprehensively. Despite environmentalist concerns about nanoparticle safety data, scientists stress the ethical importance of pursuing nanotechnology's potential in environmental remediation and health enhancement. The NanoSafety Cluster (NSC) is dedicated to identifying research needs and promoting the creation of safer nanomaterials (Cassee et al., 2024). Furthermore, fair global technology dissemination is vital to ensure widespread benefits of nanobiotechnology, preventing increased inequality and fostering inclusive sustainable development through ongoing international partnerships and technology exchange programs.

Nanobiotechnology presents a significant avenue for achieving global sustainability goals. Its application in converting waste into valuable resources like bioplastics and biofuels directly supports the principles of a circular economy, diminishing waste and encouraging the sustainable utilization of resources, thus aligning with SDG 12 (Responsible Consumption and Production). Furthermore, by improving biogas production, aiding biofuel extraction, and allowing for the real-time tracking of greenhouse gases such as methane, nanobiotechnology offers vital tools for climate change mitigation and the promotion of cleaner energy sources, directly impacting SDG 13 (Climate Action). Its contributions also span across SDG 6 (Clean Water and Sanitation) through improved pollutant removal, SDG 7 (Affordable and Clean Energy) via sustainable energy generation, and SDG 9 (Industry, Innovation, and Infrastructure) by fostering advanced industrial processes. Microbial nanotechnology, notably, holds substantial potential in contributing to various SDGs related to environmental protection (Singh et al., 2025).

The impact of nanobiotechnology goes beyond specific uses; it serves as a driver for systemic change by enabling innovative circular economy approaches, more accurate environmental monitoring, and cleaner industrial operations. Consequently, investing in nanobiotechnology signifies an investment in a future that is resilient, resource-efficient, and environmentally sound, providing a concrete way to address current environmental issues and achieve the ambitious targets outlined in global sustainability agendas.

References

Abdelaziz, O. Y., Clemmensen, I., Meier, S., Costa, C. A. E., Rodrigues, A. E., Hulteberg, C. P., & Riisager, A. (2022). On the oxidative valorization of lignin to high-value chemicals: A critical review of opportunities and challenges. *Chemsuschem, 15*(20), Article e202201232. https://doi.org/10.1002/cssc.202201232

Adak, M. K., Kumar, K., Das, A., & Dhak, D. (2023). Nano-filtration application in the textile industry for wastewater treatment. In A. Ahmad & M. B. Alshammari (Eds.), *Nanofiltration membrane for water purification* (pp. 69–82). Springer Nature Singapore. https://doi.org/10.1007/978-981-19-5315-6_5

Aguilar-Pérez, K. M., Avilés-Castrillo, J. I., Ruiz-Pulido, G., Medina, D. I., Parra-Saldivar, R., & Iqbal, H. M. N. (2021). Nanoadsorbents in focus for the remediation of environmentally-related contaminants with rising toxicity concerns. *Science of the Total Environment, 779*, Article 146465. https://doi.org/10.1016/j.scitotenv.2021.146465

Ahmad, S., Aadil, M., Ejaz, S. R., Akhtar, M. U., Noor, H., Haider, S., Alsafari, I. A., & Yasmin, G. (2022). Sol-gel synthesis of nanostructured ZnO/$SrZnO_2$ with boosted antibacterial and photocatalytic activity. *Ceramics International, 48*(2), 2394–2405. https://doi.org/10.1016/j.ceramint.2021.10.020

Ali, M. E., Hoque, M. E., Safdar Hossain, S. K., & Biswas, M. C. (2020). Nanoadsorbents for wastewater treatment: Next generation biotechnological solution. *International Journal of Environmental Science and Technology, 17*(9), 4095–4132. https://doi.org/10.1007/s13762-020-02755-4

Aliannejadi, S., Hassani, A. H., Panahi, H. A., & Borghei, S. M. (2019). Fabrication and characterization of high-branched recyclable PAMAM dendrimer polymers on the modified magnetic nanoparticles for removing naphthalene from aqueous solutions. *Microchemical Journal, 145*, 767–777. https://doi.org/10.1016/j.microc.2018.11.043

Amutha, C., Gopan, A., Pushbalatatha, I., Ragavi, M., & Reneese, J. A. (2024). Nanotechnology and governance: Regulatory framework for responsible innovation. In S. C. George & B. Tawiah (Eds.), *Nanotechnology in societal development* (pp. 481–503). Springer Nature Singapore. https://doi.org/10.1007/978-981-97-6184-5_14

Andrada Suarez, E. E., Roca Jalil, M. E., Fernandez Baldo, M. A., & Cuozzo, S. A. (2025). Nanobiotechnology approaches for the remediation of persistent and emerging organic pollutants: Strategies, interactions, and effectiveness. *Environmental Science: Nano, 12*(2), 979–1011. https://doi.org/10.1039/D4EN00424H

Antezana, P. E., Colin, V. L., Bourguignon, N., Benimeli, C. S., & Fuentes, M. S. (2023). Applied of actinobacteria consortia-based bioremediation to restore co-contaminated systems. *Research in Microbiology, 174*(4), Article 104028. https://doi.org/10.1016/j.resmic.2023.104028

Arriaga, M., Pinar, F. J., Izarra, I., Amo, J. D., Vicente, J., Fernández-Morales, F. J., & Mena, J. (2025). Valorization of agri-food waste into PHA and bioplastics: From waste selection to transformation. *Applied Sciences, 15*(3), 1008. https://doi.org/10.3390/app15031008

Arvidsson, R., Nguyen, D., & Svanström, M. (2015). Life cycle assessment of cellulose nanofibrils production by mechanical treatment and two different pretreatment processes. *Environmental Science and Technology, 49*(11), 6881–6890. https://doi.org/10.1021/acs.est.5b00888

Arya, I., Poona, A., Dikshit, P. K., Pandit, S., Kumar, J., Singh, H. N., Jha, N. K., Rudayni, H. A., Chaudhary, A. A., & Kumar, S. (2021). Current trends and future prospects of nanotechnology in biofuel production. *Catalysts, 11*(11), 1308. https://doi.org/10.3390/catal11111308

Aslam, R., Rosenkranz, A., Aslam, J., Hussain, C. M., & American Chemical Society (Eds.). (2025). *Waste derived carbon nanomaterials: Volume 1*. American Chemical Society. https://doi.org/10.1021/bk-2025-1494

Augusto, P. A., Castelo-Grande, T., Merchan, L., Estevez, A. M., Quintero, X., & Barbosa, D. (2019). Landfill leachate treatment by sorption in magnetic particles: Preliminary study. *Science of the Total Environment, 648*, 636–668. https://doi.org/10.1016/j.scitotenv.2018.08.056

Bala, S., Garg, D., Sridhar, K., Inbaraj, B. S., Singh, R., Kamma, S., Tripathi, M., & Sharma, M. (2023). Transformation of Agro-waste into value-added bioproducts and bioactive compounds: Micro/Nano formulations and application in the agri-food-pharma sector. *Bioengineering, 10*(2), 152. https://doi.org/10.3390/bioengineering10020152

Bhandari, G., Gangola, S., Dhasmana, A., Rajput, V., Gupta, S., Malik, S., & Slama, P. (2023). Nano-biochar: Recent progress, challenges, and opportunities for sustainable environmental remediation. *Frontiers in Microbiology, 14*, 1214870. https://doi.org/10.3389/fmicb.2023.1214870

Bhardwaj, P., Singh, B., & Behera, S. P. (2021). Green approaches for nanoparticle synthesis: Emerging trends. In *Nanomaterials* (pp. 167–193). Elsevier. https://doi.org/10.1016/B978-0-12-822401-4.00015-5

Bhattu, M., & Singh, J. (2023). Recent advances in nanomaterials based sustainable approaches for mitigation of emerging organic pollutants. *Chemosphere, 321*, Article 138072. https://doi.org/10.1016/j.chemosphere.2023.138072

Bian, H., Boguta, P., Huang, J., Deng, C., Kong, D., Zhou, H., & Su, X. (2024). Transforming organic solid waste management: Embracing humification for sustainable resource recovery. *ACS Sustainable Resource Management, 1*(2), 181–198. https://doi.org/10.1021/acssusresmgt.3c00080

Bippus, L., Briem, A.-K., Beck, A., Zibek, S., & Albrecht, S. (2024). Life cycle assessment for early-stage process optimization of microbial biosurfactant production using kinetic models—A case study on mannosylerythritol lipids (MEL). *Frontiers in Bioengineering and Biotechnology, 12*, 1347452. https://doi.org/10.3389/fbioe.2024.1347452

Bokare, V., Murugesan, K., Kim, J.-H., Kim, E.-J., & Chang, Y.-S. (2012). Integrated hybrid treatment for the remediation of 2,3,7,8-tetrachlorodibenzo-p-dioxin. *Science of the Total Environment, 435–436*, 563–566. https://doi.org/10.1016/j.scitotenv.2012.07.079

Booton, A., Mayer, B. K., & Zitomer, D. H. (2024). Chemical oxidation as an alternative for municipal wastewater secondary treatment: A review. *Reviews in Environmental Science and Bio/technology, 23*(1), 43–65. https://doi.org/10.1007/s11157-024-09684-5

Bouabidi, Z. B., El-Naas, M. H., & Zhang, Z. (2019). Immobilization of microbial cells for the biotreatment of wastewater: A review. *Environmental Chemistry Letters, 17*(1), 241–257. https://doi.org/10.1007/s10311-018-0795-7

Bruce, D. (2006). Ethical and social issues in nanobiotechnologies: Nano2Life provides a European ethical 'think tank' for research in biology at the nanoscale. *EMBO Reports, 7*(8), 754–758. https://doi.org/10.1038/sj.embor.7400762

Campos, A. F. C., De Oliveira, H. A. L., Da Silva, F. N., Da Silva, F. G., Coppola, P., Aquino, R., Mezzi, A., & Depeyrot, J. (2019). Core-shell bimagnetic nanoadsorbents for hexavalent chromium removal from aqueous solutions. *Journal of Hazardous Materials, 362*, 82–91. https://doi.org/10.1016/j.jhazmat.2018.09.008

Cassee, F. R., Bleeker, E. A. J., Durand, C., Exner, T., Falk, A., Friedrichs, S., Heunisch, E., Himly, M., Hofer, S., Hofstätter, N., Hristozov, D., Nymark, P., Pohl, A., Soeteman-Hernández, L. G., Suarez-Merino, B., Valsami-Jones, E., & Groenewold, M. (2024). Roadmap towards safe and sustainable advanced and innovative materials. (Outlook for 2024–2030). *Computational and Structural Biotechnology Journal, 25*, 105–126. https://doi.org/10.1016/j.csbj.2024.05.018

Ceballos, M. W., Jafari, S., Fikry, M., Shiekh, K. A., Kijpatanasilp, I., & Assatarakul, K. (2025). Changes in quality attributes of coconut water treated with UV-radiation and nisin during cold storage: Kinetics modelling and shelf-life prediction. *Food Control, 167*, Article 110801. https://doi.org/10.1016/j.foodcont.2024.110801

Chen, S., Zhao, W., Liu, W., & Zhang, S. (2008). Preparation, characterization and activity evaluation of p–n junction photocatalyst p-ZnO/n–TiO_2. *Applied Surface Science, 255*(5), 2478–2484. https://doi.org/10.1016/j.apsusc.2008.07.115

Chen, X., Wu, Z., Liu, D., & Gao, Z. (2017). Preparation of ZnO photocatalyst for the efficient and rapid photocatalytic degradation of azo dyes. *Nanoscale Research Letters, 12*(1), 143. https://doi.org/10.1186/s11671-017-1904-4

Chu, S., Liu, C., Feng, X., Wu, H., & Liu, X. (2023). Aromatic polymer dual-confined magnetic metal-organic framework microspheres enable highly efficient removal of dyes, heavy metals, and antibiotics. *Chemical Engineering Journal, 472*, Article 145159. https://doi.org/10.1016/j.cej.2023.145159

Dai, J., Ogbeide, O., Macadam, N., Sun, Q., Yu, W., Li, Y., Su, B.-L., Hasan, T., Huang, X., & Huang, W. (2020). Printed gas sensors. *Chemical Society Reviews, 49*(6), 1756–1789. https://doi.org/10.1039/C9CS00459A

Das, A. P., Mohanty, A. K., van Hullebusch, E. D., Figueiredo, G., Das, A. P., Mohanty, A. K., van Hullebusch, E. D., & Figueiredo, G. (2023). Bioremediation as an emerging technology for the removal of synthetic microplastic pollutants from marine ecosystem. *MarPB, 194*, Article 115297. https://doi.org/10.1016/J.MARPOLBUL.2023.115297

Das, A., Dey, S., & Das, A. P. (2024a). Ecotoxicological impacts of synthetic microfiber pollutants and development of sustainable mitigation strategies. *Environmental Chemistry and Ecotoxicology*. https://doi.org/10.1016/J.ENCECO.2024.12.003

Das, A. P., Nayak, B., & Behera, I. D. (2024b). Applications of microorganisms in advanced geomicrobiological and biotechnological exploration. *Geomicrobiology Journal, 41*(4), 309–311. https://doi.org/10.1080/01490451.2024.2327059/ASSET//CMS/ASSET/DE2A1F67-08C1-475A-ABC7-2B3F4C9E8BCF/01490451.2024.2327059.FP.PNG

Das, A., Dey, S., Gacem, A., Yadav, K. K., Nath, H., Ram, D. K., & Das, A. P. (2025). Revolutionizing solid hydrocarbon sourcing: Plastic's journey from waste to upcycled treasure. *Environmental Science and Engineering, Part, F218*, 331–351. https://doi.org/10.1007/978-3-031-81615-4_14

Devendrapandi, G., Balu, R., Ayyappan, K., Ayyamperumal, R., Alhammadi, S., Lavanya, M., Senthilkumar, R., & Karthika, P. C. (2024). Unearthing Earth's secrets: Exploring the environmental legacy of contaminants in soil, water, and sediments. *Environmental Research, 249*, Article 118246. https://doi.org/10.1016/j.envres.2024.118246

Dey, S., Tripathy, B., Kumar, M. S., & Das, A. P. (2023). Ecotoxicological consequences of manganese mining pollutants and their biological remediation. *Environmental Chemistry and Ecotoxicology*. https://doi.org/10.1016/J.ENCECO.2023.01.001

Dey, S., Bissoyi, A., Koteswara Reddy, G., & Das, A. P. (2024). Synthetic microfibers: An urban emerging pollutant demanding for sustainable environmental solutions. *Sustainable Microbial Technology for Synthetic and Cellulosic Microfiber Bioremediation*, 107–121. https://doi.org/10.1007/978-3-031-62600-5_7

Dey, S., Dash, A., & Das, A. P. (2024). Bioremediation as a sustainable and circular solution for mining waste management and bioenergy generation. *Sustainable Management of Mining Waste and Tailings: A Circular Economy Approach*, 107–119. https://doi.org/10.1201/9781003442455-6/BIOREMEDIATION-SUSTAINABLE-CIRCULAR-SOLUTION-MINING-WASTE-MANAGEMENT-BIOENERGY-GENERATION-SUDESHNA-DEY-AKANSHYA-DASH-ALOK-PRASAD-DAS

Dhakal, N., & Acharya, B. (2021). Syngas fermentation for the production of bio-based polymers: A review. *Polymers, 13*(22), 3917. https://doi.org/10.3390/polym13223917

Ding, Z., Sun, M., Liu, W., Meng, X., Zheng, Y., Sun, W., & Zhou, Q. (2022). In-situ growth of N–TiO_2 on delaminated N–Ti_3C_2 with highly strengthened photocatalytic activity. *International*

Journal of Hydrogen Energy, 47(49), 21204–21219. https://doi.org/10.1016/j.ijhydene.2022.04.254

Domingo, J. L., Marquès, M., Mari, M., & Schuhmacher, M. (2020). Adverse health effects for populations living near waste incinerators with special attention to hazardous waste incinerators. A review of the scientific literature. *Environmental Research, 187*, 109631. https://doi.org/10.1016/j.envres.2020.109631

Dong, D., Wang, K., Yi, M., Liang, Y., Muhammad, Y., Wei, E., Wei, Y., & Fujita, T. (2022). Preparation of TiO_2 photocatalyst microspheres by geopolymer technology for the degradation of tetracycline. *Journal of Cleaner Production, 339*, Article 130734. https://doi.org/10.1016/j.jclepro.2022.130734

Ejeian, F., Etedali, P., Mansouri-Tehrani, H.-A., Soozanipour, A., Low, Z.-X., Asadnia, M., Taheri-Kafrani, A., & Razmjou, A. (2018). Biosensors for wastewater monitoring: A review. *Biosensors and Bioelectronics, 118*, 66–79. https://doi.org/10.1016/j.bios.2018.07.019

Eker, F., Duman, H., Akdaşçi, E., Bolat, E., Sarıtaş, S., Karav, S., & Witkowska, A. M. (2024). A comprehensive review of nanoparticles: From classification to application and toxicity. *Molecules, 29*(15), 3482. https://doi.org/10.3390/molecules29153482

El Gohary, H. G., Alhagri, I. A., Qahtan, T. F., Al-Hakimi, A. N., Saeed, A., Abolaban, F., Alshammari, E. M., & Asnag, G. M. (2023). Reinforcement of structural, thermal and electrical properties and antibacterial activity of PVA/SA blend filled with hybrid nanoparticles (Ag and TiO_2 NPs): Nanodielectric for energy storage and food packaging industries. *Ceramics International, 49*(12), 20174–20184. https://doi.org/10.1016/j.ceramint.2023.03.141

El-Mansi, M. (2007). *Fermentation microbiology and biotechnology* (2nd ed.). CRC /Taylor & Francis.

El-Ramady, H., Abdalla, N., Sári, D., Ferroudj, A., Muthu, A., Prokisch, J., Fawzy, Z. F., Brevik, E. C., & Solberg, S. Ø. (2023). Nanofarming: Promising solutions for the future of the global agricultural industry. *Agronomy, 13*(6), 1600. https://doi.org/10.3390/agronomy13061600

El-sayed, M. E. A. (2020). Nanoadsorbents for water and wastewater remediation. *Science of the Total Environment, 739*, Article 139903. https://doi.org/10.1016/j.scitotenv.2020.139903

Elumalai, P., Parthipan, P., Huang, M., Muthukumar, B., Cheng, L., Govarthanan, M., & Rajasekar, A. (2021). Enhanced biodegradation of hydrophobic organic pollutants by the bacterial consortium: Impact of enzymes and biosurfactants. *Environmental Pollution, 289*, Article 117956. https://doi.org/10.1016/j.envpol.2021.117956

Emaimo, A. J., Olkhov, A. A., Iordanskii, A. L., & Vetcher, A. A. (2022). Polyhydroxyalkanoates composites and blends: Improved properties and new applications. *Journal of Composites Science, 6*(7), 206. https://doi.org/10.3390/jcs6070206

Esposito, S. (2019). "Traditional" Sol-Gel chemistry as a powerful tool for the preparation of supported metal and metal oxide catalysts. *Materials, 12*(4), 668. https://doi.org/10.3390/ma12040668

Esquivel-Castro, T. A., Robledo-Trujillo, G., Oliva, J., Rosu, H. C., & Rodríguez-González, V. (2023). A functional SiO_2–TiO_2 mesoporous assembly designed for the controlled release of carvacrol. *Applied Surface Science Advances, 13*, Article 100378. https://doi.org/10.1016/j.apsadv.2023.100378

Farhadian, N., Akbarzadeh, R., Pirsaheb, M., Jen, T.-C., Fakhri, Y., & Asadi, A. (2019). Chitosan modified N, S-doped TiO_2 and N, S-doped ZnO for visible light photocatalytic degradation of tetracycline. *International Journal of Biological Macromolecules, 132*, 360–373. https://doi.org/10.1016/j.ijbiomac.2019.03.217

Farré, M., Sanchís, J., & Barceló, D. (2011). Analysis and assessment of the occurrence, the fate and the behavior of nanomaterials in the environment. *TrAC Trends in Analytical Chemistry, 30*(3), 517–527. https://doi.org/10.1016/j.trac.2010.11.014

Fazeli-Nasab, B., Sayyed, R. Z., Mojahed, L. S., Rahmani, A. F., Ghafari, M., Antonius, S., & Sukamto. (2022). Biofilm production: A strategic mechanism for survival of microbes under stress conditions. *Biocatalysis and Agricultural Biotechnology, 42*, 102337. https://doi.org/10.1016/j.bcab.2022.102337

Flores-Contreras, E. A., González-González, R. B., Pablo Pizaña-Aranda, J. J., Parra-Arroyo, L., Rodríguez-Aguayo, A. A., Iñiguez-Moreno, M., González-Meza, G. M., Araújo, R. G., Ramírez-Gamboa, D., Parra-Saldívar, R., & Melchor-Martínez, E. M. (2024). Agricultural waste as a sustainable source for nanoparticle synthesis and their antimicrobial properties for food preservation. *Frontiers in Nanotechnology, 6*, 1346069. https://doi.org/10.3389/fnano.2024.1346069

Fu, L., You, S., Li, G., Li, X., & Fan, Z. (2023). Application of semiconductor metal oxide in chemiresistive methane gas sensor: Recent developments and future perspectives. *Molecules, 28*(18), 6710. https://doi.org/10.3390/molecules28186710

GadelHak, Y., El-Azazy, M., Shibl, M. F., & Mahmoud, R. K. (2023). Cost estimation of synthesis and utilization of nano-adsorbents on the laboratory and industrial scales: A detailed review. *Science of the Total Environment, 875*, Article 162629. https://doi.org/10.1016/j.scitotenv.2023.162629

Gagliano, E., Sgroi, M., Falciglia, P. P., Vagliasindi, F. G. A., & Roccaro, P. (2020). Removal of poly- and perfluoroalkyl substances (PFAS) from water by adsorption: Role of PFAS chain length, effect of organic matter and challenges in adsorbent regeneration. *Water Research, 171*, Article 115381. https://doi.org/10.1016/j.watres.2019.115381

Ghamarpoor, R., Fallah, A., & Jamshidi, M. (2023a). Investigating the use of titanium dioxide (TiO_2) nanoparticles on the amount of protection against UV irradiation. *Scientific Reports, 13*(1), 9793. https://doi.org/10.1038/s41598-023-37057-5

Ghamarpoor, R., Fallah, A., Jamshidi, M., & Salehfekr, S. (2023b). Using waste silver metal in synthesis of Z-scheme Ag@WO_3-CeO_2 heterojunction to increase photodegradation and electrochemical performances. *Journal of Industrial and Engineering Chemistry, 128*, 459–471. https://doi.org/10.1016/j.jiec.2023.08.010

Ghamarpoor, R., Fallah, A., & Eghbali, T. (2024). Design of bifunctional sandwich-like Co@Si/Ox-MXene nanocomposite to increase the supercapacitor properties and removal of pollutants from wastewater. *Journal of Alloys and Compounds, 983*, Article 173920. https://doi.org/10.1016/j.jallcom.2024.173920

Ghosh, S., Mohanty, S., Nayak, S., Sukla, L. B., & Das, A. P. (2016). Molecular identification of indigenous manganese solubilising bacterial biodiversity from manganese mining deposits. *Journal of Basic Microbiology, 56*(3), 254–262. https://doi.org/10.1002/JOBM.201500477

Ghosh, S., & Das, A. P. (2020). Microbial metagenomics: current advances in investigating microbial ecology and population dynamics. In S. K. Nayak & B. B. Mishra (Eds.), *Frontiers in Soil and Environmental Microbiology* (1st ed., pp. 247–254). CRC Press.

Ghosh, S., Gandhi, M., van Hullebusch, E. D., & Das, A. P. (2020). Proteomic insights into Lysinibacillus sp.-mediated biosolubilization of manganese. *Environmental Science and Pollution Research, 28*(30), 40249–40263. https://doi.org/10.1007/S11356-020-10863-4

Ghosh, S., Tripathy, B., Dey, S., & Das, A. P. (2023). Comparative investigation of fungal and bacterial manganese biomining mechanisms. *Manganese Mining Microorganisms*, 99–115. https://doi.org/10.1016/b978-0-12-822147-1.00016-9

Giese, E. C., Silva, D. D. V., Costa, A. F. M., Almeida, S. G. C., & Dussán, K. J. (2020). Immobilized microbial nanoparticles for biosorption. *Critical Reviews in Biotechnology, 40*(5), 653–666. https://doi.org/10.1080/07388551.2020.1751583

Gómez-Pastora, J., Bringas, E., & Ortiz, I. (2014). Recent progress and future challenges on the use of high performance magnetic nano-adsorbents in environmental applications. *Chemical Engineering Journal, 256*, 187–204. https://doi.org/10.1016/j.cej.2014.06.119

Grillo, R., De Jesus, M. B., & Fraceto, L. F. (2018). Editorial: Environmental impact of nanotechnology: Analyzing the present for building the future. *Frontiers in Environmental Science, 6*, 34. https://doi.org/10.3389/fenvs.2018.00034

Han, J., Liu, Y., Singhal, N., Wang, L., & Gao, W. (2012). Comparative photocatalytic degradation of estrone in water by ZnO and TiO_2 under artificial UVA and solar irradiation. *Chemical Engineering Journal, 213*, 150–162. https://doi.org/10.1016/j.cej.2012.09.066

Hauschild, M. Z., Rosenbaum, R. K., & Olsen, S. I. (Eds.). (2018). *Life cycle assessment: Theory and practice*. Springer International Publishing. https://doi.org/10.1007/978-3-319-56475-3

He, N., Li, P., Zhou, Y., Fan, S., & Ren, W. (2009). Degradation of pentachlorobiphenyl by a sequential treatment using Pd coated iron and an aerobic bacterium (H1). *Chemosphere, 76*(11), 1491–1497. https://doi.org/10.1016/j.chemosphere.2009.06.046

He, M., Wang, L., Zhang, Z., Zhang, Y., Zhu, J., Wang, X., Lv, Y., & Miao, R. (2020). Stable forward osmosis nanocomposite membrane doped with sulfonated graphene oxide@Metal–organic frameworks for heavy metal removal. *ACS Applied Materials and Interfaces, 12*(51), 57102–57116. https://doi.org/10.1021/acsami.0c17405

Hooshmand, S., Kassanos, P., Keshavarz, M., Duru, P., Kayalan, C. I., Kale, İ, & Bayazit, M. K. (2023). Wearable nano-based gas sensors for environmental monitoring and encountered challenges in optimization. *Sensors, 23*(20), 8648. https://doi.org/10.3390/s23208648

Horváthová, H., Lászlová, K., & Dercová, K. (2019b). Remediation potential of bacterial mixed cultures for polychlorinated biphenyls (PCBs) biodegradation. *Acta Chimica Slovaca, 12*(1), 1–7. https://doi.org/10.2478/acs-2019-0001

Horváthová, H., Lászlová, K., & Dercová, K. (2019a). Bioremediation versus nanoremediation: Degradation of Polychlorinated Biphenyls (PCBS) using integrated remediation approaches. *Water, Air, & Soil Pollution, 230*(8), 204. https://doi.org/10.1007/s11270-019-4259-x

Hu, W., Zhou, W., Zhang, K., Zhang, X., Wang, L., Jiang, B., Tian, G., Zhao, D., & Fu, H. (2016). Facile strategy for controllable synthesis of stable mesoporous black TiO_2 hollow spheres with efficient solar-driven photocatalytic hydrogen evolution. *Journal of Materials Chemistry A, 4*(19), 7495–7502. https://doi.org/10.1039/C6TA01928E

Huang, Z., Chen, Y., & Hu, Y. (2016). The role of nanoscale zerovalent iron particles in the biosorption and biodegradation of BDE-47 by Pseudomonas stutzeri under aerobic conditions. *International Biodeterioration and Biodegradation, 112*, 51–58. https://doi.org/10.1016/j.ibiod.2016.04.039

Huang, D., Wang, J., Ren, K., & Ji, J. (2020). Functionalized biomaterials to combat biofilms. *Biomaterials Science, 8*(15), 4052–4066. https://doi.org/10.1039/D0BM00526F

Huang, C.-W., Lin, C., Nguyen, M. K., Hussain, A., Bui, X.-T., & Ngo, H. H. (2023). A review of biosensor for environmental monitoring: Principle, application, and corresponding achievement of sustainable development goals. *Bioengineered, 14*(1), 58–80. https://doi.org/10.1080/21655979.2022.2095089

Inturi, S. N. R., Boningari, T., Suidan, M., & Smirniotis, P. G. (2014). Visible-light-induced photodegradation of gas phase acetonitrile using aerosol-made transition metal (V, Cr, Fe Co, Mn, Mo, Ni, Cu, Y, Ce, and Zr) doped TiO_2. *Applied Catalysis b: Environmental, 144*, 333–342. https://doi.org/10.1016/j.apcatb.2013.07.032

Jabbar, Z. H., Graimed, B. H., Ammar, S. H., Sabit, D. A., Najim, A. A., Radeef, A. Y., & Taher, A. G. (2024). The latest progress in the design and application of semiconductor photocatalysis systems for degradation of environmental pollutants in wastewater: Mechanism insight and theoretical calculations. *Materials Science in Semiconductor Processing, 173*, Article 108153. https://doi.org/10.1016/j.mssp.2024.108153

Jain, K., Patel, A. S., Pardhi, V. P., & Flora, S. J. S. (2021). Nanotechnology in wastewater management: A new paradigm towards wastewater treatment. *Molecules, 26*(6), 1797. https://doi.org/10.3390/molecules26061797

Jain, U., Saxena, K., Hooda, V., Balayan, S., Singh, A. P., Tikadar, M., & Chauhan, N. (2022). Emerging vistas on pesticides detection based on electrochemical biosensors—An update. *Food Chemistry, 371*, Article 131126. https://doi.org/10.1016/j.foodchem.2021.131126

Jeyakumar, R. B., & Vincent, G. S. (2022). Recent advances and perspectives of nanotechnology in anaerobic digestion: A new paradigm towards sludge biodegradability. *Sustainability, 14*(12), 7191. https://doi.org/10.3390/su14127191

Jia, Y., Hu, F., Lv, Y., Chen, Y., & Hu, Y. (2019). Biomineralization of 2′2′4′4′-Tetrabromodiphenyl ether in a Pseudomonas putida and Fe/Pd nanoparticles integrated system. *Chemosphere, 221*, 301–313. https://doi.org/10.1016/j.chemosphere.2019.01.042

Jin, J., Sun, J., Lv, K., Huang, X., Wang, J., Liu, J., Bai, Y., Guo, X., Zhao, J., Liu, J., & Hou, Q. (2021). Magnetic-responsive CNT/chitosan composite as stabilizer and adsorbent for organic contaminants and heavy metal removal. *Journal of Molecular Liquids, 334*, Article 116087. https://doi.org/10.1016/j.molliq.2021.116087

Jo, J., Kim, H., Jeong, S.-Y., Park, C., Hwang, H. S., & Koo, B. (2021). Changes in mechanical properties of polyhydroxyalkanoate with double silanized cellulose nanocrystals using different organosiloxanes. *Nanomaterials, 11*(6), 1542. https://doi.org/10.3390/nano11061542

Jönsson, A., Gustafsson, Ö., Axelman, J., & Sundberg, H. (2003). Global accounting of PCBs in the continental shelf sediments. *Environmental Science and Technology, 37*(2), 245–255. https://doi.org/10.1021/es0201404

Kalyani, N., Goel, S., & Jaiswal, S. (2021). On-site sensing of pesticides using point-of-care biosensors: A review. *Environmental Chemistry Letters, 19*(1), 345–354. https://doi.org/10.1007/s10311-020-01070-1

Kang, S., Zhao, K., Yu, D.-G., Zheng, X., & Huang, C. (2022). Advances in Biosensing and Environmental Monitoring Based on Electrospun Nanofibers. *Advanced Fiber Materials, 4*(3), 404–435. https://doi.org/10.1007/s42765-021-00129-0

Kapoor, R. T., Sillanpää, M., & Rafatullah, M. (Eds.). (2024). *Catalytic applications of biochar for environmental remediation: Valorization of lignocellulosic waste biomass into bioenergy (Vol 3)* (Vol. 1480). American Chemical Society. https://doi.org/10.1021/bk-2024-1480

Karn, B., Kuiken, T., & Otto, M. (2009). Nanotechnology and *in Situ* remediation: A review of the benefits and potential risks. *Environmental Health Perspectives, 117*(12), 1813–1831. https://doi.org/10.1289/ehp.0900793

Karthikeyan, C., Arunachalam, P., Ramachandran, K., Al-Mayouf, A. M., & Karuppuchamy, S. (2020). Recent advances in semiconductor metal oxides with enhanced methods for solar photocatalytic applications. *Journal of Alloys and Compounds, 828*, Article 154281. https://doi.org/10.1016/j.jallcom.2020.154281

Kashezheva, M., Kedakhezheva, R., Morozov, V., Kolesnikov, R., Voropaev, V., Mezentsev, S., Vasilenko, E., Vernigorova, E., Kuchuk, O., & Simonov, A. (2024). Harnessing nanocatalysts for sustainable chemical processes: Innovations in green chemistry. *Asian Journal of Green Chemistry, Online First*. https://doi.org/10.48309/AJGC.2025.483172.1564

Kasula, M., Le, T., Thomsen, A., & Rabbani Esfahani, M. (2022). Silver metal organic frameworks and copper metal organic frameworks immobilized on graphene oxide for enhanced adsorption in water treatment. *Chemical Engineering Journal, 439*, Article 135542. https://doi.org/10.1016/j.cej.2022.135542

Khan, S. Z., Zaidi, A. A., Naseer, M. N., & AlMohamadi, H. (2022). Nanomaterials for biogas augmentation towards renewable and sustainable energy production: A critical review. *Frontiers in Bioengineering and Biotechnology, 10*, Article 868454. https://doi.org/10.3389/fbioe.2022.868454

Khan, S. A., Rafiq, M. T., Feng, Y., Ali, H. M., Alsakkaf, W. A. A., Shah, F. A., Ali, A., Ahmed, T., & Aziz, R. (2025). Nanocellulose based titanium dioxide nanocomposites for the wastewater treatment of textile industry. *Frontiers in Environmental Science, 13*, 1560070. https://doi.org/10.3389/fenvs.2025.1560070

Khan, A. A., Allemailem, K. S., Almatroodi, S. A., Almatroudi, A., & Rahmani, A. H. (2020). Recent strategies towards the surface modification of liposomes: An innovative approach for different clinical applications. *3 Biotech, 10*(4), 163. https://doi.org/10.1007/s13205-020-2144-3

Khanam, Z., Gupta, S., & Verma, A. (2020). Endophytic fungi-based biosensors for environmental contaminants-a perspective. *South African Journal of Botany, 134*, 401–406. https://doi.org/10.1016/j.sajb.2020.08.007

Khatoon, H., & Rai, J. P. N. (2018). Augmentation of Atrazine biodegradation by two Bacilli immobilized on α-Fe_2O_3 magnetic nanoparticles. *Scientific Reports, 8*(1), 17831. https://doi.org/10.1038/s41598-018-36296-1

Khatoon, U. T., & Velidandi, A. (2025). An overview on the role of government initiatives in nanotechnology innovation for sustainable economic development and research progress. *Sustainability, 17*(3), 1250. https://doi.org/10.3390/su17031250

Kim, S.-S., Yum, J.-H., & Sung, Y.-E. (2005). Flexible dye-sensitized solar cells using ZnO coated TiO_2 nanoparticles. *Journal of Photochemistry and Photobiology a: Chemistry, 171*(3), 269–273. https://doi.org/10.1016/j.jphotochem.2004.10.019

Klein, J., & Wagner, F. (1983). Methods for the immobilization of microbial cells. In *Applied biochemistry and bioengineering* (Vol. 4, pp. 11–51). Elsevier. https://doi.org/10.1016/B978-0-12-041104-7.50008-7

Kokkinos, K., Karayannis, V., & Moustakas, K. (2021). Optimizing microalgal biomass feedstock selection for nanocatalytic conversion into biofuel clean energy, using fuzzy multi-criteria decision making processes. *Frontiers in Energy Research, 8*, Article 622210. https://doi.org/10.3389/fenrg.2020.622210

Kovács, Z., Molnár, C., Gyulavári, T., Magyari, K., Tóth, Z.-R., Baia, L., Pap, Z., & Hernádi, K. (2022). Solvothermal synthesis of ZnO spheres: Tuning the structure and morphology from nano- to micro-meter range and its impact on their photocatalytic activity. *Catalysis Today, 397–399*, 16–27. https://doi.org/10.1016/j.cattod.2022.03.004

Kozlova, E. A., Kurenkova, A. Y., Gerasimov, E. Y., Gromov, N. V., Medvedeva, T. B., Saraev, A. A., & Kaichev, V. V. (2021). Comparative study of photoreforming of glycerol on Pt/TiO_2 and CuOx/TiO_2 photocatalysts under UV light. *Materials Letters, 283*, Article 128901. https://doi.org/10.1016/j.matlet.2020.128901

Kumar, S. G., & Rao, K. S. R. K. (2017). Comparison of modification strategies towards enhanced charge carrier separation and photocatalytic degradation activity of metal oxide semiconductors (TiO_2, WO_3 and ZnO). *Applied Surface Science, 391*, 124–148. https://doi.org/10.1016/j.apsusc.2016.07.081

Kumar, T., Naik, S., & Jujjavarappu, S. E. (2022). A critical review on early-warning electrochemical system on microbial fuel cell-based biosensor for on-site water quality monitoring. *Chemosphere, 291*, Article 133098. https://doi.org/10.1016/j.chemosphere.2021.133098

Kumari, R., Suman, K., Karmakar, S., Mishra, V., Lakra, S. G., Saurav, G. K., & Mahto, B. K. (2023). Regulation and safety measures for nanotechnology-based agri-products. *Frontiers in Genome Editing, 5*, 1200987. https://doi.org/10.3389/fgeed.2023.1200987

Lagarde, F., & Jaffrezic-Renault, N. (2011). Cell-based electrochemical biosensors for water quality assessment. *Analytical and Bioanalytical Chemistry, 400*(4), 947–964. https://doi.org/10.1007/s00216-011-4816-7

Le, T. T., Nguyen, K.-H., Jeon, J.-R., Francis, A. J., & Chang, Y.-S. (2015). Nano/bio treatment of polychlorinated biphenyls with evaluation of comparative toxicity. *Journal of Hazardous Materials, 287*, 335–341. https://doi.org/10.1016/j.jhazmat.2015.02.001

Let, S., Dutta, S., Samanta, P., Sharma, S., & Ghosh, S. K. (2021). Magnetic nanoparticle-embedded ionic microporous polymer composite as an efficient scavenger of organic micropollutants. *ACS Applied Materials and Interfaces, 13*(43), 51474–51484. https://doi.org/10.1021/acsami.1c14819

Li, S., Zhang, C., Wang, S., Liu, Q., Feng, H., Ma, X., & Guo, J. (2018). Electrochemical microfluidics techniques for heavy metal ion detection. *The Analyst, 143*(18), 4230–4246. https://doi.org/10.1039/C8AN01067F

Li, F., Xie, J., & Wang, W. (2019). Incentivizing sustainable development: The impact of a recent policy reform on electricity production efficiency in China. *Sustainable Development, 27*(4), 770–780. https://doi.org/10.1002/sd.1942

Li, J., Chu, B., Xie, Z., Deng, Y., Zhou, Y., Dong, L., Li, B., & Chen, Z. (2022). Mechanism and DFT study of degradation of organic pollutants on rare earth ions doped TiO_2 photocatalysts prepared by sol-hydrothermal synthesis. *Catalysis Letters, 152*(2), 489–502. https://doi.org/10.1007/s10562-021-03634-4

Liang, L., Li, X., Guo, Y., Lin, Z., Su, X., & Liu, B. (2021). The removal of heavy metal cations by sulfidated nanoscale zero-valent iron (S-nZVI): The reaction mechanisms and the role of sulfur.

Journal of Hazardous Materials, 404, Article 124057. https://doi.org/10.1016/j.jhazmat.2020.124057

Liu, L., Helbling, D. E., Kohler, H.-P.E., & Smets, B. F. (2014). A model framework to describe growth-linked biodegradation of trace-level pollutants in the presence of coincidental carbon substrates and microbes. *Environmental Science and Technology, 48*(22), 13358–13366. https://doi.org/10.1021/es503491w

Louis, J., Padmanabhan, N. T., Jayaraj, M. K., & John, H. (2024). Exploring enhanced interfacial charge separation in ZnO/reduced graphene oxide hybrids on alkaline photoelectrochemical water splitting and photocatalytic pollutant degradation. *Materials Research Bulletin, 169*, Article 112542. https://doi.org/10.1016/j.materresbull.2023.112542

Luque, R., Len, C., & Triantafyllidis, K. (2018). Editorial: Nano-(Bio)Catalysis in lignocellulosic biomass valorization. *Frontiers in Chemistry, 6*, 577. https://doi.org/10.3389/fchem.2018.00577

Ma, D., Wang, D., Wang, X., Wang, Q., & Hu, Y. (2024a). Efficient treatment of old landfill leachate by peroxodisulfate assisted electro-oxidation and electro-coagulation combined system. *Chemosphere, 346*, Article 140675. https://doi.org/10.1016/j.chemosphere.2023.140675

Ma, J., Wang, Y., Zhang, Z., Wang, X., Hou, X., & Hu, Q. (2024b). Conventional and toxic pollutants removal of toxic herbal wastewater by using moving bed biofilm reactor, coagulation, and adsorption. *Journal of Environmental Chemical Engineering, 12*(3), Article 112705. https://doi.org/10.1016/j.jece.2024.112705

Maheswari, P., Sugapriya, S., Krishnaveni, N., & Senthil, T. S. (2022). Investigation of pure and CuO- and ZnO-loaded TiO_2 nanocomposites prepared by modified hydrothermal cum green synthesis (Hybanthus enneaspermus Extract) method for photocatalytic and antioxidant applications. *Brazilian Journal of Physics, 52*(4), 142. https://doi.org/10.1007/s13538-022-01112-9

Mahlambi, M. M., Ngila, C. J., & Mamba, B. B. (2015). Recent developments in environmental photocatalytic degradation of organic pollutants: The case of titanium dioxide nanoparticles—A review. *Journal of Nanomaterials, 2015*(1), Article 790173. https://doi.org/10.1155/2015/790173

Mahmoodi, N. M., Taghizadeh, M., Taghizadeh, A., Abdi, J., Hayati, B., & Shekarchi, A. A. (2019). Bio-based magnetic metal-organic framework nanocomposite: Ultrasound-assisted synthesis and pollutant (heavy metal and dye) removal from aqueous media. *Applied Surface Science, 480*, 288–299. https://doi.org/10.1016/j.apsusc.2019.02.211

Malara, A., & Frontera, P. (2022). Special issue on advanced materials and nanotechnology for sustainable energy and environmental applications. *Applied Sciences, 12*(15), 7440. https://doi.org/10.3390/app12157440

Mansur, A. A. P., & Mansur, H. S. (2025). Nanozyme-based cancer nanotheranostics: Emerging applications and challenges in brain cancer therapeutics. *Journal of Nanotheranostics, 6*(1), 4. https://doi.org/10.3390/jnt6010004

Massot, F., Bernard, N., Alvarez, L. M. M., Martorell, M. M., Mac Cormack, W. P., & Ruberto, L. A. M. (2022). Microbial associations for bioremediation. What does "microbial consortia" mean? *Applied Microbiology and Biotechnology, 106*(7), 2283–2297. https://doi.org/10.1007/s00253-022-11864-8

Massoud, A., Farid, O. M., Maree, R. M., Allan, K. F., & Tian, Z. R. (2020). An improved metal cation capture on polymer with graphene oxide synthesized by gamma radiation. *Reactive and Functional Polymers, 151*, Article 104564. https://doi.org/10.1016/j.reactfunctpolym.2020.104564

Mehrotra, T., Dev, S., Banerjee, A., Chatterjee, A., Singh, R., & Aggarwal, S. (2021). Use of immobilized bacteria for environmental bioremediation: A review. *Journal of Environmental Chemical Engineering, 9*(5), Article 105920. https://doi.org/10.1016/j.jece.2021.105920

Meng, C., Zhikun, W., Qiang, L., Chunling, L., Shuangqing, S., & Songqing, H. (2018). Preparation of amino-functionalized Fe_3O_4@$mSiO_2$ core-shell magnetic nanoparticles and their application for aqueous Fe3+ removal. *Journal of Hazardous Materials, 341*, 198–206. https://doi.org/10.1016/j.jhazmat.2017.07.062

Mohanty, J., Dey, S., Bala, K., & Das, A. P. (2024). Recent advances in microplastic pollution for its sustainable management. *Sustainable Management of Urban Plastic Waste Through Circular Economic Approaches*, 27–48. https://doi.org/10.1201/9781003465447-3

Mojiri, A., Zhou, J. L., Ratnaweera, H., Ohashi, A., Ozaki, N., Kindaichi, T., & Asakura, H. (2021). Treatment of landfill leachate with different techniques: An overview. *Journal of Water Reuse and Desalination, 11*(1), 66–96. https://doi.org/10.2166/wrd.2020.079

Momina, M., Shahadat, M., & Isamil, S. (2018). Regeneration performance of clay-based adsorbents for the removal of industrial dyes: A review. *RSC Advances, 8*(43), 24571–24587. https://doi.org/10.1039/C8RA04290J

Montalvo, D., Corro, G., Bañuelos, F., Olivares-Xometl, O., Arellanes, P., & Pal, U. (2023). Selective alcohols production through CO_2 photoreduction using Co_3O_4 /TiO_2 photocatalyst exploiting synergetic interactions between Ti3+, Co2+ and Co3+. *Applied Catalysis B: Environmental, 330*, Article 122652. https://doi.org/10.1016/j.apcatb.2023.122652

Moradi, N., Jamshidi, M., Ghamarpoor, R., & Moghbeli, M. R. (2023). Surface functionalization/silane modification of CeO_2 nanoparticles and their influences on photocatalytic activity of acrylic films for methylene blue removal. *Progress in Organic Coatings, 183*, Article 107787. https://doi.org/10.1016/j.porgcoat.2023.107787

Najafidoust, A., Allahyari, S., Rahemi, N., & Tasbihi, M. (2020). Uniform coating of TiO_2 nanoparticles using biotemplates for photocatalytic wastewater treatment. *Ceramics International, 46*(4), 4707–4719. https://doi.org/10.1016/j.ceramint.2019.10.202

Narain, R. (Ed.). (2020). *Polymer science and nanotechnology: Fundamentals and applications*. Elsevier.

Neha, R., Adithya, S., Jayaraman, R. S., Gopinath, K. P., & Arun, J. (2021). Nano-adsorbents an effective candidate for removal of toxic pharmaceutical compounds from aqueous environment: A critical review on emerging trends. *Chemosphere, 272*, 129852. https://doi.org/10.1016/j.chemosphere.2021.129852

Nelson, W. M. (2023). *Sustainable agricultural chemistry in the 21st Century: Green chemistry nexus*. Taylor and Francis.

Neto, A. S., Wainaina, S., Chandolias, K., Piatek, P., & Taherzadeh, M. J. (2024). Exploring the potential of syngas fermentation for recovery of high-value resources: A comprehensive review. *Current Pollution Reports, 11*(1), 7. https://doi.org/10.1007/s40726-024-00337-3

Ngcongco, K., Krishna, S. B. N., & Pillay, K. (2023). Biogenic metallic nanoparticles as enzyme mimicking agents. *Frontiers in Chemistry, 11*, 1107619. https://doi.org/10.3389/fchem.2023.1107619

Odobašić, A., Šestan, I., & Begić, S. (2019). Biosensors for determination of heavy metals in waters. In T. Rinken & K. Kivirand (Eds.), *Biosensors for environmental monitoring*. IntechOpen. https://doi.org/10.5772/intechopen.84139

OECD. (2016). Nanomaterials in waste streams: Current knowledge on risks and impacts. *OECD*. https://doi.org/10.1787/9789264249752-en

Okai Amu-Darko, J. N., Hussain, S., Zhang, X., Alothman, A. A., Ouladsmane, M., Nazir, M. T., Qiao, G., & Liu, G. (2023). Metal-organic frameworks-derived In_2O_3/ZnO porous hollow nanocages for highly sensitive H_2S gas sensor. *Chemosphere, 314*, Article 137670. https://doi.org/10.1016/j.chemosphere.2022.137670

Omar, S., Omar, M., Attia, N. F., El-Subruiti, G. M., & Eltaweil, A. (2024). Rational engineering and fabrication of efficient nanophotocatalysts based on ZnO–SrO–CdS for pharmaceutical pollutants based wastewater degradation. *Surfaces and Interfaces, 45*, Article 103817. https://doi.org/10.1016/j.surfin.2023.103817

Opoku, F., Kiarii, E. M., Govender, P. P., & Mamo, M. A. (2017). Metal oxide polymer nanocomposites in water treatments. In T. Akitsu (Ed.), *Descriptive inorganic chemistry researches of metal compounds*. InTech. https://doi.org/10.5772/67835

Pan, Y., Leung, P.-Y., Li, Y.-Y., Chen, J., Kong, R. Y. C., & Tam, N. F. Y. (2021). Enhancement effect of nanoscale zero-valent iron addition on microbial degradation of BDE-209 in contaminated mangrove sediment. *Science of the Total Environment, 781*, Article 146702. https://doi.org/10.1016/j.scitotenv.2021.146702

Pandit, C., Banerjee, S., Pandit, S., Lahiri, D., Kumar, V., Chaubey, K. K., Al-Balushi, R., Al-Bahry, S., & Joshi, S. J. (2023). Recent advances and challenges in the utilization of nanomaterials in transesterification for biodiesel production. *Heliyon, 9*(4), Article e15475. https://doi.org/10.1016/j.heliyon.2023.e15475

Partovinia, A., & Rasekh, B. (2018). Review of the immobilized microbial cell systems for bioremediation of petroleum hydrocarbons polluted environments. *Critical Reviews in Environmental Science and Technology, 48*(1), 1–38. https://doi.org/10.1080/10643389.2018.1439652

Patel, V., Chesmore, A., Legner, C. M., & Pandey, S. (2022). Trends in workplace wearable technologies and connected-worker solutions for next-generation occupational safety, health, and productivity. *Advanced Intelligent Systems, 4*(1), 2100099. https://doi.org/10.1002/aisy.202100099

Pathak, H. K., Seth, C. S., Chauhan, P. K., Dubey, G., Singh, G., Jain, D., Upadhyay, S. K., Dwivedi, P., & Khoo, K. S. (2024). Recent advancement of nano-biochar for the remediation of heavy metals and emerging contaminants: Mechanism, adsorption kinetic model, plant growth and development. *Environmental Research, 255*, Article 119136. https://doi.org/10.1016/j.envres.2024.119136

Patil, S. B., Basavarajappa, P. S., Ganganagappa, N., Jyothi, M. S., Raghu, A. V., & Reddy, K. R. (2019). Recent advances in non-metals-doped TiO_2 nanostructured photocatalysts for visible-light driven hydrogen production, CO_2 reduction and air purification. *International Journal of Hydrogen Energy, 44*(26), 13022–13039. https://doi.org/10.1016/j.ijhydene.2019.03.164

Patiño-Ruiz, D. A., Meramo-Hurtado, S. I., González-Delgado, Á. D., & Herrera, A. (2021). Environmental sustainability evaluation of iron oxide nanoparticles synthesized via green synthesis and the coprecipitation method: A comparative life cycle assessment study. *ACS Omega, 6*(19), 12410–12423. https://doi.org/10.1021/acsomega.0c05246

Patra, D., Haribabu, E., & McComas, K. A. (2010). Perceptions of nano ethics among practitioners in a developing country: A case of India. *NanoEthics, 4*(1), 67–75. https://doi.org/10.1007/s11569-010-0086-7

Paul, A., Dey, S., Ram, D. K., & Das, A. P. (2023). Hexavalent chromium pollution and its sustainable management through bioremediation. *Geomicrobiology Journal*, 1–11. https://doi.org/10.1080/01490451.2023.2218377

Peng, Y.-H., Chen, M., & Shih, Y. (2013). Adsorption and sequential degradation of polybrominated diphenyl ethers with zerovalent iron. *Journal of Hazardous Materials, 260*, 844–850. https://doi.org/10.1016/j.jhazmat.2013.05.057

Peralta, M. E., Mártire, D. O., Moreno, M. S., Parolo, M. E., & Carlos, L. (2021). Versatile nanoadsorbents based on magnetic mesostructured silica nanoparticles with tailored surface properties for organic pollutants removal. *Journal of Environmental Chemical Engineering, 9*(1), Article 104841. https://doi.org/10.1016/j.jece.2020.104841

Pérez-González, M., Tomás, S. A., Morales-Luna, M., Arvizu, M. A., & Tellez-Cruz, M. M. (2015). Optical, structural, and morphological properties of photocatalytic TiO_2–ZnO thin films synthesized by the sol–gel process. *Thin Solid Films, 594*, 304–309. https://doi.org/10.1016/j.tsf.2015.04.073

Pramanik, A., Chaudhary, A. A., Sinha, A., Chaubey, K. K., Ashraf, M. S., Basher, N. S., Rudayni, H. A., Dayal, D., & Kumar, S. (2023). Nanocatalyst-based biofuel generation: An update, challenges and future possibilities. *Sustainability, 15*(7), 6180. https://doi.org/10.3390/su15076180

Pranaw, K., Drewniak, L., Nain, L., & Singh, S. (2022). Editorial: Waste to wealth: A sustainable circular bioeconomy approach. *Frontiers in Bioengineering and Biotechnology, 10*, 1010811. https://doi.org/10.3389/fbioe.2022.1010811

Priyadarshini, S., Jagatee, S., & Das, A. P. (2024). Synthetic fabrics and microfiber pollution–an assessment of their global impact. *Environmental Science and Engineering, Part, F2333*, 137–157. https://doi.org/10.1007/978-3-031-51792-1_8

Pryadko, A. S., Mukhortova, Y. R., Chernozem, R. V., Shlapakova, L. E., Wagner, D. V., Romanyuk, K., Gerasimov, E. Y., Kholkin, A., Surmenev, R. A., & Surmeneva, M. A. (2022). Comprehensive study on the reinforcement of electrospun PHB scaffolds with composite magnetic Fe_3O_4–rGO fillers: Structure, physico-mechanical properties, and piezoelectric response. *ACS Omega, 7*(45), 41392–41411. https://doi.org/10.1021/acsomega.2c05184

Qiang, C., Li, N., Zuo, S., Guo, Z., Zhan, W., Li, Z., & Ma, J. (2022). Microwave-assisted synthesis of RuTe2/black TiO_2 photocatalyst for enhanced diclofenac degradation: Performance, mechanistic investigation and intermediates analysis. *Separation and Purification Technology, 283*, Article 120214. https://doi.org/10.1016/j.seppur.2021.120214

Raman, R., Lathabai, H. H., & Nedungadi, P. (2024). Sustainable development goal 12 and its synergies with other SDGs: Identification of key research contributions and policy insights. *Discover Sustainability, 5*(1), 150. https://doi.org/10.1007/s43621-024-00289-0

Ranmadugala, D., Ebrahiminezhad, A., Manley-Harris, M., Ghasemi, Y., & Berenjian, A. (2018). Magnetic immobilization of bacteria using iron oxide nanoparticles. *Biotechnology Letters, 40*(2), 237–248. https://doi.org/10.1007/s10529-017-2477-0

Rasheed, T. (2022). Magnetic nanomaterials: Greener and sustainable alternatives for the adsorption of hazardous environmental contaminants. *Journal of Cleaner Production, 362*, Article 132338. https://doi.org/10.1016/j.jclepro.2022.132338

Ray, S., Dey, S., Digal, A. K., Yadav, K. K., & Das, A. P. (2025). Sanitary waste and microplastic pollution in the Euro-Mediterranean region: Challenges and solutions. *Euro-Mediterranean Journal for Environmental Integration, 2025*, 1–23. https://doi.org/10.1007/S41207-025-00879-Y

Ray, S. K., Dey, S., Das, A. P., & John, A. (2024). Microplastic waste management through greener technologies: A review. *Sustainable Management of Urban Plastic Waste Through Circular Economic Approaches*, 237–261. https://doi.org/10.1201/9781003465447-14

Reddy Koteswara, G., Bhargavi, S., Yamuna, D., Anjani Devi, M., Jeevana Lakshmi Kumar, G., Manas, T., Adarsh, S., Bhavya Sai, V., Nath, H., Alreshidi, M. A., Togun, H., Ram, D. K., & Das, A. P. (2025). Recent advancements in the remediation of marine hydrocarbon and plastic pollutants. *Environmental Science and Engineering, Part F218*, 127–142. https://doi.org/10.1007/978-3-031-81615-4_7

Regulating Nanomedicine at the Food and Drug Administration. (2019). *AMA Journal of Ethics, 21*(4), E347–355. https://doi.org/10.1001/amajethics.2019.347

Rehman, S., Ullah, R., Butt, A. M., & Gohar, N. D. (2009). Strategies of making TiO_2 and ZnO visible light active. *Journal of Hazardous Materials, 170*(2–3), 560–569. https://doi.org/10.1016/j.jhazmat.2009.05.064

Roy, A., Sharma, A., Yadav, S., Jule, L. T., & Krishnaraj, R. (2021). Nanomaterials for remediation of environmental pollutants. *Bioinorganic Chemistry and Applications, 2021*, 1–16. https://doi.org/10.1155/2021/1764647

Saeed, M., Khan, I., Adeel, M., Akram, N., & Muneer, M. (2022). Synthesis of a CoO–ZnO photocatalyst for enhanced visible-light assisted photodegradation of methylene blue. *New Journal of Chemistry, 46*(5), 2224–2231. https://doi.org/10.1039/D1NJ05633F

Sahu, S., & Das, A. P. (2024). Microplastics degradation mechanism by aquatic plastiomes: A review. *Environmental Science and Engineering, Part, F2827*, 213–226. https://doi.org/10.1007/978-3-031-50840-0_12

Sahu, A., Dey, S., Mallick, K., Priyadarsini, S., Behera, I. D., & Das, A. P. (2024). A review on advanced technology for sustainable management of synthetic microplastic waste. *Sustainable Management of Urban Plastic Waste Through Circular Economic Approaches*, 262–279. https://doi.org/10.1201/9781003465447-15

Sajeev, E., Shekher, S., Ogbaga, C. C., Desongu, K. S., Gunes, B., & Okolie, J. A. (2023). Application of nanoparticles in bioreactors to enhance mass transfer during syngas fermentation. *Encyclopedia, 3*(2), 387–395. https://doi.org/10.3390/encyclopedia3020025

Salam, J. A., & Das, N. (2015). Degradation of lindane by a novel embedded bio-nano hybrid system in aqueous environment. *Applied Microbiology and Biotechnology, 99*(5), 2351–2360. https://doi.org/10.1007/s00253-014-6112-x

Saldaña-Ramírez, A., Cruz, M. R. A., Juárez-Ramírez, I., & Torres-Martínez, L. M. (2020). Influence of the power density and working pressure in the magnetron co-sputtering deposition of ZnO–SnO_2 thin films and their effect in photocatalytic hydrogen production. *Optical Materials, 110*, Article 110501. https://doi.org/10.1016/j.optmat.2020.110501

Sankaran, R., Markandan, K., Khoo, K. S., Cheng, C. K., Ashokkumar, V., Deepanraj, B., & Show, P. L. (2021). The expansion of lignocellulose biomass conversion into bioenergy via nanobiotechnology. *Frontiers in Nanotechnology, 3*, Article 793528. https://doi.org/10.3389/fnano.2021.793528

Sanni, S. E., Oni, B. A., Okoro, E. E., & Pandya, S. (2024). Recent advances in the use of biogenic nanomaterials and photocatalysts for wastewater treatment: Challenges and future prospects. *Frontiers in Nanotechnology, 6*, 1469309. https://doi.org/10.3389/fnano.2024.1469309

Sharifi Rad, A., Afshar, A., & Azadeh, M. (2023). Antireflection and photocatalytic single layer and double layer ZnO and ZnO–TiO_2 thin films. *Optical Materials, 136*, Article 113501. https://doi.org/10.1016/j.optmat.2023.113501

Sharma, M., Mahajan, P., Alsubaie, A. S., Khanna, V., Chahal, S., Thakur, A., Yadav, A., Arya, A., Singh, A., & Singh, G. (2025). Next-generation nanomaterials-based biosensors: Real-time biosensing devices for detecting emerging environmental pollutants. *Materials Today Sustainability, 29*, Article 101068. https://doi.org/10.1016/j.mtsust.2024.101068

Shawky, A., & Alshaikh, H. (2022). Cobalt ferrite-modified sol-gel synthesized ZnO nanoplatelets for fast and bearable visible light remediation of ciprofloxacin in water. *Environmental Research, 205*, Article 112462. https://doi.org/10.1016/j.envres.2021.112462

Shi, W., Ren, H., Huang, X., Li, M., Tang, Y., & Guo, F. (2020). Low cost red mud modified graphitic carbon nitride for the removal of organic pollutants in wastewater by the synergistic effect of adsorption and photocatalysis. *Separation and Purification Technology, 237*, Article 116477. https://doi.org/10.1016/j.seppur.2019.116477

Singh, R., Manickam, N., Mudiam, M. K. R., Murthy, R. C., & Misra, V. (2013). An integrated (nano-bio) technique for degradation of γ-HCH contaminated soil. *Journal of Hazardous Materials, 258–259*, 35–41. https://doi.org/10.1016/j.jhazmat.2013.04.016

Singh, B., Chauhan, N., Agrawal, N., & Ayyagari, A. (2025). Contribution of microbial nanotechnology to sustainable development goals. In Y. K. Mohanta, B. Mishra, & R. N. Pudake (Eds.), *Nano-microbiology for Sustainable Development* (pp. 1–23). Springer Nature Switzerland. https://doi.org/10.1007/978-3-031-78845-1_1

Singhal, N., & Kumar, U. (2017). Noble metal modified TiO_2: Selective photoreduction of CO_2 to hydrocarbons. *Molecular Catalysis, 439*, 91–99. https://doi.org/10.1016/j.mcat.2017.06.031

Solano, R. A., Herrera, A. P., Maestre, D., & Cremades, A. (2019). Fe–TiO_2 nanoparticles synthesized by green chemistry for potential application in waste water photocatalytic treatment. *Journal of Nanotechnology, 2019*, 1–11. https://doi.org/10.1155/2019/4571848

Son, B. T., Long, N. V., & Nhat Hang, N. T. (2021). Fly ash-, foundry sand-, clay-, and pumice-based metal oxide nanocomposites as green photocatalysts. *RSC Advances, 11*(49), 30805–30826. https://doi.org/10.1039/D1RA05647F

Sukhadeve, G. K., Bandewar, H., Janbandhu, S. Y., Jayaramaiah, J. R., & Gedam, R. S. (2023). Photocatalytic hydrogen production, dye degradation, and antimicrobial activity by Ag–Fe co-doped TiO_2 nanoparticles. *Journal of Molecular Liquids, 369*, Article 120948. https://doi.org/10.1016/j.molliq.2022.120948

Sun, G., Wei, X., Zhang, D., Huang, L., Liu, H., & Fang, H. (2023). Immobilization of enzyme electrochemical biosensors and their application to food bioprocess monitoring. *Biosensors, 13*(9), 886. https://doi.org/10.3390/bios13090886

Suzana, C. U. S. M., Claudia, M., A, M., Larissa Guedes Fiuacute Za, M. C. O., & Ra, T. D. S. (2013). Immobilization of microbial cells: A promising tool for treatment of toxic pollutants in industrial wastewater. *African Journal of Biotechnology, 12*(28), 4412–4418. https://doi.org/10.5897/AJB12.2677

Swain, S., Purnima Sudha, M., Dey, S., Badu, Y., Bhoi, S. R., Alreshidi, M. A., Qasim, M. T., Nath, H., Ram, D. K., & Das, A. P. (2025). Sources of micro and nano plastic hydrocarbon pollutants in marine bodies and their treatment. *Environmental Science and Engineering, Part, F218*, 45–67. https://doi.org/10.1007/978-3-031-81615-4_3

Takada, S., Nakamura, M., Matsueda, T., Kondo, R., & Sakai, K. (1996). Degradation of polychlorinated dibenzo-p-dioxins and polychlorinated dibenzofurans by the white rot fungus Phanerochaete sordida YK-624. *Applied and Environmental Microbiology, 62*(12), 4323–4328. https://doi.org/10.1128/aem.62.12.4323-4328.1996

Tavakolian, M., Wiebe, H., Sadeghi, M. A., & Van De Ven, T. G. M. (2020). Dye removal using hairy nanocellulose: Experimental and theoretical investigations. *ACS Applied Materials and Interfaces, 12*(4), 5040–5049. https://doi.org/10.1021/acsami.9b18679

Thakur, A., & Kumar, A. (2022). Recent advances on rapid detection and remediation of environmental pollutants utilizing nanomaterials-based (bio)sensors. *Science of the Total Environment, 834*, Article 155219. https://doi.org/10.1016/j.scitotenv.2022.155219

Udoh, U. G., Abodunrin, A. D., Oteh, O. C., Ameh, D. A., & Omosebi, O. B. (2025). The role of nanotechnology in enhancing biotechnology and biodiversity conservation efforts in Nigeria: Opportunities challenges and future prospect. *International Journal of Advanced Biological and Biomedical Research, 13*(2). https://doi.org/10.48309/IJABBR.2025.2041778.1544

Usman, M., Farooq, M., Wakeel, A., Nawaz, A., Cheema, S. A., Rehman, H. U., Ashraf, I., & Sanaullah, M. (2020). Nanotechnology in agriculture: Current status, challenges and future opportunities. *Science of the Total Environment, 721*, Article 137778. https://doi.org/10.1016/j.scitotenv.2020.137778

Vaidh, S., Parekh, D., Patel, D., & Vishwakarma, G. S. (2022). Leachate treatment potential of nanomaterial based assemblies: A systematic review on recent development. *Water Science and Technology, 85*(11), 3285–3300. https://doi.org/10.2166/wst.2022.168

Vannuccini, M. L., Grassi, G., Leaver, M. J., & Corsi, I. (2015). Combination effects of nano-TiO_2 and 2,3,7,8-tetrachlorodibenzo-p-dioxin (TCDD) on biotransformation gene expression in the liver of European sea bass Dicentrarchus labrax. *Comparative Biochemistry and Physiology Part C: Toxicology & Pharmacology, 176–177*, 71–78. https://doi.org/10.1016/j.cbpc.2015.07.009

Verma, M. L., Dhanya, B. S., Wang, B., Thakur, M., Rani, V., & Kushwaha, R. (2023). Bio-nanoparticles mediated transesterification of algal biomass for biodiesel production. *Sustainability, 16*(1), 295. https://doi.org/10.3390/su16010295

Vicente, D., Proença, D. N., & Morais, P. V. (2023). The role of Bacterial Polyhydroalkanoate (PHA) in a sustainable future: A review on the biological diversity. *International Journal of Environmental Research and Public Health, 20*(4), 2959. https://doi.org/10.3390/ijerph20042959

Wang, L., Liu, S., Wang, Z., Zhou, Y., Qin, Y., & Wang, Z. L. (2016). Piezotronic effect enhanced photocatalysis in strained anisotropic ZnO/TiO_2 nanoplatelets *via* thermal stress. *ACS Nano, 10*(2), 2636–2643. https://doi.org/10.1021/acsnano.5b07678

Wang, Y., Lin, K., Liu, Y., & Deng, X. (2022). Nanocomposites of functionalized Metal−Organic frameworks and magnetic graphene oxide for selective adsorption and efficient determination of Lead(II). *Journal of Solid State Chemistry, 313*, Article 123300. https://doi.org/10.1016/j.jssc.2022.123300

Wang, Q., Zhao, Y., Zhang, Z., Liao, S., Deng, Y., Wang, X., Ye, Q., & Wang, K. (2023). Hydrothermal preparation of Sn_3O_4/TiO_2 nanotube arrays as effective photocatalysts for boosting photocatalytic dye degradation and hydrogen production. *Ceramics International, 49*(4), 5977–5985. https://doi.org/10.1016/j.ceramint.2022.11.113

Wang, F., Zhou, L., Mu, D., Zhang, H., Zhang, G., Huang, X., & Xiong, P. (2024a). Current research on ecotoxicity of metal-based nanoparticles: From exposure pathways, ecotoxicological effects to toxicity mechanisms. *Frontiers in Public Health, 12*, 1390099. https://doi.org/10.3389/fpubh.2024.1390099

Wang, Z., Ma, R., Chen, B., Yu, X., Wang, X., Zuo, X., Liang, B., & Yang, J. (2024b). A transcription factor-based bacterial biosensor system and its application for on-site detection of explosives. *Biosensors and Bioelectronics, 244*, Article 115805. https://doi.org/10.1016/j.bios.2023.115805

Wei, X., Yu, Y., He, J., Yuan, Y., Wang, L., Guo, Z., Zhu, M., Yin, H., & Dang, Z. (2022a). Synergistic removal of decabromodiphenyl ether by sulfurized zero-valent iron coupling with microbacterium in the presence of heavy metals: Biochemical degradation process and mechanism. *Journal of Environmental Chemical Engineering, 10*(6), Article 108885. https://doi.org/10.1016/j.jece.2022.108885

Wei, Y., Sun, H., Zhang, S., Xie, H. Q., Li, C., Zhao, B., & Yan, B. (2022b). Multi-walled carbon nanotubes inhibit potential detoxification of dioxin-mediated toxicity by blocking the nuclear translocation of aryl hydrocarbon receptor. *Journal of Hazardous Materials, 430*, Article 128458. https://doi.org/10.1016/j.jhazmat.2022.128458

Xi, Y., Liu, Y., Ye, X., Du, J., Kong, X., Guo, D., & Xiao, Q. (2021). Enhanced anaerobic biogas production from wheat straw by herbal-extraction process residues supplementation. *Frontiers in Bioengineering and Biotechnology, 9*, Article 623594. https://doi.org/10.3389/fbioe.2021.623594

Xiang, R., Wei, W., Mei, T., Wei, Z., Yang, X., Liang, J., & Zhu, J. (2025). A review on landfill leachate treatment technologies: Comparative analysis of methods and process innovation. *Applied Sciences, 15*(7), 3878. https://doi.org/10.3390/app15073878

Xiao, X., Tu, S., Lu, M., Zhong, H., Zheng, C., Zuo, X., & Nan, J. (2016). Discussion on the reaction mechanism of the photocatalytic degradation of organic contaminants from a viewpoint of semiconductor photo-induced electrocatalysis. *Applied Catalysis b: Environmental, 198*, 124–132. https://doi.org/10.1016/j.apcatb.2016.05.042

Xiao, L., Youji, L., Feitai, C., Peng, X., & Ming, L. (2017). Facile synthesis of mesoporous titanium dioxide doped by Ag-coated graphene with enhanced visible-light photocatalytic performance for methylene blue degradation. *RSC Advances, 7*(41), 25314–25324. https://doi.org/10.1039/C7RA02198D

Xiong, J., Sun, Z., Yu, J., Liu, H., & Wang, X. (2022). Thermal self-regulatory smart biosensor based on horseradish peroxidase-immobilized phase-change microcapsules for enhancing detection of hazardous substances. *Chemical Engineering Journal, 430*, Article 132982. https://doi.org/10.1016/j.cej.2021.132982

Xu, K., Chen, X., Zheng, R., & Zheng, Y. (2020). Immobilization of multi-enzymes on support materials for efficient biocatalysis. *Frontiers in Bioengineering and Biotechnology, 8*, 660. https://doi.org/10.3389/fbioe.2020.00660

Yan, J., & Bassler, B. L. (2019). Surviving as a community: Antibiotic tolerance and persistence in bacterial biofilms. *Cell Host and Microbe, 26*(1), 15–21. https://doi.org/10.1016/j.chom.2019.06.002

Yang, Q., Wang, J., Zhang, W., Liu, F., Yue, X., Liu, Y., Yang, M., Li, Z., & Wang, J. (2017). Interface engineering of metal organic framework on graphene oxide with enhanced adsorption capacity for organophosphorus pesticide. *Chemical Engineering Journal, 313*, 19–26. https://doi.org/10.1016/j.cej.2016.12.041

Yao, N., Li, C., Yu, J., Xu, Q., Wei, S., Tian, Z., Yang, Z., Yang, W., & Shen, J. (2020). Insight into adsorption of combined antibiotic-heavy metal contaminants on graphene oxide in water. *Separation and Purification Technology, 236*, Article 116278. https://doi.org/10.1016/j.seppur.2019.116278

Yao, Y., Wang, Z., Han, Y., Xie, L., Zhao, X., Shahrokhian, S., Barsan, N., & Zhu, Z. (2023). Conductometric $Cr_2O_3/TiO_2/Ti_3C_2Tx$ gas sensor for detecting triethylamine at room temperature. *Sensors and Actuators b: Chemical, 381*, Article 133412. https://doi.org/10.1016/j.snb.2023.133412

Yıldırım, H. (2022). Excitons in nonpolar ZnO/BeZnO quantum wells: Their binding energy and its dependence on the dimensions of the structures. *Physica b: Condensed Matter, 639*, Article 413974. https://doi.org/10.1016/j.physb.2022.413974

Yildirim, R., Eskikaya, O., Keskinler, B., Karagunduz, A., Dizge, N., & Balakrishnan, D. (2023). Fabric dyeing wastewater treatment and salt recovery using a pilot scale system consisted of graphite electrodes based on electrooxidation and nanofiltration. *Environmental Research, 234*, Article 116283. https://doi.org/10.1016/j.envres.2023.116283

You, J., Wang, L., Zhao, Y., & Bao, W. (2021). A review of amino-functionalized magnetic nanoparticles for water treatment: Features and prospects. *Journal of Cleaner Production, 281*, Article 124668. https://doi.org/10.1016/j.jclepro.2020.124668

Yu, Z., Kong, W., Liang, W., Guo, Y., Cui, J., Hu, Y., Sun, Z., Elangovan, S., & Xu, F. (2025). Heterogeneously catalyzed reductive depolymerization of lignin to value-added chemicals. *Chemsuschem, 18*(3), Article e202401399. https://doi.org/10.1002/cssc.202401399

Zahra, Z., Habib, Z., Hyun, S., & Sajid, M. (2022). Nanowaste: Another future waste, its sources, release mechanism, and removal strategies in the environment. *Sustainability, 14*(4), 2041. https://doi.org/10.3390/su14042041

Zanaroli, G., Negroni, A., Vignola, M., Nuzzo, A., Shu, H., & Fava, F. (2012). Enhancement of microbial reductive dechlorination of polychlorinated biphenyls (PCBs) in a marine sediment by nanoscale zerovalent iron (NZVI) particles. *Journal of Chemical Technology and Biotechnology, 87*(9), 1246–1253. https://doi.org/10.1002/jctb.3835

Zeltins, A. (2013). Construction and characterization of virus-like particles: A review. *Molecular Biotechnology, 53*(1), 92–107. https://doi.org/10.1007/s12033-012-9598-4

Zhang, Y., Zhu, Y., Zeng, Z., Zeng, G., Xiao, R., Wang, Y., Hu, Y., Tang, L., & Feng, C. (2021). Sensors for the environmental pollutant detection: Are we already there? *Coordination Chemistry Reviews, 431*, Article 213681. https://doi.org/10.1016/j.ccr.2020.213681

Leveraging the Catalytic Microbial Organic Waste to Sustainable Diverse Valuable Benefit

Rajeswari Rath, Alok Prasad Das, and Sansuta Mohanty

1 Introduction

Waste disposal is one of the most significant concerns of present-day society, and it is emerging globally and needs to be addressed. This seems to be one of the growing concerns increasing day-by-day. Most of the organic wastes are generated from various sectors, which could be industrial, residential, and agricultural, and can be transformed into diverse, sustainable, and valuable benefits using microorganisms (Hnain et al., 2011). There is an enormous amount of organic material present in the waste streams, which are disposed off in landfills. These wastes are then involved in biological conversions, which are then liberated into the atmosphere in the form of greenhouse gases (GHG) and methane (CH_4). Instead, energy from this waste might be captured and converted into biofuels, which will decrease emissions of greenhouse gases and provide economic value.

Additional benefits include the elimination of the necessity for landfills and the substitution of fossil fuels with energy produced from organic waste. Waste comes in many forms, such as industrial, biomedical, agricultural, and municipal solid waste (Fig. 1). Municipal solid waste includes waste from food products, metal, paper, plastic, glass, and other items and is produced by both commercial and residential activities. Agro-industrial by-products, livestock waste, farm product leftovers, and garbage from tourists are all sources of agricultural waste (Bhat et al., 2012). On the other hand, industrial waste is produced in mills, mines, and factories during production or other industrial operations. By using methods like composting, anaerobic

R. Rath · S. Mohanty (✉)
Department of Molecular Biology and Biotechnology, Faculty of Agricultural Sciences, Siksha O Anusandhan (Deemed to be University), Bhubaneswar, Odisha, India
e-mail: sansutamohanty@soa.ac.in

A. P. Das
Department of Life Sciences, Rama Devi Women's University, Bhubaneswar, Odisha, India

S. Ghosh et al. (eds.), *Advances in Organic Waste Conversion Through Industrial Biotechnology*, Sustainable Environmental Waste Management Strategies,
https://doi.org/10.1007/978-3-032-11377-1_17

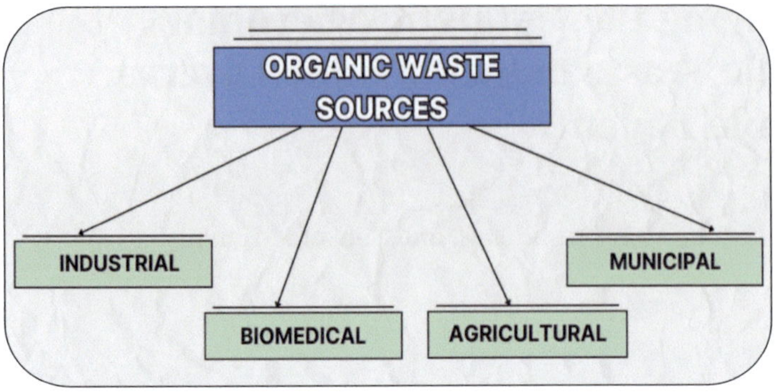

Fig. 1 Different sources of organic waste

digestion, and biogas generation, agricultural waste may be recycled and returned to the soil as organic matter and nutrients. On the other hand, industrial waste might be toxic or nontoxic (Behera & Das, 2025; Pengqi et al., 2025; Reddy Koteswara et al., 2025; Yousuf et al., 2025). Preventing environmental contamination, reducing waste, and recycling techniques like eco-design and closed-loop manufacturing systems are all part of effective industrial waste disposal.

In order to support sustainable development, the purpose of the present analysis is to look into environmental waste management techniques, with an emphasis on microbial conversion. The primary objective is to explore current waste handling techniques, understand microbial conversion, evaluate the advantages to the environment, analyse obstacles and barriers, and provide sustainable development plans. Utilizing the catalytic properties of microbial enzymes to transform complicated organic molecules into a variety of useful products, such as flavonoids, phenolics, anthocyanins, ethanol, bio-oil, essential oils, and nutritional items, represents one of the most crucial strategies. By addressing the issues of waste disposal and the advancement of ecologically sustainable procedures also improves the country's economic status by means of the circular bioeconomy (Kavya et al., 2024). This chapter highlights various microbial approaches, recent technological advances, and future prospects in the conversion of organic waste into useful chemicals, fuels, and other value-added products.

2 Microbial Approaches for Waste to Chemical Conversion

There are microbes everywhere in nature, and they serve several critical roles. Microbes contribute significantly to the organic recycling of life forms. All naturally occurring materials are biodegradable, meaning that microbes like bacteria or fungi may break them down as shown in Fig. 2. Microorganisms have shown to be quite helpful in solving a number of issues that humanity has faced in preserving

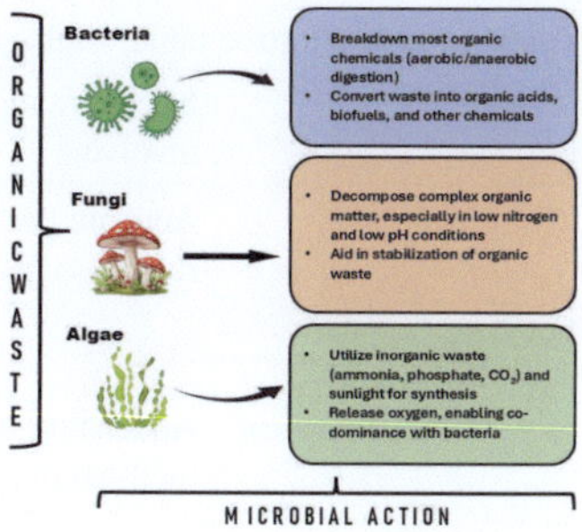

Fig. 2 Microbial roles in organic waste conversion

the environment. They have had beneficial effects on safeguarding the environment, genetic modification, the well-being of humans and animals, and the management of industrial and municipal waste (Dey & Dash et al., 2024a; Ray et al., 2025). Microorganisms have made it possible to respond in a practical and economical way that would not have been achievable using physical or chemical engineering techniques.

Furthermore, microbial advances have been effectively used to address a variety of ecological concerns, particularly those related to waste management (Husain et al., 2022). The process of collecting, moving, processing, or disposing of waste items while controlling and keeping an eye on them to reduce their negative effects on the environment and people is known as waste management. By lowering the amount of toxic effects of solid materials, solid waste treatment methods change it into an economically valuable or important form. Waste management is a process that includes waste recycling, fewer sources, monitored landfilling and burning, energy recovery from waste, and solid waste disposal in the case that none of the methods mentioned above provide an adequate response. Effective handling of solid waste involves a variety of procedures. These consist of landfilling, recycling, composting, processing, transportation, collecting, and monitoring (Nanda & Berruti, 2021). This includes various techniques like thermal treatment (in which waste products are treated using heat) like incineration, pyrolysis, gasification, and open burning; landfills and disposal sites like sanitary landfills, controlled landfills, and bioreactor landfills; and biological waste treatment like anaerobic digestion and composting (Table 1).

2.1 *Microbial Strategies for Waste Management*

Microorganisms such as bacteria, fungi, algae, protozoa, rotifers, and other higher animals are found in aerobic biological treatment methods. Different microbial agents have different roles (Fig. 3) and the chemical properties of the industrial waste, the ecological constraints of the specific waste arrangement, and the biochemical properties of microbes are all going to impact the development of any or all species of

Table 1 Comparative table: biological waste treatment techniques

Technique	Microorganisms involved	Primary output	Advantages	Limitations
Composting	Aerobic bacteria, fungi	Compost (organic manure)	Simple process, low-cost, reduces landfill load	Odour issues, requires space and aeration control
Anaerobic digestion	Anaerobic bacteria, methanogens	Biogas (CH_4 + CO_2), slurry	Renewable energy source, reduces GHG emissions	Requires careful pH/temp control, slower process
Bioreactor landfills	Mixed microbial consortia	Methane, stabilized waste	Enhanced degradation, gas capture for energy	Expensive setup, monitoring and leachate management are needed
Vermicomposting	Earthworms + decomposer microbes	Humus-rich vermicast	High-quality compost, improves soil health	Sensitive to temperature/moisture changes
Microbial fuel cells	Electroactive bacteria	Electricity	Generates clean energy directly from waste	High initial cost, lower power output, still experimental

microorganisms in a particular industrial waste management system (Mohanty et al., 2024; Priyadarshini et al., 2024; Sahu et al., 2024). Every microbe that develops in a particular industrial waste management system adds to its entire qualities, both positive and negative. Suppose the waste treatment facility is capable of being appropriately built and run for optimal efficiency. In that case, it is critical to acknowledge the contributions that each species of microorganism makes to the ultimate stabilization of the organic wastes.

2.1.1 Bacteria

The fundamental biological components of aerobic waste management systems are bacteria. Because of their unique biochemical makeup, bacteria can breakdown the majority of organic chemicals present in wastages coming from industries. All aerobic waste-disposal systems contain facultative bacteria and obligatory aerobes (Bala et al., 2024; Das et al., 2024a, 2024b, 2024c; John & Das, 2024). The competitive power of an organism to secure a portion of the method's accessible organic matter determines its growth. Two main kinds of bacteria will typically emerge from bacterial predomination: those that use the organic substances in the waste and those

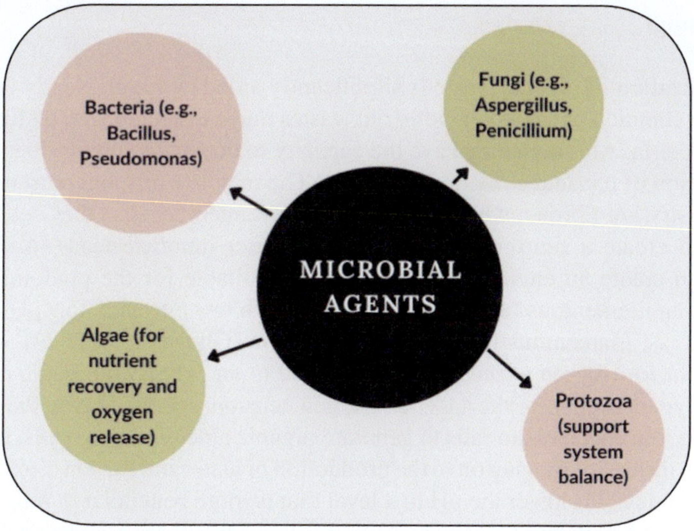

Fig. 3 Key microbial agents and their roles

that implement the lysed products produced by the initial group. The most crucial category, the bacteria that use the organic chemicals in the waste, is going to establish the features of the waste-disposal system (Husain et al., 2022). The dominant species is going to be the one that can employ most of the organic sources and grows the fastest. The timeframe of a food shortage will showcase the degree of subsequent predomination. The predominant bacteria die and lyse once the organic base is eliminated.

Other microorganisms can proliferate when the bacterial cells are released. Secondary predomination shall come into the picture because all biological processes are often overdesigned as precautionary measures. The most essential feature of the bacteria, apart from their metabolic traits, is their capacity for flocculation. The total stabilization of all aerobic biological waste management systems depends on the flocculation of the bacteria and their segregation from the liquid state. Initially, it was believed that just one type of bacteria, *Zoogloea ramigeria*, was responsible for flocculation; however, in the recent times, studies have revealed that a wide variety of bacteria may flocculate (Kaura et al., 2023). All bacteria are thought to be capable of flocculating under specific circumstances. The primary elements influencing flocculation are the bacteria's level of energy and surface charges. It has been elaborated that the electrical surface charge of bacteria cultivated in systems of diluted organic matter is less than the essential charge of 0.020 V for autoagglutination. Accordingly, when two bacteria come into contact with one another, Brownian movement offers enough energy to neutralize the electrical forces that oppose them, making the weak forces of attraction to overwhelm and keep the two bacteria united.

2.1.2 Fungi

The stabilization of organic waste is significantly aided by fungi. Nearly every kind of organic chemical included in industrial wastes may be metabolized by fungus, just like by bacteria. Although they have the capacity to do so, the fungus does not, with the exception of unusual climatic conditions. The majority of fungi present in waste from industry are fibrous, which makes them unwanted because they settle readily and fail to create a rigid, compact floc. This latter rationale leads to significant attempts to create an environment that is more suitable for the predomination of bacteria than filamentous fungus. In conditions with low nitrogen, low pH, and low-oxygen stress, filamentous fungus exceed bacteria (Ghosh et al., 2023). When the requirement for oxygen exceeds the supply due to an excessive organic content or a poor oxygen supply, a shortage of oxygen tension occurs. When the levels of oxygen are low, metabolism fails to generate organic aldehydes, alcohols, and acidic substances instead of moving on to the production of water and carbon dioxide (CO_2).

The organic acids lower the pH to a level that is more beneficial to fungal growth if there is not enough buffer in the system. It follows that there may be a relationship between low levels of oxygen and pH. While few bacteria can develop enough effectively to compete, many fungi thrive around pH 4 to 5. Compared to bacteria, fungi are in the requirement of less nitrogen per mass of protoplasm. The fungus prevails in nitrogen-deficient wastes because it can produce larger masses of protoplasm from the wastes than the bacteria can. On average, bacteria have between 10 and 12% nitrogen, whereas fungi have between 5 and 6% (Adebayo & Obiekezie, 2018). Fungi ought to be found in a typical environment and can assist with stabilizing the organic materials. However, the fungus shall not prevail and has no significance.

2.1.3 Algae

The entire stabilization of organic wastes is aided by another type of biological plant, algae. Unlike bacteria and fungus, algae lack the need to metabolize organic molecules since they get the energy they need for synthesis from sunlight. Algae mainly use inorganic waste materials, such as ammonia, phosphate, CO_2, magnesium, iron, potassium, calcium, sulphate, sodium, and other ions, to create protoplasm (Kumari et al., 2022). It is feasible for bacteria and algae to co-dominate since they lack the need for the same waste materials. Some of the inorganic substances used by the algae are released by the bacteria when they break down the organic elements of the waste. In order to achieve full aerobic stabilization of the organic materials, the bacteria absorb the oxygen released by the algae throughout protoplasm formation. Like other microorganisms, algae must rely on the metabolism of organic sources to provide the energy they need to survive in the absence of sunshine. This organic substance often originates from food that has been stored inside the cell, but in certain algae species, it may also originate from organic waste.

2.2 Microbial Waste Management

Solid waste may generally be divided into two categories: biodegradable and non-biodegradable. Biodegradables, also known as biowastes, are solid waste products that may be broken down by microorganisms and do not pose a significant threat to the environment for a long time. Paper goods, plant wastes, and animal wastes (faeces, carcasses, droppings, and poultry waste products) are among them. Even though certain types of solid waste are quickly broken down by microorganisms, they nonetheless emit unpleasant odours and are more of an annoyance to the visual environment than non-biodegradable solid waste. In Nigerian cities, they can also serve as an ideal environment for the growth of harmful microbes that have the potential to contaminate fresh food products and water supplies.

However, bacteria cannot break down solid waste that isn't biodegradable. This suggests that they are disposed of via alternative treatment methods, including recycling, landfilling, and incineration. Solid wastes from the metallurgical and smelting industries (such as abandoned cars, motorbikes, scrap metals, zinc, aluminium sheets, iron, and machine parts), the construction industry (such as sand, gravel, bitumen wastes, concrete, and waste building materials), the plastic industry (such as plastic buckets, tables, cable insulators, tyres, chairs, cellophane bags, plastic bottles, cutlery, and sachet water containments), and glass products are examples of this category of solid wastes.

Moreover, solid waste management is known to have a substantial effect on economic growth and upgraded living while minimizing or eliminating negative effects on the environment and human health (Das et al., 2023a, 2023b, 2024a, 2024b, 2024c; Dey et al., 2023). The most popular biological solid waste management technique is composting, which involves the controlled aerobic breakdown of organic waste materials by microorganisms and microscopic invertebrates. The process of composting involves breaking down organic waste items, such as food, plants, and paper, and then recycling the compost for use in landscaping and agriculture. Static pile composting, windrow composting, vermin composting, and in-vessel composting are the most often used composting methods.

2.3 Efficient Wastewater Management: Activated Sludge Versus Trickling Filters

Microorganisms play a crucial role in wastewater treatment, which employs the metabolic processes of various mixed colonies of microbes that may break down any substance that is introduced to them. The two primary goals are to eradicate every harmful microorganism found in the sewage, especially those that cause the water-borne illnesses cholera, dysentery, and typhoid. In order to create a final effluent (outflow) that can be properly released into the environment, the second goal is to break down the organic content in wastewater to primarily CO_2 and CH_4. Additionally, microbial activity can be used to bioremediation settings damaged by man-made xenobiotics and to break down these substances inside waste streams. Converting the wastewater's carbon and energy into microorganisms and removing them from the water by settling is the ultimate aim of wastewater treatment. For microorganisms, the connection between the carbon source and the energy source is crucial. The fundamental component of cell synthesis is carbon; however, for synthesis to take place, energy must be acquired from sources outside the cell. For growth, microorganisms need specific nutrients such as carbon (C), nitrogen (N), and phosphorus (P). These three fundamental nutrients are abundant in typical raw sewage. Traces of sodium (Na), potassium (K), magnesium (Mg), iron (Fe), and many other elements are needed in addition to C, N, and P (Mishra & Das, 2022; Nag et al., 2023; Pradhan et al., 2023). The majority of these nutrients are present in regular municipal wastewater. The majority of nutrient shortage issues arise when industrial waste is prevalent. The metabolism malfunctions, and a kind of bacterial fat, called slime, starts to build up around the cell when the right nutrients are not available. Due to insufficient enzyme production and insufficient nutrition penetration of the slime layer, the cell's function slows down. BOD removal slows sluggish as the sludge refuses to settle.

3 Challenges in Microbial-Assisted Waste to Chemical Conversion

One method for eliminating or purifying pollutants found in many natural habitats, including soil, water, and sediments, is remediation. When contaminants exceed permissible limits and endanger living things, remediation techniques become obsessive. Heavy metals, pesticides, insecticides, volatile hydrocarbons, radioactive substances, oil, petroleum mixes, and other poisons can persist and bioaccumulate, causing structural, physiological, neurological, and reproductive abnormalities in biological forms, such as human beings (Das & Dey, 2024). Considering all the three types of remediation—physical, chemical, and biological—the last has been more well-regarded. Utilizing living organisms such as bacteria, fungus, plants, and microorganisms to detoxify or mineralize harmful chemicals is known as bioremediation (Ghosh et al., 2020; Mishra & Das, 2023; Mishra et al., 2023). Two commonly used microbial-assisted remediation techniques are bioaugmentation, which involves adding microorganisms, and biostimulation, which involves adding nutrients to encourage native bacteria (Saleem et al., 2022).

However, some of the obstacles that hinder the degrading process include things like the environment's vulnerability and change, toxic effects, bioavailability, monitoring, and the capacity of microbial degraders to survive (Das et al., 2023a, 2023b; Dey et al., 2024b; Ghosh & Das, 2020). Techniques to support microbial persistence in field circumstances include rhizoremediation, genetic alterations, cell immobilization, and acclimatization. The implementation of biomarkers and persistent isotope probing helps evaluate the remediation process while biosparging, bioventing, surfactants, and nanoparticles speed up microbial decomposition. These adaptable methods remain difficult to carry out in field settings for microbe-assisted treatment due to the greater expense, the cooperation of the promoting matrix to microorganisms and the surroundings, alterations in the structure of the local population, the volatilization of pollutants into the environment, etc. (Phian et al., 2022).

4 Recent Advances and Technological Innovation

Recent developments in microbial technology have greatly enhanced the process of turning organic waste into useful goods and solutions, such as biofertilizers, recycling, and biogas generation. These developments address waste management issues and enhance sustainable practices by using microbial enzymes, bioremediation, and biorefinery processes.

4.1 Bioconversion Process

The need for nutritious foods, animal feed, and chemicals has surged due to rapid urbanization, which has also increased the consumption of fossil fuels and the production of organic waste. Because of its qualities, such as its great abundance, sustainability, and accessibility, organic waste might be valued as an alternate approach to waste management problems. Because of their organic and nutrient-rich makeup, a number of wastes from industry are currently used as a resource for microbial fermentation methods that produce value-added products like biofuels, biopesticides, biohydrogen, enzymes, and bioplastics (Bibi et al., 2023). The procedure entails pre-treating the waste biomass, producing the value-added product in reactors, and then recovering the product through downstream processing. The transformation of organic acids to the value-added products mentioned above is achieved through the microbial action (Fig. 4).

Thus, it is possible to carry out the viability and sustainability of producing a range of value-added products from waste materials and by-product sources (Chavan et al., 2022). Because of its harmful impacts on the environment and human well-being, environmental contamination brought on by sophisticated industrialization and human activity is a major issue in the present day (Chavan et al., 2021). In order to satisfy customer demands, the marketplace for these kinds of products has grown in tandem with the growing need for functional foods worldwide. In the end, one of the biggest problems facing the globe today is the massive creation of organic waste from many sources, including the agricultural, food, municipal, and urban sectors. Thus, by minimizing waste and lowering reliance on fossil fuels, their bioconversion into value-added goods might transform this issue into a possible solution (Awasthi et al., 2020).

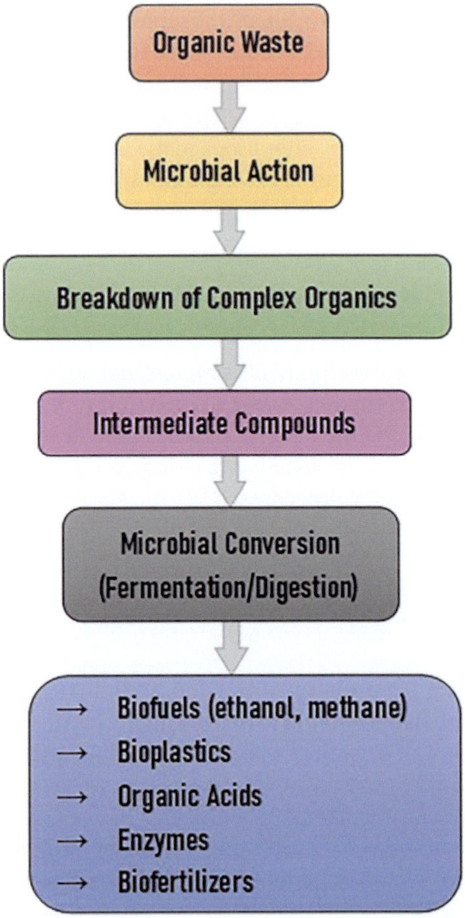

Fig. 4 Microbial bioconversion

4.2 Biogas Production

The growing global population has led to a rise in the demand for energy throughout time. The environment has been harmed by the emission generated by fossil fuels, which are running out. International researchers are concentrating on creating alternative energy production techniques (Rai & Da Silva, 2017). Any fuel derived from biomass, which comprises recently lived things and the by-products of their metabolism, can be used to generate bioenergy. All of the plant and animal matter found on the surface of the Earth is referred to as biomass. So, bioenergy is the process of gathering biomass, like dung, crops, or plants, and using it to generate heat, electricity, or motion.

The substantial volumes of biomass, livestock manure, agro-industrial waste, sludge, and moist organic materials generated nowadays provide a continuous pollution concern and might have an adverse effect on the ecosystem if not handled properly. Fuels derived from resources such as biomass or their by-products from manufacturing and conversion are known as biofuels (Ilham, 2022). Since biofuels are sustainable and environmentally beneficial energy sources, they are gaining popularity as a substitute source of energy. The organic portion of almost any type of biomass may be broken down into biogas, a combination of CO_2 and CH_4 that is regarded as a viable renewable energy source.

Since CH_4 has the greatest amount of energy of all of the elements of biogas, it is the most significant component. In turn, biogas with a high CH_4 percentage is preferred. Biogas was initially discovered to come from organic matter that was breaking down 600 years ago (Ngan et al., 2020). Biogas is combustible, smokeless, odourless, colourless, and clean, but if it isn't desulfurized, it smells like rotten eggs. Utilizing biogas technology, the organic waste is transformed into electricity. Utilizing manure and energy can result in a greener environment, social and economic advantages, and support sustainable development (Bissoyi et al., 2024a, 2024b, 2024c; Dogra et al., 2024a, 2024b; Ray et al., 2024). In addition to producing nutrient-rich organic fertilizer, the biogas process also produces an effluent slurry that is beneficial for seed germination, cultivation of fish, and algae development. Both large-scale and small-scale applications involving the generation of electricity employ the biogas technology. The total amount of this combination of gases depends on the anaerobic digestion process parameters, such as temperature, pH, and storage duration, as well as the sources. Utilizing a variety of lignocellulosic biomass and other organic wastes, including agricultural residues, plant-based wastes, animal manure, food waste, and even the organic component of municipal waste, is made viable by biogas technology. These abundant biomass sources have a great potential to meet both the current and future energy needs, particularly in the home sector. One well-established and reasonably priced method for creating biogas, a useful green primary energy source, is the anaerobic digestion of biodegradable waste components. For a variety of final applications, including transportation, energy generation, food preparation, and other power-related methods, this method of production offers a clean and effective fuel.

Anaerobic digestion is regarded as a natural process in which groups of microbes break down organic materials without the presence of oxygen. However, there are four phases: hydrolysis, acidogenesis, which is seen as creating acid, acetogenesis, that is further referred to as making acetic acid, and methanogenesis, which is the last stage and is known for generating CH_4 (Ngan et al., 2020) (Fig. 5). Purifying raw biogas and upgrading it to reach a high-quality fuel level is crucial for raising its calorie content and lowering undesirable substances like CO_2 and H_2S that harm consumption systems.

In addition to producing energy, anaerobic digestion of biomass waste has various benefits, including preventing odours, providing a disposal option, and enhancing cleanliness by eliminating germs. Additionally, the nutrient-rich digested slurry may be recycled back into the fields as fertilizer. Since CH_4 has a 21-fold greater greenhouse effect than CO_2 and would otherwise be discharged into the atmosphere, biogas also has the benefit of reducing GHG emissions overall (Prasad et al., 2017). To lessen

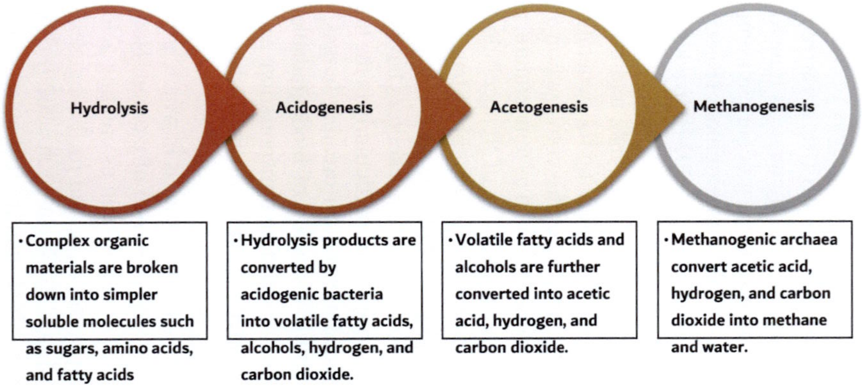

Fig. 5 Stages of anaerobic digestion

the amount of CH_4 that spontaneously dissipates into the atmosphere, it is crucial to manage organic wastes under rules and regulations.

4.3 Biochar Production

Around one-third of the food generated globally for people to consume is discarded, as reported by the Food and Agriculture Organisation (Fabi et al., 2021; Schanes et al., 2018). Regretfully, if appropriate policy reforms are not made, the percentage of food waste is predicted to double by 2050 (Barrera & Hertel, 2021). Food disposal does not remain consistent across the globe; industrialized nations produce more waste from food than undeveloped nations. The world's leading producers of food waste are Central and South Asia (Reitemeier et al., 2021). Decomposition of food waste leads to serious social and environmental problems, including resource depletion, food poverty, and GHG emissions. Among the techniques used to handle food waste are landfill accumulation, fermentation-extinction composting, pyrolysis to create solid charcoal and biogas, microbial anaerobic digestion, transformation to livestock feed, and combustion to create synthetic gas. The prospective uses of gasification and pyrolysis in the recycling of resources remain the subject of numerous recent studies (Al-Rumaihi et al., 2020).

The pyrolysis of biomass produces biochar, a black, porous, carbon-rich solid that has attracted a significant interest because of its many uses and advantages for the surroundings. In order to create biochar, biomass must be thermally broken down in an oxygen-limited environment, producing a stable carbonaceous deposit in the process (Fabi et al., 2021; Schanes et al., 2018). Biochar may be produced from a range of raw materials, including organic by-products, forest waste, and agricultural leftovers. This provides a sustainable approach to resource consumption and handling waste.

Pyrolysis refers to a thermal breakdown process that creates biochar, and bio-oil without the presence of oxygen. Because of its advantages in pollutant adsorption, enhancement of soil, and carbon storage, biochar has undergone extensively researched. To increase output and functional qualities, sophisticated biochar manufacturing techniques such microwave-assisted pyrolysis and hydrothermal carbonization have been developed. By providing more control over the circumstances of the reaction, these techniques enable biochar characteristics to be tailored according to the feed and purpose. Furthermore, the use of metal-based or biomass-derived catalysts in catalytic pyrolysis improves choice of products and energy utilization. A key component of the circular bioeconomy and agricultural sustainability, biochar has demonstrated promise in enhancing soil fertility and microbial life when combined with microbial inoculants (Nguyen et al., 2022).

Utilizing cutting-edge methods like flash carbonization, hydrothermal carbonization, and microwave pyrolysis are examples of recent developments in the creation of biochar. These techniques increase the biochar's surface area and porosity while consuming less energy. In order to enhance the quality of biochar and retain nutrients, co-pyrolysis of food particles alongside additional biomass residues has additionally been studied. Biochar's catalytic and adsorptive qualities are further improved by functionalizing it with chemical processes or microbial inoculants, which makes it suitable for use in soil remediation, treatment of wastewater, and heavy metal remediation. Additionally, the multipurpose qualities of biochar-based composites in energy and environmental uses continue to be investigated. A developing field of study that aids in the microbial breakdown of contaminants and encourages nutrient cycling is the beneficial interaction among biochar and microbial populations (Ahn et al., 2023).

Biochar has been discovered to be a practical and reasonably priced adsorbent for the removal of heavy metals from wastewater (Patra et al., 2017). Biochar activity is influenced by variables such as feedstock content and modification method. Surface modifications can optimize the potential of biochar, even if activated carbon is superior at adsorbing heavy metals. These modifications can improve the adsorption and recycling potential of biochar, making it a more practical and effective way to minimise or eliminate the heavy metals from wastewater (Sajjadi et al., 2021).

4.4 Biomedical Waste Ash (BWA)

Solid medical waste management is becoming a bigger challenge recently because of the pandemic's rapid growth of waste production. Used needles and syringes, body parts, medications, diagnostic tests, blood, synthetic materials, medical equipment, radioactive substances, and surgical masks are just a few examples of all the things that make up medical waste. It may be viewed as a subset of every waste product produced in medical institutions. Hospital waste may usually be classified as either hazardous or non-hazardous. Health care workers and the general public are exposed to physical, chemical, or microbial risks associated with the management, treatment, and disposal of hazardous waste (Mazzei & Specchia, 2023). Approximately 15% of the total amount of medical waste is considered to be harmful, and based on the waste's properties, this percentage might rise to 35%. The rest of the medical waste is considered to be harmless. There are two types of toxic waste: infectious and non-infectious. In order to eliminate the associated biohazards, infectious waste must undergo particular treatment since it includes bacteria that might cause an infectious disease in a vulnerable host (Mazzei & Specchia, 2023). Conversely, hazardous waste may result in bodily harm, intoxication, reproductive issues, or toxicity (Kumar & Das, 2016, 2017). Regular techniques can be used to treat harmless or ordinary waste; additional handling is not necessary.

The most popular heat treatment for getting rid of medical waste is incineration. The temperature range for this high-temperature procedure is 980° to 2,000 °C. To guarantee full burning process, organic substances must be burned in an environment with a plenty of oxygen. The primary benefit is the substantial decrease in material quantity, which can be as much as 90% (Datta et al., 2018). Heat, particles, fumes, and ash are the primary by-products of incineration. It offers excellent disinfection, sterilization, and pollution removal and has a steady action. In addition to controlling the flow rate and burning air, well-designed incinerators use enough dwell time to eliminate all medical waste. A primary benefit of this method is its wide range of applications; however, it is not recommended for use with pressurized gas containers, reactive waste chemicals, silver salts, PVC plastics, heavy metals, sealed containers or vials, radioactive substances, or unreliable medications, even though it is appropriate for all types of infectious medical waste (Capoor & Bhowmik, 2017).

The procedure may release carcinogenic and toxic chemicals including furans and polychlorinated dioxins. Particularly, fabrics, healthcare products, and plastic items have a significant dioxin concentration (Jiang et al., 2019). In addition to producing other harmful gases like sulphur dioxide, hydrogen fluoride, and hydrogen chloride, that call for an emissions purification system, combustion also encourages the creation of fly ash and bottom slag, which are potentially harmful by-products with a significant amount of toxic metal (Xu et al., 2020). Dioxin generation is influenced by heavy metals, fly ash's residual carbon, burning ailments waste composition, and exhaust gas's chlorine level. Since plastics make up the majority of medical waste, the primary source of elevated dioxin pollutants from incineration facilities is the chlorine found in plastics. However, inadequate combustion efficiency is mostly the reason for such emissions. In order to ensure the air quality and the ecosystem, medical waste incinerators must keep track of their emissions. If they fail to be properly handled, they can lead to technical issues such as tube corrosion and also harm to the ecosystem and human health. Consequently, employing incinerators also requires specific training on the operations.

Bioremediation is an ecologically safe method that uses microorganisms and plants (e.g. fungus, bacteria, and algae) to help restore polluted soil to its natural form, as opposed to physicochemical processes (Saha et al., 2021). By using these species' innate metabolic capacities, bioremediation transforms heavy metals into less hazardous forms, providing an environmentally responsible and long-lasting remedy for heavy metal pollution. Through a variety of methods, such as the immobilization, adhesion, processing, oxidation, and volatilization of heavy metals, microorganisms have been used to mitigate heavy metal contamination (Haider et al., 2021; Zulfiqar et al., 2022). The basic processes that control the behaviour, growth, and responses of microorganisms at polluted sites to changes in their surroundings must be identified in order to optimize the effectiveness of bioremediation. Microbe-metal interactions, biosorption, biotransformation, biomineralization, bioaccumulation, and bioleaching are examples of bioremediation approaches. They may be extracted from soil by microorganisms that rely on substances for development and growth (Mustafa et al., 2023). Microorganisms have the ability to oxidize and decrease metallic substances along with dissolving metals. Many organic solvents can damage cell membranes. To avoid harming the outer membrane, microbial cells can create defence mechanisms like solvent outflow or hydrophobic pumps.

5 Future Prospects in Microbial-Assisted Waste to Chemical Conversion

Microorganisms, among the most miniature living things on the planet, have the potential to make each of the advanced energy technologies a reality in the future. Such methods are referred to as microbial energy conversion. In microbial energy systems, microorganisms transform the chemical energy in biomass into chemical energy in the shape of, for instance, hydrogen or ethanol by producing fuels from unprocessed organic matter (Talapko et al., 2022). By breaking down organic compounds, bacteria are used in microbial energy systems to either produce fuels or directly produce power. In fact, energy is never created; instead, it may be transformed into several forms. Microorganisms convert the chemical energy in biomass into fuels that carry energy via a variety of microbial energy conversion processes. Microbial fuel cells, on the other hand, convert chemical energies into electrical energy by using bacteria to extract electrons from biomass and create a flow of electricity (Ieropoulos & Greenman, 2023).

Compared to energy derived from natural gas and petroleum, microbial energy conversion offers a number of benefits. The most significant of these advantages is that microbial energy transformations are more ecologically benign than fossil fuel technologies. For instance, there is no usage or generation of harmful compounds in microbial energy forms. More significantly, biomass is transformed into chemical or electrical energy in every microbial energy transformation mechanism discussed here, with the exception of photosynthetic hydrogen synthesis. Biomass conversion is a carbon–neutral process, which demonstrates that potentially it does not have the capability for enhancing or degrading the total amount of CO_2 (an important GHG) in the atmosphere.

Additionally, microorganisms may choose their substrate from a complicated combination of chemicals, which reduces the requirement for raw materials to be refined or purified in order to differentiate substrates from non-substrates. Lastly, energy may be produced locally, at the location of demand, using microbial energy technology. Landfills may eventually function as bioreactors, housing microbial populations and designed to improve garbage digestion and conversion into valuable products (Aromolaran & Sartaj, 2023). By encouraging the growth of microbial populations that are suited to certain waste stream types, a bioreactor method might increase the productivity and efficiency of waste-to-resource operations. Furthermore, certain metabolic reactions may be accelerated and directed by the use of genetically modified bacteria. Microbial collaborations are being used to improve the breakdown and transformation of waste materials, both organic and plastic, into bioplastics, biofuels, and other useful chemical intermediates. In order to further comprehend and optimize these microbial systems and support a biological economy model, methods from synthetic biology and biological systems are being used more and more.

Another promising field is microbial electrochemical technologies (METs), such as microbial fuel cells and microbial electrolysis cells (Ghangrekar & Nath, 2022).

These technologies can also be employed to purify wastewater and produce hydrogen gas or energy by using microorganisms to catalyze redox reactions at electrodes. Both environmental and financial advantages might result from incorporating such technologies into waste management.

6 Conclusion

The microbial conversion of organic waste into valuable products represents a transformative approach towards addressing some of the most pressing environmental, economic, and societal challenges of the twenty-first century. The diverse nature of organic waste—ranging from agricultural and industrial residues to biomedical and municipal solid waste—demands multifaceted strategies, and microbial bioconversion stands out as a sustainable, cost-effective, and environmentally friendly solution.

Microorganisms, including bacteria, fungi, and algae, possess remarkable enzymatic systems and metabolic pathways capable of degrading complex organic materials. Through processes such as composting, anaerobic digestion, biofermentation, and bioremediation, microbes facilitate the breakdown of waste and the recovery of useful compounds like biofuels, enzymes, essential oils, biochar, and bioplastics. Each microbial group plays a distinct role: bacteria dominate in aerobic treatment systems; fungi contribute under low-oxygen and acidic conditions; and algae assist in nutrient recycling and oxygen production. These coordinated microbial actions not only reduce the volume and toxicity of waste but also generate products with substantial commercial and environmental value.

Furthermore, the advent of advanced biotechnological tools—such as genetic engineering, microbial consortia design, and microbial electrochemical technologies—has enhanced the efficiency and scope of waste bioconversion. Innovations in biochar functionalization, biogas purification, and value-added product recovery from fermentation processes are accelerating the transition from waste disposal to resource recovery. However, despite these advantages, challenges persist. Environmental variability, process scalability, high setup costs, and the complexity of waste compositions can affect microbial survival and degradation efficiency in real-world settings. To ensure long-term viability, future research should focus on optimizing microbial strains, integrating waste treatment systems with circular bioeconomy models, and developing site-specific strategies tailored to regional waste characteristics. Regulatory frameworks and public awareness must also evolve to support these green technologies.

Ultimately, leveraging microbial potential for waste-to-resource conversion not only addresses environmental pollution but also supports energy independence, soil health, and economic growth. As global sustainability goals become more urgent, microbial waste valorization offers a promising, nature-inspired path forward—merging ecological wisdom with modern innovation for a cleaner, circular, and more resilient future.

References

Adebayo, F. O., & Obiekezie, S. O. (2018). Microorganisms in waste management. *Research Journal of Science and Technology, 10*(1), 28–39.

Ahn, K. H., Shin, D. C., Lee, Y. E., Jeong, Y., Jung, J., & Kim, I. T. (2023). Biochar production and demineralization characteristics of food waste for fuel conversion. *Molecules, 28*(16), 6114.

Al-Rumaihi, A., McKay, G., Mackey, H. R., & Al-Ansari, T. (2020). Environmental impact assessment of food waste management using two composting techniques. *Sustainability, 12*(4), 1595.

Aromolaran, A., & Sartaj, M. (2023). Enhancing biogas production from municipal solid waste through recirculation of blended leachate in simulated bioreactor landfills. *Biomass Conversion and Biorefinery, 13*(4), 2797–2812.

Awasthi, M. K., Sarsaiya, S., Patel, A., Juneja, A., Singh, R. P., Yan, B., & Taherzadeh, M. J. (2020). Refining biomass residues for sustainable energy and bio-products: An assessment of technology, its importance, and strategic applications in circular bio-economy. *Renewable and Sustainable Energy Reviews, 127*, Article 109876.

Bala, K., Anshul, Verma, S., Dogra, A., Das, A. P., Kumar, R., & Nandini. (2024). Sustainable Management for Urban Plastic Waste Generation. *Sustainable Management of Urban Plastic Waste Through Circular Economic Approaches*, 211–236. https://doi.org/10.1201/9781003465447-13

Bala, K., Dogra, A., Lata, P., & Das, A. P. (2024). Discussing Sources and Biological-Chemical Remediation Approaches for Microplastic Pollution. *Sustainable Microbial Technology for Synthetic and Cellulosic Microfiber Bioremediation*, 37–59. https://doi.org/10.1007/978-3-031-62600-5_3

Barrera, E. L., & Hertel, T. (2021). Global food waste across the income spectrum: Implications for food prices, production and resource use. *Food Policy, 98*, Article 101874.

Behera, I. Dipamitra., & Das, A. Prasad. (2025). *Environmental hydrocarbon pollution and zero waste approach towards a sustainable waste management* 436.

Bhat, R. A., Dar, G. H., Jehangir, A., Bhat, B. A., & Yousuf, A. R. (2012). Municipal solid waste generation and present scenario of waste management during Yatra season in Pahalgam: A tourist health resort of Kashmir valley. *International Journal of Current Research, 4*(10), 004–009.

Bibi, F., Ilyas, N., Saeed, M., Shabir, S., Shati, A. A., Alfaifi, M. Y., & Sayyed, R. Z. (2023). Innovative production of value-added products using agro-industrial wastes via solid-state fermentation. *Environmental Science and Pollution Research, 30*(60), 125197–125213.

Capoor, M. R., & Bhowmik, K. T. (2017). Current perspectives on biomedical waste management: Rules, conventions and treatment technologies. *Indian Journal of Medical Microbiology, 35*(2), 157–164.

Chavan, S., Yadav, B., Tyagi, R. D., & Drogui, P. (2021). A review on production of polyhydroxyalkanoate (PHA) biopolyesters by thermophilic microbes using waste feedstocks. *Bioresource Technology, 341*, Article 125900.

Chavan, S., Yadav, B., Atmakuri, A., Tyagi, R. D., Wong, J. W., & Drogui, P. (2022). Bioconversion of organic wastes into value-added products: A review. *Bioresource Technology, 344*, Article 126398.

Das, A. P., Mallavarapu, M., & Ghosh, S. (2023a). Ecotoxicity of mining pollutants on the environment and their remediation. *Environmental Chemistry and Ecotoxicology, 5*, 165–167. https://doi.org/10.1016/J.ENCECO.2023.08.002

Das, A. P., Mohanty, A. K., van Hullebusch, E. D., Figueiredo, G., Das, A. P., Mohanty, A. K., van Hullebusch, E. D., & Figueiredo, G. (2023b). Bioremediation as an emerging technology for the removal of synthetic microplastic pollutants from marine ecosystem. *MarPB, 194*, Article 115297. https://doi.org/10.1016/J.MARPOLBUL.2023.115297

Das, A., Dey, S., & Das, A. P. (2024b). Ecotoxicological impacts of synthetic microfiber pollutants and development of sustainable mitigation strategies. *Environmental Chemistry and Ecotoxicology*. https://doi.org/10.1016/J.ENCECO.2024.12.003

Das, A., Ray, S. K., Mohanty, M., Mohanty, J., Dey, S., & Das, A. P. (2024c). Ecotoxicity of microplastic wastes and their sustainable management: A review. *Environmental Chemistry and Ecotoxicology, 6*, 144–152. https://doi.org/10.1016/J.ENCECO.2024.05.003

Das, A., Dey, S., Bala, K., Gnanasekaran, R., & Das, A. P. (2024). Microbial bioremediation technology for sustainable treatment and management of synthetic microfiber waste. *Sustainable Microbial Technology for Synthetic and Cellulosic Microfiber Bioremediation*, 221–238. https://doi.org/10.1007/978-3-031-62600-5_12

Datta, P., Mohi, G., & Chander, J. (2018). Biomedical waste management in India: Critical appraisal. *Journal of Laboratory Physicians, 10*(01), 006–014.

Dey, S., Tripathy, B., Kumar, M. S., & Das, A. P. (2023). Ecotoxicological consequences of manganese mining pollutants and their biological remediation. *Environmental Chemistry and Ecotoxicology*. https://doi.org/10.1016/J.ENCECO.2023.01.001

Dey, S., Bissoyi, A., Koteswara Reddy, G., & Das, A. P. (2024). Synthetic microfibers: an urban emerging pollutant demanding for sustainable environmental solutions. *Sustainable Microbial Technology for Synthetic and Cellulosic Microfiber Bioremediation*, 107–121. https://doi.org/10.1007/978-3-031-62600-5_7

Dey, S., Dash, A., & Das, A. P. (2024a). Bioremediation as a sustainable and circular solution for mining waste management and bioenergy generation. *Sustainable Management of Mining Waste and Tailings: A Circular Economy Approach*, 107–119. https://doi.org/10.1201/9781003442455-6/

Fabi, C., Cachia, F., Conforti, P., English, A., & Moncayo, J. R. (2021). Improving data on food losses and waste: From theory to practice. *Food Policy, 98*, Article 101934.

Ghangrekar, M. M., & Nath, D. (2022). Microbial electrochemical technologies for wastewater treatment: insight into theory and reality. In *Clean Energy and Resource Recovery* (pp. 179–200). Elsevier.

Ghosh, S., & Das, A. P. (2020). Microbial metagenomics: current advances in investigating microbial ecology and population dynamics. In S. K. Nayak & B. B. Mishra (Eds.), *Frontiers in Soil and Environmental Microbiology* (1st ed., pp. 247–254). CRC Press.

Ghosh, S., Gandhi, M., van Hullebusch, E. D., & Das, A. P. (2020). Proteomic insights into Lysinibacillus sp.-mediated biosolubilization of manganese. *Environmental Science and Pollution Research 2020 28:30, 28*(30), 40249–40263. https://doi.org/10.1007/S11356-020-10863-4

Ghosh, S., Tripathy, B., Dey, S., & Das, A. P. (2023). Comparative investigation of fungal and bacterial manganese biomining mechanisms. *Manganese Mining Microorganisms*, 99–115. https://doi.org/10.1016/b978-0-12-822147-1.00016-9

Haider, F. U., Liqun, C., Coulter, J. A., Cheema, S. A., Wu, J., Zhang, R., & Farooq, M. (2021). Cadmium toxicity in plants: Impacts and remediation strategies. *Ecotoxicology and Environmental Safety, 211*, Article 111887.

Hnain, A. K., Cockburn, L. M., & Lefebvre, D. D. (2011). Microbiological processes for waste conversion to bioenergy products: Approaches and directions. *Environmental Reviews, 19*(NA), 214–237.

Husain, R., Vikram, N., Pandey, S., Yadav, G., Bose, S. K., Ahamad, A., & Hussain, T. (2022). Microorganisms: An eco-friendly tools for the waste management and environmental safety. In *Biotechnological Innovations for Environmental Bioremediation* (pp. 949–981). Singapore: Springer Nature Singapore.

Ieropoulos, I., & Greenman, J. (2023). The future role of MFCs in biomass energy. *Frontiers in Energy Research, 11*, 1108389.

Ilham, Z. (2022). Biomass classification and characterization for conversion to biofuels. In *Value-chain of biofuels* (pp. 69–87). Elsevier.

Jiang, X., Li, Y., & Yan, J. (2019). Hazardous waste incineration in a rotary kiln: A review. *Waste Disposal & Sustainable Energy, 1*(1), 3–37.

John, A., & Das, A. P. (2024). Government strategies for achieving circular economy for plastic waste management in Southeast Asia. *Sustainable Management of Urban Plastic Waste Through Circular Economic Approaches*, 72–82. https://doi.org/10.1201/9781003465447-6

Kaura, S., Mathur, A., & Kalra, A. (2023). Bacteria and pollutants. In: *Modern Approaches in Waste Bioremediation: Environmental Microbiology* (pp. 339–364). Cham: Springer International Publishing.

Kavya, Vashisht, M., Jain, B., & Shrivastava, S. (2024). Transforming waste into wealth: a review on microbial conversion of organic municipal wastes to value-added products. *Discover Environment*, *2*(1), 112.

Kumar, M. S., & Das, A. P. (2016). Molecular identification of multi drug resistant bacteria from urinary tract infected urine samples. *Microbial Pathogenesis, 98*, 37–44. https://doi.org/10.1016/J.MICPATH.2016.06.029

Kumar, M. S., & Das, A. P. (2017). Emerging nanotechnology based strategies for diagnosis and therapeutics of urinary tract infections: A review. *Advances in Colloid and Interface Science, 249*, 53–65. https://doi.org/10.1016/J.CIS.2017.06.010

Kumaria, N., Pandeya, S., & Pandeyb, A. K. (2022). Role of microorganisms in municipal solid waste management-A case study.

Mazzei, H. G., & Specchia, S. (2023). Latest insights on technologies for the treatment of solid medical waste: A review. *Journal of Environmental Chemical Engineering, 11*(2), Article 109309.

Mishra, S., & Das, A. P. (2022). Treatment of the wastewater polluted with synthetic microfiber released from washing machine. In B. B. Das, H. Hettiarachchi, P. K. Sahu, & S. Nanda (Eds.), *Lecture Notes in Civil Engineering* (Vol. 207, pp. 109–117). Springer Science and Business Media Deutschland GmbH. https://doi.org/10.1007/978-981-16-7509-6_9/COVER

Mishra, S., & Das, A. P. (2023). Microbial remediation of synthetic microfiber contaminated wastewater. In M. P. Shah (Ed.), *Microbial Technologies in Industrial Wastewater Treatment* (pp. 337–350). Springer, Singapore. https://doi.org/10.1007/978-981-99-2435-6_17

Mishra, S., Dash, D., & Das, A. P. (2023). Aquatic microbial diversity on plastisphere: colonization and potential role in microplastic biodegradation. *Geomicrobiology Journal*, 1–12. https://doi.org/10.1080/01490451.2023.2209750

Mohanty, M., Tripathy, B., Koteswara Reddy, G., & Das, A. P. (2024). Greener technologies for establishing restraint over microplastic pollution. *Sustainable Microbial Technology for Synthetic and Cellulosic Microfiber Bioremediation*, 239–258. https://doi.org/10.1007/978-3-031-62600-5_13

Mustafa, A., Zulfiqar, U., Mumtaz, M. Z., Radziemska, M., Haider, F. U., Holatko, J., & Brtnicky, M. (2023). Nickel (Ni) phytotoxicity and detoxification mechanisms: A review. *Chemosphere, 328*, Article 138574.

Nag, M., Lahiri, D., Ghosh, S., Sarkar, T., Pati, S., Das, A. P., Ram, D. K., Bhattacharya, D., & Ray, R. R. (2023). Application of microorganisms in biotransformation and bioremediation of environmental contaminant: A review. *Geomicrobiology Journal*. https://doi.org/10.1080/01490451.2023.2261443

Nanda, S., & Berruti, F. (2021). Municipal solid waste management and landfilling technologies: A review. *Environmental Chemistry Letters, 19*(2), 1433–1456.

Ngan, N. V. C., Chan, F. M. S., Nam, T. S., Van Thao, H., Maguyon-Detras, M. C., Hung, D. V., & Van Hung, N. (2020). Anaerobic digestion of rice straw for biogas production. *Sustainable rice straw management*, 65–92.

Nguyen, M. K., Lin, C., Hoang, H. G., Sanderson, P., Dang, B. T., Bui, X. T., & Tran, H. T. (2022). Evaluate the role of biochar during the organic waste composting process: A critical review. *Chemosphere, 299*, Article 134488.

Patra, J. M., Panda, S. S., & Dhal, N. K. (2017). Biochar as a low-cost adsorbent for heavy metal removal: A review. *Int. J. Res. Biosci, 6*(1), 1–7.

Pengqi, G., John, A., Huang, D. M., & Das, A. P. (2025). Insights to visual and analytical techniques for identification and characterization of marine micro plastic as a major source of petroleum hydrocarbon pollutants. *Environmental Science and Engineering, Part F218*, 409–436. https://doi.org/10.1007/978-3-031-81615-4_17

Phian, S., Nagar, S., Kaur, J., & Rawat, C. D. (2022). Emerging issues and challenges for microbes-assisted remediation. In *Microbes and Microbial Biotechnology for Green Remediation* (pp. 47–89). Elsevier.

Pradhan, G., Tripathy, B., Ram, D. K., Digal, A. K., & Das, A. P. (2023). Bauxite mining waste pollution and its sustainable management through bioremediation. *Geomicrobiology Journal*, 1–10. https://doi.org/10.1080/01490451.2023.2235353

Prasad, S., Rathore, D., & Singh, A. (2017). Recent advances in biogas production. *Chem Eng Process Tech, 3*(2), 1038.

Priyadarshini, S., Jagatee, S., & Das, A. P. (2024). Synthetic fabrics and microfiber pollution–an assessment of their global impact. *Environmental Science and Engineering, Part F2333*, 137–157. https://doi.org/10.1007/978-3-031-51792-1_8

Rai, M., & Da Silva, S. S. (Eds.). (2017). *Nanotechnology for bioenergy and biofuel production.* Springer.

Ray, S., Dey, S., Digal, A. K., Yadav, K. K., & Das, A. P. (2025). Sanitary waste and microplastic pollution in the Euro-Mediterranean region: Challenges and solutions. *Euro-Mediterranean Journal for Environmental Integration, 2025*, 1–23. https://doi.org/10.1007/S41207-025-00879-Y

Ray, S. K., Dey, S., Das, A. P., & John, A. (2024). Microplastic waste management through greener technologies: A review. *Sustainable Management of Urban Plastic Waste Through Circular Economic Approaches*, 237–261. https://doi.org/10.1201/9781003465447-14

Reddy Koteswara, G., Bhargavi, S., Yamuna, D., Anjani Devi, M., Jeevana Lakshmi Kumar, G., Manas, T., Adarsh, S., Bhavya Sai, V., Nath, H., Alreshidi, M. A., Togun, H., Ram, D. K., & Das, A. P. (2025). Recent advancements in the remediation of marine hydrocarbon and plastic pollutants. *Environmental Science and Engineering, Part F218*, 127–142. https://doi.org/10.1007/978-3-031-81615-4_7

Reitemeier, M., Aheeyar, M., & Drechsel, P. (2021). Perceptions of food waste reduction in Sri Lanka's commercial capital, Colombo. *Sustainability, 13*(2), 838.

Saha, L., Tiwari, J., Bauddh, K., & Ma, Y. (2021). Recent developments in microbe–plant-based bioremediation for tackling heavy metal-polluted soils. *Frontiers in Microbiology, 12*, Article 731723.

Sahu, A., Dey, S., Mallick, K., Priyadarsini, S., Behera, I. D., & Das, A. P. (2024). A review on advanced technology for sustainable management of synthetic microplastic waste. *Sustainable Management of Urban Plastic Waste Through Circular Economic Approaches*, 262–279. https://doi.org/10.1201/9781003465447-15

Sajjadi, B., Shrestha, R. M., Chen, W. Y., Mattern, D. L., Hammer, N., Raman, V., & Dorris, A. (2021). Double-layer magnetized/functionalized biochar composite: Role of microporous structure for heavy metal removals. *Journal of Water Process Engineering, 39*, Article 101677.

Saleem, A., Naureen, I., Tasleem, G., Anwar, R., Mairaj, M., Muddassar, H., & Rana, N. J. (2022). Microbial assisted bioremediation of polluted water. *Haya: Saudi J Life Sci*, *7*(4), 116–127.

Schanes, K., Dobernig, K., & Gözet, B. (2018). Food waste matters-A systematic review of household food waste practices and their policy implications. *Journal of Cleaner Production, 182*, 978–991.

Talapko, J., Talapko, D., Matić, A., & Škrlec, I. (2022). Microorganisms as new sources of energy. *Energies, 15*(17), 6365.

Xu, L., Dong, K., Zhang, Y., & Li, H. (2020, December). Comparison and analysis of several medical waste treatment technologies. In *IOP conference series: Earth and Environmental Science* (Vol. 615, No. 1, p. 012031). IOP Publishing.

Yousuf, M., Das, A. P., Masurkar, P., Bag, M. K., Mitra, D., & Ray, A. (2025). Recent advances and research in biodiesel as a sustainable, eco-friendly approach toward the green economy. *Environmental Science and Engineering, Part F218*, 305–329. https://doi.org/10.1007/978-3-031-81615-4_13

Zulfiqar, U., Jiang, W., Xiukang, W., Hussain, S., Ahmad, M., Maqsood, M. F., & Mustafa, A. (2022). Cadmium phytotoxicity, tolerance, and advanced remediation approaches in agricultural soils; a comprehensive review. *FrontIers in Plant Science, 13*, Article 773815.

Microbial Fermentation Strategy for Converting Organic Waste to Value-Added Product

Abdulhalim Musa Abubakar, Hamadou Mamoudou, Rashid Shamsuddin, and Saroj Raj Kafle

1 Introduction

Several microorganisms have been used for the production of biogas (Christy et al., 2014; Lebuhn & Gronauer, 2009; Malakahmad et al., 2009). Among them, diverse type of Clostridium spp. available took part: namely, *Clostridium chauvoei*, *Clostridium sordellii*, *Clostridium parapurificum*, *Clostridium butyricum*, *Clostridium haemolyticum*, *Clostridium tertium*, *Clostridium difficile*, *Clostridium bifermentans*, *Clostridium subterminale*, *Clostridium botulinum*, *Clostridium tyrobutyricum*, *Clostridium cadaveris*, *Clostridium novyi*, *Clostridium baratii*, *Clostridium welchii*, *Clostridium septicum*, *Clostridium glycolicum*, and *Clostridium sporogenes* (Neuhaus et al., 2015a, 2015b; Pulvirenti et al., 2015). *Clostridium welchii*, now more commonly referred to as *Clostridium perfringens*, can be isolated from a variety of sources, including soil contaminated with fecal matter, human and animal feces, decaying vegetation, stagnant water, improperly cooked or stored meat, and infected

A. M. Abubakar (✉)
Department of Chemical Engineering, Faculty of Engineering, Modibbo Adama University, Yola, Adamawa, Nigeria
e-mail: abdulhalim@mau.edu.ng

H. Mamoudou
Department of Biological Sciences, Faculty of Science, University of Maroua, Maroua, Cameroon
e-mail: mamoudhmd@gmail.com

R. Shamsuddin
Department of Chemical Engineering, Faculty of Engineering, Islamic University of Madinah, Madinah 42311, Saudi Arabia
e-mail: r.shamsuddin@iu.edu.sa

S. R. Kafle
Department of Chemical Engineering, Chungbuk National University, Cheongju 28644, South Korea
e-mail: sarojrajkafle@gmail.com

S. Ghosh et al. (eds.), *Advances in Organic Waste Conversion Through Industrial Biotechnology*, Sustainable Environmental Waste Management Strategies,
https://doi.org/10.1007/978-3-032-11377-1_18

wounds or tissues (Derongs et al., 2020). It is commonly found in goat feces as part of the normal gut flora in many animals, including poultry litter (Popova et al., 2014), goats (Uzal, 2004), sheep (Uzal & Songer, 2008), cattle (Klein et al., 2011), pigs (Syrchina et al., 2025), and humans (Hashimoto et al., 2016). However, under certain conditions, it can become pathogenic and cause diseases such as enterotoxemia (commonly known as "pulpy kidney" or "overeating disease") in goats and other livestock. *Clostridium welchii* obtains nutrients through heterotrophic metabolism, meaning it feeds on organic material rather than producing its own food, as in photosynthetic organisms. Because it thrives in environments devoid of oxygen (anaerobic conditions), such as deep wounds, intestines, and decaying organic matter, it can be a good candidate for biogas production. It does that by breaking down proteins (proteolysis) and carbohydrates (saccharolysis) in the organic matter for energy, through the secretion of collagenase, proteases, and glycosidases enzymes. Findings of Watcharasukarn et al. (2009) and Lorine et al. (2021) show that *Clostridium welchii* is one of the most heat resistant microorganism. Sahlström et al. (2008) proved this by showing that *Clostridia* spores in biowaste for biogas production can survive up to a temperature close to 70 °C but not above. Hence, during a thermophilic AD of piggery wastewater by Skillman et al. (2009), there was an inverse relationship between the levels of *Clostridium welchii* and the biogas production observed. Such is attributable to the pH condition as Salsali et al. (2008) discovered that inhibitory effects is felt around $pH = 4.5–5.5$ as a result of volatile fatty acid, also mentioned in Delmon et al. (2023). In many biogas plants, *Clostridium welchii* is often found in the digestate or the plants' waste byproduct (BIOHAZ, 2009; Derongs et al., 2020; Neuhaus et al., 2015a, 2015b). Reports by Fröschle et al. (2015) and Paroha et al. (2020) shows the effectiveness of using *Clostridium botulinum* and *Clostridium welchii* for AD biogas production. Apart from biogas, *Clostridium welchii* are used to produce bio-hydrogen (Wang et al., 2011, 2014; Wong et al., 2018; Yu et al., 2013).

Organic waste such as cattle manure was previously employed by Dohrmann et al. (2015) to produce biogas from *Clostridium botulinum*. Romanazzi et al. (2016) used *Clostridium difficile* in sludge to produce biogas. Novel microorganisms application have the capability to optimize production rate, since it is among the most important factors influencing gas production (Adetunji et al., 2022). Without augmenting a biogas system with any of the *Clostridium* specie, manures contains most of them in desired quantity (Bagge et al., 2009, 2010; Neuhaus Schrödl et al., 2015a, 2015b). Biogas is a value-added byproduct because it converts organic waste into a renewable energy source, thereby reducing environmental pollution and providing sustainable energy. It also yields digestate, a nutrient-rich fertilizer, and further enhances agricultural productivity. Other byproducts produced via microbial fermentation include ethanol (Banerjee et al., 2014), lactic acid (Silveira et al., 2019), butanol (Lee et al., 2008), citric acid (Thippareddi et al., 2003), and biosurfactants (Arora et al., 2019). In this study, we focused on the production of biogas using *Clostridium welchii* added to an equal ratio blend of cassava peel, sugarcane bagasse, and poultry manure. The use of Fourier Transform Infrared Spectroscopy (FTIR), Energy Dispersive X-ray Fluorescence (EDXRF), Scanning Electron Microscopy (SEM), and UV spectroscopy in analyzing a feedstock for biogas production is very important. Presence of compounds that affects biodegradability can be spotted using FTIR, while EDXRF spots metallic element or nutrient compositions that either support or inhibit microbial activity during anaerobic digestion. Based on the surface structure and texture of the feedstock, SEM allow us to know how easily microbes could attach and break down the material. As it concerns UV spectroscopy, aromatic compounds and certain organic acids that influence the chemical reactions during digestion are revealed. The objective of the study is to run the aforementioned characterization techniques on the biogas feedstock selected and employ existing kinetic models to analyze the performance of the biogas produced from two setups containing varying concentrations of *Clostridium welchii*. The estimation of study of kinetic parameters from First-Order, Modified First-Order, Transfert, Richards and Transference biogas models, represent a novel contribution of this work to existing knowledge. By employing feedstock under the same category (i.e., cattle manure, food waste & wheat straw), Transference and First-Order kinetic models were reported to fit well to the measured biogas yield result (Rahmani et al., 2022). Hence, models that more accurately incorporate the biological contributions of *C. welchii*; especially its kinetics and environmental sensitivity, might improve predictive performance in future simulations. Another unique contribution of this study is the incorporation of error analysis.

2 Methodology

2.1 Feedstock Characterization

For the FTIR analysis, approximately 2 mg of the homogenized mixture was mixed with 200 mg of spectroscopic-grade potassium bromide (KBr) and pressed into a transparent pellet using a hydraulic press at 10 tons pressure for 2 min. The pellet was placed in the FTIR spectrometer (model: PerkinElmer Spectrum Two), and the spectrum was recorded in the range of 4000–400 cm^{-1} at a resolution of 4 cm^{-1} with 32 scans. During EDXRF, 5 g of the powdered sample were pressed into a pellet using a 25 mm die set and a pressure of 15 tons for 5 min. The pellet was then analyzed using an EDXRF spectrometer (model: Shimadzu EDX-7000) with an Rh anode X-ray tube operating at 50 kV and 1 mA. The spectrum was collected for 300 s to determine the elemental composition of the sample. In SEM analysis, a small quantity of the sample powder was mounted on aluminum stubs using double-sided carbon tape. The mounted samples were then sputter-coated with gold using a sputter coater (model: Quorum Q150R ES) for 60 s to improve conductivity. The surface morphology was observed using a SEM (model: JEOL JSM-6610LV) at an accelerating voltage of 15 kV and a magnification range of 500 × to 5000x. For UV-Vis spectroscopy, 0.5 g of the powdered mixture was dissolved in 100 mL of distilled water and sonicated for 15 min to extract soluble components. The solution was then filtered using Whatman No. 1 filter paper. The filtrate was scanned using a UV-Vis spectrophotometer (model: UV-1800, Shimadzu) in the wavelength range of 200–800 nm at 1 nm intervals. The absorbance peaks were recorded and used to identify chromophoric groups present in the mixture. The four analytical procedures were slightly in line with the study conducted by Ranganathan and Arunodayaraj (2024), Kirawa et al. (2020) and Ormaechea et al. (2018).

2.2 Isolation of *Clostridium Perfringens*

Goat fecal samples were freshly collected from healthy adult goats using sterile spatulas and transferred into sterile, screw-capped containers. Approximately 10 g of feces were suspended in 90 mL of sterile phosphate-buffered saline (PBS, pH 7.2) to form a 10% (w/v) fecal slurry (Xue et al., 2023). The mixture was homogenized by vortexing for 5 min and allowed to settle for 10 min at room temperature. From the supernatant, 1 mL was serially diluted tenfold up to 10^{-6} using sterile PBS. Aliquots of 0.1 mL from appropriate dilutions (10^{-3} to 10^{-6}) were aseptically spread on Perfringens Agar Base (OPSP agar) supplemented with selective agents including TSC (Tryptose Sulfite Cycloserine) and egg yolk emulsion to enhance the selective recovery of *Clostridium perfringens* (Hauschild & Hilsheimer, 1974). Plates were incubated anaerobically at 37 °C for 24–48 h using an anaerobic jar containing GasPak™ anaerobic system envelopes. Colonies with characteristic black centers (due to sulfite reduction) and lecithinase activity (opaque halo) were presumptively identified as *C. perfringens.* Suspected colonies were subcultured in accordance with Bagge et al. (2010) and confirmed by Gram staining (showing Gram-positive rods), heat resistance test (survival at 80 °C for 10 min), and biochemical tests including lactose fermentation, nitrate reduction, and gelatin liquefaction. Figure 1 presents the steps as followed.

To prepare an inoculum of 200 mg/L *C. perfringens* for feeding into an anaerobic reactor, the biomass concentration was estimated using a pre-established standard curve relating optical density at 600 nm (OD600) to dry cell weight. Cultures were grown in Reinforced Clostridial Medium (RCM) under anaerobic conditions at 37 °C for 24 h until mid-log phase (Andrighetto et al., 2023). The culture was centrifuged at 6000 rpm for 10 min, and the pellet was washed and resuspended in sterile PBS. A dry weight equivalent of *C. perfringens* was adjusted to 200 mg/L by gravimetric quantification. This was achieved by concentrating the harvested cells, drying at 105 °C to constant weight, and suspending the required mass in sterile anaerobic buffer. The suspension was then introduced into the anaerobic digester "A" under strict anaerobic conditions to initiate biogas production. Similar procedural route was followed to obtain a 300 mg/L of *C. perfringens* to be fed into a different reactor "B," as shown in Fig. 2.

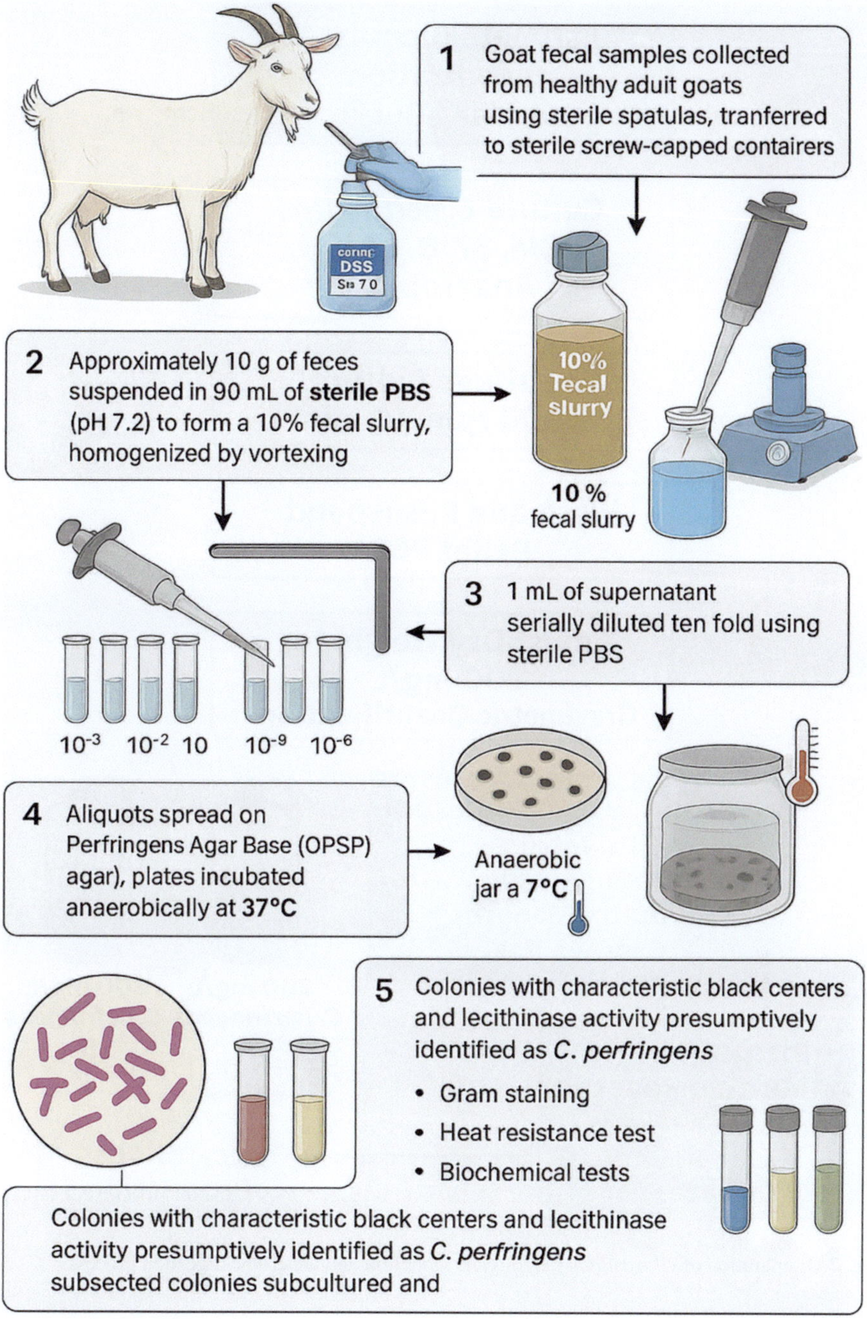

Fig. 1 Flowchart illustrating the isolation and identification procedure of *Clostridium perfringens* from foat fecal samples

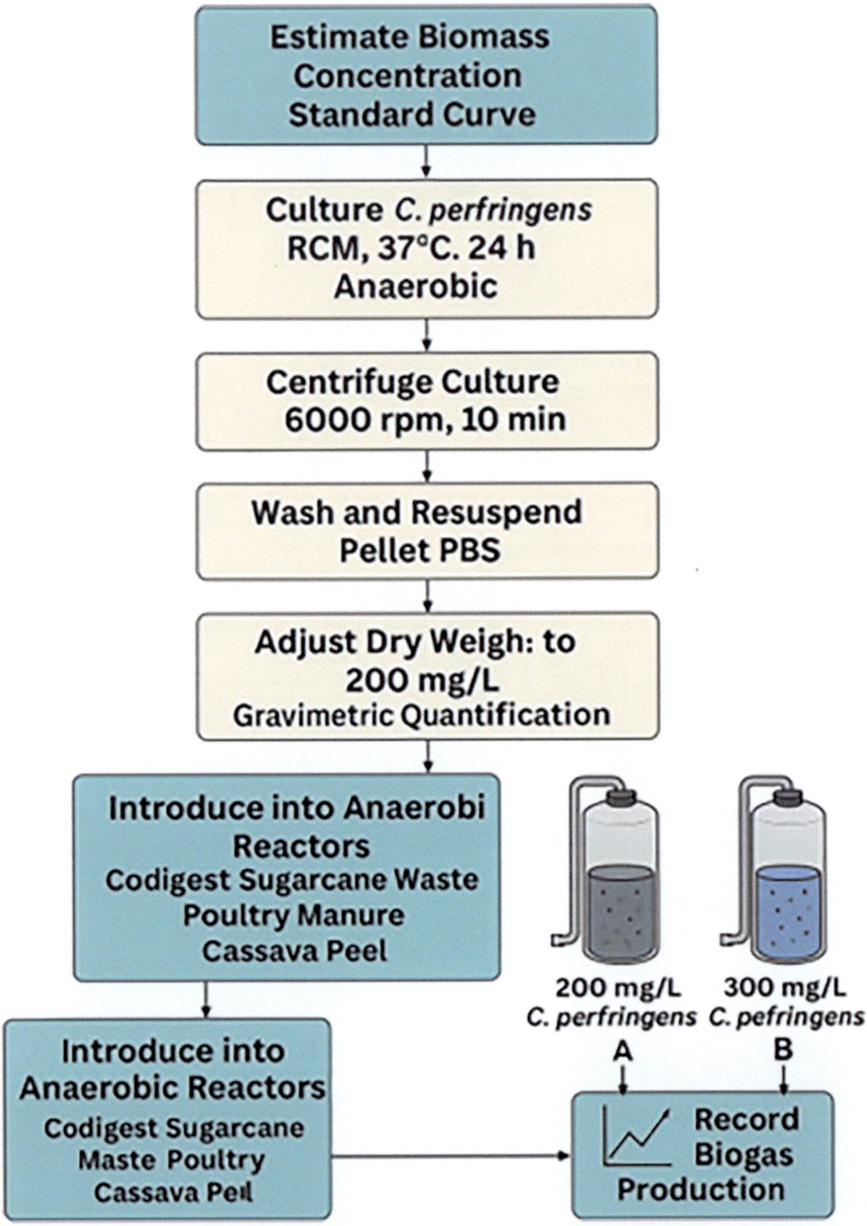

Fig. 2 Preparation of *Clostridium perfringens* inoculum and anaerobic digestion process

2.3 Bioaugmented Biogas Production

Two reactors, A and B, were set up with different concentrations of a specific microorganism to observe their effects on gas production. To avoid limited mass transfer due to low moisture content highlighted in Avinash and Mishra (2023), sufficient amount of water (2 L) was added. It was setup to codigest sugarcane waste, poultry manure, and cassava peel. Reactors A and B were bioaugmented with 200 mg/L and 300 mg/L of *Clostridium welchii*, respectively. The biogas produced from each reactor was recorded over time.

2.4 Kinetic Model Fitting and Error Analysis

The biogas data was later fitted to the main equations: First-Order, Modified First-Order, Transfert, Transference and Richards biogas models enumerated in Table 1. This was done to determine their parameters. In those models, the cumulative biogas yield (CBY) is the dependent variable set as a function of retention time (t). In Excel software, the retention time (RT) values were entered into one column and the measured CBY data in the next column. Then, a biogas kinetic model equation such as the First-Order model was selected and set initial guesses for its parameters (e.g., BP and k) in separate cells. Next, a formula was typed in Excel using the model equation to calculate predicted CBY based on the time and parameter values. By dragging through the column, the formula was copied down for each time point and the predicted CBY was determined.

To estimate the best-fit parameters, the Solver Tool in Excel was employed. Subsequently, the objective was set to minimize the Sum of Squared Errors (SSEs) by changing the parameter cells (BP and k). After running Solver, Excel returned the optimal parameters that gave the best fit between measured and predicted CBY values. The steps were repeated for each model and for the other biogas setup, i.e., "B," in order to determine their parameters in a separate sheet.

For each A and B and biogas model measured (CBY_i^{meas}) and predicted CBY data (CBY_i^{calc}), 10 error metrics were determined using expressions in Table 2. An Excel table of values with two columns was first prepared: one for the experimental (measured) CBY and the other for the model-predicted CBY values. Additional columns were used to compute the differences between measured and predicted values. Initially, ERRSQ was calculated by subtracting predicted values from measured values, squaring the result, and summing all the squared differences using a typed function formula. Others were also calculated using several built-in functions such as SUM, AVERAGE, ABS, IF, and SQRT.

It is worthy of note that MRE expresses the average error as a proportion of the actual values, while the mean absolute error (MAE) calculated in Mariam and Safar (2024), gives the average of the absolute differences between predicted and actual values without considering their relative scale. After obtaining the values of

Table 1 Selected biogas kinetic models for bioaugmentation performance assessment (Abubakar et al., 2022a, 2022b)

Model name	Equation	Parameters	No. of model parameters (p)
First-order	$CBY = BP\left(1 - e^{-kt}\right)$	BP and k	2
Modified first-order	$CBY = BP\left[(1 - \beta) - (1 - \beta)e^{-kt}\right]$	BP, β and k	3
Transfert	$CBY = BPe^{-e^{\left[1 - \frac{k.e}{BP}(LP - t)\right]}}$ where, $e = 2.718282$	BP, LP and k	3
Richards	$CBY = BP\left\{1 + SF\,e^{(1+SF)}.e^{\left[\frac{k}{BP}(1+SF)\left(1+\frac{1}{SF}\right)(LP-t)\right]}\right\}^{1/SF}$	BP, SF, LP and k	4
Transference	$CBY = BP\left\{1 - e^{\left[-\frac{k(t-LP)}{BP}\right]}\right\}$	BP, LP and k	3

where BP = maximum biogas production potential (m^3/kg), k = biogas production rate (day^{-1}), β = non-degradable fraction of the substrate, LP = lag phase (day) and SF = shape factor (dimensionless)

all the error functions in Table 2 for reactor A data, a new sheet was created to run the same analysis for reactor B data. Excel is easy to use, does not require programming knowledge, and allows for quick visualization of results. It supports real-time updates, where changing one value automatically updates the entire model. With built-in Solver, complex models can be fitted by non-specialists without coding. Graphs were instantly generated to assess the quality of fit.

3 Results and Discussion

3.1 Derivations from Feedstock Characterization

Oleszek and Krzemińska (2021) reported a shift at 1620 cm^{-1}, attributable to protein aggregation, based on FTIR analysis. Obviously, the key functional groups in the current study feedstock are shown in Fig. 3a. Namely, the peak at 3278.85 cm^{-1} indicates O–H stretching vibrations, common in cellulose, hemicellulose, and lignin; 2919.07 cm^{-1} peak corresponds to C–H stretching, often found in aliphatic structures of polysaccharides and proteins; 2278.83 cm^{-1} band suggests the presence of nitriles or isocyanate groups, which may be from protein decomposition or manure content; 1415.23 cm^{-1} peak signifies bending vibrations of $–CH_2$ or carboxylate

Table 2 List of error measurement models

S/No	Error metric	Expression
1	Sum square of errors (ERRSQ)	$ERRSQ = \sum_{i=1}^{n}(CBY_i^{calc} - CBY_i^{meas})^2$
2	Hybrid fractional error function (HYBRID)	$HYBRID = \frac{100}{n-p}\sum_{i=1}^{n}\left[\frac{(CBY_i^{meas}-CBY_i^{calc})^2}{CBY_i^{meas}}\right]$
3	Average relative error (ARE)	$ARE = \frac{100}{n}\sum_{i=1}^{n}\left[\frac{CBY_i^{calc}-CBY_i^{meas}}{CBY_i^{meas}}\right]$
4	Marquardt's percent standard deviation (MPSD)	$MPSD = \sqrt{\frac{1}{n-p}\sum_{i=1}^{n}\left[\frac{(CBY_i^{meas}-CBY_i^{calc})}{CBY_i^{meas}}\right]^2}$
5	Sum of absolute errors (EABS)	$EABS = \sum_{i=1}^{n}(CBY_i^{meas} - CBY_i^{calc})$
6	Sum of normalized errors (SNE)	$SNE = \sum_{i=1}^{n}\left\lvert\frac{CBY_i^{meas}-CBY_i^{calc}}{CBY_i^{meas}}\right\rvert$
7	Mean relative error (MRE)	$MRE = \frac{1}{n}\sum_{i=1}^{n}\left(\frac{CBY_i^{meas}-CBY_i^{calc}}{CBY_i^{meas}}\right)$
8	Standard deviation of relative errors (SRE)	$SRE = \sqrt{\frac{1}{n}\sum_{i=1}^{n}\left(\frac{CBY_i^{meas}-CBY_i^{calc}}{CBY_i^{meas}} - MRE\right)^2}$
9	Nonlinear chi-square test (χ^2)	$\chi^2 = \sum_{i=1}^{n}\frac{(CBY_i^{calc}-CBY_i^{meas})^2}{CBY_i^{meas}}$
10	Coefficient of nondetermination (COND)	$COND = 1 - R^2$

where n = number of points in the data set (precisely = 21) and p = the number of model parameters used in the calculation (i.e., degrees of freedom used in the model)

groups, indicating degraded lignin or organic acids, and; 1027.05 cm^{-1} band corresponds to C–O stretching in alcohols and polysaccharides, confirming the presence of hemicellulose and cellulose. Cellulose and hemicellulose are fermentable by *Clostridium welchii*, a hydrolytic and acidogenic bacterium that breaks them down into volatile fatty acids (Salsali et al., 2008). Lignin, though less degradable, may still be partially decomposed to facilitate long-term digestion. As a result, the presence of the aforementioned functional groups observed in Fig. 3a, ensures a steady supply of fermentable substrates, capable of boosting biogas yield during hydrolysis and acidogenesis phases. In Fig. 3b, the EDXRF spectrum (intensity versus energy) revealed the presence of essential elements such as calcium (Ca), magnesium (Mg), potassium (K), phosphorus (P), iron (Fe), and possibly trace heavy metals. Similarly, an EDXRF of liquid manure reveals the presence of P, K, Mg, Ca, sulfur (S), manganese (Mn), Fe, copper (Cu), and zinc (Zn) (Horf et al., 2021). Elements such as Ca, Mg, K, and P are beneficial—they act as micronutrients that support microbial enzyme activity

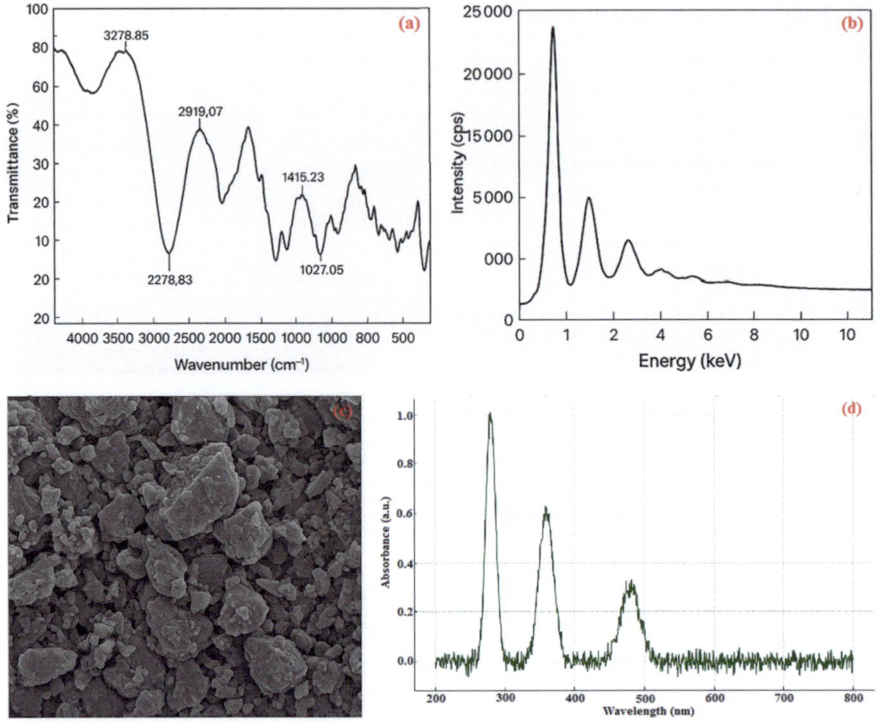

Fig. 3 **a** FTIR, **b** EDXRF, **c** SEM, and (d) UV spectroscopy of the feedstock

and buffer system pH during anaerobic digestion. Due to the absence of inhibitory elements allows *Clostridium welchii* and methanogens to function optimally.

It is observed that the SEM micrograph showed a rough, porous, and heterogeneous surface composed of fragmented biomass particles. The varied texture, with cracks, voids, and granules, increases the surface area available for microbial colonization (Onu et al., 2022). Figure 3c shows the morphological structure of the combined feedstock that makes it easier for *Clostridium welchii* to adhere to the surface and hydrolyze the organic matter. The roughness enhances substrate-microbe interaction and promotes efficient enzymatic breakdown. Thus, the structural morphology seen under SEM supports the effective penetration of enzymes and accelerates the hydrolysis stage during biogas production (Ormaechea et al., 2018). In Fig. 3d, the UV-Vis spectrum exhibited absorbance peaks primarily between 270 and 480 nm. Notable peaks occurred around 280 nm (associated with phenolic compounds or proteins) and 360–480 nm (suggesting the presence of lignin derivatives and humic substances), in close agreement with findings of Waltham and Örmeci (2020). Generally, the peaks confirmed the presence of aromatic compounds and organic acids, which influence the chemical pathways during digestion. Such compounds serve as precursors in microbial fermentation and may enhance acidogenesis and acetogenesis stages. However, in very high concentrations, some aromatics can be toxic to methanogens. In this feedstock, the concentration appears balanced.

3.2 Daily Biogas Production

The gradual increase in CBY over time in both datasets (A and B) can be partially attributed to the enzymatic activity of *Clostridium welchii*, which accelerates the conversion of organic substrates during the early and mid-phases of digestion. Its activity could also explain the steady biogas production observed between day 1 and day 40, as the bacterium supports continuous breakdown and substrate availability for methanogens. Table 3 show that, before a full 24 h, a minimum of 0.001 m^3 of biogas per kilogram was achieved in both biogas production setups.

Table 3 Daily record of biogas production rate

t (day)	CBY (m^3/kg)	
	A	B
1	0.001	0.001
3	0.008696	0.021739
5	0.052174	0.086957
7	0.16087	0.173913
9	0.28	0.265217
11	0.408261	0.373913
13	0.536522	0.484348
15	0.664783	0.601739
17	0.816957	0.71913
19	0.97087	0.84
21	1.124783	0.961739
23	1.278696	1.083478
25	1.422174	1.215652
27	1.558696	1.346087
29	1.693478	1.467826
31	1.814348	1.58087
33	1.901304	1.65913
35	1.975217	1.726087
37	2.04913	1.793043
39	2.097826	1.836522
40	2.13	1.853913

Between the setups, Setup A performed better, producing 2.13 m^3/kg compared to 1.85 m^3/kg from Setup B.

3.3 Implications of the Biogas Kinetic Parameters Obtained

Alongside the kinetic parameters, the coefficient of determination (COD) or R^2 is displayed for dataset A and B in Table 4. BP represents the maximum biogas that can be generated from the feedstock. The values differ widely across the models and setups. Some models (First-Order and Modified First-Order) give very high BP values, such as 677.38 m^3/kg for Setup A and 575.13 m^3/kg for Setup B, while Richards and Transference models estimate very small values such as 0.51 m^3/kg. This wide difference suggests that some models may be overestimating or underestimating due to poor fitting. When compared to the actual maximum biogas produced at day 40—which is 2.13 m^3/kg for Setup A and 1.85 m^3/kg for Setup B, it becomes clear that the BP values from First-Order and Modified First-Order are unrealistic and do not reflect experimental values, indicating overprediction. In contrast, Richards align more closely with the experimental data and is possibly the most reliable estimation. Transference model showed the highest k values (0.09 for A and 0.1 for B), suggesting the fastest biogas production. However, Transfert model have $k = 0$, which implies no change over time, meaning the model failed to capture the dynamic behavior of digestion. A zero-rate constant makes the model unreliable because it assumes that biogas production is static, which contradicts the increasing trend seen in the actual data (Lower et al., 2025). Thus, models with $k = 0$ in Table 4 are not good choices for representing real digestion kinetics. Values for β were very small (0.000158 and 0.0001596), which suggests that almost all the feedstock is degradable, and is a reasonable result for organic wastes used herein (i.e., cassava peel, sugarcane bagasse, and poultry manure). This estimate makes sense, especially considering the good digestibility of the mixture. Both First-Order and Modified First-Order models gave similar trends in k and high BP values, which suggests these models are mathematically similar, but the Modified First-Order includes a correction for non-degradable material via β. Their k values are lower compared to 0.1123 day^{-1} reported for First-Order AD of corn stover by Shitophyta et al. (2023).

LP represents the delay before biogas production starts. In some models, LP is 0, indicating no lag phase (Akbay et al., 2021); in others, it is less than 1 day or slightly above, such as LP = 4.99 days in Setup A (Transfert Model). However, these values are not very reliable because the R^2 values for these models are very low, especially for Transfert and Richards models (near 0 or below 0.4). Hence, LP estimates may not reflect actual process behavior. LP in Transfert model in setup A and B is 65.18 and 68.23% less than the value reported by Abubakar et al., (2022a, 2022b). Still, conceptually, an LP > 1 day could indicate a delay in microbial adaptation, while LP = 0 could point to immediate microbial activity due to bioaugmentation with *Clostridium welchii*. But during co-digestion of cow manure and food waste,

Table 4 Estimated biogas kinetic parameters

Model	Parameter	Setup A	Setup B
First-order	BP (m^3/kg)	677.3788	197.6732
	k (day^{-1})	0.000048	0.000164
	R^2	0.7081	0.7477
Modified first-order	BP (m^3/kg)	606.2642	575.1316
	β	0.000158	0.0001596
	k (day^{-1})	0.000054	0.000055
	R^2	0.7081	0.7480
Transfert	BP (m^3/kg)	7.7639	7.5114
	LP (day)	4.9936	4.5553
	k (day^{-1})	0	0
	R^2	0	0.00034335
Richards	BP (m^3/kg)	0.5123	0.4649
	SF	85.8263	85.1901
	LP (day)	0	0
	k (day^{-1})	0.6164	0.6
	R^2	-1.0125×10^{-9}	0.000343
Transference	BP (m^3/kg)	10	10.8
	LP (day)	0.05	0.0567
	k (day^{-1})	0.09	0.1
	R^2	0.4708	0.3871

Mohammadianroshanfekr et al. (2024) reported an LP < 1 in the range of 0.42–0.91 day, which still greater than those obtained in this study from Transference model. In consonance with the study herein, an LP $= 0.05$ day from Transference model have been reported by Li et al. (2012) after pretreating grass using microwave. High SF (85.8 & 85.2) and poor R^2 suggest the model is overfitted or too rigid for the dataset. Based on the fit quality and reliability of estimates, the models can be ranked as follows: First-Order Model (best overall with good R^2 and low error values) > Modified First-Order Model (similar trend, but slightly higher errors) > Transference Model (good k value, moderate reliability) > Richards Model (poor R^2 and uncertain parameter fit) > Transfert Model (worst, with zero rate and extremely poor fit). Interestingly, Setup B had better model fitting statistics (lower errors, slightly higher R^2). The amount of *Clostridium welchii* bioaugmented played a significant role. Setup B had 300 mg/L, while Setup A had 200 mg/L, and the increased dose may have helped achieve more controlled fermentation, even if the total biogas yield was slightly lower. Richards model gave vital predictions in Zhang et al. (2021), but the worst one herein.

3.4 Calculated Error Metrics and Fit Plots

Each model estimates CBY based on the corresponding kinetic model's parameters determined. Experimentally, Case A reaches a CBY of 2.13 m^3/kg at 40 days, whereas Case B reaches a slightly lower CBY value of 1.853913 m^3/kg at the same time. Thus, this Case A reactor setup performed better in terms of biogas production. First-Order and Modified First-Order Models steadily approach the maximum CBY but slightly underpredict early values and overpredict later ones, especially in Case A. But the Modified First-Order model fits better than First-Order, particularly at higher retention times. Calculated CBY at the specified span of time are constant across all time points (0.5123 m^3/kg for A and 0.4957 m^3/kg for B), implying either: the model assumes an asymptotic value very early or parameter fitting resulted in poor flexibility or over-simplification. In that case, both models are less suitable in this context if CBY is clearly time-varying. On the other hand, Transference model tracks the trend quite closely, increasing as RT increases. It offers the best fit visually, especially at higher CBY ranges for both A and B.

Since almost all the model performed poorly, it is important to pinpoint possible causative issues. In subsequent studies, it is preferable/recommended to examine the experimental conditions (such as temperature, pH, or catalyst dosage), as inconsistencies can lead to noisy data that misaligns with theoretical predictions. Applying data smoothing or transformation techniques might help reduce such noise. Often, Levenberg–Marquardt optimization algorithm is used to refine the parameter fitting (Huang et al., 2023). In addition, combining models in a hybrid or semi-empirical form might capture complex behaviors more effectively than a single model alone. Nonlinear regression is incomplete when error metrics or statistical analysis is not conducted (Karichappan et al., 2014). Despite the increasing trend of the predicted CBY in Transference model, corresponding to the behavior of the measured CBY, the model R^2 values of 0.4708 and 0.3871 for "A" and "B" are lower than those of the First-Order model. First-Order Model consistently shows the lowest error values across most metrics (ERRSQ, HYBRID, ARE, MPSD, EABS, χ^2, etc.) in both Setup A and B. Arguably, it is the most accurate model for predicting CBY under the conditions tested. The Modified First-Order Model yields extremely high error values, especially for HYBRID, ARE, SNE, and χ^2. It is consistent with severe overfitting or instability, likely due to poor parameter estimation or inappropriate model structure for the dataset. Transfert and Transference Models show moderate to high errors. While not as poor as the Modified First-Order model, they still fail to match the accuracy of the First-Order model. Notably, the Transfert model has infinite (∞) χ^2 values as shown in Table 5, suggesting model breakdown or non-convergence during fitting. Lastly, Richards Model performs better than the Transfert and Transference models, but still worse than the First-Order model. It could offer more flexibility in curve shape but could be too complex for this dataset, especially if the digestion process follows simpler first-order kinetics. Table 6 is the key to these error metrics. Setup B consistently shows lower errors than Setup A across all models, suggesting better experimental conditions or more stable digestion performance.

Table 5 Values of the errors determined from each setup dataset

Error metric	Setup "A"					Setup "B"				
	First-order	Modified first-order	Transfert	Richards	Transference	First-order	Modified first-order	Transfert	Richards	Transference
ERRSQ	6.285	7,321,026	36.779	18.49	912.899	3.019	6,590,339	27.867	13.486	1218.479
HYBRID	585.97	2.64×10^{10}	105.26	17,018.22	410,105.15	75.37	4.03×10^{9}	105.26	2186.77	109,780.51
ARE	93.53	480,728.4	100	383.51	2419.46	46.42	240,463.4	100	175.49	1801.87
MPSD	24.21	162,537	10.26	130.45	640.39	8.68	63,520.9	10.26	46.76	331.33
EABS	9.238	12,100.43	22.945	15.97	131.753	6.352	11,480.71	20.091	13.569	152.314
SNE	0.601	700,202.6	3.518	1.768	87.312	0.393	857,667.4	3.627	1.755	158.573
MRE	−0.344	− 4,807.28	1	− 2.962	− 24.195	− 0.010	−2,404.63	1	− 0.904	− 18.019
SRE	2.334	15,095.15	0	12.365	57.538	0.846	5705.2	0	4.467	26.8
χ^2	6.437	12,077.54	∞	36.117	112.792	3.093	11,460.66	∞	29.029	134.968
COND	0.2919	0.2919	1	1	0.5292	0.2523	0.252	0.9997	0.9997	0.6129

Table 6 Predicted CBY from the selected biogas kinetic model

		First-order	Modified-first order	Transfert	Richards	Transference
		A	A	A	A	A
RT (day)	CBY (Meas.) (m^3/kg)	CBY (Calc.) (m^3/kg)	CBY (Calc.) (m^3/kg)	CBY (Calc.) (m^3/kg)	CBY (Calc.) (m^3/kg)	CBY (Calc.) (m^3/kg)
1	0.001	0.032611574	0.032700111	0.512322246	0.512322177	0.085135527
3	0.008696	0.097830011	0.098095041	0.512322246	0.512322173	0.262006473
5	0.052174	0.163042169	0.163482916	0.512322246	0.512322173	0.435722225
7	0.16087	0.228248048	0.228863737	0.512322246	0.512322173	0.606339067
9	0.28	0.293447648	0.294237503	0.512322246	0.512322173	0.77391228
11	0.408261	0.358640971	0.359604217	0.512322246	0.512322173	0.93849616
13	0.536522	0.423828016	0.424963878	0.512322246	0.512322173	1.100144034
15	0.664783	0.489008786	0.490316487	0.512322246	0.512322173	1.258908277
17	0.816957	0.554183279	0.555662046	0.512322246	0.512322173	1.414840329
19	0.97087	0.619351496	0.621000555	0.512322246	0.512322173	1.567990715
21	1.124783	0.684513439	0.686332015	0.512322246	0.512322173	1.718409056
23	1.278696	0.749669108	0.751656426	0.512322246	0.512322173	1.866144088
25	1.422174	0.814818503	0.816973789	0.512322246	0.512322173	2.011243681
27	1.558696	0.879961626	0.882284105	0.512322246	0.512322173	2.153754846
29	1.693478	0.945098476	0.947587375	0.512322246	0.512322173	2.29372376
31	1.814348	1.010229054	1.0128836	0.512322246	0.512322173	2.431195772
33	1.901304	1.075353362	1.07817278	0.512322246	0.512322173	2.566215426
35	1.975217	1.140471398	1.143454916	0.512322246	0.512322173	2.698826468
37	2.04913	1.205583165	1.208730009	0.512322246	0.512322173	2.829071867
39	2.097826	1.270688663	1.273998059	0.512322246	0.512322173	2.956993822
40	2.13	1.303239061	1.306629444	0.512322246	0.512322173	3.020096489
		First-order	Modified-first order	Transfert	Richards	Transference
		B	B	B	B	B
RT (day)	CBY (Meas.) (m^3/kg)	CBY (Calc.) (m^3/kg)	CBY (Calc.) (m^3/kg)	CBY (Calc.) (m^3/kg)	CBY (Calc.) (m^3/kg)	CBY (Calc.) (^{m3}/kg)
1	0.001	0.032611574	0.031520599	0.495659389	0.495659394	0.093919245
3	0.021739	0.097830011	0.094556615	0.495659389	0.495659394	0.290355532
5	0.086957	0.163042169	0.15758572	0.495659389	0.495659394	0.483187585
7	0.173913	0.228248048	0.220607915	0.495659389	0.495659394	0.672481535
9	0.265217	0.293447648	0.283623202	0.495659389	0.495659394	0.858302301
11	0.373913	0.358640971	0.34663158	0.495659389	0.495659394	1.040713607

(continued)

Table 6 (continued)

		First-order	Modified-first order	Transfert	Richards	Transference
		B	B	B	B	B
RT (day)	CBY (Meas.) (m^3/kg)	CBY (Calc.) (m^3/kg)	CBY (Calc.) (m^3/kg)	CBY (Calc.) (m^3/kg)	CBY (Calc.) (m^3/kg)	CBY (Calc.) (^{m3}/kg)
13	0.484348	0.423828016	0.409633051	0.495659389	0.495659394	1.219778012
15	0.601739	0.489008786	0.472627615	0.495659389	0.495659394	1.395556924
17	0.71913	0.554183279	0.535615274	0.495659389	0.495659394	1.568110627
19	0.84	0.619351496	0.598596027	0.495659389	0.495659394	1.737498296
21	0.961739	0.684513439	0.661569876	0.495659389	0.495659394	1.903778023
23	1.083478	0.749669108	0.724536821	0.495659389	0.495659394	2.067006832
25	1.215652	0.814818503	0.787496864	0.495659389	0.495659394	2.227240701
27	1.346087	0.879961626	0.850450005	0.495659389	0.495659394	2.384534583
29	1.467826	0.945098476	0.913396244	0.495659389	0.495659394	2.53894242
31	1.58087	1.010229054	0.976335583	0.495659389	0.495659394	2.690517167
33	1.65913	1.075353362	1.039268021	0.495659389	0.495659394	2.839310804
35	1.726087	1.140471398	1.102193561	0.495659389	0.495659394	2.985374359
37	1.793043	1.205583165	1.165112203	0.495659389	0.495659394	3.128757926
39	1.836522	1.270688663	1.228023947	0.495659389	0.495659394	3.269510676
40	1.853913	1.303239061	1.259477232	0.495659389	0.495659394	3.338915614

In Table 6 and Fig. 5a, the predicted CBY values from the First-Order model in Setup B closely follow the experimental data. Because it captures the biogas production trend without large deviations or overestimation. First-order kinetics assumes a simple exponential approach to maximum biogas yield, which is often suitable for anaerobic digestion when a single dominant microbial pathway is present. The poor performance of some models, as revealed by Figs. 4 and 5, may result from inaccurate initial guesses or non-optimized fitting routines. It could be a severe case of error made; the type explained in Parajuli (2011) and Walker et al. (2009). Using global optimization techniques (e.g., genetic algorithms, particle swarm) will possibly yield better parameter sets. Scaling or transforming the data may reduce fitting errors, especially for models sensitive to magnitudes. Also, incorporating the activity of *Clostridium welchii* into the models will help capture real biochemical behavior, especially during the lag and exponential phases of digestion. Again, cross-validation or test on an external dataset ensures that the model isn't overfitting the training data (Yildirim & Ozkaya, 2023). Ten fit plots are presented in Figs. 4 and 5 for the setups and biogas models. Red trending line is indicative of the predicted or calculated CBY behavior and the black is the measured or experimentally obtained CBY data. Transfert and Richards predicted CBY cases for A and B reactor data are constant at 0.5123 and 0.4957 m^3/kg, respectively, as shown in Figs. 4c–d and 5c–d. Trending

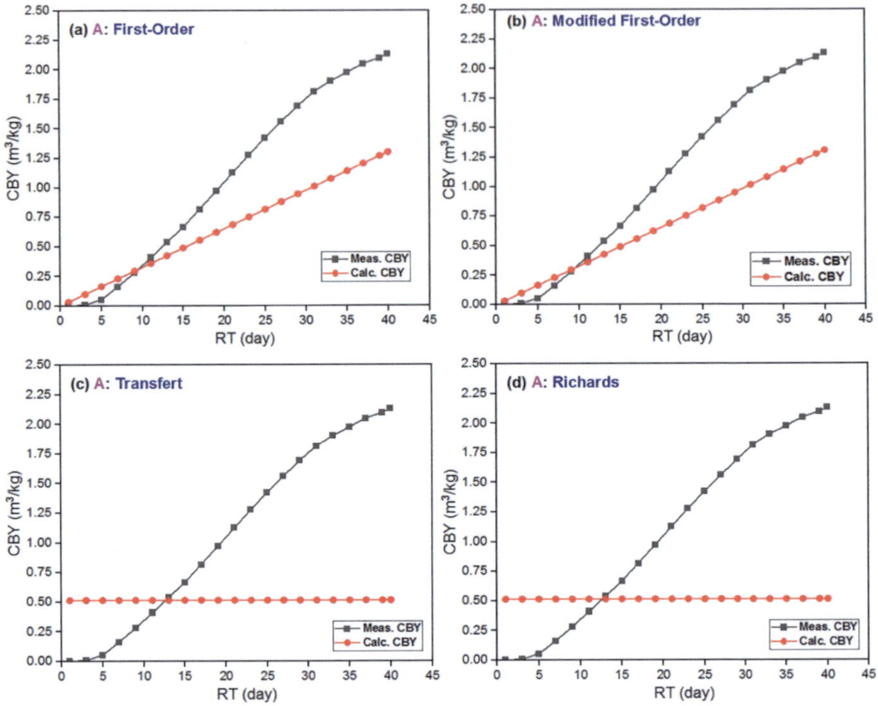

Fig. 4 Fitted plots from setup "A" CBY data based on **a** First-order, **b** Modified first-order, **c** Transfert, and **d** Richards Biogas kinetic models

line for computed CBY for First-Order and Modified First-Order kinetic model in Figs. 4a–b and 5a–b is linear, which is not as expected. The extent of fit has already been ascertain using the error estimates conducted earlier.

The SF value in the Richards model was very high (> 85), which may indicate sharp transitions in biogas production rate over time. However, the R^2 value for this model is also very low, meaning the curve fit is poor. So, although the SF gives a theoretical shape to the growth curve, its reliability in this study is questionable due to poor model performance. Among the tested models, the Transference kinetic model demonstrated the best agreement with experimental data for both Case A and Case B, as it closely followed the increasing trend of CBY with retention time (i.e., Fig. 6). While the Modified First-Order model provided a reasonable fit, particularly at mid to high RT values, the Transfert and Richards models failed to capture the dynamic behavior of CBY, likely due to their asymptotic assumptions. Quantitative error metrics further confirmed the potentials of the Transference model in describing the cumulative biogas production kinetics. The predicted data curve in Fig. 6 with less curvature in its path, is almost similar to fit plots showcased in Nweke et al. (2022).

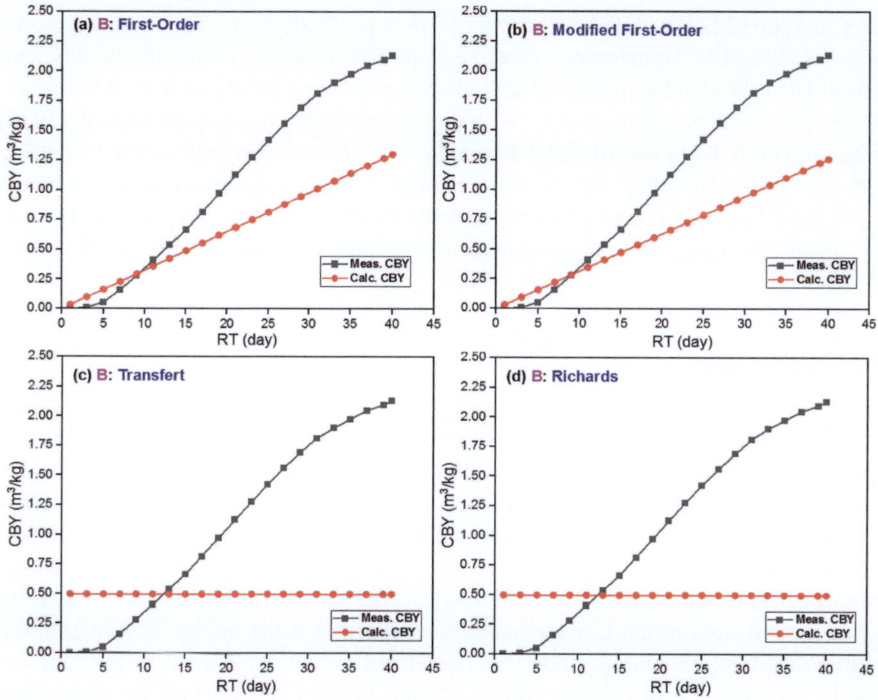

Fig. 5 Fitted plots from setup "B" CBY data based on **a** First-order, **b** Modified first-order, **c** Transfert, and **d** Richards Biogas kinetic Models

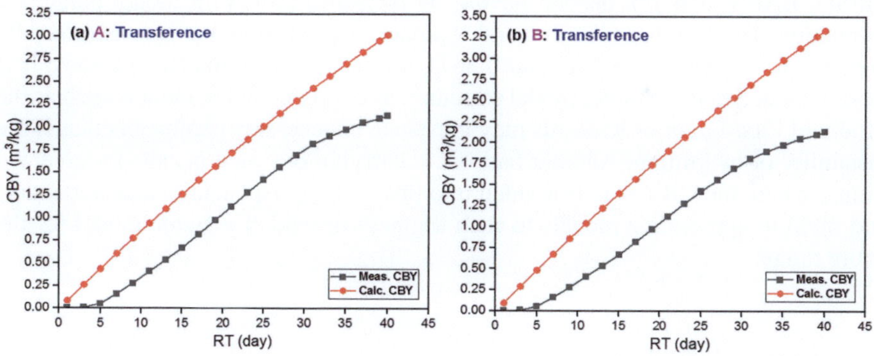

Fig. 6 Correlated plots based on transference model for **a** setup "A" and **b** setup "B" CBY data

To improve the fit, we recommend the use of global optimization tools such as genetic algorithms (Ibro et al., 2024), ANFIS (Yang et al., 2021), and artificial neural networks (ANNs) (Onu et al., 2022) in Excel Solver, MATLAB employed by Avinash and Mishra (2024), or Python software that could yield better fitting parameters than basic local optimization (Iweka et al., 2021). Without estimating the

LP, Clarkson (2023) fitted his experimental data perfectly to the Transference model ($R^2 = 0.999$). The Transference model assumes that biogas production begins with a short LP, followed by an exponential growth phase and finally reaches a stationary phase. But an LP in the range of 6.2–7.6 days reported by Manu and Clarkson (2022) is way too high. It assumes that the digestion rate is governed by microbial adaptation and substrate availability. The model also implies that the process is monotonic and asymptotic, meaning biogas production increases over time and then plateaus. Herein, it is almost the case, but analytical inaccuracies are presumed to be the fault.

4 Conclusion

EDXRF, FTIR, SEM, and UV spectroscopy of the blend (i.e., poultry manure, cassava peels, and sugarcane bagasse) helped ascertain the possibility or quality of the feed for biogas production. UV reveals the presence of phenolic compounds, SEM describe it as having a porous structure, FTIR fetched the cellulose, hemicellulose and lignin, polysaccharide and protein content of the feedstock, while EDXRF show that it contains Mg, K, Ca, Fe and P which will allow the growth of the AD microorganism called *Clostridium welchii*. Since anaerobic digestion is the microbial fermentation approach employed, kinetic study was conducted, in which none of the BP estimate was in agreement with the experimental value of 2.13 and 1.85 m^3/kg from setup A and B, respectively. Bioaugmenting the digesters A and B with different levels of *Clostridium welchii* proved significant. The best model is the First-Order kinetic model having lower error estimates (ERRSQ, HYBRID, ARE, COND, SRE, MRE, MPSD, EABS, and χ^2), despite having its own discrepancy in model parameter estimation. Transference model fit and estimates is promising, but is constrained by poor fit ($R^2 = 0.4708$ for "A" and 0.3871 for "B"). For predictive accuracy, it is recommended to use simpler model structure, as overly complex models such as the Modified First-Order or Richards may introduce unnecessary parameters that cause instability or overfitting. Another bail-out is carrying out what is called segmented fitting, where the CBY data is divided into phases (lag, exponential and stationary) and fitted to appropriate models to each segment instead of a single model for the entire range.

References

Abubakar, A. M., Umdagas, L. B., Waziri, A. Y., & Itamah, E. I. (2022b). Estimation of biogas potential of liquid manure from kinetic models at different temperature. *International Journal of Scientific Research in Computer Science and Engineering (IJSRCSE), 10*(2), 46–63. https://doi.org/10.5281/zenodo.6835863

Abubakar, A. M., Silas, K., Aji, M. M., Taura, U. H., & Undiandeye, J. (2022). Biogas production from chicken manure: Characterization and kinetic models. *Bayero Journal of Engineering Technology (BJET), 17*(3), 1–13. https://bjet.ng/index.php/jet/article/view/123

Adetunji, C. O., Olaniyan, O. T., & Varma, A. (2022). Novel microorganisms involved in the production of sustainable biogas production. In Mahajan, S., & Varma, A. (Eds.), Animal Manure (pp. 123–130). Springer, Cham. https://doi.org/10.1007/978-3-030-97291-2_7

Akbay, H. E. G., Dizge, N., & Kumbur, H. (2021). Enhancing biogas production of anaerobic co-digestion of industrial waste and municipal sewage sludge with mechanical, chemical, thermal, and hybrid pretreatment. *Bioresource Technology, 340*(125688). https://doi.org/10.1016/j.biortech.2021.125688

Andrighetto, C., Spolaor, D., De Dea, P., Fioravanzo, E., Brazzale, P., De Battista, M., Dolci, E., Maccà, F., Sperotto, D., & Lombardi, A. (2023). Comparison of two analytical methods for detecting Clostridium spores in milk. *International Dairy Journal, 142*(105649). https://doi.org/10.1016/j.idairyj.2023.105649

Arora, P., Kshirsagar, P. R., Rana, D. P., & Dhakephalkar, P. K. (2019). Hyperthermophilic Clostridium sp. N-4 produced a glycoprotein biosurfactant that enhanced recovery of residual oil at 96 °C in lab studies. *Colloids and Surfaces B: Biointerfaces, 182*(110372). https://doi.org/10.1016/j.colsurfb.2019.110372

Avinash, L. S., & Mishra, A. (2023). Enhancing biogas production in anaerobic digestion of MSW with addition of bio-solids and various moisture sources. *Fuel, 354*(129414). https://doi.org/10.1016/j.fuel.2023.129414

Avinash, L. S., & Mishra, A. (2024). Comparative evaluation of Artificial intelligence based models and kinetic studies in the prediction of biogas from anaerobic digestion of MSW. *Fuel, 367*(131545), 1–15. https://doi.org/10.1016/j.fuel.2024.131545

Bagge, E., Persson, M., & Johansson, K.-E. (2010). Diversity of spore-forming bacteria in cattle manure, slaughterhouse waste and samples from biogas plants G. *Journal of Applied Microbiology, 109*(5), 1549–1565. https://doi.org/10.1111/j.1365-2672.2010.04790.x

Bagge, E., Lewerin, S. S., & Johansson, K.-E. (2009). Detection and identification by PCR of Clostridium chauvoei in clinical isolates, bovine faeces and substrates from biogas plant. *Acta Veterinaria Scandinavica, 51*(8). https://doi.org/10.1186/1751-0147-51-8

Banerjee, A., Leang, C., Ueki, T., Nevin, K. P., & Lovley, D. R. (2014). Lactose-inducible system for metabolic engineering of Clostridium ijungdahlii. *Applied and Environmental Microbiology, 80*(8), 2410–2416. https://doi.org/10.1128/AEM.03666-13

BIOHAZ. (2009). Statement on technical assistance related to the EFSA opinion on transformation of animal by-products into biogas and compost. *EFSA Journal, 7*(11), 1–8. https://doi.org/10.2903/j.efsa.2009.1370

Christy, P. M., Gopinath, L. R., & Divya, D. (2014). A review on anaerobic decomposition and enhancement of biogas production through enzymes and microorganisms. *Renewable and Sustainable Energy Reviews, 34*, 167–173. https://doi.org/10.1016/j.rser.2014.03.010

Clarkson, M. A. (2023). A simulation of biogas production from lignocellulosic biomass co-digested with agricultural waste under the influence bentonite catalyst. *African Journal of Environmental Sciences & Renewable Energy, 12*(1), 87–96. https://publications.afropolitanjournals.com/index.php/ajesre/article/view/629/457

Delmon, C., Ouk, C., Casellas, M., & Prorot, A. (2023). Evaluation of the additive effects of volatile fatty acids and moderate heat treatment for enhancing the inactivation of vegetative cells and spores of Clostridium perfringens by flow cytometry. *Anaerobe, 84*(102802). https://doi.org/10.1016/j.anaerobe.2023.102802

Derongs, L., Druilhe, C., Ziebal, C., Maréchal, C. Le, & Pourcher, A.-M. (2020). Characterization of Clostridium perfringens isolates collected from three agricultural biogas plants over a one-year period. *International Journal of Environmental Research and Public Health, 17*(15). https://doi.org/10.3390/ijerph17155450

Dohrmann, A. B., Walz, M., Löwen, A., & Tebbe, C. C. (2015). Clostridium cluster I and their pathogenic members in a full-scale operating biogas plant. *Applied Microbiology and Biotechnology, 99*, 3585–3598. https://doi.org/10.1007/s00253-014-6261-y

Fröschle, B., Messelhäusser, U., Höller, C., & Lebuhn, M. (2015). Fate of Clostridium botulinum and incidence of pathogenic clostridia in biogas processes. *Journal of Applied Microbiology, 119*(4), 936–947. https://doi.org/10.1111/jam.12909

Hashimoto, A., Tsuchioka, H., Higashi, K., Ota, N., & Harada, H. (2016). Distribution of enterotoxin gene-positive Clostridium perfringens spores among human and livestock samples and its potential as a human fecal source tracking indicator. *Journal of Water Environment Technology, 14*(6), 447–454. https://doi.org/10.2965/jwet.16-022

Hauschild, A. H. W., & Hilsheimer, R. (1974). Enumeration of food-borne Clostridium perfringens in egg yolk-free Tryptose-Sulfite-Cycloserine agar. *Applied and Environmental Biology, 27*(3). https://doi.org/10.1128/am.27.3.521-526.1974

Horf, M., Gebbers, R., Vogel, S., Ostermann, M., Piepel, M.-F., & Olfs, H.-W. (2021). Determination of nutrients in liquid manures and biogas digestates by portable energy-dispersive x-ray fluorescence spectrometry. *Sensors, 21*(11). https://doi.org/10.3390/s21113892

Huang, X., Cao, H., & Jia, B. (2023). Optimization of Levenberg Marquardt algorithm applied to nonlinear systems. *Processes, 11*(6). https://doi.org/10.3390/pr11061794

Ibro, M. K., Ancha, V. R., & Lemma, D. B. (2024). Biogas production optimization in the anaerobic codigestion process: A critical review on process parameters modeling and simulation tools. *Journal of Chemistry*. https://doi.org/10.1155/2024/4599371

Iweka, S. C., Owuama, K. C., Chukwuneke, J. L., & Falowo, O. A. (2021). Optimization of biogas yield from anaerobic co-digestion of corn-chaff and cow dung digestate: RSM and python approach. *Heliyon, 7*(11), 1–8. https://doi.org/10.1016/j.heliyon.2021.e08255

Karichappan, T., Venkatachalam, S., & Jeganathan, P. M. (2014). Investigation on biogas production process from chicken processing industry wastewater using statistical analysis: Modelling and optimization. *Journal of Renewable and Sustainable Energy, 6*(4), 1–9. https://doi.org/10.1063/1.4892604

Kirawa, L. S., Gitau, A. N., & Gatari, M. J. (2020). Comparative evaluation of the chemical composition of biogas digestate slurries under varying feedstocks. *Journal of Engineering in Agriculture and the Environment, 6*(1), 83–97. https://doi.org/10.37017/jeae-volume6-no1.2020-6

Klein, M., Brown, L., Ashbolt, N. J., Stuetz, R. M., & Roser, D. J. (2011). Inactivation of indicators and pathogens in cattle feedlot manures and compost as determined by molecular and culture assays. *FEMS Microbiology Ecology, 77*(1), 200–210. https://doi.org/10.1111/j.1574-6941.2011.01098.x

Lebuhn, M., & Gronauer, A. (2009). Microorganisms in the biogasprocess—the unknown beings. *Landtechnik, 64*(2), 127–130. https://www.landtechnik-online.eu/ojs-2.4.5/index.php

Lee, S. Y., Park, J. H., Jang, S. H., Nielsen, L. K., Kim, J., & Jung, K. S. (2008). Fermentative butanol production by clostridia. *Biotechnology & Bioengineering, 101*(2), 209–228. https://doi.org/10.1002/bit.22003

Li, L., Kong, X., Yang, F., Li, D., Yuan, Z., & Sun, Y. (2012). Biogas production potential and kinetics of microwave and conventional thermal pretreatment of grass. *Applied Biochemistry and Biotechnology, 166*, 1183–1191. https://doi.org/10.1007/s12010-011-9503-9

Lorine, D., Céline, D., Caroline, L. M., Frédéric, B., Lorette, H., Julie, B., Laure, M., Christine, Z., Typhaine, P., Sandra, R., Emmanuelle, H., Rabab, S. Z., Jeanne, C., & Anne-Marie, P. (2021). Influence of operating conditions on the persistence of E. coli, enterococci, Clostridium perfringens and Clostridioides difficile in semi-continuous mesophilic anaerobic reactors. *Waste Management, 134*, 32–41. https://doi.org/10.1016/j.wasman.2021.08.003

Lower, L., Qiu, Y., Sartor, R. C., Sagues, W. J., & Cheng, J. J. (2025). Kinetic modeling of thermophilic anaerobic digestion of Lemnaceae for biogas production. *BioEnergy Research, 18*(23), 1–12. https://doi.org/10.1007/s12155-025-10824-0

Malakahmad, A., Zain, S. M., Basri, N. E. A., Kutty, S. R. M., & Isa, M. H. (2009). Identification of anaerobic microorganisms for converting kitchen waste to biogas. *International Journal of Environmental and Ecological Engineering, 3*(12), 410–413. http://scholar.waset.org/1307-6892/12462

Manu, G., & Clarkson, M. A. (2022). Modelling the effects of dilute alkaline pretreatment of lignocellulosic biomass on biogas production. *African Journal of Environmental Sciences & Renewable Energy, 5*(1), 49–57. https://publications.afropolitanjournals.com/index.php/ajesre/article/view/188

Mariam, A., & Safar, A. B. (2024). Modeling of biogas production of camel and sheep manure using tomato and rumen as co-substrate via kinetic models. *Journal of Ecological Engineering (JEE), 25*(8), 10–23. https://doi.org/10.12911/22998993/189231

Mohammadianroshanfekr, M., Pazoki, M., Pejman, M. B., Ghasemzadeh, R., & Pazoki, A. (2024). Kinetic modeling and optimization of biogas production from food waste and cow manure co-digestion. *Results in Engineering, 24*(103477), 1–11. https://doi.org/10.1016/j.rineng.2024.103477

Neuhaus, J., Schrödl, W., Shehata, A. A., & Krüger, M. (2015a). Detection of Clostridium botulinum in liquid manure and biogas plant wastes. *Folia Microbiologica, 60*, 451–456. https://doi.org/10.1007/s12223-015-0381-3

Neuhaus, J., Shehata, A. A., & Krüger, M. (2015b). Detection of pathogenic clostridia in biogas plant wastes. *Folia Microbiologica, 60*, 15–19. https://doi.org/10.1007/s12223-014-0334-2

Nweke, C. N., Nwabanne, J. T., Onu, C. E., & Ewuzie, N. V. (2022). Biogas production kinetic studies from yam peels. *Journal of Engineering and Applied Sciences (JEAS), 20*(1), 753–766. https://d1wqtxts1xzle7.cloudfront.net/121132381/YP

Oleszek, M., & Krzemińska, I. (2021). Biogas production from high-protein and rigid cell wall microalgal biomasses: Ultrasonication and FT-IR evaluation of pretreatment effects. *Fuel, 296*(120676). https://doi.org/10.1016/j.fuel.2021.120676

Onu, C. E., Nweke, C. N., & Nwabanne, J. T. (2022). Modeling of thermo-chemical pretreatment of yam peel substrate for biogas energy production: RSM, ANN, and ANFIS comparative approach. *Applied Surface Science Advances, 11*(100299), 1–13. https://doi.org/10.1016/j.apsadv.2022.100299

Ormaechea, P., Castrillón, L., Suárez-Peña, B., Megido, L., Fernández-Nava, Y., Negral, L., Marañón, E., & Rodríguez-Iglesias, J. (2018). Enhancement of biogas production from cattle manure pretreated and/or co-digested at pilot-plant scale. Characterization by SEM. *Renewable Energy, 126*, 897–904. https://doi.org/10.1016/j.renene.2018.04.022

Parajuli, P. (2011). Biogas measurement techniques and the associated errors. In: Rintala, J., Kaparaju, P., Jagadabhi, P. S., Koistinen, M., Siitonen, L., Ahonen, J., Karusalmi, A., & Bursiewicz, P. (Eds.), [Masters' Thesis, Department of Biological and Environmental Science, Renewable Energy Programme, University of Jyväskylä]. https://d1wqtxts1xzle7.cloudfront.net/32201768/water_displacement-libre.pdf

Paroha, S., Kachhwaha, A., & Singh, S. (2020). Diversity of micro-flora in different combinations of press-mud for biogas production. *International Journal of Microbiology Research, 12*(4), 1821–1827. https://www.bioinfopublication.org/jouarchive.php?opt=&jouid=BPJ0000234

Popova, T. P., Kirov, V. K., Kaleva, M. D., & Baykov, B. D. (2014). Comparative study of microflora of new energy crops for biogas production. *International Journal of Current Microbiology and Applied Sciences, 3*(11), 539–548. http://www.ijcmas.com/

Pulvirenti, A., Ronga, D., Zaghi, M., Tomasselli, A. R., Mannella, L., & Pecchioni, N. (2015). Pelleting is a successful method to eliminate the presence of Clostridium spp. from the digestate of biogas plants. *Biomass and Bioenergy, 81*, 479–482. https://doi.org/10.1016/j.biombioe.2015.08.008

Rahmani, A. M., Tyagi, V. K., Ahmed, B., Kazmi, A. A., Ojha, C. S. P., & Singh, R. (2022). Critical insights into anaerobic co-digestion of wheat straw with food waste and cattle manure: Synergistic effects on biogas yield and kinetic modeling. *Environmental Research, 212*(113382). https://doi.org/10.1016/j.envres.2022.113382

Ranganathan, T., & Arunodayaraj, I. S. (2024). Disintegration of lignocellulosic material through visible light SiO2/g-C3N4 photocatalyst for biogas generation. *Journal of the Indian Chemical Society, 101*(11). https://doi.org/10.1016/j.jics.2024.101371

Romanazzi, V., Bonetta, S., Fornasero, S., De Ceglia, M., Gilli, G., & Traversi, D. (2016). Assessing Methanobrevibacter smithii and Clostridium difficile as not conventional faecal indicators in effluents of a wastewater treatment plant integrated with sludge anaerobic digestion. *Journal of Environmental Management, 184*, 170–177. https://doi.org/10.1016/j.jenvman.2016.09.081

Sahlström, L., Bagge, E., Emmoth, E., Holmqvist, A., Danielsson-Tham, M.-L., & Albihn, A. (2008). A laboratory study of survival of selected microorganisms after heat treatment of biowaste used in biogas plants. *Bioresource Technology, 99*(16), 7859–7865. https://doi.org/10.1016/j.biortech.2007.09.071

Salsali, H., Parker, W. J., & Sattar, S. A. (2008). The effect of volatile fatty acids on the inactivation of Clostridium perfringens in anaerobic digestion. *World Journal of Microbiology and Biotechnology, 24*, 659–665. https://doi.org/10.1007/s11274-007-9514-4

Shitophyta, L. M., A, A., & Wulansari, H. D. A. (2023). Evaluation and modelling of biogas production from batch anaerobic digestion of corn stover with oxalic acid. *Research in Agricultural Engineering, 69*(3), 151–157. https://doi.org/10.17221/98/2022-RAE

Silveira, V. A. I., Nishio, E. K., Freitas, C. A. U. Q., Amador, I. R., Kobayashi, R. K. T., Caretta, T., Macedo, F., & Celligoi, M. A. P. C. (2019). Production and antimicrobial activity of sophorolipid against Clostridium perfringens and Campylobacter jejuni and their additive interaction with lactic acid. *Biocatalysis and Agricultural Biotechnology, 21*(101287). https://doi.org/10.1016/j.bcab.2019.101287

Skillman, L. C., Bajsa, O., Ho, L., Santhanam, B., Kumar, M., & Ho, G. (2009). Influence of high gas production during thermophilic anaerobic digestion in pilot-scale and lab-scale reactors on survival of the thermotolerant pathogens Clostridium perfringens and Campylobacter jejuni in piggery wastewater. *Water Research, 43*(14), 3281–3291. https://doi.org/10.1016/j.watres.2009.04.031

Syrchina, N. V., Pilip, L. V., Kolevatykh, E. P., & Ashikhmina, T. Y. (2025). Effect of different processing methods on the number of bacteria of the genus Clostridium (Clostridia, bacteria) in animal by-products. *Biology Bulletin, 51*, 3075–3082. https://doi.org/10.1134/S1062359024700845

Thippareddi, H., Juneja, V. K., Phebus, R. K., Marsden, J. L., & Kastner, C. L. (2003). Control of Clostridium perfringens germination and outgrowth by buffered sodium citrate during chilling of roast beef and injected pork. *Journal of Food Protection, 66*(3), 376–381. https://doi.org/10.4315/0362-028X-66.3.376

Uzal, F. A. (2004). Diagnosis of Clostridium perfringens intestinal infections in sheep and goats. *Anaerobe, 10*(2), 135–143. https://doi.org/10.1016/j.anaerobe.2003.08.005

Uzal, F. A., & Songer, J. G. (2008). Diagnosis of Clostridium perfringens intestinal infections in sheep and goats. *Journal of Veterinary Diagnostic Investigation, 20*(3), 253–265. https://doi.org/10.1177/104063870802000301

Walker, M., Zhang, Y., Heaven, S., & Banks, C. (2009). Potential errors in the quantitative evaluation of biogas production in anaerobic digestion processes. *Bioresource Technology, 100*(24), 6339–6346. https://doi.org/10.1016/j.biortech.2009.07.018

Waltham, B., & Örmeci, B. (2020). Fluorescence intensity, conductivity, and UV–vis absorbance as surrogate parameters for real-time monitoring of anaerobic digestion of wastewater sludge. *Journal of Water Process Engineering, 37*(101395), 1–9. https://doi.org/10.1016/j.jwpe.2020.101395

Wang, R., Zong, W., Qian, C., Wei, Y., Yu, R., & Zhou, Z. (2011). Isolation of Clostridium perfringens strain W11 and optimization of its biohydrogen production by genetic modification. *International Journal of Hydrogen Energy, 36*(19), 12159–12167. https://doi.org/10.1016/j.ijhydene.2011.06.105

Wang, H., Ma, S., & Bu, H. (2014). Fermentative hydrogen production by newly isolated Clostridium perfringens ATCC 13124. *Journal of Renewable Sustainable Energy, 6*(013130). https://doi.org/10.1063/1.4863085

Watcharasukarn, M., Kaparaju, P., Steyer, J.-P., Krogfelt, K. A., & Angelidaki, I. (2009). Screening Escherichia coli, Enterococcus faecalis, and Clostridium perfringens as indicator organisms in evaluating pathogen-reducing capacity in biogas plants. *Environmental Microbiology, 58*, 221–230. https://doi.org/10.1007/s00248-009-9497-9

Wong, Y. M., Wu, T. Y., Ling, T. C., Show, P. L., Lee, S. Y., Chang, J.-S., Ibrahim, S., & Juan, J. C. (2018). Evaluating new bio-hydrogen producers: Clostridium perfringens strain JJC, Clostridium bifermentans strain WYM and Clostridium sp. strain Ade.TY. *Journal of Bioscience and Bioengineering, 125*(5), 590–598. https://doi.org/10.1016/j.jbiosc.2017.12.012

Xue, Y., Huang, X., Ou, Z., Wu, Y., Li, Q., Huang, X., Wen, M., Yang, Y., Situ, B., & Zheng, L. (2023). Isolation and purification of bacterial extracellular vesicles from human feces using density gradient centrifugation. *JoVE Journal, 199*(e65574), 1–24. https://doi.org/10.3791/65574

Yang, Y., Zheng, S., Ai, Z., & Jafari, M. M. M. (2021). On the prediction of biogas production from vegetables, fruits, and food wastes by ANFIS- and LSSVM-based models. *BioMed Research International*. https://doi.org/10.1155/2021/9202127

Yildirim, O., & Ozkaya, B. (2023). Prediction of biogas production of industrial scale anaerobic digestion plant by machine learning algorithms. *Chemosphere, 335*(138976). https://doi.org/10.1016/j.chemosphere.2023.138976

Yu, R., Wang, R., Bi, T., Sun, W., & Zhou, Z. (2013). Blocking the butyrate-formation pathway impairs hydrogen production in Clostridium perfringens. *Acta Biochimica Et Biophysica Sinica (ABBS), 45*(5), 408–415. https://doi.org/10.1093/abbs/gmt024

Zhang, H., An, D., Cao, Y., Tian, Y., & He, J. (2021). Modeling the methane production kinetics of anaerobic co-digestion of agricultural wastes using sigmoidal functions. *Energies, 14*(2), 1–12. https://doi.org/10.3390/en14020258

The Role of Green Chemistry in Achieving Sustainability Goals

Ch. Venkateswara Rao, V. L. N. Balaji Gupta Tiruveedhi, Sri Sudewi, Hui-Fen Wu, Myung-Geol Pang, and Gangaraju Gedda

1 Introduction: Green Chemistry as a Driver of Sustainability

1.1 Overview of Green Chemistry and Its 12 Principles

Green chemistry, often referred to as sustainable chemistry, represents a paradigm shift in chemical sciences by promoting environmentally benign practices that minimize hazardous substances and waste. Introduced by Paul T. Anastas and John C. Warner in 1998, the 12 principles of green chemistry serve as a blueprint for designing

Ch. Venkateswara Rao
Department of Basic Sciences and Humanities, Swarnandhra College of Engineering and Technology, Seetharamapuram, Andhra Pradesh, India

V. L. N. B. G. Tiruveedhi
Department of Basic Science, Vishnu Institute of Technology, Bhimavaram, Andhra Pradesh, India

S. Sudewi
Department of Pharmacy, Faculty of Mathematics and Natural Science, Universitas Sam Ratulangi, Manado, Indonesia

H.-F. Wu
Department of Chemistry, National Sun Yat-Sen University, Kaohsiung, Kaohsiung, Taiwan

M.-G. Pang · G. Gedda (✉)
Department of Animal Science and Technology and BET Research Institute, Chung-Ang University, Anseong, Gyeonggi-Do, Republic of Korea
e-mail: raju.analy@gmail.com

G. Gedda
Central Research Laboratory, NITTE (Deemed to Be University), K S Hegde Medical Academy, Deralakatte, Mangaluru, Karnataka, India

S. Ghosh et al. (eds.), *Advances in Organic Waste Conversion Through Industrial Biotechnology*, Sustainable Environmental Waste Management Strategies,
https://doi.org/10.1007/978-3-032-11377-1_19

safer, more sustainable chemical processes (Anastas and Warner, 2000). These principles emphasize atom economy, waste prevention, and the use of non-toxic reagents, aiming to reduce the environmental footprint of industrial and laboratory practices.

The 12 principles of green chemistry are as follows:

1. Prevention—Avoiding waste generation rather than treating it post-production.
2. Atom Economy—Maximizing the incorporation of reactants into the final product.
3. Less Hazardous Chemical Syntheses—Designing safer substances and reactions.
4. Designing Safer Chemicals—Minimizing toxicity while maintaining efficacy.
5. Safer Solvents and Auxiliaries—Reducing or eliminating hazardous solvents.
6. Energy Efficiency—Conducting reactions at ambient temperature and pressure.
7. Renewable Feedstocks—Sourcing materials from renewable rather than depletable sources.
8. Reduce Derivatives—Avoiding unnecessary modifications that require additional reagents.
9. Catalysis—Using selective catalysts to enhance efficiency and minimize waste.
10. Design for Degradation—Ensuring products break down into non-toxic components.
11. Real-Time Analysis for Pollution Prevention—Monitoring and controlling hazardous substances.
12. Safer Chemistry for Accident Prevention—Designing chemicals and processes to minimize risks such as explosions and toxic releases.

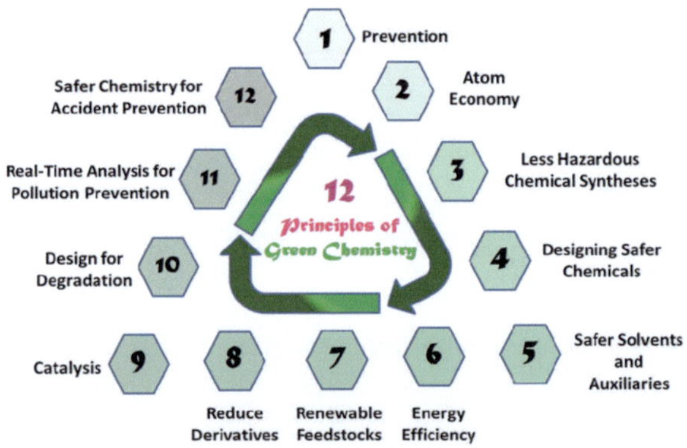

The implementation of these principles has significantly influenced modern chemistry by fostering the development of sustainable industrial practices. From solvent-free reactions to bio-based catalysts, the widespread adoption of green chemistry strategies aligns with sustainability goals by reducing hazardous byproducts and energy consumption in chemical manufacturing (Matlack, 2010).

1.2 Importance of Green Chemistry in Reducing Environmental Impact

Green chemistry plays a critical role in mitigating environmental pollution by addressing fundamental issues related to hazardous chemical production, excessive energy consumption, and industrial waste disposal (Behera & Das, 2025; Das et al., 2023). Traditional chemical processes often rely on toxic reagents and generate significant amounts of hazardous waste, contributing to ecological degradation and climate change (Clark & Macquarrie, 2002). A major environmental challenge addressed by green chemistry is the elimination of persistent pollutants. Many conventional industrial chemicals, such as polychlorinated biphenyls (PCBs) and perfluorinated compounds, persist in the environment for decades, leading to bioaccumulation and toxicity in aquatic and terrestrial ecosystems (Tundo & Griguol, 2018). By shifting toward biodegradable materials and eco-friendly synthesis routes, green chemistry prevents long-term ecological damage.

Furthermore, green chemistry promotes the replacement of hazardous solvents with bio-based or aqueous alternatives. Organic solvents like benzene, toluene, and dichloromethane are widely used in industrial processes but pose severe health risks due to their carcinogenic and neurotoxic properties. The introduction of ionic liquids, deep eutectic solvents, and supercritical CO_2 as greener alternatives has significantly reduced solvent-related toxicity while enhancing reaction efficiency (Anon., 2004). Additionally, catalytic efficiency is a key aspect of green chemistry's impact on sustainability. Traditional stoichiometric reactions often require excess reagents, leading to excessive waste. Heterogeneous and enzymatic catalysts enable selective transformations with minimal byproduct formation, improving atom economy and reducing energy demands (Ghosh et al., 2020; Nag et al., 2023; Sheldon, 2018).

The integration of bio-catalysis into industrial processes has facilitated the development of waste-free pharmaceutical synthesis, eco-friendly polymerization, and biofuel production. Moreover, the energy savings achieved through green chemistry strategies are profound. Conventional chemical manufacturing relies on high-temperature and high-pressure conditions, consuming vast amounts of fossil fuel-derived energy. Green chemistry emphasizes mild reaction conditions, such as microwave-assisted and photochemical synthesis, which significantly lower energy consumption and greenhouse gas emissions (Horváth, 2018). In essence, by minimizing hazardous waste, replacing toxic solvents, improving catalytic efficiency, and reducing energy consumption, green chemistry actively contributes to climate change mitigation, pollution prevention, and resource conservation—all essential components of global sustainability efforts (Pradhan et al., 2023).

1.3 Connection Between Green Chemistry and Industrial Biotechnology

The intersection of green chemistry and industrial biotechnology represents a powerful synergy for advancing sustainability goals, particularly in the production of bio-based chemicals, pharmaceuticals, and materials. Industrial biotechnology harnesses biological systems—such as enzymes, microorganisms, and genetically engineered biocatalysts to facilitate environmentally friendly chemical transformations (Straathof et al., 2019). One of the most significant applications of this synergy is the valorization of organic waste through biochemical conversion. Traditional chemical industries depend heavily on petrochemical feedstocks, which are finite and environmentally harmful. Green chemistry, in conjunction with biotechnology, enables the transformation of agricultural residues, food waste, and lignocellulosic biomass into valuable platform chemicals like lactic acid, succinic acid, and bioethanol (Werpy & Petersen, 2004). This approach not only reduces landfill waste but also lowers dependency on fossil-based resources.

Another crucial aspect is the role of enzymatic catalysis in green chemical synthesis. Unlike conventional catalysts, which often require harsh conditions and toxic reagents, biocatalysts operate under ambient temperatures and aqueous media, minimizing energy consumption and hazardous waste production (Tao & Kazlauskas, 2011a). Enzymes such as lipases, oxidoreductases, and hydrolases have been successfully employed in bio-based polymer production, pharmaceutical synthesis, and fine chemical industries. Moreover, industrial biotechnology contributes to green solvent innovation. Many green chemistry processes integrate bio-based solvents derived from plant sources or engineered microbial fermentation. For instance, glycerol, bioethanol, and terpenes serve as sustainable alternatives to volatile organic compounds (VOCs), reducing air pollution and workplace hazards (Schuur et al., 2019).

The synergy between green chemistry and biotechnology also extends to waste valorization in the circular economy. Traditional linear economic models rely on single-use manufacturing, leading to excessive waste. However, biorefinery concepts integrate green chemistry strategies with biotechnological processing, transforming waste streams into biodegradable plastics, biofuels, and bio-based surfactants (Kamm et al., 2007).

2 Industrial Biotechnology and the Role of Green Chemistry

2.1 *Definition and Significance of Industrial Biotechnology*

Industrial biotechnology, often referred to as white biotechnology, involves the application of biological systems such as enzymes, microorganisms, and genetically modified organisms (GMOs) to develop sustainable industrial processes (Chen, 2012). Unlike conventional industrial methods, which rely on petrochemicals and energy-intensive synthesis, industrial biotechnology integrates bio-based feedstocks and eco-friendly catalysts to produce biofuels, bioplastics, fine chemicals, and pharmaceuticals (Das et al., 2025; Reddy Koteswara et al., 2025; Renewable Energy Generation and Value Addition From Environmental, 2024).

The significance of industrial biotechnology lies in its environmental and economic advantages. Traditional chemical industries contribute to environmental degradation through high greenhouse gas (GHG) emissions, toxic waste generation, and resource depletion. By contrast, industrial biotechnology promotes carbon–neutral production methods, replacing fossil-based raw materials with renewable biomass(Fröhling & Hiete, 2020). Notably, the biotechnological production of biopolymers like polylactic acid (PLA) and polyhydroxyalkanoates (PHAs) reduces reliance on petroleum-derived plastics, mitigating plastic pollution and advancing circular economy principles. Another key aspect is energy efficiency. Industrial biotechnology employs mild reaction conditions, often operating at ambient temperatures and pressures, which significantly lowers energy consumption compared to traditional chemical synthesis (Clomburg et al., 2017).For instance, the enzymatic synthesis of bio-based surfactants and detergents requires fewer processing steps, leading to higher atom economy and lower waste production.

Moreover, industrial biotechnology fosters sustainability in pharmaceuticals and agrochemicals. The application of microbial fermentation and synthetic biology has enabled the production of biopharmaceuticals, including insulin, antibiotics, and monoclonal antibodies, with greater specificity and fewer toxic byproducts (Choi et al., 2023). Similarly, bio-based pesticides and fertilizers derived from microbial sources reduce agricultural dependence on synthetic chemicals, promoting eco-friendly farming practices.

Thus, the integration of industrial biotechnology with green chemistry principles offers an innovative pathway toward reducing environmental impact, enhancing resource efficiency, and developing sustainable alternatives to petrochemical-based products.

2.2 How Microbial and Enzymatic Processes Contribute to Green Chemical Synthesis

Microbial and enzymatic processes play a transformative role in green chemical synthesis by catalyzing reactions with unparalleled specificity, efficiency, and eco-compatibility. Unlike traditional chemical catalysts, which often require high temperatures, extreme pH conditions, and hazardous reagents, enzymes function under mild and aqueous conditions, making them ideal for sustainable industrial applications. (Bornscheuer & Buchholz, 2005).

One of the most significant contributions of microbial and enzymatic processes is their ability to convert biomass into biofuels and bio-based chemicals. For example:

(1) Fermentative production of bioethanol and biodiesel utilizes microbes such as *Saccharomyces cerevisiae* and *Escherichia coli*, reducing the dependence on fossil fuels (Lange & Huang, 2020).
(2) Enzymatic hydrolysis of lignocellulosic biomass by cellulases and hemicellulases facilitates the production of platform chemicals like furfural, levulinic acid, and 5-hydroxymethylfurfural (HMF), which serve as sustainable building blocks for bio-based materials (Dutta et al., 2022).

Moreover, microbial and enzymatic catalysis have revolutionized pharmaceutical and fine chemical synthesis. Traditional synthetic pathways for active pharmaceutical ingredients (APIs) often involve hazardous solvents and reagents, generating toxic waste. In contrast, enzymatic biotransformations enable selective functionalization with minimal byproducts, improving both yield and environmental safety. For instance:

(1) Lipase-mediated transesterification is employed in the synthesis of pharmaceutical intermediates and biodegradable polyesters (Sheldon & Woodley, 2018).
(2) Oxidoreductase enzymes, such as laccases and peroxidases, facilitate the detoxification of industrial effluents by degrading persistent organic pollutants(Singh et al., 2025).

The efficiency of enzyme immobilization techniques further enhances the sustainability of biocatalysis. Immobilized enzymes, such as glucose oxidase and nitrilases, demonstrate reusability and prolonged catalytic activity, reducing waste and operational costs in industrial-scale processes (Rodríguez-Restrepo & Orrego, 2020). Thus, microbial and enzymatic processes are at the forefront of sustainable chemical synthesis, offering environmentally benign alternatives to petrochemical-based manufacturing while maintaining high selectivity and efficiency.

2.3 Synergies Between Green Chemistry and Biotechnology in Waste Valorization

The integration of green chemistry with industrial biotechnology presents a powerful solution for waste valorization, transforming organic waste streams into valuable bio-based chemicals. This approach not only minimizes environmental pollution but also contributes to resource recovery and circular economy strategies (Dey et al., 2024; John & Das, 2024; Koutinas et al., 2014)). A key area of synergy is the conversion of agro-industrial and municipal waste into biofuels and platform chemicals. Traditional waste disposal methods, such as landfilling and incineration, generate greenhouse gas emissions and leach toxic substances into the environment. In contrast, biochemical and green chemistry techniques enable waste valorization by:

(1) Microbial fermentation of food waste and agricultural residues to produce bioethanol, biogas, and organic acids (Ashokkumar et al., 2022).
(2) Enzymatic hydrolysis of lignocellulose for the production of bio-based polymers like polylactic acid (PLA) and polyhydroxyalkanoates (PHAs), reducing plastic waste accumulation (Song et al., 2009).

The utilization of biocatalysts in green chemical processes further enhances waste valorization. For instance:

(1) Lipase-catalyzed biodiesel synthesis from waste cooking oil provides a sustainable alternative to petroleum diesel while reducing oil disposal issues (Thangaraj et al., 2019).
(2) Microbial degradation of electronic waste (e-waste) using bioleaching bacteria (*Acidithiobacillus ferrooxidans*) facilitates the recovery of valuable metals such as gold and copper, minimizing electronic pollution (Singh et al., 2020).

Additionally, waste valorization through green chemistry and biotechnology supports the development of biorefineries, where waste biomass serves as a raw material for multiple bio-based products. The integration of bio-based solvents, enzymatic transformations, and sustainable extraction methods ensures that waste valorization processes remain energy-efficient and environmentally responsible (Clasen et al., 2024). Ultimately, the synergies between green chemistry and biotechnology not only contribute to waste minimization and resource efficiency but also drive innovation in bio-based industries, fostering a more sustainable and resilient global economy (Mohanty et al., 2024; Ray et al., 2024).

3 Valorization of Organic Waste Through Green Chemistry

3.1 The Concept of Waste-To-Wealth Transformation

The waste-to-wealth concept is a fundamental pillar of sustainability, aligning with the principles of green chemistry and the circular economy. It revolves around the conversion of waste materials into valuable products, minimizing environmental pollution while maximizing resource efficiency (Orooji et al., 2022). Unlike traditional waste disposal methods such as incineration or landfilling, which contribute to greenhouse gas (GHG) emissions and ecological degradation, the waste-to-wealth strategy leverages biochemical, thermochemical, and catalytic processes to extract useful compounds from organic waste streams (Mishra & Das, 2023; Mishra et al., 2024).

A crucial driver of waste-to-wealth initiatives is the integration of green chemistry principles, particularly those emphasizing waste prevention, the use of renewable feedstocks, and catalytic efficiency (Anastas & Eghbali, 2010). Instead of treating waste as an environmental burden, green chemistry enables its transformation into bio-based chemicals, biofuels, biodegradable polymers, and fertilizers. One of the most promising applications of this concept is biorefineries, which utilize waste biomass as a raw material for multiple value-added products, similar to petroleum refineries (Ioannidou et al., 2022). For example, lignocellulosic agricultural residues, such as corn stover, wheat straw, and sugarcane bagasse, can be converted into biofuels and bio-based chemicals through enzymatic hydrolysis and microbial fermentation.

Furthermore, waste-to-wealth transformation is supported by government policies and industrial collaborations that promote sustainable waste management and green chemistry innovations (Clark, 2019). Many companies and research institutions are investing in catalytic depolymerization, enzymatic degradation, and bioelectrochemical processes to extract valuable compounds from food waste, municipal solid waste (MSW), and industrial byproducts.

Thus, the concept of waste-to-wealth is not only an environmental necessity but also an economic opportunity, fostering sustainable industrial development, energy security, and climate resilience.

3.2 Organic Waste as a Sustainable Feedstock for Platform Chemical Production

Organic waste, derived from agriculture, food processing, and municipal sources, represents an abundant and renewable feedstock for producing platform chemicals—key intermediates used in the synthesis of biofuels, bioplastics, pharmaceuticals, and specialty chemicals Platform chemicals are fundamental building blocks in the chemical industry, traditionally sourced from petroleum. However, green chemistry and biotechnological advancements have enabled the conversion of organic waste into sustainable alternatives. Examples of bio-based platform chemicals include:

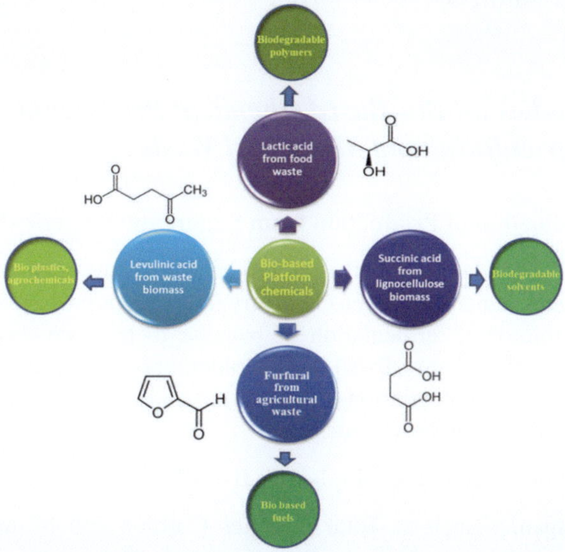

1. Lactic acid—Used in biodegradable plastics (polylactic acid, PLA), pharmaceuticals, and food preservatives. Produced via microbial fermentation of food waste and agricultural residues (Cho et al., 2020).
2. Succinic acid—A precursor for bioplastics, polyesters, and biodegradable solvents. Derived from lignocellulosic biomass through enzymatic hydrolysis and fermentation (Kim et al., 2025).
3. Furfural—A key intermediate in the production of resins, solvents, and bio-based fuels, obtained from hemicellulose-rich agricultural waste (Zhang et al., 2022)
4. Levulinic acid—Used in pharmaceuticals, agrochemicals, and bioplastics, synthesized from cellulose-containing waste biomass (Santos et al., 2021).

The adoption of organic waste as a sustainable feedstock is driven by multiple factors:

(a) Environmental Benefits—Reduces landfill waste, lowers carbon emissions, and minimizes pollution.
(b) Economic Viability—Provides a cost-effective alternative to fossil-based raw materials.
(c) Technological Advancements—Improvements in catalytic conversion, enzymatic hydrolysis, and microbial fermentation enhance efficiency and scalability (Tuck et al., 2012).

Industries and research institutions are increasingly investing in waste valorization projects, developing biorefineries that integrate biochemical and thermochemical processes to optimize resource recovery. For instance, integrated biorefineries in Europe and North America utilize municipal and agricultural waste to produce bioethanol, biopolymers, and bio-based surfactants, contributing to a greener and more circular economy (Clark et al., 2012).

3.3 Case Studies on Bio-Based Chemical Production from Agricultural and Municipal Waste

Case Study 1: Bioethanol Production from Sugarcane Bagasse (Brazil)

Brazil is a global leader in bioethanol production using sugarcane bagasse, a byproduct of sugar production. Green chemistry innovations have enabled enzymatic hydrolysis and microbial fermentation of bagasse to produce second-generation bioethanol, reducing reliance on fossil fuels (Goldemberg et al., 2014). This approach significantly reduces GHG emissions compared to traditional gasoline, contributing to Brazil's low-carbon economy strategy.

Case Study 2: Lactic Acid and Bioplastics from Food Waste (Europe)

In Europe, companies such as Total Energies Corbion and Nature Works have developed commercial-scale lactic acid production from food waste. The fermentation of starch-rich waste materials, such as potato peels and expired bakery products, yields lactic acid, which is further polymerized into polylactic acid (PLA), a biodegradable plastic (Hafid et al., 2017). PLA-based materials are widely used in packaging, textiles, and medical applications, offering an eco-friendly alternative to petroleum-derived plastics.

Case Study 3: Biogas and Biofertilizers from Municipal Solid Waste (India)

In India, the Indore Biogas Plant utilizes municipal solid waste (MSW) for biogas production through anaerobic digestion. Organic fractions of MSW, such as food scraps and vegetable waste, are decomposed by methanogenic bacteria, producing biogas (a mixture of methane and carbon dioxide (Kumar, Singh, et al., 2023a, 2023b). The residual slurry is converted into biofertilizer, reducing chemical fertilizer dependency in agriculture.

Case Study 4: Succinic Acid Production from Corn Stover (USA)

In the USA, bio-based succinic acid production from corn stover (a lignocellulosic agricultural residue) has gained industrial significance. Companies like BioAmber and Myriant employ engineered microbial strains to ferment hemicellulosic sugars, yielding high-purity succinic acid (Bechthold et al., 2008). This bio-based succinic acid serves as a precursor for biodegradable polymers and specialty chemicals, enhancing sustainability in chemical manufacturing.

4 Green Chemistry Innovations in Bioprocessing

4.1 Solvent-Free and Bio-Based Solvent Applications in Industrial Biotechnology

Traditional chemical processes rely heavily on volatile organic solvents (VOCs) such as toluene, benzene, and dichloromethane, which contribute to air pollution, toxicity, and environmental degradation. Green chemistry seeks to eliminate or replace these hazardous solvents through solvent-free processes and bio-based solvent applications, which align with sustainability goals by minimizing waste and energy consumption (Clarke et al., 2018).

Solvent-Free Processes

Solvent-free processes eliminate the use of traditional solvents by enabling direct solid-state, melt-phase, or gas-phase reactions. These methods offer multiple advantages, such as:

(1) Reduced environmental impact by eliminating solvent waste and VOC emissions.
(2) Higher atom economy, ensuring that most reactants are incorporated into the final product.
(3) Improved energy efficiency, as reactions often proceed at ambient temperature and pressure.

One example is mechanochemistry, a solvent-free technique where mechanical energy (e.g., ball milling) drives chemical transformations. This approach has been widely used in pharmaceutical and polymer synthesis, significantly reducing hazardous waste production (Guzmán et al., 2021).

Bio-Based Solvent Applications

Bio-based solvents, derived from renewable plant-based feedstocks, provide sustainable alternatives to petroleum-derived solvents. Examples include:

1. Ionic Liquids (ILs)—Non-volatile, tunable solvents that improve selectivity in enzymatic and catalytic reactions (Zhao et al., 2017).
2. Deep Eutectic Solvents (DESs)—Low-cost, biodegradable solvents synthesized from natural compounds such as choline chloride and glycerol, which have been applied in biomass processing and enzymatic biotransformations (Tan & Dou, 2020).
3. Supercritical CO_2 ($scCO_2$)—A non-toxic, recyclable solvent that enhances extraction, polymerization, and pharmaceutical synthesis while reducing energy consumption (Ardila-Fierro & Hernández, 2021).

The combination of solvent-free techniques and bio-based solvents in industrial biotechnology leads to cleaner, more efficient production processes, contributing to sustainability and green chemistry advancements.

4.2 Role of Biocatalysts and Green Catalysts in Sustainable Chemical Transformations

Biocatalysts and green catalysts are essential in reducing the environmental impact of chemical transformations, offering high selectivity, energy efficiency, and reduced waste generation. Unlike conventional metal catalysts, which often require harsh reaction conditions and toxic reagents, biocatalysts (enzymes) and green catalysts (non-toxic or bio-derived catalysts) facilitate reactions under mild and eco-friendly conditions (Sheldon & Woodley, 2018).

Biocatalysts in Sustainable Synthesis

Enzymatic catalysis has emerged as a powerful tool in green chemistry due to its:

(1) High specificity, reducing unwanted side reactions.
(2) Mild reaction conditions, lowering energy consumption.
(3) Biodegradability, ensuring minimal environmental toxicity (Tao & Kazlauskas, 2011b).

Common enzymatic transformations in industrial biotechnology include:

1. Lipase-catalyzed transesterification—Used in biodiesel production, replacing traditional acid or base catalysts (Rodríguez-Restrepo & Orrego, 2020).
2. Oxidoreductase-driven oxidations—Applied in the synthesis of pharmaceutical intermediates and biopolymers, reducing hazardous byproducts (Singh et al., 2021).
3. Protease-mediated peptide synthesis—Enabling pharmaceutical and biopolymer production under aqueous conditions (Choi et al., 2024).

Green Catalysts in Chemical Transformations

Green catalysts, including bio-based, heterogeneous, and recyclable catalysts, offer an eco-friendly alternative to toxic transition-metal catalysts. Notable advancements include:

(1) Heterogeneous catalysis—Solid-state catalysts that can be easily separated and reused, reducing waste. Metal–organic frameworks (MOFs) and biochar-supported catalysts are widely applied in biofuel synthesis and fine chemical production (Darmayanti et al., 2024).
(2) Bio-derived catalysts—Catalysts derived from renewable sources such as plant extracts, agricultural waste, and enzymatic modifications, reducing reliance on rare metals (Kumar et al., 2023b).

These catalytic innovations contribute to more sustainable and cost-effective industrial processes, reinforcing the role of green chemistry in achieving sustainability goals.

4.3 *Process Intensification Techniques for Enhanced Efficiency and Scalability*

Process intensification (PI) aims to enhance efficiency, reduce waste, and minimize resource consumption in industrial biotechnology. By adopting novel reactor designs, energy-efficient technologies, and integrated reaction-separation methods, PI ensures greener and more economically viable bioprocessing (Dimian et al., 2014).

Key Process Intensification Strategies.

1. Microreactor Technology—Microfluidic reactors enable faster heat and mass transfer, improving reaction selectivity and reducing solvent usage. These reactors are widely used in pharmaceutical and fine chemical manufacturing (Jensen, 2017).
2. Membrane Bioreactors (MBRs)—Combining enzymatic catalysis with membrane separation enhances product recovery and solvent recycling in biochemical and biofuel production (Khambhati et al., 2019).
3. Ultrasound and Microwave-Assisted Processing—These energy-efficient techniques accelerate reaction rates, enhance catalytic efficiency, and reduce energy demand, widely applied in extraction, polymerization, and biofuel synthesis (Khadhraoui et al., 2021).

Scalability and Industrial Applications

Process intensification is critical in scaling up green bioprocessing techniques from lab to industrial scale. Companies implementing PI strategies have reported:

(1) 50–70% reduction in energy consumption compared to conventional methods.
(2) Increased product yields and reduced raw material usage, improving economic feasibility.
(3) Lower environmental impact through minimized solvent waste and carbon emissions (Tóth et al., 2023).

By integrating process intensification with green chemistry principles, industrial biotechnology can achieve higher efficiency, lower costs, and enhanced environmental sustainability.

5 Economic and Environmental Impact of Green Chemistry in Industrial Biotechnology

5.1 *Comparative Analysis of Traditional Versus Green Chemical Processes*

Traditional chemical processes have long been associated with high environmental costs, including resource depletion, hazardous waste generation, and significant carbon emissions. Green chemistry, by contrast, seeks to design products and processes that minimize toxicity, enhance efficiency, and reduce waste, aligning with sustainability objectives (Anastas and Warner, 2000).

Key Differences Between Traditional and Green Chemical Processes

Aspect	Traditional chemical processes	Green chemical processes
Raw materials	Petroleum-based, non-renewable	Bio-based, renewable
Energy consumption	High energy demand, fossil fuel reliance	Low energy input, renewable sources
Waste Generation	Toxic byproducts, hazardous waste	Minimal waste, byproduct valorization
Solvent use	Organic solvents (e.g., benzene, toluene)	Water, supercritical CO_2, ionic liquids
Catalysts	Heavy metal catalysts, non-recyclable	Biocatalysts, recyclable green catalysts

The transition to green chemistry in industrial biotechnology has led to significant reductions in pollution and energy use. For example, enzyme-based bioprocesses in pharmaceutical synthesis have replaced toxic chemical reagents, improving product purity and yield while minimizing hazardous waste (Sheldon, 2014). Another notable example is bioethanol production. Traditional petroleum-derived fuel production generates high greenhouse gas (GHG) emissions, whereas bioethanol derived from lignocellulosic biomass significantly reduces carbon footprints and dependency on fossil fuels (Lugani et al., 2020). The adoption of green chemical processes in industrial biotechnology is thus not only an environmental necessity but also a key driver for achieving long-term economic sustainability.

5.2 *Life Cycle Assessment and Carbon Footprint Reduction*

Life cycle assessment (LCA) is a comprehensive tool used to evaluate the environmental impact of a product or process from raw material extraction to disposal. In the context of green chemistry, LCA helps quantify energy consumption, waste generation, and greenhouse gas emissions, allowing industries to make informed decisions for sustainable process improvements (Bruijn et al., 2004).

Carbon Footprint Reduction Through Green Chemistry

One of the primary goals of green chemistry in industrial biotechnology is reducing the carbon footprint of chemical manufacturing. Key strategies include:

(1) Utilization of renewable feedstocks—Transitioning from petroleum-derived chemicals to bio-based alternatives (e.g., bioplastics) significantly cuts CO_2 emissions (Zimmerman et al., 2020).
(2) Energy-efficient processing—Technologies such as enzymatic catalysis and supercritical CO_2 extraction require lower temperatures and pressures, reducing overall energy consumption (Clark & Macquarrie, 2002).
(3) Carbon capture and utilization (CCU)—The integration of CO_2 conversion into valuable chemicals (e.g., methanol, carbonates) minimizes industrial carbon emissions and enhances circular economy approaches (Aresta, 2019).

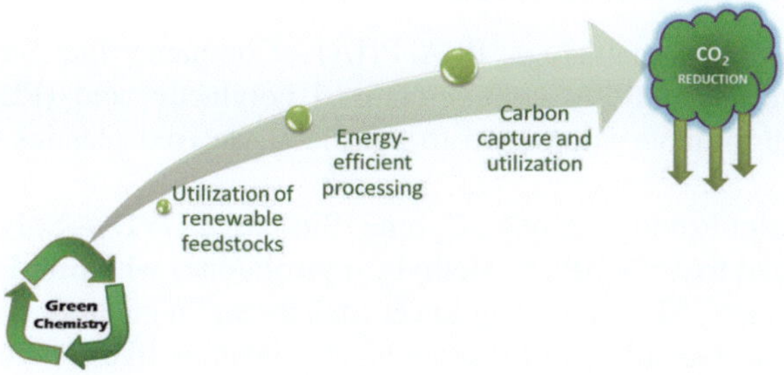

For instance, the bioplastic industry has shown a 50–80% reduction in carbon emissions compared to traditional petroleum-based plastics, highlighting the role of green chemistry in decarbonization efforts (Narayan, 2020). While LCA confirms the environmental benefits of green chemistry, continuous technological innovations, policy support, and industrial adoption are necessary to optimize sustainability outcomes in biotechnology sectors.

5.3 *Economic Feasibility and Commercialization of Green Chemical Technologies*

The economic viability of green chemistry innovations in industrial biotechnology depends on multiple factors, including production costs, scalability, market demand, and policy incentives. While green chemical processes often have higher initial costs, long-term benefits such as resource efficiency, reduced regulatory burdens, and enhanced market competitiveness make them financially attractive (Patel et al., 2020).

Key Economic Considerations in Green Chemistry Adoption

1. Raw Material Costs—Renewable feedstocks, such as agricultural residues and algae-derived biopolymers, are often cheaper than petroleum-based chemicals but require efficient supply chain management (Chen, 2012).
2. Process Efficiency and Yield—Green catalysts and biocatalysts have demonstrated higher selectivity and conversion rates, leading to lower energy consumption and reduced waste disposal costs (Sheldon, 2020).
3. Regulatory and Consumer Demand—Stricter environmental regulations and consumer preference for sustainable products have driven the commercialization of green chemicals, bioplastics, and bio-based solvents (Clark et al., 2012).

Successful Commercialization Cases

Several green chemical technologies have reached commercial success, demonstrating both economic feasibility and environmental impact reduction:

(1) Biodegradable Polymers (e.g., PLA, PHA)—Companies like NatureWorks and BASF have successfully commercialized polylactic acid (PLA) from corn starch, a sustainable alternative to petroleum-derived plastics (Cywar et al., 2021)
(2) Green Solvent Production (e.g., Cyrene, Bio-based GVL)—Solvent companies have replaced toxic NMP (N-Methyl-2-pyrrolidone) with bio-derived gamma-valerolactone (GVL), achieving lower toxicity and regulatory compliance
(3) Bio-based Surfactants and Detergents—Companies like Ecover and Seventh Generation use plant-derived surfactants, reducing synthetic detergent pollution and enhancing biodegradability (Hayes, 2019)

Challenges and Future Outlook

Despite these successes, barriers to widespread commercialization include:

(1) High initial investment costs for green technology development.
(2) Limited infrastructure for bio-based feedstock processing.
(3) Market competition with traditional chemical industries.

To address these challenges, increased government incentives, research funding, and cross-industry collaboration are essential. The adoption of circular economy principles and carbon pricing mechanisms can further drive the scalability and profitability of green chemical innovations in industrial biotechnology.

6 Policy and Regulatory Frameworks Driving Green Chemical Adoption

6.1 Global Sustainability Policies and Incentives for Green Chemistry Adoption

The adoption of green chemistry is significantly influenced by global sustainability policies and incentive mechanisms designed to encourage industries to transition toward environmentally friendly chemical processes. Various international frameworks, such as the United Nations Sustainable Development Goals (SDGs), emphasize sustainable industrialization and responsible consumption. The European Green Deal also provides regulatory support and financial incentives for green chemistry initiatives, aiming to make Europe climate-neutral by 2050. Furthermore, the U.S. Environmental Protection Agency (EPA) has implemented programs such as the Green Chemistry Challenge Awards, which encourage innovation in eco-friendly chemical synthesis.

Financial incentives, including tax benefits, subsidies, and grants, have played a crucial role in stimulating research and development in green chemistry. For example, the BioPreferred Program in the United States prioritizes government procurement of bio-based products, driving industrial investment in sustainable chemical alternatives (USDA, 2022). Similar schemes exist in Canada and Japan, where policies support bio-based manufacturing and eco-friendly supply chain integration (OECD, 2023). These incentives create a favorable ecosystem for industries to adopt sustainable practices while reducing reliance on fossil fuel-based chemicals.

6.2 Role of Government Regulations in Promoting Bio-Based Chemical Production

Government regulations serve as a fundamental driver for the adoption of bio-based chemical production by setting legally binding environmental standards. The European Union's REACH (Registration, Evaluation, Authorization, and Restriction of Chemicals) regulation ensures that industries assess and mitigate the risks associated with chemical substances before their commercialization Similar legislation, such as the Toxic Substances Control Act (TSCA) in the United States, mandates chemical safety evaluations and promotes the substitution of hazardous materials with safer alternatives.

In Asia, countries like China and India have strengthened their regulatory frameworks to encourage the use of biodegradable and non-toxic chemical processes. China's "Made in China 2025" initiative includes a strong focus on green manufacturing and sustainable chemical production. Additionally, India's National Policy on Biofuels provides a roadmap for scaling up bio-based industrial processes, fostering green chemistry applications in energy and material sciences. These regulations not only promote industrial biotechnology but also accelerate the global shift toward sustainable manufacturing.

6.3 Industry Best Practices and Guidelines for Sustainable Industrial Applications

The implementation of best practices in the industry is essential for the widespread adoption of green chemistry principles. Many multinational corporations and small-scale enterprises have integrated green chemistry into their operations to align with sustainability targets and regulatory requirements. Companies such as BASF and Dow Chemical have pioneered innovative bio-based product lines, reducing the environmental impact of their supply chains.

Certification schemes and voluntary industry standards also play a significant role in guiding sustainable practices. The International Organization for Standardization (ISO) has established standards such as ISO 14001, which provides a framework for environmentally responsible manufacturing. Additionally, green labeling programs, such as the USDA Certified Bio-based Product label, help consumers identify and support bio-based alternatives (USDA, 2022). Collaborative initiatives between academia, industry, and government agencies further enhance the implementation of green chemistry in industrial biotechnology. The Ellen MacArthur Foundation, for instance, promotes the circular economy concept, which integrates waste valorization and bio-based materials into industrial production systems. Such collaborations ensure that sustainable industrial applications continue to evolve and contribute to global sustainability goals.

By integrating global policies, stringent regulations, and industry best practices, green chemistry adoption can be significantly accelerated, paving the way for a more sustainable future in industrial biotechnology.

7 Future Prospects and Challenges in Green Chemistry for Biotechnology

7.1 Emerging Trends in Green Chemistry and Industrial Biotechnology

The landscape of green chemistry and industrial biotechnology is evolving rapidly, driven by innovations in synthetic biology, precision fermentation, and computational chemistry. One of the most significant trends is the use of artificial intelligence (AI) and machine learning to design environmentally benign chemical pathways. These tools enable researchers to predict molecular behavior, optimize enzymatic reactions, and reduce experimental iterations, thereby minimizing resource use and waste (Peng et al., 2022).

Another critical development is the emergence of biofoundries-integrated facilities that combine high-throughput experimentation, automation, and data analytics to accelerate the design-build-test-learn cycle in synthetic biology. These facilities are increasingly used to engineer microorganisms for the production of high-value chemicals from renewable feedstocks, aligning with green chemistry principles (Carbonell et al., 2020).

Green solvents, including ionic liquids and deep eutectic solvents, continue to gain traction due to their low volatility and recyclability. Additionally, photoredox and electrochemical synthesis methods are emerging as sustainable alternatives to traditional redox reactions, offering energy efficiency and reduced hazardous byproducts (Yan et al., 2020).

7.2 Challenges in Large-Scale Implementation and Technological Limitations

Despite the promise of green chemistry and biotechnology, several challenges hinder their large-scale industrial implementation. One of the primary barriers is the high initial capital investment required for developing bioprocessing infrastructure and green chemical technologies. Many industries, especially in developing regions, lack access to the funding and expertise needed to transition from conventional methods to greener alternatives (Sheldon, 2016).

Technological limitations also pose significant hurdles. Enzyme stability under industrial conditions, such as high temperatures and variable pH, remains a challenge. Similarly, the scalability of microbial fermentation processes often encounters bottlenecks related to oxygen transfer, product toxicity, and low yields (Wandrey et al., 2000).

Moreover, the lack of standardization and benchmarking in green chemistry metrics complicates comparative assessments of sustainability. While tools like life cycle assessment (LCA) are valuable, their application requires robust datasets, which are often unavailable for emerging biotechnologies(Parisi et al., 2020).

Regulatory hurdles further complicate deployment. Navigating diverse environmental and safety regulations across regions can delay commercialization and increase compliance costs. The dynamic nature of policy landscapes also creates uncertainty for investors and industrial stakeholders.

7.3 The Role of Interdisciplinary Research in Advancing Green Chemical Solutions

Interdisciplinary collaboration is crucial for advancing green chemistry and biotechnology. The convergence of chemistry, biology, materials science, engineering, and data science fosters holistic approaches to sustainability challenges. For instance, integrating systems biology with process engineering allows for the design of more robust microbial cell factories capable of converting complex biomass into platform chemicals (Nielsen & Keasling, 2016). Collaborations between computational scientists and experimental chemists are yielding advanced modeling tools that can simulate reaction mechanisms and predict toxicity profiles. This integrative approach reduces trial-and-error experimentation and enhances the safety and efficiency of chemical processes (Gómez-Bombarelli et al., 2018).

Academic-industry partnerships are also playing a pivotal role. These collaborations facilitate knowledge exchange, access to infrastructure, and the alignment of research objectives with commercial needs. Initiatives such as the Bioenergy Technologies Office (BETO) in the U.S. and the European Bio-Based Industries Joint Undertaking (BBI-JU) exemplify how cross-sector partnerships can accelerate the development and adoption of sustainable technologies.

8 Conclusion: The Path Forward for Sustainable Industrial Chemistry

8.1 Summary of Key Takeaways and Their Relevance to Global Sustainability Goals

This chapter has thoroughly examined how green chemistry serves as a cornerstone for advancing sustainability within industrial biotechnology. The 12 principles of green chemistry emphasizing waste prevention, atom economy, energy efficiency, and the use of renewable feedstocks have been systematically integrated into biotechnological practices to reduce the environmental and economic costs of chemical manufacturing. Key insights include the valorization of organic waste into high-value chemicals, the transformative role of microbial and enzymatic catalysis in green synthesis, and the integration of solvent-free and bio-based solvent systems for cleaner production methods. These strategies are directly aligned with global sustainability goals, particularly those outlined in the United Nations Sustainable Development Goals (SDGs), such as climate action (SDG 13), responsible consumption and production (SDG 12), and affordable and clean energy (SDG 7). Life cycle assessments have shown marked reductions in carbon footprints, and successful case studies across Brazil, Europe, India, and the U.S. confirm the feasibility of green chemistry at scale. Together, these findings affirm that green chemistry not only mitigates environmental damage but also contributes to economic resilience and industrial innovation.

8.2 The Imperative of Green Chemistry in Circular Economy Strategies

The shift toward a circular economy one that emphasizes reuse, recycling, and the regenerative use of resources demands a fundamental transformation of industrial processes. Green chemistry offers the scientific and technological framework necessary for this transformation. From waste-to-wealth initiatives and the use of biodegradable feedstocks to the development of recyclable solvents and sustainable catalysts, green chemistry enables industries to close material loops and reduce reliance on non-renewable inputs. Biorefineries exemplify this circular approach by converting agricultural and municipal waste into multiple product streams, including biofuels, platform chemicals, and biodegradable polymers. The synergistic application of green chemistry and industrial biotechnology ensures that waste is not an endpoint but a starting point for new value creation. This circularity also enhances supply chain resilience and resource security, particularly in light of growing environmental regulations and consumer demand for sustainable products.

By embedding green chemistry into circular economy frameworks, industries can achieve systemic sustainability minimizing waste, maximizing resource efficiency, and promoting environmental stewardship.

8.3 Call for Collaboration Between Researchers, Industries, and Policymakers

To realize the full potential of green chemistry in sustainable industrial development, robust collaboration across sectors is imperative. Researchers must continue to innovate in areas such as enzyme engineering, green solvent development, and process intensification. At the same time, industries must commit to adopting these innovations through pilot programs, infrastructure investment, and life cycle-based decision-making. Policymakers play a crucial role in creating an enabling environment through regulation, financial incentives, and educational initiatives. Global and national policies such as REACH in the EU, TSCA in the U.S., and bioeconomy strategies in China and India are vital for catalyzing this transition. The involvement of international agencies, such as the OECD and UNEP, in setting sustainability benchmarks and supporting knowledge transfer further strengthens global momentum. Moreover, interdisciplinary collaboration must extend beyond the laboratory to include economists, environmental scientists, engineers, and sociologists, ensuring that green chemical innovations are not only technically feasible but also socially and economically inclusive.

In conclusion, the future of sustainable industrial chemistry lies in our collective ability to integrate green chemistry principles into every stage of research, production, and policy. Through coordinated action, we can forge a path toward a cleaner, circular, and more sustainable global economy.

References

Anastas, P. T., & Warner, J. C. (2000). *Green chemistry: Theory and practice* [online]. Oxford University PressOxford. https://academic.oup.com/book/53104

Anastas, P., & Eghbali, N. (2010). Green chemistry: Principles and practice. *Chemical Society Reviews, 39*(1), 301–312.

Anon., 2004. Green solvents for chemistry: perspectives and practice. *Choice Reviews Online, 41*(06), 41–3457.

Ardila-Fierro, K. J., & Hernández, J. G. (2021). Sustainability assessment of mechanochemistry by using the twelve principles of green chemistry. *Chemsuschem, 14*(10), 2145–2162.

Aresta, M. (2019). Carbon dioxide utilization: The way to the circular economy. *Greenhouse Gases: Science and Technology, 9*(4), 610–612.

Ashokkumar, V., Flora, G., Venkatkarthick, R., SenthilKannan, K., Kuppam, C., Mary Stephy, G., Kamyab, H., Chen, W.-H., Thomas, J., & Ngamcharussrivichai, C. (2022). Advanced technologies on the sustainable approaches for conversion of organic waste to valuable bioproducts: Emerging circular bioeconomy perspective. *Fuel, 324*, Article 124313.

Bechthold, I., Bretz, K., Kabasci, S., Kopitzky, R., & Springer, A. (2008). Succinic acid: A new platform chemical for biobased polymers from renewable resources. *Chemical Engineering & Technology, 31*(5), 647–654.

Behera, I., Dipamitra., & Das, A. P. (2025). *Environmental hydrocarbon pollution and zero waste approach towards a sustainable waste management*. p. 436.

Bornscheuer, U. T., & Buchholz, K. (2005). Highlights in biocatalysis—Historical landmarks and current trends. *Engineering in Life Sciences, 5*(4), 309–323.

Bruijn, H., Duin, R., Huijbregts, M. A. J., Guinee, J. B., Gorree, M., Heijungs, R., Huppes, G., Kleijn, R., Koning, A., Oers, L., & Wegener Sleeswijk, A. (2004). *Handbook on life cycle assessment: Operational guide to the ISO Standards*. Kluwer Academic Publishers.

Carbonell, P., Le Feuvre, R., Takano, E., & Scrutton, N. S. (2020). *In silico* design and automated learning to boost next-generation smart biomanufacturing. *Synthetic Biology, 5*(1), ysaa020.

Chen, G.-Q. (2012). New challenges and opportunities for industrial biotechnology. *Microbial Cell Factories, 11*(1), 111.

Cho, E. J., Trinh, L. T. P., Song, Y., Lee, Y. G., & Bae, H.-J. (2020). Bioconversion of biomass waste into high value chemicals. *Bioresource Technology, 298*, Article 122386.

Choi, S. Y., Lee, Y., Yu, H. E., Cho, I. J., Kang, M., & Lee, S. Y. (2023). Sustainable production and degradation of plastics using microbes. *Nature Microbiology, 8*(12), 2253–2276.

Choi, J.-H., You, S.-M., Ahn, M.-R., Jung, C.-D., Seong, H., Kim, Y., Cha, H. G., & Kim, H. (2024). Simultaneous lignin extraction and fibrillation: A Novel approach for a continuous biorefinery. *ACS Sustainable Chemistry & Engineering, 12*(6), 2324–2333.

Clark, J. H. (2019). Green biorefinery technologies based on waste biomass. *Green Chemistry, 21*(6), 1168–1170.

Clark, J. H., Luque, R., & Matharu, A. S. (2012). Green Chemistry, Biofuels, and Biorefinery. *Annual Review of Chemical and Biomolecular Engineering, 3*(1), 183–207.

Clark, J. H., & Macquarrie, D. (Eds.). (2002). *Handbook of green chemistry and technology* [online] (1st ed.). Wiley. https://doi.org/10.1002/9780470988305

Clarke, C. J., Tu, W.-C., Levers, O., Bröhl, A., & Hallett, J. P. (2018). Green and sustainable solvents in chemical processes. *Chemical Reviews, 118*(2), 747–800.

Clasen, A. P., Agostinho, F., Almeida, C. M. V. B., Liu, G., & Giannetti, B. F. (2024). Advancing towards circular economy: Environmental benefits of an innovative biorefinery for municipal solid waste management. *Sustainable Production and Consumption, 46*, 571–581.

Clomburg, J. M., Crumbley, A. M., & Gonzalez, R. (2017). Industrial biomanufacturing: The future of chemical production. *Science, 355*(6320), aag0804.

Cywar, R. M., Rorrer, N. A., Hoyt, C. B., Beckham, G. T., & Chen, E.Y.-X. (2021). Bio-based polymers with performance-advantaged properties. *Nature Reviews Materials, 7*(2), 83–103.

Darmayanti, M. G., Tuck, K. L., & Thang, S. H. (2024). Carbon dioxide capture by emerging innovative polymers: Status and perspectives. *Advanced Materials, 36*(30), 2403324.

Das, A. P., Mohanty, A. K., van Hullebusch, E. D., Figueiredo, G., Das, A. P., Mohanty, A. K., van Hullebusch, E. D., & Figueiredo, G. (2023). Bioremediation as an emerging technology for the removal of synthetic microplastic pollutants from marine ecosystem. *MarPB, 194*, Article 115297. https://doi.org/10.1016/J.MARPOLBUL.2023.115297

Das, A., Dey, S., Gacem, A., Yadav, K. K., Nath, H., Ram, D. K., & Das, A. P. (2025). Revolutionizing solid hydrocarbon sourcing: Plastic's journey from waste to upcycled treasure. *Environmental Science and Engineering, Part, F218*, 331–351. https://doi.org/10.1007/978-3-031-81615-4_14

Dey, S., Dash, A., & Das, A. P. (2024). Bioremediation as a Sustainable and Circular Solution for Mining Waste Management and Bioenergy Generation. *Sustainable Management of Mining Waste and Tailings: A Circular Economy Approach*, 107–119. https://doi.org/10.1201/9781003442455-6/BIOREMEDIATION-SUSTAINABLE-CIRCULAR-SOLUTION-MINING-WASTE-MANAGEMENT-BIOENERGY-GENERATION-SUDESHNA-DEY-AKANSHYA-DASH-ALOK-PRASAD-DAS

Dimian, A. C., Bildea, C. S., & Kiss, A. A. (2014). Process intensification. In *Computer aided chemical engineering* [online] (pp. 397–448). Elsevier. https://linkinghub.elsevier.com/retrieve/pii/B9780444627001000103

Dutta, N., Usman, M., Luo, G., & Zhang, S. (2022). An insight into valorization of lignocellulosic biomass by optimization with the combination of hydrothermal (HT) and biological techniques: A review. *Sustainable Chemistry, 3*(1), 35–55.

Fröhling, M., & Hiete, M. (2020). Sustainability and life cycle assessment in industrial biotechnology: A review of current approaches and future needs. In M. Fröhling, & M. Hiete (Eds.), *Sustainability and life cycle assessment in industrial biotechnology* [online] (pp. 143–203). Springer International Publishing. https://doi.org/10.1007/10_2020_122

Ghosh, S., Gandhi, M., van Hullebusch, E. D., & Das, A. P. (2020). Proteomic insights into Lysinibacillus sp.-mediated biosolubilization of manganese. *Environmental Science and Pollution Research 2020 28:30, 28*(30), 40249–40263. https://doi.org/10.1007/S11356-020-10863-4

Goldemberg, J., Mello, F. F. C., Cerri, C. E. P., Davies, C. A., & Cerri, C. C. (2014). Meeting the global demand for biofuels in 2021 through sustainable land use change policy. *Energy Policy, 69*, 14–18.

Gómez-Bombarelli, R., Wei, J. N., Duvenaud, D., Hernández-Lobato, J. M., Sánchez-Lengeling, B., Sheberla, D., Aguilera-Iparraguirre, J., Hirzel, T. D., Adams, R. P., & Aspuru-Guzik, A. (2018). Automatic chemical design using a data-driven continuous representation of molecules. *ACS Central Science, 4*(2), 268–276.

Guzmán, H., Salomone, F., Batuecas, E., Tommasi, T., Russo, N., Bensaid, S., & Hernández, S. (2021). How to make sustainable CO_2 conversion to methanol: Thermocatalytic versus electrocatalytic technology. *Chemical Engineering Journal, 417*, Article 127973.

Hafid, H. S., Rahman, N. A. A., Shah, U. K. M., Baharuddin, A. S., & Ariff, A. B. (2017). Feasibility of using kitchen waste as future substrate for bioethanol production: A review. *Renewable and Sustainable Energy Reviews, 74*, 671–686.

Hayes, D. G. (Ed.). (2019). *Biobased surfactants and detergents: Synthesis, properties, and applications*. Academic Press.

Horváth, I. T. (2018). Introduction: Sustainable chemistry. *Chemical Reviews, 118*(2), 369–371.

Ioannidou, S. M., Galanopoulos, C., Ladakis, D., & Koutinas, A. (2022). Techno-economic evaluation and life-cycle assessment of integrated biorefineries within a circular bioeconomy concept. *In*: *Biomass, biofuels, biochemicals* [online] (pp. 541–556). Elsevier. https://linkinghub.elsevier.com/retrieve/pii/B9780323898553000157

Jensen, K. F. (2017). Flow chemistry—Microreaction technology comes of age. *AIChE Journal, 63*(3), 858–869.

John, A., & Das, A. P. (2024). Government strategies for achieving circular economy for plastic waste management in Southeast Asia. In *Sustainable management of urban plastic waste through circular economic approaches* (pp. 72–82). https://doi.org/10.1201/9781003465447-6

Kamm, B., Gruber, P. R., & Kamm, M. (2007). Biorefineries—Industrial processes and products. In V. C. H. Wiley (ed.), *Ullmann's encyclopedia of industrial chemistry* [online]. Wiley. https://doi.org/10.1002/14356007.l04_l01

Khadhraoui, B., Ummat, V., Tiwari, B. K., Fabiano-Tixier, A. S., & Chemat, F. (2021). Review of ultrasound combinations with hybrid and innovative techniques for extraction and processing of food and natural products. *Ultrasonics Sonochemistry, 76*, Article 105625.

Khambhati, K., Bhattacharjee, G., Gohil, N., Braddick, D., Kulkarni, V., & Singh, V. (2019). Exploring the potential of cell-free protein synthesis for extending the abilities of biological systems. *Frontiers in Bioengineering and Biotechnology, 7*, 248.

Kim, S.-Y., Choi, J.-B., Kim, D.-S., An, D.-H., Jeong, J.-H., Cho, H.-M., & Cha, Y.-L. (2025). Succinic acid production from sweet potato via separate hydrolysis and fermentation. *Biotechnology and Bioprocess Engineering* [online]. https://doi.org/10.1007/s12257-025-00177-y

Koutinas, A. A., Vlysidis, A., Pleissner, D., Kopsahelis, N., Lopez Garcia, I., Kookos, I. K., Papanikolaou, S., Kwan, T. H., & Lin, C. S. K. (2014). Valorization of industrial waste and by-product streams via fermentation for the production of chemicals and biopolymers. *Chemical Society Reviews, 43*(8), 2587.

Kumar, A., Singh, E., Mishra, R., Lo, S. L., & Kumar, S. (2023a). Global trends in municipal solid waste treatment technologies through the lens of sustainable energy development opportunity. *Energy, 275*, Article 127471.

Kumar, V., Saha, R., Chatterjee, S., & Mishra, V. (2023b). From traditional to greener alternatives: Potential of plant resources as a biotransformation tool in organic synthesis. *Reaction Chemistry & Engineering, 8*(11), 2677–2688.

Lange, L., & Huang, Y. (2020). The potential of integrated bio- and chemical-engineering—for a more sustainable world. *Green Chemical Engineering, 1*(1), 9–15.

Lugani, Y., Rai, R., Prabhu, A. A., Maan, P., Hans, M., Kumar, V., Kumar, S., Chandel, A. K., & Sengar, R. S. (2020). Recent advances in bioethanol production from lignocelluloses: A comprehensive review with a focus on enzyme engineering and designer biocatalysts. *Biofuel Research Journal, 7*(4), 1267–1295.

Matlack, A. (2010). *Introduction to green chemistry* [online]. CRC Press. https://www.taylorfrancis.com/books/9781439882115

Mishra, S., Dash, D., & Das, A. P. (2024). Aquatic microbial diversity on plastisphere: Colonization and potential role in microplastic biodegradation. *Geomicrobiology Journal, 41*(4), 312–323. https://doi.org/10.1080/01490451.2023.2209750

Mishra, S., & Das, A. P. (2023). Microbial remediation of synthetic microfiber contaminated wastewater. In M. P. Shah (Ed.), *Microbial technologies in industrial wastewater treatment* (pp. 337–350). Springer. https://doi.org/10.1007/978-981-99-2435-6_17

Mohanty, M., Tripathy, B., Koteswara Reddy, G., & Das, A. P. (2024). greener technologies for establishing restraint over microplastic pollution. In *Sustainable microbial technology for synthetic and cellulosic microfiber bioremediation* (pp. 239–258). https://doi.org/10.1007/978-3-031-62600-5_13

Morone, P., Caferra, R., D'Adamo, I., Falcone, P. M., Imbert, E., & Morone, A. (2021). Consumer willingness to pay for bio-based products: Do certifications matter? *International Journal of Production Economics, 240*, 108248. https://doi.org/10.1016/j.ijpe.2021.108248

Nag, M., Lahiri, D., Ghosh, S., Sarkar, T., Pati, S., Das, A. P., Ram, D. K., Bhattacharya, D., & Ray, R. R. (2023). Application of microorganisms in biotransformation and bioremediation of environmental contaminant: A review. *Geomicrobiology Journal*. https://doi.org/10.1080/01490451.2023.2261443

Narayan, R. (2020). 16. Principles, drivers, and analysis of biodegradable and biobased plastics. In C. Bastioli (ed.), *Handbook of biodegradable polymers* [online] (pp. 455–474). De Gruyter. https://doi.org/10.1515/9781501511967-016/html

Nielsen, J., & Keasling, J. D. (2016). Engineering cellular metabolism. *Cell, 164*(6), 1185–1197.

OECD. (2023). OECD Economic Outlook. 2023(1). OECD Publishing.

Orooji, Y., Han, N., Nezafat, Z., Shafiei, N., Shen, Z., Nasrollahzadeh, M., Karimi-Maleh, H., Luque, R., Bokhari, A., & Klemeš, J. J. (2022). Valorisation of nuts biowaste: Prospects in sustainable bio(nano)catalysts and environmental applications. *Journal of Cleaner Production, 347*, Article 131220.

Parisi, M. L., Dessì, A., Zani, L., Maranghi, S., Mohammadpourasl, S., Calamante, M., Mordini, A., Basosi, R., Reginato, G., & Sinicropi, A. (2020). Combined LCA and green metrics approach for the sustainability assessment of an organic dye synthesis on lab scale. *Frontiers in Chemistry, 8*, 214.

Patel, A. D., Zabeti, M., Seshan, K., & Patel, M. K. (2020). Economic and environmental assessment of catalytic and thermal pyrolysis routes for fuel production from lignocellulosic biomass. *Processes, 8*(12), 1612.

Peng, J., Schwalbe-Koda, D., Akkiraju, K., Xie, T., Giordano, L., Yu, Y., Eom, C. J., Lunger, J. R., Zheng, D. J., Rao, R. R., Muy, S., Grossman, J. C., Reuter, K., Gómez-Bombarelli, R., & Shao-Horn, Y. (2022). Human- and machine-centred designs of molecules and materials for sustainability and decarbonization. *Nature Reviews Materials, 7*(12), 991–1009.

Pradhan, G., Tripathy, B., Ram, D. K., Digal, A. K., & Das, A. P. (2023). Bauxite mining waste pollution and its sustainable management through bioremediation. *Geomicrobiology Journal*, 1–10. https://doi.org/10.1080/01490451.2023.2235353

Ray, S. K., Dey, S., Das, A. P., & John, A. (2024). Microplastic waste management through greener technologies: A review. In *Sustainable management of urban plastic waste through circular economic approaches* (pp. 237–261). https://doi.org/10.1201/9781003465447-14

Reddy Koteswara, G., Bhargavi, S., Yamuna, D., Anjani Devi, M., Jeevana Lakshmi Kumar, G., Manas, T., Adarsh, S., Bhavya Sai, V., Nath, H., Alreshidi, M. A., Togun, H., Ram, D. K., & Das, A. P. (2025). Recent advancements in the remediation of marine hydrocarbon and plastic pollutants. *Environmental Science and Engineering, Part, F218*, 127–142. https://doi.org/10.1007/978-3-031-81615-4_7

Renewable Energy Generation and Value Addition from Environmental. (2024).

Rodríguez-Restrepo, Y. A., & Orrego, C. E. (2020). Immobilization of enzymes and cells on lignocellulosic materials. *Environmental Chemistry Letters, 18*(3), 787–806.

Santos, L., Fraga, G., Pontes, D., Campos, L., Pontes, L., & Teixeira, L. (2021). Alternatives for the production of levulinic acid obtained from biomass. *Química Nova* [online]. http://quimicanova.sbq.org.br/audiencia_pdf.asp?aid2=9325&nomeArquivo=RV2021-0061.pdf

Schuur, B., Brouwer, T., Smink, D., & Sprakel, L. M. J. (2019). Green solvents for sustainable separation processes. *Current Opinion in Green and Sustainable Chemistry, 18*, 57–65.

Sheldon, R. A. (2014). Green and sustainable manufacture of chemicals from biomass: State of the art. *Green Chemistry, 16*(3), 950–963.

Sheldon, R. A. (2016). Green chemistry and resource efficiency: Towards a green economy. *Green Chemistry, 18*(11), 3180–3183.

Sheldon, R. A. (2018). Metrics of green chemistry and sustainability: Past, present, and future. *ACS Sustainable Chemistry & Engineering, 6*(1), 32–48.

Sheldon, R. A. (2020). Biocatalysis and biomass conversion: Enabling a circular economy. *Philosophical Transactions of the Royal Society a: Mathematical, Physical and Engineering Sciences, 378*(2176), 20190274.

Sheldon, R. A., & Woodley, J. M. (2018). Role of biocatalysis in sustainable chemistry. *Chemical Reviews, 118*(2), 801–838.

Singh, R., Das, R., Sangwan, S., Rohatgi, B., Khanam, R., Peera, S. K. P. G., Das, S., Lyngdoh, Y. A., Langyan, S., Shukla, A., Shrivastava, M., & Misra, S. (2021). Utilisation of agro-industrial waste for sustainable green production: A review. *Environmental Sustainability, 4*(4), 619–636.

Singh, A., Srivastava, S., Rathore, D., & Pant, D. (2020). *Environmental microbiology and biotechnology: Volume 1: Biovalorization of solid wastes and wastewater treatment*. Springer Singapore Pte. Limited.

Singh, S., Singh, A., Gupta, A., Tripathy, D. B., & Kesarwani, S. (2025). Biological and enzymatic approach for different applications. *In*: *Green Chemistry* [online] (pp. 197–223). Elsevier. https://linkinghub.elsevier.com/retrieve/pii/B9780443219900000092

Song, J. H., Murphy, R. J., Narayan, R., & Davies, G. B. H. (2009). Biodegradable and compostable alternatives to conventional plastics. *Philosophical Transactions of the Royal Society b: Biological Sciences, 364*(1526), 2127–2139.

Straathof, A. J. J., Wahl, S. A., Benjamin, K. R., Takors, R., Wierckx, N., & Noorman, H. J. (2019). Grand research challenges for sustainable industrial biotechnology. *Trends in Biotechnology, 37*(10), 1042–1050.

Tan, J.-N., & Dou, Y. (2020). Deep eutectic solvents for biocatalytic transformations: Focused lipase-catalyzed organic reactions. *Applied Microbiology and Biotechnology, 104*(4), 1481–1496.

Tao, J. (Alex), & Kazlauskas, R., (Eds.). (2011a). *Biocatalysis for green chemistry and chemical process development* [online] (1st ed.). Wiley. https://doi.org/10.1002/9781118028308

Tao, J., & Kazlauskas, R. J. (Eds.). (2011b). *Biocatalysis for green chemistry and chemical process development*. John Wiley & Sons.

Thangaraj, B., Solomon, P. R., Muniyandi, B., Ranganathan, S., & Lin, L. (2019). Catalysis in biodiesel production—A review. *Clean Energy, 3*(1), 2–23.

Tóth, A. J., Fózer, D., Mizsey, P., Varbanov, P. S., & Klemeš, J. J. (2023). Physicochemical methods for process wastewater treatment: Powerful tools for circular economy in the chemical industry. *Reviews in Chemical Engineering, 39*(7), 1123–1151.

Tuck, C. O., Pérez, E., Horváth, I. T., Sheldon, R. A., & Poliakoff, M. (2012). Valorization of biomass: Deriving more value from waste. *Science, 337*(6095), 695–699.

Tundo, P., & Griguol, E. (2018). Green chemistry for sustainable development. *Chemistry International, 40*(1), 18–24.

Wandrey, C., Liese, A., & Kihumbu, D. (2000). Industrial biocatalysis: Past, present, and future. *Organic Process Research & Development, 4*(4), 286–290.

Werpy, T., & Petersen, G. (2004). *Top value added chemicals from biomass: Volume I—Results of screening for potential candidates from sugars and synthesis gas* [online]. No. DOE/GO-102004–1992, 15008859. http://www.osti.gov/servlets/purl/15008859/

Yan, Z., Ji, M., Xia, J., & Zhu, H. (2020). Recent advanced materials for electrochemical and photoelectrochemical synthesis of ammonia from dinitrogen: One step closer to a sustainable energy future. *Advanced Energy Materials, 10*(11), 1902020.

Zhang, T., Li, W., Xiao, H., Jin, Y., & Wu, S. (2022). Recent progress in direct production of furfural from lignocellulosic residues and hemicellulose. *Bioresource Technology, 354*, Article 127126.

Zhao, Y., Gani, R., Afzal, R. M., Zhang, X., & Zhang, S. (2017). Ionic liquids for absorption and separation of gases: An extensive database and a systematic screening method. *AIChE Journal, 63*(4), 1353–1367.

Zimmerman, J. B., Anastas, P. T., Erythropel, H. C., & Leitner, W. (2020). Designing for a green chemistry future. *Science, 367*(6476), 397–400.

Microbial Genomics for a Stable Future: Tackling Strategies for Heavy Metal and Organic Pollutant Remediation

Subhasmita Mallik, Ellojita Rout, Subham Jyoti Sahoo, Swayamprabha Sahoo, and Jatindra Nath Mohanty

1 Introduction

The rapid pace of industrialization and anthropogenic development has significantly contributed to environmental degradation, primarily through the excessive exploitation of natural resources and the persistent release of pollutants into soil, water, and air (Medfu Tarekegn et al., 2020, Reddy Koteswara et al., 2025; Sahu & Das, 2024; Swain et al., 2025). Some of the industries like mining, chemical processing, textiles, and petroleum refining generate extensive amounts of toxic waste that includes heavy metals and organic pollutants, which pose long-term risks to ecological and human health (Das et al., 2023a; Dey et al., 2023). If we see about the cause, will find the lack of effective waste discharge protocols, which often leads to the accumulation of hazardous substances with genotoxic and mutagenic effects (Prabakar et al., 2018). The old traditional methods for remedies like chemical precipitation, incineration, and adsorption though widely practiced are there but they often come with

S. Mallik · S. J. Sahoo
Department of Botany, School of Applied Sciences, Centurion University of Technology and Management, Jatni, Bhubaneswar, India

E. Rout
Department of Microbiology, Malla Reddy Medical College for Women, Quthbullapur, Hyderabad, Telangana, India

S. Sahoo
Department of Biotechnology, School of Pharmaceutical Sciences, SOA Deemed to Be University, Bhubaneswar, India

J. N. Mohanty (✉)
Assistant Professor, School of Biotechnology, Centurion University of Technology and Management, Jatni, Bhubaneswar, India
e-mail: jatindranathmohanty@gmail.com

S. Ghosh et al. (eds.), *Advances in Organic Waste Conversion Through Industrial Biotechnology*, Sustainable Environmental Waste Management Strategies,
https://doi.org/10.1007/978-3-032-11377-1_20

high operational costs, limited efficiency and the possibility of secondary contamination as well. Now a days, we have the sustainable alternatives like bioremediation that utilizes biological systems to degrade or neutralize environmental contaminants (Ramrakhiani et al., 2016).

The microorganisms plays a vital role as a biological agents found having enormous potential towards bioremediation which is due to their adaptability, resistance, and metabolic versatility. They are tolerant extreme condition and dealing with different types of contaminant (Poursat et al., 2019), also we found microbes are great candidate for decontamination process. Microorganisms play a major role in ecosystem processes, particularly nutrient cycling and ecological stability (Mohanty et al., 2024a; Priyadarshini et al., 2024). They are considered the most important players either in biotransformation, immobilization, or degradation of organic substances such as polycyclic aromatic hydrocarbons (PAHs) and to a lesser extent for inorganic pollutants such as cadmium, lead, and arsenic type compounds (Ghosh & Das, 2020; Ghosh et al., 2016, 2020; Joutey et al., 2013). The microorganisms used in remediation can be exogenous, brought in from other ecosystems where they exhibit higher degradative capabilities, or indigenous, living naturally in contaminated areas (Das et al., 2025; Allam et al., 2025; Ray et al., 2025). In this regard, microbial bioremediation has gained increased recognition as an environmentally benign, economical, and sustainable substitute. The metabolic capacity of some microbial species to detoxify, break down, change, or immobilize toxic substances is innate (Das et al., 2023b, 2024a). Yan et al. (2020a, 2020b), reported about the diverse microbial groups like bacteria, fungi, and archaea that plays critical roles in contaminant degradation through mechanisms such as enzymatic transformation, biosorption, redox reactions as well as bioaccumulation. Still, the natural efficacy of these microbial communities in polluted environments is often limited. The major causes are found as slow growth rates, high sensitivity to environmental stress and narrow metabolic capabilities (Das et al., 2024a; Dey et al., 2024a, 2024b). If we see the recent advancements in molecular biology and microbial genomics, we get promising solutions to overcome these limitations. In bioremediation, the use of next-generation sequencing (NGS), functional genomics and comparative genomic analyses enables researchers to unravel the genetic architecture of microorganism (Mishra et al., 2023; Mishra & Das, 2022, 2023). By using these tools we can get key resistance genes, stress response pathways, and biodegradative enzymes responsible for the detoxification of organic pollutants and heavy metals (Tang et al., 2024). Besides, the whole genome sequencing (WGS) techniques provide deeper insights into microbial adaptation and also ecological function within contaminated ecosystems as well (Sharma et al., 2021). These days, identifying certain metal resistance gene and pollutant degradation pathways is made possible by genomic insights only. The best example is enzymatically converting Hg (II) to the less hazardous elemental mercury, the mer operon that includes genes like merA, merB, merP, and merT. These genes are mostly found associated with the mercury detoxification. Likewise if we see the ars operon, which codes for proteins mostly involved in cellular efflux and arsenate reduction, controls arsenic resistance (Singh et al., 2023). In certain bacterial strains also both respiratory and detoxification pathways are used to reduce arsenate, showcasing the genetic adaptability of microbes in highly toxic environments (Ray et al., 2024; Sahu et al., 2024). The advanced metagenomic analysis has recognized Methylocystis HL18, from methane-fed bioreactors as a promising candidate for heavy metal cleanup due to its mer and ars gene clusters (Shi et al., 2019). Other than that strains like Alcaligenes aquatilis possess genes associated with biosurfactant production and aromatic hydrocarbon degradation, supporting the breakdown of complex pollutants like diclofenac (Mohamed et al., 2023, Das et al., 2023b, Das et al., 2024b). Furthermore, the discovery of metal efflux operons, such as czc (conferring resistance to cadmium, zinc, and cobalt) and cop (copper homeostasis), explains details on microbial adaptations to heavy metal stress. The functional characterization study of these systems offers valuable molecular targets for metabolic engineering and synthetic biology. One of the examples is *Bacillus cereus* BC4, that has shown strong resistance to chromate and can detoxify polluted soils (Liu et al., 2025). Due to the presence of advance tools like transposon mutagenesis and RNA sequencing help identify the genes and pathways involved in sensing and breaking down pollutants. These techniques allow detailed understanding of microbial responses, aiding the creation of improved microbes for bioremediation.

The model microorganism, Pseudomonas putida and Escherichia coli are genetically modified for expression of express arsenic-methylating genes (arsM and arsA), facilitating the transformation of inorganic arsenic into less toxic, volatile forms (Yang et al., 2013). Similar report also there with Bacillus subtilis that has been engineered to express thermostable arsenite methyltransferases, enhancing its utility in the treatment of organic and mixed waste streams (Huang et al., 2015). By combining genomics with metabolic engineering, scientists can now add or enhance pathways in microbes to boost their tolerance, speed up pollutant breakdown, and target a wider range of pollutants, including drugs and pesticides (Shah et al., 2024). Recent breakthroughs in CRISPR-Cas genome editing have further transformed microbial engineering for environmental applications. This technology enables precise, site-specific modifications, such as the insertion, deletion, or regulation of genes involved in metal resistance and organic contaminant degradation. CRISPR-based modifications have been employed to enhance the expression of metal-binding proteins and catabolic enzymes in Pseudomonas and *E. coli*, improving their capacities to degrade cadmium, lead, and hydrocarbons. Unlike conventional mutagenesis, CRISPR offers high specificity with minimal off-target effects, making it ideal for designing tailored microbial strains for site-specific remediation. Moreover, CRISPR interference (CRISPRi) and activation (CRISPRa) systems allow reversible and tunable control of gene expression, enabling fine adjustment of microbial functions in contaminated settings without permanent genomic alteration.

Altogether, microbial genomics has revolutionized our approach to environmental remediation. By bridging the gap between the intrinsic capabilities of native microbes and practical application, it enables the strategic design and optimization of microbial strains for targeted bioremediation. As environmental contamination grows in complexity, the integration of omics-based technologies including CRISPR, metagenomics, and systems biology that will be pivotal in developing next-generation, sustainable solutions for the remediation of heavy metals and organic pollutants. In this chapter, we aim to highlight these recent advancements in microbial genomics and bioengineering approaches, with a particular focus on how cutting-edge tools are driving innovation in the field of bioremediation.

2 Microbial Genomics and Bioremediation Strategies

With day by day, due to excess human interference, organic pollutants, and widespread presence of heavy metals increasing tremendously in terrestrial ecosystems, which has raised the urgent need for effective and sustainable remediation strategies. Apart from various industrial, effective yet expensive chemical remediation strategies (Gong et al., 2018, Mohanty et al., 2024b; Pradhan et al., 2023), bioremediation stands out as the most sustainable, economically and environmentally friendly one to clean out the heavy metals from both land and water resources. Microbes with their ability to sustain in extreme environments and conditions and with the ability to accumulate and metabolize toxic compounds make them the key candidate of the Bioremediation process (Yan et al., 2020a, 2020b). Despite of their potential, some significant properties like slow growth rate and low metabolic activity under harsh conditions stands as pillar of limitation infront of the measures of microbial remediation (Fig. 1), which could hamper the overall efficiency of these microbes in large scale application and long run. (Liu et al., 2023). With recent development of the microbial genomics field, now researchers can interpret the genetic and functional blueprints of these organisms, enabling the strategic enhancement of their detoxification capabilities (Tang et al., 2024).

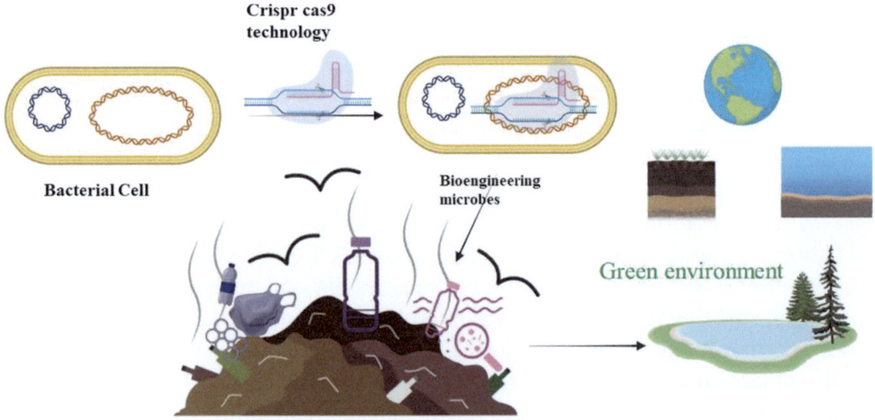

Fig. 1 Microbial bioremediation process using the strategies like CRISPR-Cas 9 to have a healthy green environment

Having a better understanding about the genomic tools such as high throughput sequencing, functional genome annotation and comparative genomics have revolutionized our understanding of key resistance genes and respective enzymes responsible for degradation of different types of pollutants and stress response elements. This understanding can emphasize knowing the pathways involved in the uptake and utilisation of heavy metals. While the whole genome sequencing of microbes involved in bioremediation can give an insight to the genetic blueprint of that microbe which enables those microbe to acquire these kind of characteristics (Ghosh et al., 2023; Paul et al., 2023). In the traditional way of studying microbial genomics, it often failed to capture the full microbial diversity, but advancements in the next-generation sequencing have revolutionized this field of research by lowering costs and offering deeper insight into microbial genomics, proteomics and metabolomics (Sharma et al., 2021). The potential of the microbial remediation now can be modified by researchers with the knowledge of metagenomics. Illustration of the potential of microbes in bioremediation is highlighted in several case studies which also defines their remarkable genetic adaptability. Remediation or conversion of toxic mercury ions [Hg(II)]into a less toxic mercuric form is done by both Gram positive and Gram negative bacteria possessing mer operon, which synthesize mercuric reductase enzyme and this operon includes several genes like *merA, merB, merP, merT,* and others, though the exact composition varies among strains (Singh et al., 2023). Similarly, *Shewanella* sp. strain ANA-3 is capable of reducing arsenate [As (V)] through two systems: the arr respiratory pathway and the ars detoxification operon, which includes the core genes ars RBC, coding for a repressor, a reductase, and a transmembrane efflux pump. In a later study, the metagenomic research uncovers a novel *Methylocystis* species (HL18) in a methane-fed bioreactor. This strain possesses both the gene clusters for merA and ars operons, for which it can managed to clean up waters with both mercury and arsenic contamination (Shi et al., 2019). On the other side, further genomic analysis revealed another bacteria *Alcaligenes aquatilis* having multiple aromatic hydrocarbon pathways and biosurfactant genes, which enables complete degradation of some inorganic compounds like diclofenac (Mohamed et al., 2023). *Bacillus cereus* BC4 strain have shown chromate resistance and detoxify chromate contaminated soils (Liu et al., 2025). Similar to this some other operons like CZC (or cadmium, zinc, and cobalt efflux), and cop (for copper homeostasis), has also characterized and discovery of such genes supports our understanding regarding microbial adaptation to polluted environments with providing molecular targets for genetic engineering. Functional genomics approaches, which includes transposon mutagenesis and RNA-seq, further enable the identification of gene networks and regulatory circuits involved in pollutant sensing, transport, and breakdown. It can be said that the genome sequencing serves as the foundational step in identifying and characterizing efficient microbial candidates for bioremediation and opens the way for metabolic engineering and synthetic biology interventions for boosting their bioremediation.

By identifying and characterizing these mechanisms, metabolic engineers can modify key enzymes to enhance microbial efficiency in breaking down specific heavy metals, while also expanding their capability to target a broader range of

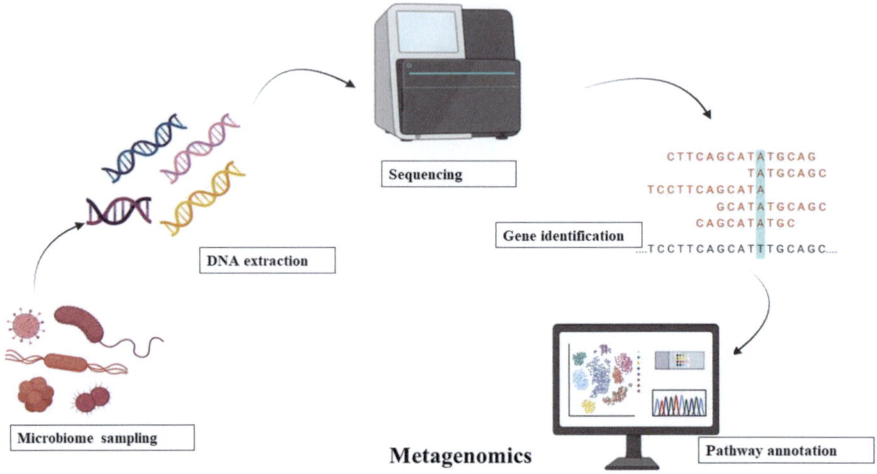

Fig. 2 Workflow of metagenomics approach for bioremediation using microbial sequencing data

contaminants. Recent advances have enabled the introduction or overexpression of genes targeting specific pollutants, enabling microbes to operate more efficiently under harsh conditions (Fig. 2). For example, *Pseudomonas putida* and *E. coli* has been genetically modified to express the arsM and arsA gene, respectively (Table 1), enabling it to methylate and volatilize inorganic arsenic into less toxic forms—a key study demonstrating arsenic detox potential in water treatment (Han, 2024; Yang et al., 2013). Similarly, Huang et al. (2015) engineered *Bacillus subtilis* with heat-resistant arsenite methyltransferase for bioremediation of arsenic from contaminated organic wastes. So integration of omics data with metabolic engineering accelerates bioremediation of complex pollutants with a range from heavy metals to unwanted contaminants like pharmaceutical chemicals and pesticides (Shah et al., 2024).

3 Metagenomics Approaches for Discovering Novel Bioremediators

Functional metagenomics seeks to perform the identification of protein roles that are encoded by the collective genomes of microbial communities in totality, foregoing any need for culturing or isolating individual organisms. The enormous pool of uncultured microbes makes this approach uncover new biocatalysts in an effective way. Different functional metagenomic strategies have been used in recent years to experimentally validate gene functions as well as discover enzymes plus metabolic pathways involved in the breakdown of pollutants, including both those that involve cloning metagenomic fragments into expression hosts alongside those that do not (Berini et al., 2017). Bioremediation takes advantage of this because

Table 1 Key genes involved in microbial heavy metal resistance and degradation

Sl no	Gene	Function	Pollutant targeted	Microbial example	References
1	merA	Mercuric reductase—reduces Hg(II) to Hg(0)	Mercury (Hg)	*Pseudomonas*, *Bacillus*, *Methylocystis*	Zheng et al. (2018)
2	arsC	Arsenate reductase—reduces As(V) to As(III)	Arsenic (As)	*Shewanella*, *Escherichia coli*	Saltikov et al. (2005)
3	copA	Cu-efflux ATPase—pumps out excess copper	Copper (Cu)	*Enterococcus*, *Cupriavidus*	Andrei et al. (2020)
4	czcA	Cobalt-zinc-cadmium resistance protein	Cobalt-zinc-cadmium	*Ralstonia*, *Alcaligenes*	Remenar et al. (2018)
5	catA	Catechol 1,2-dioxygenase—breaks aromatic rings	Hydrocarbons	*Pseudomonas putida, Achromobacter xylosoxidans*	Guzik et al. (2011); Liu et al. (2023)
6	alkB	Alkane monooxygenase—oxidizes alkanes	Alkanes	*Alcanivorax borkumensis*	Bukliarevich et al. (2023)
7	chrA, chrR	Chromate reductase-reduce Cr(VI) to Cr(III)	Chromate (Cr)	*Bacillus cereus, Escherichia coli FACU*	Ayangbenro and Babalola (2020)
8	RutE, NfsA, NfsB	Nitroreductase	Chromate (Cr)	*Serratia sp.*	Dong et al. (2018)
9	cadA, cadB	Cadmium efflux ATPase—removes cadmium ions from the cytoplasm	Cadmium (Cd)	*Serratia marcescens, Staphylococcus aureus*	Zhu et al. (2022); Hussain et al. (2022)
10	nccA	Efflux transporter for nickel–cobalt-cadmium	Nickel (Ni), Cobalt (Co), Cadmium (Cd)	*Alcaligenes xylosoxidans 31A*	Siunova et al. (2025)

(continued)

Table 1 (continued)

Sl no	Gene	Function	Pollutant targeted	Microbial example	References
11	zntA	$Zn^{2+}/Pb^{2+}/Cd^{2+}$ P-type ATPase—exports excess heavy metals from cells	Zinc (Zn), Lead (Pb), Cadmium (Cd)	*Staphylococcus aureus, Escherichia coli, Ralstonia metallidurans*	Hussain et al. (2022)
12	ycdH operon yciC operon		Zn	*Bacillus subtilis*	Hussain et al. (2022)

environmental microbes detoxify naturally. Bioremediation is thus an emerging method within bioengineering. It endeavours to eradicate or counteract contaminants inside remediation architectures by dint of microbial metabolic mechanisms. To achieve efficacious bioremediation, pollutants should be organic and biodegradable, also microbes must exhibit the requisite metabolic capabilities (Ghosh et al., 2023; Paul et al., 2023). Furthermore, environmental conditions must strengthen degradation coupled with microbial activity. Metagenomic techniques may assist with discovering revolutionary functional microbes plus genes that allow the creation of microbial strains exhibiting elevated degradation efficiency, broad adaptability, also stable performance (Zhang et al., 2021). Combined approaches using metagenomics as well as metatranscriptomics have been employed as researchers probe the microbial communities along with genes responsible for metabolic activities throughout nitrifying sludge. We have discerned various bacterial groups, such as Bacteroidetes, Alphaproteobacteria, and Betaproteobacteria. Additionally, metatranscriptomic scrutiny markedly augmented the articulation of genes implicated in the decomposition of aromatic substances (Sun et al., 2019). Fang et al. reported that activated sludge from wastewater treatment facilities has genes identified therein that are associated with biodegradation and pesticide breakdown. Environment-related factors are changing operational conditions, exhibiting a positive effect on microbial diversity, which changed between 50.4 and 76.8%.The wide range of microorganisms responsible for the bioremediation of arsenic (As) and antimony (Sb) degradation has been analysed by metagenomics approaches (Fang et al., 2018). According to Luo et al., five microbial phyla, Actinobacteria, Nitrospirae, Gemmatimonadetes, Firmicutes, and Tenericutes are actively involved in disinfection of As and Sb-contaminated soils, may be utilized at other pollutant regions (Luo et al., 2014). It has become necessary to recognize the physical characteristics and chemical components of the pollutant in the sample taken from the polluted area. In recent years, the selection of microbial species has had a significant effect on the extent to which bioremediation strategies are essential. Metagenomics has evolved through various phases, from early sequencing techniques to advanced next-generation sequencing (NGS), which reduces chimeric sequences that may affect data quality along with providing more accurate microbial DNA profiles (Heidari & Panico, 2020; Kumar et al., 2020).Sequence-based metagenomics involves communities extracting whole genomic DNA from microbes, then researchers sequence shotgun and assemble the sequencing reads afterward. For the analysis of assembled sequences, bioinformatics tools are being used. This analysis determines the diversity in functional roles of the present microbes. In the past decade, numerous sequencing efforts have focused on highly contaminated environments so that they may explore the microbial community structure within these ecosystems. (Thomas et al., 2012). Genetic circuits are expected to then regulate microbial functions, and also mathematical tools such as dynamic and agent-based models forecast results then help simulate community interactions. These stable and effective methods support microbial consortia development for biotechnology and environmental applications (Han et al., 2023; He et al., 2024). Promoters together with transcription factors serve as components that are reliable yet flexible. They act as activators or repressors, responding to chemical inducers as they build strong genetic circuits inside synthetic microbial communities (Bittihn et al., 2018; Selberg et al., 2018). Additionally, these helpful tools can develop a synthetic microbial community with CRISPR technology (Jeong et al., 2023). Viruses as well as their hosts exist in a continuous evolutionary conflict. Various resistance and counter-defence strategies have developed from this conflict. The genome of Haloferax mediterranei was where Clustered Regularly Interspaced Short Palindromic Repeats (CRISPR), a defence system, was first identified (Hutinet et al., 2019; Mojica et al., 1993). Several microbial genes, including *atz*, *psb*, *ndo*, *tfd*, *puh*, *tri*, and *trz*, produce several kind of enzymes such as oxidoreductases (like oxygenases, laccases, and peroxidases) and hydrolytic enzymes (such as lipases, proteases, and cellulases), which play an important role in the degradation of herbicides (Aislabie et al., 2005; Clausen et al., 2002). The whole genome of several bacteria, like Rhodococcus species and Pseudomonas putida, which is known for their ability to bio-mediated pesticides, these bacteria have been sequenced. Identifying the bacterial genotypes and phenotypes related to biodegradation has improved due to proteomics, metabolomics, and transcriptomics. The development of genome-scale models (GEMs) of genomes, which determined the most promising microorganism for breaking down the pesticides and other xenobiotics. For example, lysine pathway in Corynebacterium glutamicum ATCC has been significantly identified by provided significantly identify metabolomics and transcriptome data analysis of 13,284 points of information (Kromer et al., 2004). Under specific conditions, omics studies comprising genomic, transcriptomics, and proteomics, help in identifying alterations in gene or protein expression. Transcriptional analysis *Pseudomonas bharatica* CSV86T revealed the activation of benzoate (benE, benK) and glucose (gbp, oprB, glcG) metabolism linked genes (Choudhary et al., 2017). In Methylorubrum sp. ZY-1, combined transcriptomic and metabolomic analyses, assisted in identifying the molecular mechanisms involved in the degradation of pentachlorobiphenyl (PCB 118). This evidences gives a detailed overview on how omics approaches are applied in the bioremediation of contaminants from pharmaceuticals, cyanotoxins, plasticizers, and pesticides of emerging concern (CECs) (Wu et al., 2024). The metagenomic studies of wastewater samples contaminated with testosterone revealed that *Comamonas* and *Pseudomonas* have major contribution to the breakdown of proteins. Additionally, the tesB gene, which encodes meta-cleavage dioxygenase, showed a significant increase in expression after 48 h of incubation (Chen et al., 2016). In aquatic ecosystems, Proteobacteria and Actinobacteria play a significant role in the breakdown of oestrogen as determined by comparative genomics and metagenomics studies. Genes *aedJ* and *edcC* were identified as promising biomarkers for the breakdown of oestrogen in *Proteobacteria* and *Actinobacteria* providing key devices for monitoring the environment (Hsiao et al., 2023).

4 Synthetic Biology and CRISPR-Based Microbial Engineering

In recent years, gene editing techniques using CRISPR-Cas 9 have emerged as the most successful approach for enhancing tolerance to heavy metals through efflux pump mechanisms, promoting the enzymatic pathway to break down the complex organic compounds, and removing gene regulatory restriction to improve pollutant uptake (Chen et al., 2024; Ren et al., 2025).The editing of regulated genes through CRISPR-based approaches is control the central metabolism, redirected carbon flux into contaminant degradation routes (Bhanu et al., 2024). Similarly editing the metal-binding gene and regulators of stress response to enhance microbial resistance to heavy metals (Ren et al., 2025). For the degradation of organic pollutants, CRISPR can be used to insert the catabolic gene, thereby enhancing the enzymatic pathway (Chen et al., 2024).The breakdown of harmful compounds including aromatic compounds, pesticides, microplastics, and greenhouse gases can be achieved using synthetic microbes engineered through advance CRISPR-Cas technology (Bhatt et al., 2021; Miri et al., 2022; Xiang et al., 2021). If we talk about synthetic bacteria can be designed to target heavy metals by gene editing metal specific uptake gene enabling the removal these pollutants from environment (Thai et al., 2023; Das et al., 2023b; Das et al., 2024b).

Jia et al., 2020 reported that *E. coli* BL21 which has three Pb binding proteins PbrR, PbrD and PbrR691, upon fusing to the N-terminal domain of the ice-nucleation protein enabled to improved degradation of Pb from 200 μm to 942.1,754.3, and 864.8 μmol for soil contaminants (Jia et al., 2020). Similarly, *B. subtilis PY79* with the P cotB Promoters, which expressed the fusion of cotB protein and His18 peptide, effectively removed Ni from 3 ppm to 30.66 nmol/mg (Hinc et al., 2010). Other than efflux pump mechanism and enzymatic modification, researcher have been able to improve targeted genetic modification using CRISPR-Cas 9,which can precisely edit the specific genes and operon that enhance their expression such as mer (Mercury), ars (arsenic), and czc (cobalt, zinc, cadmium) (Verma & Kuila, 2019; Akoijam &Joshi, 2024). If we see the pollutant resistance to our environment increasing, day by day it is a serious threat for us due to chemical stability. To enhance the degradation of harmful compounds, scientists have been able to design a catabolic pathway through synthetic biology. Nowadays, following this approach, bacteria like *E. coli* and *Pseudomonas putida* are suitable for insertion of an engineered gene that break down particular pollutants such as industrial dyes, pesticides and polycyclic aromatic hydrocarbons (Nazir et al., 2024).By the use of oxidation, methylation and reduction, synthetic bacteria have incorporated detoxification pathway that chemical transform harmful metals into less toxic forms. Synthetic bacteria have been able to produce metal-binding peptides which enhance the metal accumulation (Thai et al., 2023).

Jeong et al. (2023) reported bioengineered bacteria have promising approaches towards restoring polluted environments by specifically remediating complex organic metal pollutants. In the case of synthetic microbes, scientists can engineer bacteria to detect contaminants using biosensors that is responsible for heavy metals like copper, mercury, and lead (Jeong et al., 2023). Similarly, researchers can edit the microbes' genetic circuit to regulate and monitor the concentration of contaminants in the environment, and adjust their metabolic activity (Mao et al., 2024). Another study by Aminian-Dehkordi et al. (2023), reported that bioengineered bacteria have efficacy to target and degrade complex organic metals, pharmaceutical residues and plastic through engineering their metabolic pathway. These bacteria also increasing their capacity to produce enzyme when concentration of contaminant are high.

5 Microbial Consortia and Horizontal Gene Transfer in Bioremediation

The use of microbial communities in the breakdown of pollutants during the bioremediation process has gained worldwide attention due to its sustainability and eco-friendliness towards the environment. If we talk about the bioremediation process, microbial consortia play a significant role in degrading organic pollutants like petroleum hydrocarbons (Martínez-Cuesta et al., 2023). *Bacillus* and *Pseudomonas* species break down the petroleum compounds using their enzyme hydroxylases and dioxygenases, which degrade the hydrocarbons (Aziz et al., 2024; Dell'Anno et al., 2021). Recent studies have shown the several microbial species are capable to degrading heavy metals through synthetic microorganisms and microbial consortia, which remove complex metals via processes such as biomineralization, adsorption, and biological leaching (Pande et al., 2022; Das et al., 2024a; Dey et al., 2024a, 2024b). In we see the case of cadmium degradation in the aquatic environment, *Vibrio harveyi* has shown more potential for bioremediation (Abd-Elnaby et al., 2011). Microbial consortia have also demonstrated efficiency in degrading mercury ions through microbial detoxification mechanisms (Radojević et al., 2024). In another mechanism, metal mobilization and immobilization microorganisms exhibit the ability to transform, tolerate, and biosorb the heavy metals like lead, copper, zinc, and cadmium. Several environmental factors significantly influence the efficacy of oxygen availability, concentration of pollutant, temperature, and pH (Igiri et al., 2018; Jeyakumar et al., 2023). Bioengineered microorganisms have addressed the limitations of natural microbial habitats by enabling the cleanup of hazardous chemicals, pesticides, and xenobiotics (Pant et al., 2021). Curiel-Alegre et al. reported that in microbial horizontal gene transfer mechanisms, bacteria such as Pseudomonas and Acinetobacter played a key role in the degradation of petroleum hydrocarbons (Curiel-Alegre et al., 2024).

Similarly, *Delftia lacustris* plays a significant role in expelling the harmful metals such as lead, mercury, cadmium, and chromium through the membrane via the efflux pump mechanism (Wu et al., 2016). In the biostimulation process, the *Enterobacter kobai* has the ability to degrade lead enhanced by nutrients, pH, and oxygen level (Adeyi et al., 2017).The more effective mechanism of HGT allow microorganisms to acquire gene responsible for degradation of complex pollutants such as aromatic hydrocarbons, heavy metals (Kim et al., 2022).Similarly, Zhu et al. reported that heavy metals like mercury, lead, and arsenic which are resistance to the environment due to the HGT mechanism, where bacteria acquired genes that enable them to detoxify these metals (Zhu et al., 2020). In the microbial communities, HGT improves bioremediation potential by enabling the spread of degradative pathways across microbial consortia (Li et al., 2022). The degradation of large hydrocarbon into small molecules which can then be further degraded by other species allows the mobilization of these pollutants from one microbial consortium to another (Yadav et al., 2021).

Researchers have focused on genetically modified organisms (GMO) within microbial consortia to enhance the HGT capabilities, enabling the consortia to degrade specific contaminants in the environment (Srivastava, 2021). Recently advanced technologies such as CRISPR-Cas 9, prime editing and TALENs have played a significant role in breakdown of heavy metals, oil hydrocarbons and organic pollutants through the use of genetically modified microbes (Qattan, 2025). Synthetic microbes such as *Cupriavidus metallidurans*, *Pseudomonas putida*, and *Shewanella oneidensis* have shown the ability to degrade arsenic, cadmium, mercury, (Saravanan et al., 2022) and lead through the overexpressed enzymes that facilitating uptake and detoxification (Table 2).

6 Omics Technologies for Monitoring and Optimizing Bioremediation

To understand the function and behaviour of microbes in a contaminated site and the interaction in a microbiome is observed by molecular approaches such as genomics, proteomics, transcriptomics, metabolomics, and fluxomics. To get a clearer view of how microbes function during the process of bioremediation, scientists combines data from different omics techniques and a complete road map of genes presents in microbes to help in remediation to how they are expressing and how they respond to different toxic substances to how they interact with other microbes in soil microbiome can be understood (Aziz et al., 2021). With the help of these tools it is easier to explore how microbes respond to pollutants. With proteomics, we can track which proteins are involved in breaking down harmful compounds. Metabolomics helps us detect and measure the by-products formed during this process, while transcriptomics reveals how gene expression shifts when microbes are exposed to toxic substances (Rodríguez et al., 2020).

This omics approaches not only help in studying the process of bioremediation, it also help in getting a brief idea about the extent and history of contamination. For that scientists have to analyse the entire microbial community or focus upon specific genes of interest to trace contamination sources precisely. Techniques like quantitative PCR (qPCR) are especially useful for this. qPCR can quickly detect particular genes or microbial species within just a few hours of collecting a sample.

Table 2 List of bioengineered microbes and their involvement in different bioremediation related function

Bioengineered microbes	Target pollutant	Mechanism	Description of the mechanism	References
Bacillus cereus	Pb(II), As(V) and Cr(VI)	Adsorption	Bacteria can attach heavy metals to their cell surfaces through functional groups like carboxyl, hydroxyl, and phosphate	Zhang et al. (2024a, 2024b); Zhang et al. (2024a, 2024b)
Cupriavidus metallidurans	Cadmium and mercury	Bioleaching	Bacteria secrete enzymes or organic acids that help dissolve metal ions, improve their availability for uptake and removal	Alviz-Gazitua et al. (2022)
Bacillus species	Cadmium and lead	Biomineralization	Bacteria promote the transformation of heavy metals into insoluble mineral compounds, thereby reducing their mobility and harmful effects	Zhang et al. (2024a, 2024b); Zhang et al. (2024a, 2024b)
Vibrio harveyi	Cadmium	Cellular deposition	The uptake and accumulation of heavy metals within bacterial cells help decrease their harmful impact	Abd-Elnaby et al. (2011)
Pseudomonas aeruginosa	Polycyclic aromatic hydrocarbons	Enzymatic oxidation	Bacterial enzymes oxidize pollutants, leading to the degradation of harmful substances	Gangola et al. (2018)
Halomonas sp.	Arsenite, arsenate	Biotransformation	Metal ions undergo oxidation state changes by bacterial action, resulting in reduced toxicity and better removal	Zhang et al. (2024a, 2024b); Zhang et al. (2024a, 2024b)

(continued)

Table 2 (continued)

Bioengineered microbes	Target pollutant	Mechanism	Description of the mechanism	References
Pseudoxanthomonas mexicana	Mercury and arsenic	Bioaugmentation introducing specific	Bioremediation can be optimized by inoculating polluted sites with selected microbial species	El-Beltagi et al. (2024)
Enterobacter kobei	Lead	Biostimulation	Bacterial efficiency can be enhanced by optimizing environmental parameters, including pH, oxygen levels, and essential nutrients	Adeyi et al. (2017)
Delftia lacustris	Cr(VI), Hg(II), Cd(II), and Pb(II)	Efflux pump mechanism	Through specialized membrane transport mechanisms, bacteria eliminate toxic metals to adapt to harsh environments	Wu et al. (2016)
Pseudomonas aeruginosa	n-alkanes and PAHs	Hydrocarbon degradation	Bacteria utilize specific metabolic pathways to degrade petroleum hydrocarbons	Nizamutdinov et al. (2022)

It has been successfully used to monitor nitrate pollution (Carrey et al., 2021), track microbial denitrification in groundwater (Kim, 2020), and follow the activity of oil-degrading microbes during events like the Deepwater Horizon oil spill. Additionally, qPCR has proven effective in identifying sources of pathogens (Vadde et al., 2019). Examining RNA from active cells through meta-transcriptomic analysis has also helped in identifying key genes that act like biosensors, signalling the presence of organic pollutants and heavy metals in soil and water environments.

Apart from DNA, genes and RNA, proteomics and metabolomics offer valuable insight into the actual metabolic and functional state of a microbial community in a particular time period at a particular environmental condition. These fields give valuable insight about the interaction of different microbes in a different ways under specific conditions (Kleiner, 2019). E.g., metaproteomics has been widely applied in bioremediation studies to detect proteins linked to microbes stress response, helping to know how microbes tolerate and break down pollutants like metals and organic compounds (Khatiwada et al., 2020). Analysis of phospholipid fatty acids (PLFAs), which are components of bacterial membranes, can be an another helpful tool as its profiling provides a quick insight about the health and stress of microbial community and it has proven useful in monitoring the impact of oil spills (Hazen, 2020). When data from proteomics and metabolomics are combined with genomic information, it offers a much deeper understanding of not just who is present in the microbial community, but what they're doing and how they respond to pollutants.

Individual omics approaches gives different insight about the microbial community's action regarding the bioremediation of contaminants like how they get affected and how they adapt to that situation (Paradis et al., 2022). Studies in metal, solvent, nitric acid, and hydrocarbon-contaminated environments have consistently revealed a decline in overall microbial diversity. When the environmental stress increases the dominant microbial populations often change, with the ecosystem favouring stress resilient species (Wu et al., 2017). Over time, this stress-driven selection naturally enhances the presence of microbes with traits that mirror those intentionally introduced in engineered bioremediation approaches. Essentially, contamination pressures select for hardy, functionally capable microbes that are already equipped to deal with toxic environments, offering insight into how natural adaptation can inform and complement human-led remediation strategies.

7 Conclusion

In conclusion, microbial genomics has emerged as a transformative approach in addressing the growing threat of heavy metals and organic pollutants. By unlocking the genetic potential of microorganisms through advanced tools like metagenomics, synthetic biology, and CRISPR-Cas systems, we can design and optimize strains capable of efficiently degrading heavy metals and organic pollutants. The integration of omics technologies like transcriptomics, proteomics, and metabolomics may further enhance our understanding of microbial behaviour under stress conditions, enabling the development of targeted and adaptive bioremediation strategies. While challenges such as biosafety concerns, regulatory hurdles and environmental unpredictability persist, continued research and interdisciplinary collaboration hold the key to overcoming these barriers. Moving forward, coupling genomics with AI-driven modelling and systems biology will pave the way for next-generation, sustainable remediation solutions tailored to the complexity of real world contamination scenarios. To sum up, the microbial genomics has evolved as changer in devising strategies against mounting problem of heavy metals and organic pollutants. We will realize the genetic potential of microorganisms with powerful tools such as metagenomics, synthetic biology, and CRISPR-Cas for designing and improving strains to degrade efficiently heavy metals or organic pollutants. The combination of omics, including transcriptomics, proteomics, and metabolomic approaches should further improve our understanding of microbial response to stress conditions that simultaneously lead enhancing targeted or adaptive bioremediation. Despite challenges including biosafety issues, regulatory obstacles and uncertain environmental effects of such applications progress in research as well as interdisciplinary cooperation is likely to help address these limitations. In the future, genomics will be combined with AI-based modelling and systems biology to open new doors for future.

References

Abd-Elnaby, H., Abou-Elela, G. M., & El-Sersy, N. A. (2011). Cadmium resisting bacteria in Alexandria Eastern Harbor (Egypt) and optimization of cadmium bioaccumulation by Vibrio harveyi. *African Journal of Biotechnology, 10*(17), 3412–3423.

Adeyi, V. A., Adeyi, A. A., Giwa, A., & Solomon, B. O. (2017). Comparative bioremediation of diesel oil bearing soil by Pseudomonas aeruginosa and Pseudomonas fluorescence. *International Journal of Engineering Research in Africa, 30*, 134–143.

Aislabie, J., Bej, A. K., Ryburn, J., Lloyd, N., & Wilkins, A. (2005). Characterization of Arthrobacter nicotinovorans HIM, an atrazine-degrading bacterium, from agricultural soil New Zealand. *FEMS Microbiology Ecology, 52*(2), 279–286.

Akoijam, N., & Joshi, S. R. (2024). Genome editing and genetically engineered bacteria for bioremediation of heavy metals. In *Genome editing in bacteria (Part 2)* (pp. 184–221). Bentham Science Publishers.

Allam, S. S. T., Gacem, A., Govind, G., Tummalacharla, P., Ramakrishnan, G., Yadav, K. K., & Das, A. P. (2025). Zero waste initiatives in industrial hydrocarbons and waste management. *Environmental Science and Engineering, Part, F218*, 105–126. https://doi.org/10.1007/978-3-031-81615-4_6

Alviz-Gazitua, P., Durán, R. E., Millacura, F. A., Cárdenas, F., Rojas, L. A., & Seeger, M. (2022). Cupriavidus metallidurans CH34 possesses aromatic catabolic versatility and degrades benzene in the presence of mercury and cadmium. *Microorganisms, 10*(2), 484.

Aminian-Dehkordi, J., Rahimi, S., Golzar-Ahmadi, M., Singh, A., Lopez, J., Ledesma-Amaro, R., & Mijakovic, I. (2023). Synthetic biology tools for environmental protection. *Biotechnology Advances, 68*, Article 108239.

Andrei, A., Öztürk, Y., Khalfaoui-Hassani, B., Rauch, J., Marckmann, D., Trasnea, P. I., & Koch, H. G. (2020). Cu homeostasis in bacteria: The ins and outs. *Membranes, 10*(9), 242.

Ayangbenro, A. S., & Babalola, O. O. (2020). Genomic analysis of Bacillus cereus NWUAB01 and its heavy metal removal from polluted soil. *Scientific Reports, 10*(1), 19660.

Aziz, Z. S., Jazza, S. H., Dageem, H. N., Banoon, S. R., Balboul, B. A., & Abdelzaher, M. A. (2024). Bacterial biodegradation of oil-contaminated soil for pollutant abatement contributing to achieve sustainable development goals: A comprehensive review. *Results in Engineering, 22*, Article 102083.

Aziz, A., Yasmeen, T., Tariq, M., Arif, M. S., Shahzad, S. M., Riaz, M., & Rizwan, M. (2021). Genomics in understanding bioremediation of inorganic pollutants. In *Handbook of bioremediation* (pp. 397–410). Academic press.

Berini, F., Casciello, C., Marchetto, D., & Sosio, M. (2017). Engineering of microbial communities through functional metagenomics for pollutant degradation. *Trends in Biotechnology, 35*(9), 715–726. https://doi.org/10.1016/j.tibtech.2017.05.001

Bhanu, L. M., Kataki, S., & Chatterjee, S. (2024). CRISPR: New promising biotechnological tool in wastewater treatment. *Journal of Microbiological Methods, 227*, Article 107066.

Bhatt, P., Bhatt, K., Sharma, A., Zhang, W., Mishra, S., & Chen, S. (2021). Biotechnological basis of microbial consortia for the removal of pesticides from the environment. *Critical Reviews in Biotechnology, 41*(3), 317–338.

Bittihn, P., Din, M. O., Tsimring, L. S., & Hasty, J. (2018). Rational engineering of synthetic microbial systems: From single cells to consortia. *Current Opinion in Microbiology, 45*, 92–99.

Bukliarevich, H. A., Gurinovich, A. S., Filonov, A. E., & Titok, M. A. (2023). Molecular genetic and functional analysis of the genes encoding alkane 1-monooxygenase synthesis in members of the genus Rhodococcus. *Microbiology, 92*(2), 242–255.

Carrey, R., Ballesté, E., Blanch, A. R., Lucena, F., Pons, P., López, J. M., & Otero, N. (2021). Combining multi-isotopic and molecular source tracking methods to identify nitrate pollution sources in surface and groundwater. *Water Research, 188*, Article 116537.

Chen, Y. L., Wang, C. H., Yang, F. C., Ismail, W., Wang, P. H., Shih, C. J., & Chiang, Y. R. (2016). Identification of Comamonas testosteroni as an androgen degrader in sewage. *Scientific Reports, 6*(1), 35386.

Chen, F., Chen, L., Yan, Z., Xu, J., Feng, L., He, N., & Liu, C. (2024). Recent advances of CRISPR-based genome editing for enhancing staple crops. *Frontiers in Plant Science, 15*, 1478398.

Choudhary, A., Modak, A., Apte, S. K., & Phale, P. S. (2017). Transcriptional modulation of transport-and metabolism-associated gene clusters leading to utilization of benzoate in preference to glucose in Pseudomonas putida CSV86. *Applied and Environmental Microbiology, 83*(19), e01280-e1317.

Clausen, G. B., Larsen, L., Johnsen, K., Radnoti de Lipthay, J., Aamand, J. (2002). Quantification of the atrazine-degrading Pseudomonas sp. strain ADP in aquifer sediment by quantitative competitive polymerase chain reaction. *FEMS Microbiology Ecology 41*, 221–229. [Google Scholar] [CrossRef]

Curiel-Alegre, S., de la Fuente-Vivas, D., Khan, A. H. A., García-Tojal, J., Velasco-Arroyo, B., Rumbo, C., & Barros, R. (2024). Unveiling the capacity of bioaugmentation application, in comparison with biochar and rhamnolipid for TPHs degradation in aged hydrocarbons polluted soil. *Environmental Research, 252*, Article 118880.

Das, A. P., Mallavarapu, M., & Ghosh, S. (2023a). Ecotoxicity of mining pollutants on the environment and their remediation. *Environmental Chemistry and Ecotoxicology, 5*, 165–167. https://doi.org/10.1016/J.ENCECO.2023.08.002

Das, A. P., Mohanty, A. K., van Hullebusch, E. D., Figueiredo, G., Das, A. P., Mohanty, A. K., van Hullebusch, E. D., & Figueiredo, G. (2023b). Bioremediation as an emerging technology for the removal of synthetic microplastic pollutants from marine ecosystem. *MarPB, 194*, Article 115297. https://doi.org/10.1016/J.MARPOLBUL.2023.115297

Das, A., Dey, S., & Das, A. P. (2024a). Ecotoxicological impacts of synthetic microfiber pollutants and development of sustainable mitigation strategies. *Environmental Chemistry and Ecotoxicology*. https://doi.org/10.1016/J.ENCECO.2024.12.003

Das, A. P., Nayak, B., & Behera, I. D. (2024b). Applications of Microorganisms in Advanced Geomicrobiological and Biotechnological Exploration. *Geomicrobiology Journal,*

41(4), 309–311. https://doi.org/10.1080/01490451.2024.2327059/ASSET//CMS/ASSET/DE2A1F67-08C1-475A-ABC7-2B3F4C9E8BCF/01490451.2024.2327059.FP.PNG

Das, A., Dey, S., Gacem, A., Yadav, K. K., Nath, H., Ram, D. K., & Das, A. P. (2025). Revolutionizing solid hydrocarbon sourcing: Plastic's journey from waste to upcycled treasure. *Environmental Science and Engineering, Part, F218*, 331–351. https://doi.org/10.1007/978-3-031-81615-4_14

Dell'Anno, F., Rastelli, E., Sansone, C., Brunet, C., Ianora, A., & Dell'Anno, A. (2021). Bacteria, fungi and microalgae for the bioremediation of marine sediments contaminated by petroleum hydrocarbons in the omics era. *Microorganisms, 9*(8), 1695.

Dey, S., Tripathy, B., Kumar, M. S., & Das, A. P. (2023). Ecotoxicological consequences of manganese mining pollutants and their biological remediation. *Environmental Chemistry and Ecotoxicology*. https://doi.org/10.1016/J.ENCECO.2023.01.001

Dey, S., Bissoyi, A., Koteswara Reddy, G., & Das, A. P. (2024a). Synthetic microfibers: An urban emerging pollutant demanding for sustainable environmental solutions. In *Sustainable microbial technology for synthetic and cellulosic microfiber bioremediation* (pp. 107–121). https://doi.org/10.1007/978-3-031-62600-5_7

Dey, S., Dash, A., & Das, A. P. (2024b). Bioremediation as a sustainable and circular solution for mining waste management and bioenergy generation. In *Sustainable management of mining waste and tailings: A circular economy approach* (pp. 107–119). https://doi.org/10.1201/9781003442455-6/BIOREMEDIATION-SUSTAINABLE-CIRCULAR-SOLUTION-MINING-WASTE-MANAGEMENT-BIOENERGY-GENERATION-SUDESHNA-DEY-AKANSHYA-DASH-ALOK-PRASAD-DAS

Dong, L., Zhou, S., He, Y., Jia, Y., Bai, Q., Deng, P.,... & Xiao, H. (2018). Analysis of the genome and chromium metabolism-related genes of Serratia sp. S2. *Applied Biochemistry and Biotechnology, 185*(1), 140–152.

El-Beltagi, H. S., Halema, A. A., Almutairi, Z. M., Almutairi, H. H., Elarabi, N. I., Abdelhadi, A. A., & Abdelhaleem, H. A. (2024). Draft genome analysis for Enterobacter kobei, a promising lead bioremediation bacterium. *Frontiers in Bioengineering and Biotechnology, 11*, 1335854.

Fang, H., Zhang, H., Han, L., Mei, J., Ge, Q., Long, Z., & Yu, Y. (2018). Exploring bacterial communities and biodegradation genes in activated sludge from pesticide wastewater treatment plants via metagenomic analysis. *Environmental Pollution, 243*, 1206–1216.

Gangola, S., Sharma, A., Bhatt, P., Khati, P., & Chaudhary, P. (2018). Presence of esterase and laccase in Bacillus subtilis facilitates biodegradation and detoxification of cypermethrin. *Scientific Reports, 8*(1), 12755.

Ghosh, S., Mohanty, S., Nayak, S., Sukla, L. B., & Das, A. P. (2016). Molecular identification of indigenous manganese solubilising bacterial biodiversity from manganese mining deposits. *Journal of Basic Microbiology, 56*(3), 254–262. https://doi.org/10.1002/JOBM.201500477

Ghosh, S., & Das, A. P. (2020). Microbial metagenomics: current advances in investigating microbial ecology and population dynamics. In S. K. Nayak & B. B. Mishra (Eds.), *Frontiers in soil and environmental microbiology* (1st ed., pp. 247–254). CRC Press.

Ghosh, S., Gandhi, M., van Hullebusch, E. D., & Das, A. P. (2020). Proteomic insights into Lysinibacillus sp.-mediated biosolubilization of manganese. *Environmental Science and Pollution Research, 28*(30), 40249–40263. https://doi.org/10.1007/S11356-020-10863-4

Ghosh, S., Tripathy, B., Dey, S., & Das, A. P. (2023). Comparative investigation of fungal and bacterial manganese biomining mechanisms. *Manganese Mining Microorganisms*, 99–115. https://doi.org/10.1016/b978-0-12-822147-1.00016-9

Gong, Y., Zhao, D., & Wang, Q. K. (2018). An overview of field-scale studies on remediation of soil contaminated with heavy metals and metalloids: Technical progress over the last decade. *Water research, 147*, 440–460.

Guzik, U., Greń, I., Hupert-Kocurek, K., & Wojcieszyńska, D. (2011). Catechol 1, 2-dioxygenase from the new aromatic compounds–degrading Pseudomonas putida strain N6. *International Biodeterioration & Biodegradation, 65*(3), 504–512.

Han, Y. H., Kim, G., & Seo, S. W. (2023). Programmable synthetic biology tools for developing microbial cell factories. *Current Opinion in Biotechnology, 79*, Article 102874.

Han, L. (2024). Beyond traditional bioremediation: The potential of engineered syncoms in tackling complex environmental pollutants. *Genomics and Applied Biology, 15*.

Hazen, T. C. (2020). Lessons from the 2010 Deepwater Horizon accident in the Gulf of Mexico. *Hydrocarbons, oils and lipids: Diversity, origin, chemistry and fate* (pp. 847–864). Springer International Publishing.

He, Y., Yun, H., Peng, L., Ji, J., Wang, W., & Li, X. (2024). Deciphering the potential role of quorum quenching in efficient aerobic denitrification driven by a synthetic microbial community. *Water Research, 251*, Article 121162.

Heidari, P., & Panico, A. (2020). Sorption mechanism and optimization study for the bioremediation of Pb (II) and Cd (II) contamination by two novel isolated strains Q3 and Q5 of Bacillus sp. *International Journal of Environmental Research and Public Health, 17*(11), 4059.

Hinc, K., Ghandili, S., Karbalaee, G., Shali, A., Noghabi, K. A., Ricca, E., & Ahmadian, G. (2010). Efficient binding of nickel ions to recombinant Bacillus subtilis spores. *Research in Microbiology, 161*(9), 757–764.

Hsiao, T. H., Chen, P. H., Wang, P. H., Brandon-Mong, G. J., Li, C. W., Horinouchi, M., & Chiang, Y. R. (2023). Harnessing microbial phylum-specific molecular markers for assessment of environmental estrogen degradation. *Science of the Total Environment, 896*, Article 165152.

Huang, K., Chen, C., Shen, Q., Rosen, B. P., & Zhao, F. J. (2015). Genetically engineering Bacillus subtilis with a heat-resistant arsenite methyltransferase for bioremediation of arsenic-contaminated organic waste. *Applied and Environmental Microbiology, 81*(19), 6718–6724.

Hussain, S., Khan, M., Sheikh, T. M. M., Mumtaz, M. Z., Chohan, T. A., Shamim, S., & Liu, Y. (2022). Zinc essentiality, toxicity, and its bacterial bioremediation: A comprehensive insight. *Frontiers in Microbiology, 13*, Article 900740.

Hutinet, G., Kot, W., Cui, L., Hillebrand, R., Balamkundu, S., Gnanakalai, S., & de Crécy-Lagard, V. (2019). 7-Deazaguanine modifications protect phage DNA from host restriction systems. *Nature Communications, 10*(1), 5442.

Igiri, B. E., Okoduwa, S. I., Idoko, G. O., Akabuogu, E. P., Adeyi, A. O., & Ejiogu, I. K. (2018). Toxicity and bioremediation of heavy metals contaminated ecosystem from tannery wastewater: A review. *Journal of Toxicology, 2018*(1), 2568038.

Jeong, S. H., Lee, H. J., & Lee, S. J. (2023). Recent advances in CRISPR-Cas technologies for synthetic biology. *Journal of Microbiology, 61*(1), 13–36.

Jeyakumar, P., Debnath, C., Vijayaraghavan, R., & Muthuraj, M. (2023). Trends in bioremediation of heavy metal contaminations. *Environmental Engineering Research, 28*(4).

Jia, X., Ma, Y., Bu, R., Zhao, T., & Wu, K. (2020). Directed evolution of a transcription factor PbrR to improve lead selectivity and reduce zinc interference through dual selection. *AMB Express, 10*(1), 67.

Joutey, N. T., Bahafid, W., Sayel, H., & El Ghachtouli, N. (2013). Biodegradation: Involved microorganisms and genetically engineered microorganisms. *Biodegradation-Life of Science, 1*, 289–320.

Khatiwada, B., Hasan, M. T., Sun, A., Kamath, K. S., Mirzaei, M., Sunna, A., & Nevalainen, H. (2020). Proteomic response of Euglena gracilis to heavy metal exposure–identification of key proteins involved in heavy metal tolerance and accumulation. *Algal Research, 45*, Article 101764.

Kim, H. (2020). Comparison of PCR primers for analyzing denitrifying microorganisms in the hyporheic zone. *Applied Sciences, 10*(12), 4172.

Kim, D. W., Ahn, J. H., & Cha, C. J. (2022). Biodegradation of plastics: Mining of plastic-degrading microorganisms and enzymes using metagenomics approaches. *Journal of Microbiology, 60*(10), 969–976.

Kleiner, M. (2019). Metaproteomics: much more than measuring gene expression in microbial communities.*mSystems, 4*, e00115–19.

Kromer, J. O., Sorgenfrei, O., Klopprogge, K., Heinzle, E., & Wittmann, C. (2004). In-depth profiling of lysine-producing Corynebacterium glutamicum by combined analysis of the transcriptome, metabolome, and fluxome. *Journal of Bacteriology, 186*(6), 1769–1784.

Kumar, V., Thakur, I. S., Singh, A. K., & Shah, M. P. (2020). Application of metagenomics in remediation of contaminated sites and environmental restoration. In *Emerging technologies in environmental bioremediation* (pp. 197–232). Elsevier.

Li, X., Sun, M., Zhang, L., Finlay, R. D., Liu, R., & Lian, B. (2022). Widespread bacterial responses and their mechanism of bacterial metallogenic detoxification under high concentrations of heavy metals. *Ecotoxicology and Environmental Safety, 246*, Article 114193.

Liu, Z., Cai, Y., Chen, X., Cang, Y., Yu, J., Shaaban, M., Cai, Y., & Peng, Q. A. (2025). Functional genomic analysis of Bacillus cereus BC4 strain for chromium remediation in contaminated soil. *Current Research in Microbial Sciences, 8*, 100388.

Liu, C. J., Deng, S. G., Hu, C. Y., Gao, P., Khan, E., Yu, C. P., & Ma, L. Q. (2023). Applications of bioremediation and phytoremediation in contaminated soils and waters: CREST publications during 2018–2022. *Critical Reviews in Environmental Science and Technology, 53*(6), 723–732.

Luo, J., Bai, Y., Liang, J., & Qu, J. (2014). Metagenomic approach reveals variation of microbes with arsenic and antimony metabolism genes from highly contaminated soil. *PLoS ONE, 9*(10), Article e108185.

Mao, S., Jiang, J., Xiong, K., Chen, Y., Yao, Y., Liu, L., & Li, X. (2024). Enzyme engineering: Performance optimization, novel sources, and applications in the food industry. *Foods, 13*(23), 3846.

Martínez-Cuesta, M. C., Lasa, A. V., Garrido-Sanz, D., Manzano, J., & Martín, M. (2023). Co-culture of hydrocarbon-degrading bacteria with complementary catabolic traits improves the degradation of petroleum hydrocarbons. *Journal of Hazardous Materials, 441*, Article 129959. https://doi.org/10.1016/j.jhazmat.2022.129959

Medfu Tarekegn, M., Zewdu Salilih, F., & Ishetu, A. I. (2020). Microbes used as a tool for bioremediation of heavy metal from the environment. *Cogent Food & Agriculture, 6*(1), 1783174.

Miri, S., Saini, R., Davoodi, S. M., Pulicharla, R., Brar, S. K., & Magdouli, S. (2022). Biodegradation of microplastics: Better late than never. *Chemosphere, 286*, Article 131670.

Mishra, S., & Das, A. P. (2022). Treatment of the wastewater polluted with synthetic microfiber released from washing machine. In B. B. Das, H. Hettiarachchi, P. K. Sahu, & S. Nanda (Eds.), *Lecture notes in civil engineering* (vol. 207, pp. 109–117). Springer Science and Business Media Deutschland GmbH. https://doi.org/10.1007/978-981-16-7509-6_9/COVER

Mishra, S., & Das, A. P. (2023). Microbial remediation of synthetic microfiber contaminated wastewater. In M. P. Shah (Ed.), *Microbial technologies in industrial wastewater treatment* (pp. 337–350). Springer. https://doi.org/10.1007/978-981-99-2435-6_17

Mishra, S., Dash, D., & Das, A. P. (2023). Aquatic microbial diversity on plastisphere: Colonization and potential role in microplastic biodegradation. *Geomicrobiology Journal*, 1–12. https://doi.org/10.1080/01490451.2023.2209750

Mohamed, M. S., Asair, A. A., Fetyan, N. A., & Elnagdy, S. M. (2023). Complete biodegradation of diclofenac by new bacterial strains: Postulated pathways and degrading enzymes. *Microorganisms, 11*(6), 1445.

Mohanty, M., Mohanty, J., Dey, S., Dutta, K., Shah, M. P., & Das, A. P. (2024b). The face mask: A tale from protection to pollution and demanding sustainable solution. *Emerging Contaminants, 10*(2), Article 100298. https://doi.org/10.1016/J.EMCON.2023.100298

Mohanty, J., Dey, S., Bala, K., & Das, A. P. (2024a). Recent advances in microplastic pollution for its sustainable management. In *Sustainable management of urban plastic waste through circular economic approaches* (pp. 27–48). https://doi.org/10.1201/9781003465447-3

Mojica, F. J., Juez, G., & Rodriguez-Valera, F. (1993). Transcription at different salinities of Haloferax mediterranei sequences adjacent to partially modified PstI sites. *Molecular Microbiology, 9*(3), 613–621.

Nazir, M., Khan, S., Nassani, A. A., Naseem, I., Zaman, K., & Haffar, M. (2024). Do regulatory affairs, political stability, and renewable energy demand improve environmental quality? Perceptive from the South Asian Association for Regional Cooperation region. *Energy & Environment*, 0958305X241236270.

Nizamutdinov, T. I., Suleymanov, A. R., Morgun, E. N., Dinkelaker, N. V., & Abakumov, E. V. (2022). Ecotoxicological analysis of fallow soils at the Yamal experimental agricultural station.

Pande, V., Pandey, S. C., Sati, D., Bhatt, P., & Samant, M. (2022). Microbial interventions in bioremediation of heavy metal contaminants in agroecosystem. *Frontiers in Microbiology, 13*, Article 824084.

Pant, G., Garlapati, D., Agrawal, U., Prasuna, R. G., Mathimani, T., & Pugazhendhi, A. (2021). Biological approaches practised using genetically engineered microbes for a sustainable environment: A review. *Journal of Hazardous Materials, 405*, Article 124631.

Paradis, C. J., Miller, J. I., Moon, J. W., Spencer, S. J., Lui, L. M., Van Nostrand, J. D., & Hazen, T. C. (2022). Sustained ability of a natural microbial community to remove nitrate from groundwater. *Groundwater, 60*(1), 99–111.

Paul, A., Dey, S., Ram, D. K., & Das, A. P. (2023). Hexavalent chromium pollution and its sustainable management through bioremediation. *Geomicrobiology Journal*, 1–11. https://doi.org/10.1080/01490451.2023.2218377

Poursat, B. A., van Spanning, R. J., de Voogt, P., & Parsons, J. R. (2019). Implications of microbial adaptation for the assessment of environmental persistence of chemicals. *Critical Reviews in Environmental Science and Technology, 49*(23), 2220–2255.

Prabakar, J., John, J., Arumugham, I. M., Kumar, R. P., & Sakthi, D. S. (2018). Comparing the effectiveness of probiotic, green tea, and chlorhexidine-and fluoride-containing dentifrices on oral microbial flora: A double-blind, randomized clinical trial. *Contemporary Clinical Dentistry, 9*(4), 560–569.

Pradhan, G., Tripathy, B., Ram, D. K., Digal, A. K., & Das, A. P. (2023). Bauxite mining waste pollution and its sustainable management through bioremediation. *Geomicrobiology Journal*, 1–10. https://doi.org/10.1080/01490451.2023.2235353

Priyadarshini, S., Jagatee, S., & Das, A. P. (2024). Synthetic fabrics and microfiber pollution–An assessment of their global impact. *Environmental Science and Engineering, Part, F2333*, 137–157. https://doi.org/10.1007/978-3-031-51792-1_8

Qattan, S. Y. (2025). Harnessing bacterial consortia for effective bioremediation: Targeted removal of heavy metals, hydrocarbons, and persistent pollutants. *Environmental Sciences Europe, 37*(1), 85.

Radojević, I., Jakovljević, V., Grujić, S., Ostojić, A., & Ćirković, K. (2024). Biofilm formation by selected microbial strains isolated from wastewater and their consortia: Mercury resistance and removal potential. *Research in Microbiology, 175*(3), Article 104092.

Ramrakhiani, L., Ghosh, S., Sarkar, S., & Majumdar, S. (2016). Heavy metal biosorption in multi component system on dried activated sludge: Investigation of adsorption mechanism by surface characterization. *Materials Today: Proceedings, 3*(10), 3538–3552.

Ray, S., Dey, S., Digal, A. K., Yadav, K. K., & Das, A. P. (2025). Sanitary waste and microplastic pollution in the Euro-Mediterranean region: Challenges and solutions. *Euro-Mediterranean Journal for Environmental Integration, 2025*, 1–23. https://doi.org/10.1007/S41207-025-008 79-Y

Ray, S. K., Dey, S., Das, A. P., & John, A. (2024). Microplastic waste management through greener technologies : A review. In *Sustainable management of urban plastic waste through circular economic approaches* (pp. 237–261). https://doi.org/10.1201/9781003465447-14

Reddy Koteswara, G., Bhargavi, S., Yamuna, D., Anjani Devi, M., Jeevana Lakshmi Kumar, G., Manas, T., Adarsh, S., Bhavya Sai, V., Nath, H., Alreshidi, M. A., Togun, H., Ram, D. K., & Das, A. P. (2025). Recent advancements in the remediation of marine hydrocarbon and plastic pollutants. *Environmental Science and Engineering Part, F218*, 127–142. https://doi.org/10.1007/978-3-031-81615-4_7

Remenar, M., Kamlarova, A., Harichova, J., Zámocký, M., & Ferianc, P. (2018). The heavy-metal resistance determinant of newly isolated bacterium from a nickel-contaminated soil in Southwest Slovakia. *Polish Journal of Microbiology, 67*(2), 191.

Ren, X., Zhao, L., Shen, J., Zhou, P., Zhao, K., Yuan, C., & Yan, X. (2025). Engineered microbial platform confers resistance against heavy metals via phosphomelanin biosynthesis. *Nature Communications, 16*(1), 4836.

Rodríguez, A., Castrejón-Godínez, M. L., Salazar-Bustamante, E., Gama-Martínez, Y., Sánchez-Salinas, E., Mussali-Galante, P., & Ortiz-Hernández, M. L. (2020). Omics approaches to pesticide biodegradation. *Current Microbiology, 77*(4), 545–563.

Sahu, S., & Das, A. P. (2024). Microplastics degradation mechanism by aquatic plastiomes: A review. *Environmental Science and Engineering, Part, F2827*, 213–226. https://doi.org/10.1007/978-3-031-50840-0_12

Sahu, A., Dey, S., Mallick, K., Priyadarsini, S., Behera, I. D., & Das, A. P. (2024). A review on advanced technology for sustainable management of synthetic microplastic waste. In *Sustainable management of urban plastic waste through circular economic approaches* (pp. 262–279). https://doi.org/10.1201/9781003465447-15

Saltikov, C. W., Wildman Jr, R. A., & Newman, D. K. (2005). Expression dynamics of arsenic respiration and detoxification in Shewanella sp. strain ANA-3. *Journal of Bacteriology, 187*(21), 7390–7396.

Saravanan, A., Kumar, P. S., Ramesh, B., & Srinivasan, S. (2022). Removal of toxic heavy metals using genetically engineered microbes: Molecular tools, risk assessment and management strategies. *Chemosphere, 298*, Article 134341.

Selberg, J., Gomez, M., & Rolandi, M. (2018). The potential for convergence between synthetic biology and bioelectronics. *Cell Systems, 7*(3), 231–244.

Shah, B. A., Malhotra, H., Papade, S. E., Dhamale, T., Ingale, O. P., Kasarlawar, S. T., & Phale, P. S. (2024). Microbial degradation of contaminants of emerging concern: Metabolic, genetic and omics insights for enhanced bioremediation. *Frontiers in Bioengineering and Biotechnology, 12*, 1470522.

Sharma, V. (2022). Microbial genomics: Sequencing and comparative analysis. In S. Sridhar & U. Farooq (Eds.), *Microbial Genetics* (Revised ed., pp. 55–61). Wisdom Press, New Delhi. ISBN: 978-93-82006-21-3.

Sharma, K. V., Sarvalingam, B. K., & Marigoudar, S. R. (2021). A review of mesocosm experiments on heavy metals in marine environment and related issues of emerging concerns. *Environmental science and pollution research, 28*(2), 1304–1316.

Shi, L. D., Chen, Y. S., Du, J. J., Hu, Y. Q., Shapleigh, J. P., & Zhao, H. P. (2019). Metagenomic evidence for a methylocystis species capable of bioremediation of diverse heavy metals. *Frontiers in Microbiology, 9*, 3297.

Singh, K., Tripathi, S., & Chandra, R. (2023). Bacterial assisted phytoremediation of heavy metals and organic pollutants by Cannabis sativa as accumulator plants growing on distillery sludge for ecorestoration of polluted site. *Journal of Environmental Management, 332*, Article 117294.

Siunova, T. V., Filonov, A. E., Gorovtsov, A. V., Akhmetov, L. I., Ivanov, F. D., Rajput, V. D., & Biswas, J. K. (2025). Potential of nickel and cobalt resistant microorganisms for effective phytoremediation of heavy metal contaminated soils. *Discover Environment, 3*(1), 99.

Srivastava, N. (2021). Remediation of heavy metals through genetically engineered microorganism. In *Environmental pollution and remediation* (pp. 315–366). Springer Singapore.

Sun, H., Narihiro, T., Ma, X., Zhang, X. X., Ren, H., & Ye, L. (2019). Diverse aromatic-degrading bacteria present in a highly enriched autotrophic nitrifying sludge. *Science of the Total Environment, 666*, 245–251.

Swain, S., Purnima Sudha, M., Dey, S., Badu, Y., Bhoi, S. R., Alreshidi, M. A., Qasim, M. T., Nath, H., Ram, D. K., & Das, A. P. (2025). Sources of micro and nano plastic hydrocarbon pollutants in marine bodies and their treatment. *Environmental Science and Engineering, Part, F218*, 45–67. https://doi.org/10.1007/978-3-031-81615-4_3

Tang, H., Xiang, G., Xiao, W., Yang, Z., & Zhao, B. (2024). Microbial mediated remediation of heavy metals toxicity: Mechanisms and future prospects. *Frontiers in Plant Science, 15*, 1420408.

Thai, T. D., Lim, W., & Na, D. (2023). Synthetic bacteria for the detection and bioremediation of heavy metals. *Frontiers in Bioengineering and Biotechnology, 11*, 1178680.

Thomas, T., Gilbert, J., & Meyer, F. (2012). Metagenomics—a guide from sampling to data analysis. *Microbial Informatics and Experimentation, 2*(1), 3. https://doi.org/10.1186/2042-5783-2-3

Vadde, K. K., McCarthy, A. J., Rong, R., & Sekar, R. (2019). Quantification of microbial source tracking and pathogenic bacterial markers in water and sediments of Tiaoxi River (Taihu Watershed). *Frontiers in Microbiology, 10*, 699.

Verma, S., & Kuila, A. (2019). Bioremediation of heavy metals by microbial process. *Environmental Technology & Innovation, 14*, Article 100369.

Wu, W., Huang, H., Ling, Z., Yu, Z., Jiang, Y., Liu, P., & Li, X. (2016). Genome sequencing reveals mechanisms for heavy metal resistance and polycyclic aromatic hydrocarbon degradation in Delftia lacustris strain LZ-C. *Ecotoxicology, 25*(1), 234–247.

Wu, Y., Zeng, J., Zhu, Q., Zhang, Z., & Lin, X. (2017). PH is the primary determinant of the bacterial community structure in agricultural soils impacted by polycyclic aromatic hydrocarbon pollution. *Scientific Reports, 7*(1), 40093.

Wu, Y., Zhu, M., Ouyang, X., Qi, X., Guo, Z., Yuan, Y., & Yin, H. (2024). Integrated transcriptomics and metabolomics analyses reveal the aerobic biodegradation and molecular mechanisms of 2, 3′, 4, 4′, 5-pentachlorodiphenyl (PCB 118) in Methylorubrum sp. ZY-1. *Chemosphere, 356*, 141921.

Xiang, L., Li, G., Wen, L., Su, C., Liu, Y., Tang, H., & Dai, J. (2021). Biodegradation of aromatic pollutants meets synthetic biology. *Synthetic and Systems Biotechnology, 6*(3), 153–162.

Yadav, J. S., Sharma, R. K., Yadav, J., Tiwari, S., Kumar, U., Singh, P. K., & Kumar, I. (2021). Effects of bacterial consortium on heavy metal accumulation and growth performance of Zea mays L. Plants.

Yan, Y., Chang, L., & Wang, L. (2020a). Laboratory testing of SARS-CoV, MERS-CoV, and SARS-CoV-2 (2019-nCoV): Current status, challenges, and countermeasures. *Reviews in Medical Virology, 30*(3), Article e2106.

Yan, A., Wang, Y., Tan, S. N., Mohd Yusof, M. L., Ghosh, S., & Chen, Z. (2020b). Phytoremediation: A promising approach for revegetation of heavy metal-polluted land. *Frontiers in Plant Science, 11*, Article 513099.

Yang, T., Liu, J. W., Gu, C., Chen, M. L., & Wang, J. H. (2013). Expression of arsenic regulatory protein in Escherichia coli for selective accumulation of methylated arsenic species. *ACS Applied Materials & Interfaces, 5*(7), 2767–2772.

Zhang, L., Chen, F., Zeng, Z., Xu, M., Sun, F., Yang, L., & Xie, Y. (2021). Advances in metagenomics and its application in environmental microorganisms. *Frontiers in Microbiology, 12*, Article 766364.

Zhang, H., Wang, K., Liu, X., Yao, L., Chen, Z., & Han, H. (2024a). Exopolysaccharide-producing bacteria regulate soil aggregates and bacterial communities to inhibit the uptake of cadmium and lead by lettuce. *Microorganisms, 12*(11), 2112.

Zhang, Z., Zhang, Q., Chen, B., Yu, Y., Wang, T., Xu, N., & Qian, H. (2024b). Global biogeography of microbes driving ocean ecological status under climate change. *Nature Communications, 15*(1), 4657.

Zheng, R., Wu, S., Ma, N., & Sun, C. (2018). Genetic and physiological adaptations of marine bacterium Pseudomonas stutzeri 273 to mercury stress. *Frontiers in Microbiology, 9*, 682.

Zhu, Y., Liu, K., Muhammad, Y., Zhang, H., Tong, Z., Yu, B., & Sahibzada, M. (2020). Effects of divalent copper on tetracycline degradation and the proposed transformation pathway. *Environmental Science and Pollution Research, 27*(5), 5155–5167.

Zhu, G., Xie, L., Tan, W., Ma, C., & Wei, Y. (2022). Cd2+ tolerance and removal mechanisms of Serratia marcescens KMR-3. *Journal of Biotechnology, 359*, 65–74.

Correction to: Advances in Organic Waste Conversion Through Industrial Biotechnology

Shreya Ghosh, Alok Prasad Das, Chandra Mohan, and Eric van Hullebusch

Correction to:
S. Ghosh et al. (eds.), *Advances in Organic Waste Conversion Through Industrial Biotechnology*, Sustainable Environmental Waste Management Strategies, https://doi.org/10.1007/978-3-032-11377-1

The original version of this book was inadvertently published with an incorrect title in Chapter 9 which has now been corrected, and also a new affiliation for the co-author (Hadiya Amin Kantroo) has been added in Chapter 14. The book and the chapters have been updated with the changes.

The updated versions of these chapters can be found at
https://doi.org/10.1007/978-3-032-11377-1_9
https://doi.org/10.1007/978-3-032-11377-1_14

S. Ghosh et al. (eds.), *Advances in Organic Waste Conversion Through Industrial Biotechnology*, Sustainable Environmental Waste Management Strategies,
https://doi.org/10.1007/978-3-032-11377-1_21

Index

S. Ghosh et al. (eds.), *Advances in Organic Waste Conversion Through Industrial Biotechnology*, Sustainable Environmental Waste Management Strategies,
https://doi.org/10.1007/978-3-032-11377-1

MIX
Papier aus verantwortungsvollen Quellen
Paper from responsible sources
FSC® C105338

If you have any concerns about our products,
you can contact us on
ProductSafety@springernature.com

In case Publisher is established outside the EU,
the EU authorized representative is:
Springer Nature Customer Service Center GmbH
Europaplatz 3, 69115 Heidelberg, Germany

Printed by Libri Plureos GmbH
in Hamburg, Germany